IMPERFECT BALANCE

THE HISTORICAL ECOLOGY SERIES | WILLIAM BALÉE
CAROLE L. CRUMLEY
EDITORS

This series explores complex links between people and the landscapes that both mold individuals and societies and are fashioned by them. Drawing on scientific and humanistic scholarship, books in the series focus on cognition and on temporal and spatial change. The series offers examples, explores issues, and develops concepts that preserve experience and derive lessons from other places and times.

THE HISTORICAL ECOLOGY SERIES

William Balée, Editor, *Advances in Historical Ecology*

DAVID L. LENTZ, *Editor*

IMPERFECT BALANCE | LANDSCAPE TRANSFORMATIONS IN THE PRECOLUMBIAN AMERICAS

Columbia University Press / NEW YORK

Columbia University Press
Publishers Since 1893
New York   Chichester, West Sussex

LIBRARY OF CONGRESS CATALOGING-IN-PUBLICATION DATA
Imperfect balance : landscape transformations in the Precolumbian
Americas / David L. Lentz, editor.
        p. cm. — (The historical ecology series)
    Includes bibliographical references and index.
    ISBN 0–231–11156–8 (cloth : alk. paper) —
    ISBN 0–231–11157–6 (pbk. : alk. paper)
    1. Human ecology—America—History.   2. Landscape changes—
America—History.   3. Indians—Land tenure.   4. Indians—
Antiquities.   5. America—Antiquities.   I. Lentz, David
Lewis.   II. Series.
GF500 .A5 2000
304.3'097—dc21   00–022681

⊗

Casebound editions of Columbia University Press books are printed on
permanent and durable acid-free paper.
Printed in the United States of America

c 10 9 8 7 6 5 4 3 2 1
p 10 9 8 7 6 5 4 3 2 1

To the women who made a difference
WILMA *and* VORSILA

CONTENTS

TABLES AND FIGURES

TABLES

FIGURES

CONTRIBUTORS

TIMOTHY BEACH is an Assistant Professor in the Department of Geography and Environmental Science at Georgetown University, Program in Science, Technology, & International Affairs, Washington, D.C.

MARK BRENNER is an Assistant Research Professor in the Department of Fisheries and Aquatic Sciences, College of Agriculture, University of Florida, Gainesville, Florida.

STEVEN P. CHURCHILL is an Associate Curator at the Missouri Botanical Garden, St. Louis, Missouri.

JASON H. CURTIS is an Assistant in Geochemistry, Department of Geology, University of Florida, Gainesville, Florida.

TERENCE N. D'ALTROY is an Associate Professor in the Department of Anthropology, Columbia University, New York, New York.

DOUGLAS C. DALY is the B. A. Krukoff Curator of Amazonian Botany, Institute of Systematic Botany, The New York Botanical Garden, Bronx, New York.

WILLIAM M. DENEVAN is a Professor Emeritus in the Department of Geography at the University of Wisconsin, Madison, Wisconsin.

NICHOLAS DUNNING is an Associate Professor in the Department of Geography at the University of Cincinnati, Cincinnati, Ohio.

CLARK L. ERICKSON is an Associate Professor in the Department of Anthropology at the University of Pennsylvania, Philadelphia, Pennsylvania.

SUZANNE K. FISH is an Assistant Curator of Archaeology at the Arizona State Museum, University of Arizona, Tucson.

GAYLE J. FRITZ is an Associate Professor in the Department of Anthropology at Washington University, St. Louis, Missouri.

ANDREW M. GRELLER is a Professor Emeritus in the Department of Biology, Queens College, City University of New York, Flushing, New York.

DAVID A. HODELL is a Professor in the Department of Geology, University of Florida, Gainesville, Florida.

DAVID L. LENTZ is Director of the Graduate Studies Program at The New York Botanical Garden, Bronx, New York.

JAMES L. LUTEYN is a Senior Curator in the Institute of Systematic Botany at The New York Botanical Garden, Bronx, New York.

EMILY MCCLUNG DE TAPIA is a Professor in the Instituto de Investigaciones Antropológicas, Universidad Nacional Autonoma de Mexico, Mexico City, D.F.

JOHN D. MITCHELL is an Honorary Curator in the Institute of Systematic Botany at The New York Botanical Garden, Bronx, New York.

CHARLES M. PETERS is the Kate E. Tode Curator of Botany, Institute of Economic Botany, The New York Botanical Garden, Bronx, New York.

ANNA C. ROOSEVELT is a Professor in the Department of Anthropology at the University of Illinois–Chicago Circle, and Curator of Anthropology at the Field Museum of Natural History, Chicago, Illinois.

CHARLES S. SPENCER is the Curator of Mexican and Central American Archaeology, Department of Anthropology, American Museum of Natural History, New York, New York.

FOREWORD

A new scholarly debate has emerged at the end of the twentieth century, although its roots are much earlier. The magnitude of human impacts on the "natural" environment today is staggering, life threatening for many species, even for groups of our own kind. But what about in the past? Did people substantially change the environment in pre-European or preindustrial times? For the Western Hemisphere, some scholars believe that such changes prior to 1492 were relatively insignificant, except locally. Others believe that most of the continent was no longer pristine, no longer a wilderness, at the time of Columbus, and that, instead, "humanization" of the environment was pervasive to varying degrees, even though not always observable to the untrained eye. Curiously, there is more acceptance of early human impacts for Europe, Asia, the Pacific, and Africa than there is for the New World, where there is, of course, a briefer time depth.

A catalyst for debates in the New World context was the publication of several independent articles during the observance of the 500th anniversary of the discovery of the Americas (Bowden 1992; Denevan 1992; Diamond 1992; Gómez-Pompa and Kaus 1992; Simms 1992; Turner and Butzer 1992; Wilson 1992). These scholars reopened the discussion of early New World human-land interactions. Most landscapes in the Americas, they argue, were altered to some degree by Native Americans prior to 1492. Others (e.g., Russell 1997; Krech 1999) have since come to similar conclusions.

There has been a backlash, however, to the study of human environmental impacts by indigenous peoples, particularly from advocates of wilderness preservation (see Callicott and Nelson 1998). How can we argue for protecting wilderness if there is no real wilderness? This is, in part, a matter of semantics. Most environmentalists will agree, when presented with good evidence, that people have long been present in and have had some impact on environments. They will say, however, that the "wilderness" they are concerned about appears natural, is primarily the result of nonhuman processes, and has minimal or no current human presence, even though there may have been human disturbances in the past which have some residual manifestations. Possibly we should differentiate between such "wilderness" and "pristine wilderness" which has not been disturbed in any way by humans or at least not for a very long time.

There is another dimension of the backlash, and this is the argument that proponents of prehistoric landscape change by humans have seriously exaggerated their case. Thomas Vale (1998: 231) says, "We have simply replaced the old [pristine] myth with a new one: the myth of the humanized landscape." Certainly the expression "humanized" is misleading if it conveys an image of man-made structures rather than a continuum from a slightly modified forest to a fully constructed landscape of buildings, roads, and fields. And, of course, "humanized" does not apply to unpopulated regions.

Better definitions will improve communication, but they will not resolve the basic issue of the degree to which environments were modified by human action in the distant past. As the authors in this book show, this can be achieved by rigorous local research which identifies past environmental change, dates its occurrence, and attempts to evaluate potential causes of that change, which may have been natural, human, or, as often was the case, a combination of both.

We can discover environmental change in the Americas, especially for climate and vegetation, from the time of the early human record up to the availability of written records after 1492, by examining pollen and ice cores, lake levels, sediment accumulation, and other biophysical evidence, and by archaeological study of human presence. Until recently, it was often assumed that much of what change there was during this period could be explained by human activity, that the environment was otherwise relatively stable. However, a new perspective, the "new ecology" (cf. Zimmerer 1994) advocated by Lentz and his colleagues, emphasizes the instability of the biophysical environment and the dynamic flux that characterizes human interactions with their surroundings. Hence there are renewed questions regarding the distinction between natural and human impacts. It may be very difficult to differentiate one from the other.

This volume brings together natural scientists, archaeologists, and cultural ecologists to examine the physical environment and human modifications thereof, a rare combination that we need to see more often. There are treatments of climatic change and vegetation patterns, along with regional and local land-use studies for North America, Mexico, Central America, the Andes, and Amazonia. The most significant type of environmental change brought about by Precolumbian human activity was the modification of vegetation. Other changes were of soil (erosion, structure, loss of fertility, creation of fertility), wildlife (genetics, numbers, distribution), microclimate, hydrology, and the land surface itself. This collection of articles provides examples of all of these influences.

Vegetation was primarily altered by the clearing of forest and by intentional burning. Natural fires certainly occurred but varied in frequency and strength in different habitats. Anthropogenic fires, for which there is ample documentation, tended to be more frequent but weaker, with a different seasonality than natural fires, and thus had a different type of influence on vegetation. The result of clearing and burning was, in many regions, the conversion of forest to grassland, savanna, scrub, open woodland, and forest with grassy openings. Although all of these vegetation types occurred naturally, their extent was increased at the expense of forest by added burning episodes. Also, some forests remained forest but their basic composition was changed; for example, mixed hardwoods were converted to pines in some regions. Forest composition was altered more subtly by the harvesting of useful species or plant products and by intentional or inadvertent introductions and activities, including agroforestry and fallow management.

Authors here are concerned not only with natural ecosystems and human impacts on them but also with agroecosystems—ecosystems which are primarily human artifacts constructed and controlled by farmers. Agricultural landscapes are the most dominant form of complete environmental change, being even more extensive than urbanization. In light of the backlash discussed above, editor David Lentz states that "it is also a goal [of this book] to reconsider effective agricultural technologies used by early complex societies." Hence discussions are included of raised fields, irrigation, terraces, agroforestry, and the origins and evolution of agriculture itself. One is impressed by the diversity of early farming systems, their ecological sophistication, their productivity, and the degree and extent to which they reshaped the landscape.

The scope of this volume on human impacts is large—the Western Hemisphere, 10,000 years or more, the environment when people first appeared, natural change, and various forms of land use. The studies included represent the type of knowledge that is needed before we can arrive at confident generalizations regarding the degree, extent, and persistence of early human impacts on the environment. I hope this collection will encourage more scholars to pursue greater understanding of the Precolumbian aspects of what is one of the great issues of our time: the human modification of the earth.

WILLIAM M. DENEVAN

REFERENCES

Bowden, M. J. 1992. The invention of American tradition. *J. Historical Geography* 18: 3–26.

Callicott, J. B., and M. P. Nelson, eds. 1998. *The Great New Wilderness Debate*. Athens: University of Georgia Press.

Denevan, W. M. 1992. The pristine myth: The landscape of the Americas in 1492. *Annals of the Association of American Geographers* 82: 369–385.

Diamond, J. 1992. The Golden Age that never was. In J. Diamond, ed., *The Third Chimpanzee*, pp. 317–338. New York: Harper Collins.

Gómez-Pompa, A., and A. Kaus. 1992. Taming the wilderness myth. *BioScience* 42: 271–279.

Humboldt, A. von. 1869 [1808]. *Views of Nature: or, Contemplations on the Sublime Phenomena of Creation*, E. C. Otté and H. G. Bohn, trans. London: Bell & Daldy.

Krech III, S. 1999. *The Ecological Indian: Myth and History.* New York: W. W. Norton.

Russell, E. W. B. 1997. *People and the Land Through Time: Linking Ecology and History.* New Haven: Yale University Press.

Simms, S. R. 1992. Wilderness as a human landscape. In S. I. Zeveloff, L. M. Vause, and W. H. McVaugh, eds., *Wilderness Tapestry,* pp. 183–201. Reno: University of Nevada Press.

Turner II, B. L., and K. W. Butzer. 1992. The Columbian encounter and land-use change. *Environment* 43(8): 16–20, 37–44.

Vale, T. R. 1998. The myth of the humanized landscape: An example from Yosemite National Park. *Natural Areas J.* 18: 231–236.

Wilson, S. M. 1992. "That unmanned countrey": Native Americans both conserved and transformed New World environments. *Natural History* 69: 16–17.

Zimmerer, K. S. 1994. Human geography and the "New Ecology": The prospect and promise of integration. *Annals of the Association of American Geographers* 84: 108–125.

ACKNOWLEDGMENTS

This is to acknowledge those whose names may not be listed among the authors on the pages below, but who nevertheless made substantial contributions to this volume. First, I would like to recognize the contribution of the Center for Environmental Research and Conservation at Columbia University, directed by Donald Melnick, who provided initial funds in the form of a course preparation grant. The grant helped me to accumulate and organize much of the data, first presented at Columbia University as a course entitled "Precolumbian New World Ecosystems." Many reviewers have read the following essays and offered comments that were extremely helpful. Sincere thanks go to Mark Bush, Scott Stine, Dennis Stevenson, Steven McLaughlin, Lawrence Kaplan, Rita Wright, Cynthia Robin, Richard Diehl, Kent Flannery, Joyce Marcus, Payson Sheets, Christine Padoch, Margaret Scarry, Roseann Cohen, Chelsea Specht, Christine Hastorf, Hans ter Steege, and Scott Mori. Marlene Bellengi made most of the illustrations, formatted all of the tables, and helped in many other ways.

IMPERFECT BALANCE

DAVID L. LENTZ

1

INTRODUCTION
Definitions and Conceptual Underpinnings

*To a virgin continent
where prairie grass
waved tall as a man
and vast forests
perfumed the air for
miles off shore came
Spanish adventurer,
French trapper, Dutch
sailor, and doughty
Englishman.*
—JULES B.
BILLARD(1975:20)

The myth that Europeans "discovered" America and arrived upon a virgin land has lingered in the Western psyche for centuries. For many early explorers and writers, the Native American imprint was transparent; the forests were undisturbed, the land unplowed, and the mountains untraversed. Somehow through the veil of ethnic hubris, Western scholars have resisted providing an accurate appraisal of the effect of indigenous cultural developments on the American biota. This underestimation of Native American culture has a long history in Western thought, as first evidenced by the Mound Builder debate of the eighteenth and nineteenth centuries (Fagan 1998:184–187; Silverberg 1986). The numerous large tumuli of eastern North America posed an unsolvable conundrum to early American academics (Thomas Jefferson was a notable exception). They concocted every kind of speculative explanation for who, or even what, might have built the earthen mounds of the Eastern Woodlands. This was a reflection of the disbelief that the Native Americans could have constructed these impressive structures, and it was a debate that was not resolved, at least in the minds of the recognized scientific community, until the last decade of the nineteenth century (Willey and Sabloff 1974:49). To add to the misconception that the Native Americans were as ecologically unimpressive as they were perceived culturally, often areas were depopulated before the arrival of European colonists. Not only did the Native American population crash rapidly after 1492, but it was also a long time before the continents were repopulated to precontact levels. For example, it has been estimated that by 1650, the hemispheric population was only 10 percent of what it had been in 1492 (Denevan 1992a:xxix). The environmental impact of this population crash was to allow trees to reclaim cleared land (Denevan 1992b:377–378). Although it may have appeared to early colonists that the Americas were unpopulated and the forests primeval, the Western Hemisphere indeed had substantial populations before the advent of the Europeans, and in several regions the landscape was unquestionably an anthropogenic one. This treatise is not intended as an indictment of Native American land stewardship; in many instances the Precolumbian Americans developed efficient and highly productive agricultural systems that conserved resources and soil fertility. However, the concept of the "ecologically noble savage" can be safely assigned to the realm of mythology because there are many examples of negative impacts on the landscape by indigenous Americans (e.g., Abrams et al. 1996; Alvard 1993, 1994; Redford 1991; Redman 1999; Krech 1999). Through an objective evaluation of precontact land use practices, we can learn from past successes as well as failures.

In terms of conceptual orientation, this volume follows the postulate that humans are components of a dynamic biosphere, constantly adjusting and manipulating the other biotic components to garner a greater share of the nutrient output. These concepts parallel "new ecology" ideas (Botkin 1990:33; Zimmerer 1994) that object to the prevailing tendency of twentieth-century ecologists to view nature, when undisturbed, as a constant. From this viewpoint, a natural population in the absence of disturbance would achieve a stable level at or oscillating about its carrying capacity until disturbed by some outside force. When disturbing factors are removed, populations return to their previous abundance, according to the old ecological paradigm. Systems in nature, however, are not stable, but dynamic and remain in flux as a result of the constant physical and evolutionary changes that characterize life on earth. Human influence has often been regarded as a disturbing factor rather than an integral component in the natural scheme of things. Certainly people are immersed in the biosphere as much as any other organism and should not be viewed as distinct. Even though humans have the ability to influence the environment in ways that exceed other species through extrasomatic innovations coupled with willful intent, we nevertheless remain dependent on the biosphere for needed nutrients, air, and water which are, in the ultimate analysis, finite.

In an approach related to the new ecology, historical ecology attempts to examine the multifaceted historical interactions between humans and the ecosystems they inhabit. It is a conceptual apparatus that examines the "relationships between humans and the biosphere in specific temporal, regional, cultural, and biotic contexts" (Balée 1998:3). Humans do not merely have a negative impact on the natural environment but are integral to its processes. Balée sees the interaction between humankind and the biosphere ("nature") as a dialectical phenomenon, whereby the relationships between culture and nature take on the form of an unspoken dialogue, constantly interacting and promoting change as a result (ibid.). Crumley sees historical ecology, encompassing the perspectives of natural scientists and anthropologists, as "the study of past ecosystems by charting the change in landscapes over time" (1994:6). As Kirch points out, historical ecological studies should be more than just a list of the various impacts of humans in a landscape through time, but "its aim is ecological understanding, including the complex and reciprocal connections linking human populations with the myriad other life-forms that share their world" (1997:18–19).

This volume attempts to arrive at an ecological understanding of the Precolumbian New World by creating evaluations of prehuman vegetation and then offering discussions as to how humans became involved in local ecosystems. In this process, environmental evaluations, both climatic and biotic, are furnished by natural scientists (botanists, geologists, and ecologists), while human environmental interactions from prehistoric times are presented by social scientists (archaeologists and geographers).

The region selected for study is the Western Hemisphere. Part of the reason for doing so is a reflection of my own personal history, but the Americas offer several advantages for historical ecological studies. First, the Western Hemisphere represents a discrete and easily defined geographic unit with a large landmass uninhabited by humans for most of its evolutionary history. Second, we have a reasonably good idea when humans arrived in the Americas, probably somewhere between 12,000 and 15,000 years ago. When humans appeared on the horizon in the first wave of migration, the impact was gradual but observable even in our earliest archaeological records. For example, some have suggested that human hunters contributed to the mass extinction event of the Pleistocene megafauna (Martin and Klein 1989). Third, we understand something of how and where complex societies developed sui generis. Finally, we know when the indigenous cultures in different regions were overshadowed by incursions of Western culture. At the early end of human occupation, we have a "pristine" region taken over by early Precolumbians who rapidly spread their numbers across the landscape and later developed efficient trophic systems based on domesticates with the capability of supporting complex societies. At the more recent end of the Precolumbian occupation, we have an imposing cultural system that dramatically altered, and in many cases terminated, the indigenous cultural/ecological systems that were in place at the time of contact. To provide a chronological guidepost for the reader, table 1.1 lists some of the major cultural developments in each of the regions addressed, beginning in the late Pleistocene and ending with the age of discovery.

Of course, a key factor in the degree of landscape modification and the effect of human interactions with Precolumbian ecosystems was tempered by the number of humans occupying an area in any given time period. Population estimates for the precontact New World have always been controversial, mostly because the accounts of the earliest explorers were vague, at times exaggerated, and based on limited or nonexistent data. To confound the problem, the introduction of European diseases shortly after contact decimated Native American populations (Newson 1998), perhaps 90 percent or more in many areas (Lovell 1992). The impact of this phenomenon was so dramatic that seventeenth-century colonists arrived to what seemed like virgin forests but in fact were remnant landscapes that had been depopulated decades and even centuries before their arrival. For example, when first explored by Verazano in the early 1500s, New England had a population of somewhere around 90,000 to 100,000. By the time the Pilgrims landed at Plymouth Rock in 1620, the population of New England had been reduced by European diseases, such as smallpox and chicken pox, to

TABLE 1.1 NEW WORLD CULTURAL DEVELOPMENT FROM THE LATE PLEISTOCENE TO THE AGE OF DISCOVERY

Time scale	Maya/Central America[a]	Basin of Mexico[b]	Oaxaca[c]	Tehuacán Valley[d]	Andes Region[e]	Amazonia[f]	Southwest U.S.[g]	Eastern U.S.[h]	Time scale
A.D. 1500	European contact	European contact	European contact	European contact	European contact / Late horizon *Inka*	European contact / Complex societies *Santarém*	European contact	European contact	1500
1250	Postclassic	*Aztec* / Late horizon	Monte Alban V	Venta Salada	Late Intermediate period	*Santarém*	Late Classic Hohokam	*Moundville* / Mississippian	1250
1000		*Mazapan* / *Toltec*					Early Classic Hohokam	*Cahokia*	1000
750	*Copan* / *Tikal*	Coyotlatelco	Monte Alban IV		Middle horizon *Tiwanaku*	*Marajoara*		Late Woodland	750
500	Classic	*Teotihuacán* Middle horizon					Preclassic Hohokam		500
250			Monte Alban III	Palo Blanco	Early Intermediate period	Ranked societies		Middle Woodland *Hopewell*	250
A.D. / B.C.	Formative	Tzacualli / Patlachique	Monte Alban II			*Ananatuba* first Mound Builders	Pre-Hohokam Ceramic phases		A.D. / B.C.
500		Ticoman / Zacatenco	Monte Alban I	Santa Maria	Early horizon			Early Woodland	500
1000	*first ceramics*	Early horizon / *first ceramics*	Rosario / San José / Tierras Largas / Espiridion / *first ceramics*	Ajalpan	Initial period / *first ceramics*	Formative	Early Agricultural	*Poverty Point*	1000
2000				Purrón / Abejas	Preceramic period	*first ceramics*	Archaic	*first ceramics*	2000
4000	Archaic	Zohapilco	Archaic	*first maize* Coxcatlan	*first beans, Chilca caves*	*Tapeinha*		Archaic	4000
6000		Playa I & II		El Riego	Lithic period	Archaic	Paleoindian		6000
8000 / end of Pleistocene	Paleoindian	*Iztapan*	*first squash at Guila Naquitz*	Ajuerado	*Quebrada Jaguay*	Paleoindian / *Pedra Pintada*	*Clovis*	*Hester* / Paleoindian / *Thunderbird*	8000 / end of Pleistocene
10000			*Cueva Blanca*		*Monte Verde*				10000
12000									12000

[a] Sharer 1994
[b] Blanton et al. 1981; Hughes 1985
[c] Marcus and Flannery 1996
[d] MacNeish 1992, Byers 1967
[e] D'Altroy 1997
[f] Price and Feinman 1993
[g] Sebastian 1992
[h] Meltzer 1993

only a few thousand (Wessels 1997:52). In addition to ethnohistorical approaches, paleodemographers have relied on archaeological data as a basis for population estimates (Ubelaker 1988).

Notwithstanding the complexities, in the last decade or so population counters have begun to narrow the range of estimate, with most in the 40 to 70 million range for the entire hemisphere at the time of contact. Various recently published precontact estimates include: 43 million by Whitmore (1991:483), 40 million by Lord and Burke (1991), 40 to 50 million by Cowley (1991), 72 million by Thornton (1990), 44 million by Crawford (1998:39), and 43 to 65 million by Denevan (1992a:xxix). The last two authors have broken down their population estimates into regions, and these are listed in table 1.2. Results shown in the table reveal a considerable variability and reflect an ongoing debate that no doubt will continue into the next century and beyond. For the purposes of this discussion, we can see that although the exact number is unknowable, there is general agreement that Precolumbian populations were significant in portions of the hemisphere, especially in Mesoamerica and the altiplano and intermontane valleys of the Andes.

It stands to reason that the human ecological impact was greatest in areas where populations were most concentrated. For example, the Maya Lowlands, which included parts of southern Mexico, northern Guatemala, most of Belize, and western Honduras, were intensively occupied during Late Classic times (A.D. 600–900). The core zone of the Maya Lowlands, made up of the population centers of Tikal, Calakmul, and Río Bec, comprised about 22,715 km² in the heart of the Maya realm. The population of the core zone has been estimated at 3 million with a density of 150/km² at its zenith around A.D. 800 (Turner 1990).

If this information is combined with ethnohistorical data from the Puuc region of southern Yucatán (which is adjacent to the Río Bec area), we can approximate the amount of land that would have been cleared in the core zone. Smyth calculates the amount of shelled corn used by a family of five at 3.0 kg daily, or on average 219 kg/person/year (1991:39). Land in the Puuc region produces an average of 1,000 kg of maize per hectare under shifting cultivation (Smyth 1991:61); this is lower than some

TABLE 1.2. NEW WORLD PRECONTACT POPULATION ESTIMATES (IN MILLIONS)

North America	Crawford (1998)	Denevan (1992a)
North America	2	3.8
Central America and Mexico	25	22.8*
Caribbean	7	3.0
South America	10	24.3**
Total	44	53.9

* Subdivided into 17.2 million for Mexico and 5.6 million for Central America.
** Subdivided into 15.7 million for the Andes and 8.6 million for lowland South America.

areas in Mesoamerica but higher than others (see Loker 1989 for regional compari-sons). Using these figures, it takes about 0.219 ha/year to provide enough maize for the average person in a Puuc household. Other major crops, such as beans and squash, probably were intercropped in the same fields, a standard practice in Mesoamerica. Accordingly, to feed 3 million residents in any given year, it would take approximately 6,570 km² of cultivated land. Most land in a shifting cultivation scenario would be allowed to fallow for at least 5 (Hanks 1990:360) to 7 years (Redfield and Villa Rojas 1990:43) and more commonly for 10, 15, or even 20 years (Ewell and Merrill-Sands 1987). A recently recorded average fallow time for the state of Yucatán is about 10 years (Ewell and Merrill-Sands 1987). To take a step further, if we calculate 5 people per household with a solar or dooryard garden equal to about 2,500 m² (see Alcorn 1984:370), then a population of 3 million would require about 1,500 km² in inten-sively used, highly productive solar space. If we add this to the total land space required for outfields (6,570 km²), we arrive at a figure of 8,070 km². This does not include land for roadways, civic-ceremonial centers, or other public space, not to mention fallow land or land that was not arable. Even the most casual observer will have noted that in light of the large population, the land could not be fallowed for 10 years because there were only 22,715 km² in the whole area. With that much land in production, only a 2- to 3-year fallow cycle would be possible.

Up to this point I have relied on the assumption that the Maya were dependent only on shifting agriculture, as currently practiced in the Puuc, yet we know this was un-likely because of substantial evidence that various forms of intensive agriculture were practiced among the ancient Maya (Harrison and Turner 1978), viz., the use of ter-races (Dunning 1996), check dams (Scarborough 1996), raised fields (Siemens 1998), drained fields (Pohl and Bloom 1996; Pope et al. 1996), and other techniques. These innovations undoubtedly helped the Late Classic Maya sustain a growing population for several centuries. I suspect the Maya practiced shifting cultivation until fallow cycles became too short to replenish the soil, then farmers were obliged to adopt ag-ricultural approaches that would conserve water and restore nutrients. Paleoecological evidence from the Three Rivers area in northern Belize supports this assertion (see Dunning and Beach, this volume). Given the population dynamics and the options available to the populace, it seems likely that the entire core zone of the Maya Low-lands was heavily impacted and probably cleared of its forests at some time during the Late Classic period. It is inconceivable that virgin forests, rooted in fertile soil, would have been left standing in an area of high population density unless there was a specific cultural reason to have set aside such land, and there is no evidence to suggest they did. Pollen evidence for the region shows extensive forest clearance by Maya inhabitants (Vaughan et al. 1985; Deevey 1978). The productivity calculations coupled with the paleodemographic and palynological data demonstrate clearly that the Maya dramati-cally altered their environment through forest clearance to meet the demands of a bur-geoning population.

The Lowland Maya are but an example of a dense Precolumbian population coupled with intensive land use. By A.D. 1500 the Aztecs in the Basin of Mexico, dependent, no doubt, on their 12,000 ha of chinampas, or raised fields (Sanders et al. 1979:390), supported a population density of 180.4 persons/km², even greater than the Lowland

Maya (Turner 1990). On the Bolivian side of Lake Titicaca in the central Andes, 19,000 ha of raised fields were developed to support the Tiwanaku empire (Kolata 1991:109). In total, Tiwanakan raised fields surrounding Lake Titicaca extended to as much as 122,000 ha (Erickson, personal communication). More than 600,000 ha of prehistoric terracing survives in the Peruvian highlands (Denevan 1988:20). Surveys in the eastern Amazon Basin revealed mounded habitation sites with 10,000 or more occupants (Roosevelt 1991:31 and this volume). The settlement of Cahokia, built by Mississippian Mound Builders near the confluence of the Mississippi and Missouri rivers, housed approximately 30,000 inhabitants (Fowler 1989:90). Garcilaso de la Vega, one of De Soto's lieutenants, observed "great fields" of corn and other crops "spread out as far as the eye could see" during their travels through the Timucan lands of the Florida panhandle in 1539 (Dobyns 1983:138). The Salt River Valley in Arizona had enough prehistoric canals to irrigate more land than is currently practiced in modern times (Denevan 1992; Fish this volume). All of these data indicate there were areas of intensive land use in the Precolumbian Americas. Moreover, the population of 40–70 million inhabitants of the Americas was concentrated in pockets, and with that came a dispossession of the natural biota followed by a supplanting of organisms more useful to humans.

This book has been organized to examine the effects of Precolumbian societies, like those in the Maya region, on the landscape of the Americas. Following the premise that high population densities create areas of greatest impact, the essays that follow will concentrate on areas that had aggregated settlements, had the most environmental impact, and, perhaps not surprisingly, also showed the most elaborate cultural developments. Hodell, Brenner, and Curtis, all geologists, discuss the impact of climate change beginning with the Pleistocene-Holocene transition and how these changes affected the biotic environment as well as cultural developments (essay 2). They use the complex interplay of culture, climate, and environment during the Classic Maya period as an example of the potential impact of human activities and climate change on the lowland tropical forest of the Yucatán Peninsula. They allude to the growing body of evidence that reveals how deforestation and other human actions have not only immediate effects on the local landscape but also more far-reaching effects, as on climate, for example. Hodell, Brenner, and Curtis focus mostly on changes in Central America but also extrapolate to other areas of the hemisphere.

The third essay describes the vegetation of both North and Central America as it might have existed without human influence. Greller divides this huge area into two floristic kingdoms: the Holarctic and the Neotropical, each with its own geological history. He takes an evolutionary phytogeographic approach that relates the flora in each region to other regions and how these phytochoria may have come about through long-term processes. This key essay provides an important backdrop for the following seven essays.

The fourth essay examines the nature of human-centered food webs, the interconnected metapopulations of humans, other animals, plants, and microorganisms that developed through time into tightly managed systems where one species (humans) gained control over the life cycles of dozens of plants and animals (domesticates) and channeled the stored energy reserves for their own exclusive consumption. The development and refinement of trophic systems based largely on domesticated plants in the Americas provided the surpluses that allowed the cultural diversification and elaboration that coincided with the growth of chiefdoms and state organizations. The obvious effect of these habitat manipulations by humans was to expand the distribution of certain species at the expense of others.

The essay by McClung de Tapia focuses on the precontact agricultural systems in the Basin of Mexico, perhaps the most densely populated portion of the New World. She outlines previous work that characterizes the agricultural systems of the Aztecs, which include: sloped piedmont cultivation, terraces, floodplain cultivation, floodwater irrigation, drained fields, permanent irrigation, and chinampas. Results of her own paleoethnobotanical research in Teotihuacán, a site which antedates the Tenochtitlán capital of the Aztecs, are also presented.

In essay 6, Spencer provides details of Precolumbian water management and agricultural intensification systems in Mexico and Venezuela. He offers three archaeological cases, the Purrón Dam site in Tehuacán, the Cañada de Cuicatlán in Oaxaca, and the La Tigra site in Venezuela, that define how prehispanic agriculturalists were able to generate workable, technical, and sociological solutions to water control problems.

Dunning and Beach present evidence (essay 7) for agricultural intensification through the construction of terraces and check dams and soil conservation as practiced by the Late Classic Maya at La Milpa and in the Petexbatun region. Their results indicate that the Maya were able to maintain a system of sustainable agriculture complete with canals, reservoirs, terraces, and check dams well into the Late Classic period. Theirs was a well-engineered landscape that provided sustenance and minimized degradation during an era of high population.

Indigenous agroforestry techniques in both Mesoamerica and the Amazon are compared and contrasted by Peters in essay 8.

Precontact agricultural systems and the use of locally developed cultigens in the Mississippi Valley are assessed by Fritz in essay 9. She divides the valley into three regions (lower, central, and upper) and discusses agricultural developments in each. Today this area is a highly productive agricultural region in the great heartland of North America, and so it was in the past, as well.

In the final essay on North America, Fish sketches a portrait of the Southwest and how the Hohokam developed elaborate hydraulic agricultural systems in an arid zone turned lush by careful water management. Topics in these essays point to the main regions of population density of North America during Precolumbian times and represent areas of maximum anthropogenic impact.

South America also had its areas of extensive landscape modification. This section begins with a description of the vegetation types of the Andes and the Pacific coastal zone by Luteyn and Churchill. They describe an area of dramatic elevational change and concomitant environmental variation. Vegetation types discussed by Luteyn and Churchill include: montane forest along the eastern flanks of the Andes, puna in the altiplano (now mostly grassland but once tree covered), the alpine tundralike páramo just below the snowline, the Pacific coastal areas that range from desert to savanna and scrub forest, and the tropical moist forest of the Chocó along the northwestern coast.

Within a sometimes harsh and often dynamic setting, the occupants of the Lake

Titicaca basin developed agricultural systems that were both flexible and productive in ways that scientists are only beginning to appreciate (Erickson, essay 12). Inhabitants of the Tiwanaku Valley used all sorts of intensive cultivation mechanisms to enhance their productivity, including raised fields, sunken gardens, agroforestry, terraces, irrigated pastures, river modifications, canals, and many others that still leave their traces on the modern landscape. D'Altroy explores the variety of agricultural adaptations of Andean inhabitants during protohistoric times. He points out how human populations expanded through an economic infrastructure dependent on production of staple crops as well as animal husbandry. The structural component of this agropastoral network was immense, and large-scale farming enterprises included irrigation, terracing, and land reclamation.

On the other side of the continent, the South American lowland areas represent an incredibly diverse mosaic of vegetation types as described by Daly and Mitchell. Roosevelt, in turn, examines how humans exploited, manipulated, and modified the habitats of the lower Amazon in late prehistoric times. She presents data on discernible impacts by the occupants of the Monte Alegre site near the Tapajos-Amazon confluence and contrasts these with the environmental history of the region. Her assertions are based on years of archaeological fieldwork in eastern Brazil.

From these papers, a picture of Precolumbians as integral and influential components in the American landscape arises. Certainly, the influence was greater in some regions than others, but the impact in many areas was indelible. Not only is it the intent of this volume to inventory human-induced landscape changes during Precolumbian times, but it is also a goal to reconsider effective agricultural technologies used by early complex societies. By a thorough examination of the landscape changes wrought by Precolumbian Americans, we can better understand the nature of today's New World environment. Finally, we can learn from the ecological successes of the past as well as the mistakes to help shape our land use practices in the future in ways that are both wise and sustainable.

REFERENCES

Abrams, E. M., A. Freter, D. J. Rue, and J. D. Wingard. 1996. The role of deforestation in the collapse of the Late Classic Copán Mayan state. In L. E. Sponsel, T. H. Headland, and R. C. Bailey, eds., *Tropical Deforestation*, pp. 55–75. New York: Columbia University Press.

Alcorn, J. B. 1984. *Huastec Mayan Ethnobotany*. Austin: University of Texas Press.

Alvard, M. S. 1993. Testing the "ecologically noble savage" hypothesis. *Human Ecology* 21(4): 355–387.

———. 1994. Conservation by native peoples: Prey choice in a depleted habitat. *Human Nature* 5: 127–154.

Balée, W., ed. 1998. Introduction in *Advances in Historical Ecology*. New York: Columbia University Press.

Billard, J. B. 1975. *National Geographic Atlas of the World*. 4th ed. Washington, D.C.: National Geographic Society.

Blanton, R. E., S. A. Kowalewski, G. Feinman, and J. Appel. 1981. *Ancient Meso-America: A Comparison in Three Regions*. New York: Cambridge University Press.

Botkin, D. B. 1990. *Discordant Harmonies*. Oxford: Oxford University Press.

Byers, D. S., ed. 1967. *The Prehistory of the Tehuacan Valley, 1: Environment and Subsistence*. Austin: University of Texas Press.

Cowley, G. 1991. The great disease migration. In *1492–1992, When Worlds Collide: How Columbus's Voyages Transformed both East and West. Newsweek,* Special Issue, Fall/Winter, pp. 54–56.

Crawford, M. H. 1998. *The Origins of Native Americans: Evidence from Anthropological Genetics.* Cambridge: Cambridge University Press.

Crumley, C. L. 1994. *Historical Ecology: Cultural Knowledge and Changing Landscapes.* Santa Fe, N.Mex.: School of American Research Press.

D'Altroy, T. N. 1997. Recent research on the central Andes. *J. Archaeological Research* 5: 3–73.

Deevey, E. S. 1978. Holocene forests and Maya disturbance near Quexil Lake, Petén, Guatemala. *Polskie Archiwum Hydrobiologii* 25: 117–129.

Denevan, W. M. 1988. Measurement of abandoned terracing from air photos: Colca Valley, Peru. *Yearbook, Conference of Latin Americanist Geographers* 14: 20–30.

———. 1992a. *The Native Population of the Americas in 1492.* Madison: University of Wisconsin Press.

———. 1992b. The pristine myth: The landscape of the Americas in 1492. *Annals of the Association of American Geographers* 82(3): 369–385.

Dobyns, H. F. 1983. *Their Number Became Thinned: Native American Population Dynamics in Eastern North America.* Knoxville, Tenn.: University of Tennessee Press.

Dunning, N. P. 1996. A reexamination of regional variability in the prehistoric agricultural landscape. In S. L. Fedick, ed., *The Managed Mosaic: Ancient Maya Agriculture and Resource Use,* pp. 53–68. Salt Lake City: University of Utah Press.

Ewell, P. T., and D. Merrill-Sands. 1987. Milpa in Yucatán: A long fallow maize system and its alternatives in the Maya peasant economy. In B. L. Turner II and S. B. Brush, *Comparative Farming Systems,* pp. 95–129. New York: Guilford Press.

Fagan, B. 1998. *From Black Land to Fifth Sun: The Science of Sacred Sites.* Reading, Mass.: Addison-Wesley.

Fowler, M. 1989. *The Cahokia Atlas: A Historical Atlas of Cahokia Archaeology.*

Studies in Illinois Archaeology 6. Springfield: Illinois Historic Preservation Agency.

Hanks, W. F. 1990. *Referential Practice: Language and Lived Space among the Maya.* Chicago: University of Chicago Press.

Harrison, P. D., and B. L. Turner II, eds. 1978. *Pre-Hispanic Maya Agriculture.* Albuquerque: University of New Mexico Press.

Hughes, J., ed. 1985. *The World Atlas of Archaeology.* New York: Portland House.

Kirch, P. V. 1997. Introduction: The environmental history of Oceanic Islands. In P. V. Kirch and T. L. Hunt, eds., *Historical Ecology in the Pacific Islands,* pp. 1–21. New Haven: Yale University Press.

Kolata, A. L. 1991. The technology and organization of agricultural production in the Tiwanaku state. *Latin American Antiquity* 2: 99–125.

Krech, S. III. 1999. *The Ecological Indian: Myth and History.* New York: Norton.

Loker, W. 1989. Contemporary land use and prehistoric settlement: An ethnoarchaeological approach. In K. G. Hirth, G. Lara Pinto, and G. Hasemann, eds., *Archaeological Research in the El Cajon Region, 1: Prehistoric Cultural Ecology,* pp. 135–187. Memoirs in Latin American Archaeology 1. Pittsburgh: University of Pittsburgh.

Lord, L., and S. Burke. 1991. America before Columbus. *U.S. News and World Report,* July 8, pp. 22–37.

Lovell, W. G. 1992. Heavy shadows and black night: Disease and depopulation in colonial Spanish America. *Annals of the Association of American Geographers* 82(3): 426–436.

MacNeish, Richard S. 1992. *The Origins of Agriculture and Settled Life.* Norman: University of Oklahoma Press.

Marcus, J., and K. V. Flannery. 1996. *Zapotec Civilization: How Urban Society Evolved in Mexico's Oaxaca Valley.* New York: Thames & Hudson.

Martin, P. S., and R. G. Klein. 1989. *Quaternary Extinctions: A Prehistoric Revolution.* Tucson: University of Arizona Press.

Meltzer, David. 1993. *Search for the First Americans.* Montreal: St. Remy Press.

Newson, L. 1998. A historical-ecological perspective on epidemic disease. In W. Balée, ed., *Advances in Historical Ecology*. New York: Columbia University Press.

Pohl, M. D., and P. Bloom. 1996. Prehistoric Maya farming in the wetlands of northern Belize: More data from Albion Island and beyond. In S. L. Fedick, ed., *The Managed Mosaic: Ancient Maya Agriculture and Resource Use,* pp. 145–164. Salt Lake City: University of Utah Press.

Pope, K. O., Pohl, M. D., and J. S. Jacobs. 1996. Formation of ancient Maya wetland fields: Natural and anthropogenic processes. In S. L. Fedick, ed., *The Managed Mosaic: Ancient Maya Agriculture and Resource Use,* pp. 165–176. Salt Lake City: University of Utah Press.

Price, Douglas T., and Gary M. Feinman. 1993. *Images of the Past.* Mountain View, Calif.: Mayfield.

Redfield, R., and A. Villa Rojas. 1990. *Chan Kom: A Maya Village.* Prospect Heights, Ill.: Waveland Press.

Redford, K. H. 1991. The ecologically noble savage. *Cultural Survival Quarterly* 15(1): 46–48.

Redman, C. L. 1999. *Human Impacts on Ancient Environments.* Tucson: University of Arizona Press.

Roosevelt, A. C. 1991. *Moundbuilders of the Amazon: Geophysical Archaeology on Marajo Island, Brazil.* New York: Academic Press.

Sanders, W. T., J. R. Parsons, and R. S. Santley. 1979. *The Basin of Mexico: Ecological Processes in the Evolution of a Civilization.* New York: Academic Press.

Scarborough, V. 1996. Reservoirs and watersheds in the central Maya Lowlands. In S. L. Fedick, ed., *The Managed Mosaic: Ancient Maya Agriculture and Resource Use,* pp. 304–314. Salt Lake City: University of Utah Press.

Sebastian, L. 1992. *The Chaco Anasazi: Sociopolitical Evolution in the Prehistoric Southwest.* New York: Cambridge University Press.

Sharer, R. J. 1994. *The Ancient Maya.* 5th ed. Stanford, Calif.: Stanford University Press.

Siemens, A. H. 1998. *A Favored Place: San Juan River Wetlands, Central Veracruz,* A.D. 500 to the Present. Austin: University of Texas Press.

Silverberg, R. 1986. *The Mound Builders.* Athens: Ohio University Press.

Smyth, M. P. 1991. *Modern Maya Storage Behavior: Ethnoarchaeological Examples from the Puuc Region of Yucatán.* Memoirs in Latin American Archaeology 3. Pittsburgh: University of Pittsburgh.

Thornton, R. 1990. *American Indian Holocaust and Survival: A Population History since 1492.* Norman: University of Oklahoma Press.

Turner, B. L. II. 1990. Population reconstruction of the central Maya Lowlands: 1000 B.C. to A.D. 1500. In P. Culbert and D. S. Rice, eds., *Precolumbian Population History in the Maya Lowlands,* pp. 301–324. Albuquerque: University of New Mexico Press.

Ubelaker, D. H. 1988. North American Indian population size, A.D. 1500 to 1985. *American J. Physical Anthropology* 77: 289–294.

Vaughan, H. H., E. S. Deevey Jr., and S. E. Garrett-Jones. 1985. Pollen stratigraphy of two cores from the Petén lake district, with an appendix of two deep-water cores. In M. Pohl, ed., *Prehistoric Lowland Maya Environment and Subsistence Economy,* pp. 73–89. Cambridge, Mass.: Peabody Museum of Archaeology and Ethnology, Harvard University.

Wessels, T. 1997. *Reading the Forested Landscape: A Natural History of New England.* Woodstock, Vt.: Countryman Press.

Whitmore, T. M. 1991. A simulation of the sixteenth-century population collapse in the Basin of Mexico. *Annals of the Association of American Geographers* 81: 464–487.

Willey, G. R., and J. A. Sabloff. 1974. *A History of American Archaeology.* San Francisco, Calif.: W. H. Freeman & Co.

Zimmerer, K. S. 1994. Human geography and the "new ecology": The prospect and promise of integration. *Annals of the Association of American Geographers* 84(1): 108–125.

DAVID A. HODELL
MARK BRENNER
JASON H. CURTIS

2 | CLIMATE CHANGE IN THE NORTHERN AMERICAN TROPICS AND SUBTROPICS SINCE THE LAST ICE AGE
Implications for Environment and Culture

Climate change during the past 20,000 years affected both the environments and civilizations of the Americas. The last glaciation marked the arrival of humans into the New World, and in the last 3,500 years, many advanced cultures arose, flourished, and ultimately collapsed within the context of Holocene climate variability. Here we review the history of climate change in the northern American tropics and subtropics between ca. 10 and 30°N from the last glacial maximum (ca. 20,000 years) to the present. We cite the Classic Maya civilization as an example of the complex interplay among culture, climate, and environment and discuss the potential impact of human activities and climate in lowland tropical Mesoamerica.

Because instrumental records of climate are limited to the last couple of centuries and historical documents extend at most over the last four centuries, we must rely on natural sources of information to reconstruct paleoclimate and paleoenvironment during older time periods. For this purpose, lacustrine sediments are valuable archives of paleoclimatic and paleoecologic information. Lake sediments accumulate in an ordered manner through time and record changes in the hydrologic and material fluxes within a lake and its drainage basin (Binford et al. 1987). Paleoenvironmental changes can be inferred by stratigraphic analysis of environmental "proxies" that are well preserved in lacustrine deposits. These proxies are natural materials, such as pollen grains, minerals, diatoms, or animal microfossils (e.g., gastropod or ostracod shells), that are affected by environmental variables and can therefore be used to infer past conditions. For example, pollen grains composed of sporopollenin are highly resistant to decay and can be used to infer past vegetation shifts. Because both natural climate change and human agricultural practices can lead to changes in vegetation, however, differentiation of these effects using pollen alone is difficult (Bradbury et al. 1990). In many areas of Mesoamerica, for instance, changes in the relative abundance of pollen types after 3000 B.P. were highly influenced by human-induced land clearance (Vaughan et al. 1985).

Geochemical proxies, such as stable isotope and trace element ratios of fossil shell carbonate, can be used to reconstruct past changes in temperature and the hydrologic budget of lakes (see Holmes 1996). In tropical closed-basin lakes, the relative ratio of ^{18}O to ^{16}O (expressed as $\delta^{18}O$)* in shell calcite is controlled mainly by the ratio of evaporation to precipitation (E/P) over the lake basin (Fontes and Gonfiantini 1967). This proxy can be used to reconstruct relative changes in precipitation, which can profoundly influence natural vegetation, agriculture, and cultural development. Although it is generally accepted that oxygen isotopes reflect dominantly climatic changes, the construction of dams, canals, and dikes by humans may alter the water budgets of lakes (e.g., Lake Coba, Leyden et al. 1998), making paleohydrologic proxies such as $\delta^{18}O$ difficult to interpret. Furthermore, it is possible that human disturbance can influence climate; for example, widespread deforestation may lead to changes in a lake's hydrologic budget by changes in regional precipitation or runoff (Lean and Warrilow 1989; Brenner et al. 1990; Rosenmeier et al. submitted).

Distinguishing between the effects of human disturbance and climate change on environment is difficult, but the use of multiple proxy variables in lake sediment cores can potentially differentiate between these environmental stresses. For example, changing pollen assemblages unaccompanied by shifts in oxygen isotope ratios might point to human disturbance, rather than climate, as the cause of vegetation change. Here we present multiple proxy data from lake sediment cores to elucidate the complex interactions that existed among culture, climate, and environment in Precolumbian Mesoamerica.

MODERN CLIMATE OF MIDDLE AMERICA AND THE INTRA-AMERICAS SEA

The region of study we consider here is the area from the southern United States to northern South America and includes the Gulf of Mexico and Caribbean, referred to jointly as the Intra-Americas Sea (figure 2.1a). The climate of the region is subtropical to tropical, and climate variability is most notably expressed by temporal and spatial changes in precipitation. Much of the area is under the influence of the northeast trade winds, and precipitation is highly seasonal. The rainy season coincides with Northern Hemisphere summer when the intertropical convergence zone (ITCZ), a band of low pressure usually located near the equator, moves northward (Hastenrath 1966, 1967,

---

* $\delta^{18}O = \left(\dfrac{(^{18}O/^{16}O) \text{ sample}}{(^{18}O/^{16}O) \text{ reference}} - 1 \right) \cdot 1000$; reference is PDB and units are 0/00.

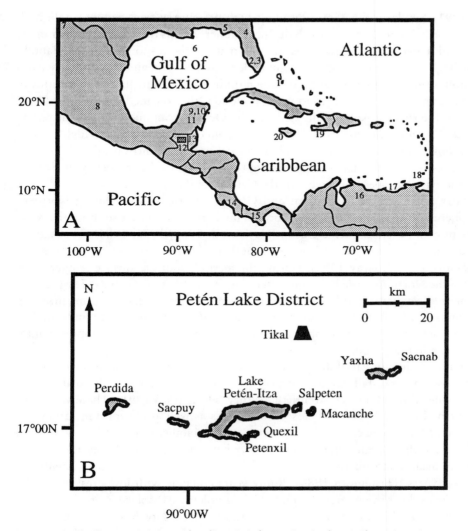

FIGURE 2.1. A: Sites containing paleoclimatic information in the northern American tropics and subtropics. 1, Church's Blue Hole, Andros Island, Bahamas. 2, Lake Tulane, Florida. 3, Lake Annie, Florida. 4, Lake Sheelar, Florida. 5, Camel Lake, Florida. 6, Orca Basin, Gulf of Mexico. 7, Chihuahua sites (Metcalfe et al. 1997). 8, Basin of Mexico (Pátzcuaro Basin; La Piscina de Yuriria; Metcalfe et al. 1989). 9, Lake Coba, Mexico. 10, Lake Punta Laguna, Mexico. 11, Lake Chichancanab, Mexico. 12, Petén Lake District, Guatemala (see B for detailed map). 13, Laguna de Cocos and Cobweb Swamp, Belize. 14, Laguna Bonilla and Laguna Bonillita, Costa Rica. 15, Lakes El Valle and La Yeguada, Panama. 16, Lake Valencia, Venezuela. 17, Caiaco Basin (Hughen et al. 1996). 18, Barbados (Guilderson et al. 1994). 19, Lake Miragoane, Haiti. 20, Wallywash Pond, Jamaica. B: Petén Lake District, Guatemala (detail of site 12).

1976, 1991). Similarly, the Azores-Bermuda high-pressure system, centered in the mid-latitude North Atlantic, moves north during summer. At this time, surface water in the tropical-subtropical North Atlantic is warm and atmospheric moisture is abundant. In the Caribbean and Central America region, the rainy season begins slowly in late April and is followed by high rainfall during the months of May and June. The months of July and August are typically drier and less cloudy (the so-called *canicula,* or "little dry"), and the rainy season usually ends by late October. Summer rains in the region often occur as violent thunderstorms resulting from convection of air to high altitude in the atmosphere. Tropical storms and hurricanes can also contribute greatly to rainfall averages for a particular year, and long-term trends in the frequency of tropical cyclone activity have been observed for the Caribbean (Gray 1987, 1993).

The dry season in the Caribbean and Central America occurs during Northern Hemisphere winter when the ITCZ is located farther to the south over both the western Atlantic and eastern Pacific Ocean (Hastenrath 1966, 1976, 1984). The Azores-Bermuda high-pressure system also moves south and dominates in the Intra-Americas Sea (Nieuwolt 1977; Gray 1993). During this period, cool sea surface temperatures prevail in the North Atlantic in a band extending from about 10 to 20°N. The pressure gradient on the equatorial side of the Azores-Bermuda high is steep, resulting in strong trade winds. A strong temperature inversion associated with the trade winds inhibits convective activity in the atmosphere. All of these factors conspire to suppress winter precipitation in the region.

Interannual climate variability in the Caribbean–Central American region is primarily reflected in rainfall quantity. Runs of predominantly wet years have alternated with dry ones throughout the twentieth century (Hastenrath 1991). These variations affect the hydrologic budgets of closed-basin lakes as well as natural vegetation and agriculture. Meteorological studies of interannual variability of rainfall in the Central America–Caribbean region during this century have identified conditions (i.e., departures from mean atmospheric and oceanic fields) that coincide with exceptionally dry versus wet years (Hastenrath 1984). Rainy years coincide with low pressure on the equatorward side of the Azores-Bermuda high, a poleward displacement of the North Atlantic trade winds, warm sea surface temperatures in the North Atlantic between 10 and 20°N, and enhanced convergence and cloudiness (Hastenrath 1984). Wet years in the region appear to result, in part, from an enhancement of the annual cycle when the ITCZ and Bermuda-Azores high move farther north during summer. In contrast, dry years are associated with the opposite of the aforementioned conditions, resulting from a reduction in the intensity of the annual cycle.

Rainfall in the Caribbean and Central American region is also highly variable spatially and related to local circulation features that are affected by coastline orientation, topography, and other factors. For example, the coastlines of Belize, Guatemala, and Honduras form a right angle in the Gulf of Honduras. This geometry together with the location of the Maya Mountains tends to direct moisture-laden winds into Belize and northern Guatemala (Wilson 1984). Summer precipitation in the Central America–Caribbean region is predominantly of Atlantic origin, but winter rainfall in northern Mexico and the northern portion of the Gulf of Mexico can originate from the Pacific when the wintertime subtropical westerly jet is accelerated (Hastenrath 1991). Northern Mexico and the southern United States tend to receive abundant precipitation during winters of El Niño years.

The seasonal cycle of precipitation exerts a strong control on native vegetation and agriculture in Mesoamerica. Natural vegetation may be less susceptible to variability in rainfall than domesticated crops, such as maize (Leyden et al. 1998). Much of the ancient agricultural activity in Mesoamerica was maize-based, swidden (slash-and-burn, shifting) agriculture. In the swidden cycle, forest is felled during the dry season (January–May) and burned (approximately April), prior to the onset of rains and planting during spring (May). The timing of rainfall can be as important as the amount for milpa productivity because a dry season is necessary for burning and the onset of the rainy season is critical following planting. Any disruption in the timing or amount of rainfall can result in poor crop yields by adversely affecting burning of cleared fields, seed germination, or crop maturation (Leyden et al. 1998).

PALEOCLIMATIC HISTORY

THE LAST ICE AGE

Climate conditions in the study region during the last glaciation were significantly different from today. Sea surface temperature (SST) in the tropical Atlantic was 4 to 5°C lower than today (Guilderson et al. 1994). Growth of large continental ice sheets reduced sea level by ca. 120 m and altered the geography of the Caribbean Sea and Gulf of Mexico. Water tables in karst regions surrounding the Intra-Americas Sea were lowered in aquifers where the freshwater lies atop seawater. Deep-water circulation in the North Atlantic changed during the last glaciation when the production of North Atlantic deep water (NADW) was reduced and glacial North Atlantic intermediate water (GNAIW) was enhanced. This change has implications for the Caribbean Sea because intermediate waters fill the basin from the North Atlantic, flowing over the sill at depths of 1,600 to 1,800 m. The glacial reduction of NADW also changed the heat and salt balance of the North Atlantic and may have reduced the cross-equatorial transport of surface waters from the South Atlantic to the North Atlantic. Lastly, the Laurentide Ice Sheet contributed seasonal meltwater to the Gulf of Mexico, thereby lowering salinity and SST (Spero and Williams 1990; Overpeck et al. 1989).

Evidence from lake sediment records in the circum-Caribbean and southeast United States indicates that climate was cooler and drier during the last ice age compared to the Holocene. Most shallow lake basins were dry (in some cases, the result of lowered sea level), and evidence for glacial-age lacustrine sedimentation is found in only a few deep lakes. For example, pollen evidence indicates extremely arid conditions surrounding Lake Quexil, Guatemala, and temperatures between 6.5 and 8°C colder than today (Leyden et al. 1993). Glacial-age vegetation was dominated by savanna scrub and juniper around the lake, and gypsum precipitated from the shallow lake water under arid conditions (figure 2.2). In Panama, palynological records from lakes El Valle and La Yeguada suggest drier conditions during the last glaciation with temperature depression of 4–6°C (Bush and Colinvaux 1990; Piperno et al. 1990). This area remained

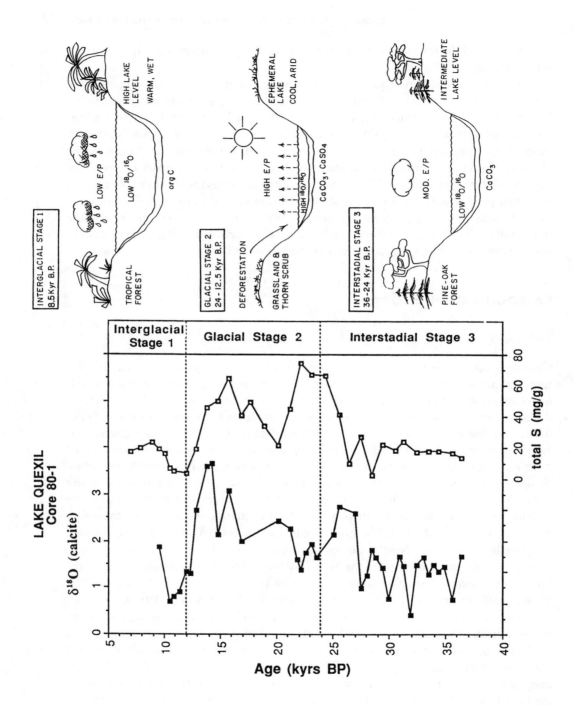

forested during the last glaciation, however, and supported an oak-rich cloud forest that does not exist in the region today. In Venezuela, pollen assemblages in a sediment core from Lake Valencia indicate extreme aridity during the last glaciation (Leyden 1985). Many insular Caribbean lakes were dry during the last glacial period, such as Wallywash Pond, Jamaica, which was desiccated between 93,500 and 10,000 B.P. (Street-Perrott et al. 1993; Holmes 1998). Lake Miragoane, Haiti, was also dry prior to 10,500 B.P. (Hodell et al. 1991). In Florida, most lakes were dry during the last glaciation owing to reduced precipitation and/or lowered sea level (Watts and Hansen 1994). Nevertheless, some deep solution basins held water during the last glacial period, including Lakes Annie (Watts 1975), Sheelar (Watts and Stuiver 1980), Camel (Watts et al. 1992), and Tulane (Grimm et al. 1993). In Lake Tulane, abrupt shifts from pine forest to oak-savanna or grassland occurred during the last glaciation, indicating rapid shifts in moisture (Grimm et al. 1993). These abrupt changes have been correlated to the North Atlantic "Heinrich events" that represent massive ice discharges from the eastern margin of the Laurentide Ice Sheet (Bond et al. 1992).

Late Pleistocene aridity throughout the circum-Caribbean is consistent with lower SST (Emiliani and Ericson 1991) and the occurrence of high-salinity surface waters in the Caribbean Sea (Prell and Hays 1976). A reduction of tropical SST by 5 to 6°C would have significantly decreased evaporation rates and atmospheric water vapor. The Caribbean may have also experienced less rainfall during the last glaciation because the ITCZ was located farther south than today. Colinvaux (1972) reported that climate was drier on the Galapagos Islands during the late Pleistocene, and Newell (1973) suggested this aridity may have been caused by a southward shift of the ITCZ.

The Laurentide Ice Sheet had a profound effect on glacial-age circulation in the North Atlantic. The polar front was displaced to the south with a very strong westerly flow (jet stream) in the North Atlantic (Kutzbach and Webb 1993). The trade winds were probably stronger during glacial time (Broccoli and Manabe 1987; Bradbury 1997), and both the ITCZ and Azores-Bermuda high were located farther south than today because of an enhanced temperature gradient in the mid-latitudes of the Northern Hemisphere. Southward displacement of the Azores-Bermuda high would have

---

FIGURE 2.2. Oxygen isotopic composition of bulk carbonate and total sulfur records from Lake Quexil core 80-1, along with cartoon depicting climate conditions between ~36 and 5 kyr B.P. (data from Leyden et al. 1993). Higher oxygen isotopic ratios are interpreted as representing increases in the ratio of evaporation to precipitation (i.e., drier climate), and high sulfur concentration indicates gypsum precipitation under evaporative conditions. Vegetation is inferred from the pollen spectrum (data from Leyden et al. 1993). During interstadial stage 3 (36 to 24 kyr B.P.), vegetation was more temperate than today and marked by a montane pine-oak forest, suggesting cool-moist conditions. During the last glaciation (glacial stage 2), climate was extremely dry and the pine-oak forest of stage 3 was replaced by sparse savanna scrub with juniper, indicating increased aridity and temperatures 6.5 to 8°C colder than today. Dry climatic conditions began to ameliorate with the onset of deglaciation, and lowland tropical forests developed in the watershed during the early Holocene, ca. 8.5 radiocarbon kyr B.P.

resulted in a stronger temperature inversion associated with the trade winds over the Caribbean–Central America region, thereby reducing convection and summer precipitation. Summer melting of ice would have delivered meltwater down the Mississippi and cooled SSTs in the Gulf of Mexico. Together with overall cooler glacial-age SSTs in the tropics (Guilderson et al. 1994), evaporation would have been reduced in the Intra-Americas Sea, resulting in less precipitation during the summer rainy season (Overpeck et al. 1989).

Whereas most of Mesoamerica and the Intra-Americas Sea experienced reduced summer precipitation during the last glaciation, the region west of 95° W was marked by increased winter precipitation (Bradbury 1997). This pattern probably resulted from a split of the jet stream with the southern limb located at about 30° N during winter, directing storm tracks into the American southwest (Thompson et al. 1993) and into northwestern Mexico as far south as 20° S (Bradbury 1997; Metcalfe et al. 1997).

PLEISTOCENE–HOLOCENE TRANSITION

The transition from the last glacial maximum to the Holocene (i.e., Termination I) represents a profound change relative to the smaller amplitude of climatic variability during the last 10,000 years. The transition was not smooth and was marked by abrupt changes in North Atlantic climate. Deglaciation occurred in two distinct steps marked by meltwater pulses to the ocean and sea level rise (Fairbanks 1989). The early meltwater of the Laurentide Ice Sheet was directed south down the Mississippi River and resulted in cold, low-salinity surface waters in the Gulf of Mexico. Meltwater delivery to the Gulf is recorded by low oxygen isotopic values of planktonic foraminifera in the Orca Basin starting by 16,000 B.P. and culminating in meltwater pulse (MWP) 1A at 12,000 B.P. (Leventer et al. 1982; Flower and Kennett 1990; Spero and Williams 1990). This was followed by a return to glacial-like conditions corresponding to the Younger Dryas period between ca. 11,000 and 10,000 B.P. The second step in the deglaciation process was marked by MWP 1B, centered on 9500 B.P., when meltwater input to the Gulf of Mexico was curtailed and drainage of the Laurentide Ice Sheet was diverted to the east via the St. Lawrence Seaway.

The best record of the last deglaciation in the Caribbean comes from studies of laminations in deep-sea cores retrieved from the Cariaco Basin off the coast of Venezuela (figure 2.3; Hughen et al. 1996). These laminations consist of alternating dark- (mineral-rich) and light-colored (plankton-rich) bands that are produced by runoff and seasonal upwelling in the Cariaco Basin. An abrupt decrease in gray-scale intensity of Cariaco sediments during the Younger Dryas Event has been interpreted as indicating increased wind-induced upwelling, implying that the northeast trade winds intensified as the North Atlantic cooled. The similarity of the Cariaco gray-scale record and the Greenland (GRIP) ice core (see figure 2.3) suggests that climate in the Caribbean and North Atlantic must have been tightly coupled during the last deglaciation until about 8000 B.P. (Hughen et al. 1996).

Few complete lacustrine records of the last deglaciation are available from the region because most shallow lakes were dry throughout this period. The Younger Dryas is recorded in pollen records from Costa Rica at 2,310 m by a 2–3°C decrease in

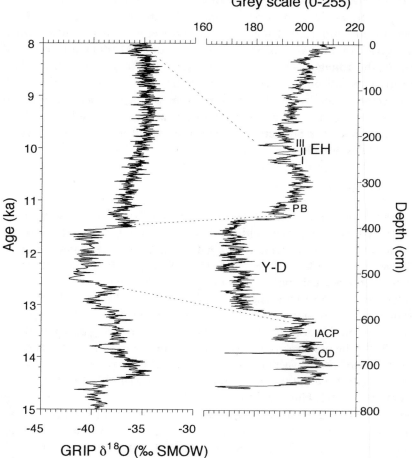

Cariaco Basin Core PL07-56PC
Grey scale (0-255)

FIGURE 2.3. Comparison of gray scale from Cariaco Basin core PL07-56PC taken just north of Venezuela to oxygen isotopes in the Greenland (GRIP) ice core. The similarity of signals suggests a tight coupling between climate of the high-latitude North Atlantic and Caribbean. Times of lower temperature (lower $\delta^{18}O$) in Greenland correlate with brighter (low) values of gray scale indicating increased wind-induced upwelling in the Cariaco Basin. Climatic events identified include Older Dryas (OD), inter-Allerod cold period (IACP), Younger Dryas (Y-D), Preboreal (PB), and Early Holocene (EH) events I, II, and III. *Source:* After figure 4a of Hughen et al. 1996.

temperature and a drier climate (Islebe et al. 1995). Leyden et al. (1994) found a series of events in Lake Quexil, Guatemala, associated with deglaciation, but lack of high-resolution dating hampered direct correlation with other paleoclimatic records. The deglacial sequence in Lake Quexil included a change from cold and dry conditions during the last glaciation, to warmer and moister climate during early deglaciation, followed by cold and moist conditions during the Younger Dryas, and a cool, wet climate after 10,300 B.P. This sequence does not completely agree with climate modeling results that predict: colder and drier during meltwater delivery to the Gulf of Mexico between 14,000 and 12,000 B.P., warmer and wetter climate starting when meltwater was diverted into the Gulf of St. Lawrence (12,000–11,000 B.P.), decreased temperature during the Younger Dryas Chronozone (ca. 11,000–10,000 B.P.), and increased precipitation and warming at about 9000 B.P. in the early Holocene (Overpeck et al. 1989; Maasch and Oglesby 1990). Continuous lacustrine sediment records and good age control spanning the period of the last deglaciation are needed to test fully the sequence of climatic events predicted by modeling experiments (Overpeck et al. 1989; Maasch and Oglesby 1990).

EARLY TO MIDDLE HOLOCENE

Pollen evidence throughout the circum-Caribbean and southeast United States indicates an abrupt switch from arid to humid conditions near the Pleistocene-Holocene boundary (Piperno et al. 1990; Markgraf 1989; Leyden 1984, 1985; Leyden et al. 1993, 1994; Deevey et al. 1983; Bradbury et al. 1981; Watts 1975; Islebe et al. 1996; among others). In northern Guatemala, for example, a mesic tropical forest developed in the early Holocene at ca. 8500 B.P. (figure 2.2; Leyden 1984; Islebe et al. 1996). A similar vegetation history has been inferred for Lake Valencia, Venezuela, suggesting that the tropical forests in this area are youthful and developed during the early Holocene (Leyden 1985). In the northeastern Caribbean, sediments from Lake Miragoane, Haiti, record a switch from dry vegetation dominated by xeric palms and montane shrubs between ca. 10,400 and 8200 B.P., to mesic forests between 8200 and 2500 B.P. (figure 2.4; Hodell et al. 1991; Higuera-Gundy et al. 1999).

Oxygen isotope data in sediment cores from Lakes Miragoane and Valencia suggest relatively dry conditions during the earliest Holocene (10,500 to 8500 B.P.) followed by abundant moisture after ca. 8500 B.P. (figure 2.4; Hodell et al. 1991; Curtis et al. 1999). Early Holocene δ^{18}O records from lakes on the Yucatán Peninsula are difficult to interpret because the basins were in the early stages of filling (Curtis 1997; Curtis et al. 1998). If interpreted simply as reflecting changes in evaporation/precipitation (E/P), they would suggest relatively drier climate during the early Holocene until ca. 6000 B.P., which is at odds with pollen evidence suggesting widespread tropical forest and hence moist conditions after 8500 B.P. (figure 2.5). Perhaps there was sufficient rainfall to support tropical dry forest vegetation during the early Holocene, but high δ^{18}O values simply reflect the annual loss of a significant fraction of the lake volume to evaporation from these initially shallow basins. Alternatively, dense early Holocene vegetation may have decreased runoff of surface and groundwater into the lake (Rosenmeier et al. submitted). In marine sediment cores from the Cariaco Basin, the earliest Holocene is marked by three distinct oscillations (EH I, II, and III) in gray scale from 9000 to 7500 calendar years B.P. (figure 2.3; Hughen et al. 1996). These same oscillations appear to be recorded near the base of the oxygen isotopic record in a sediment core from Lake Valencia, indicating that rapidly alternating dry and wet

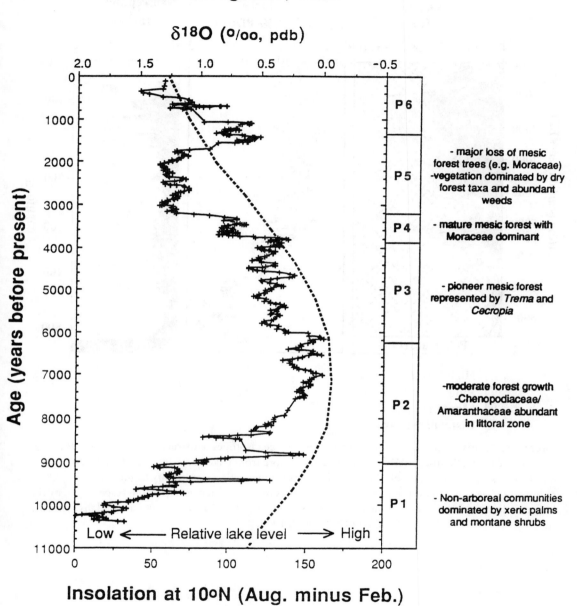

FIGURE 2.4. Oxygen isotope measurements of the ostracod *Candona* sp. in a core from Lake Miragoane, Haiti. The isotope record, indicating changing evaporation/precipitation and lake level, is compared to the difference in insolation (Langleys) at the top of the atmosphere at 10°N between the months of August and February, which is a measure of the intensity of the annual cycle. Pollen zonation and interpretation is provided to the right of figure. *Source:* After figure 2 of Hodell et al. 1991.

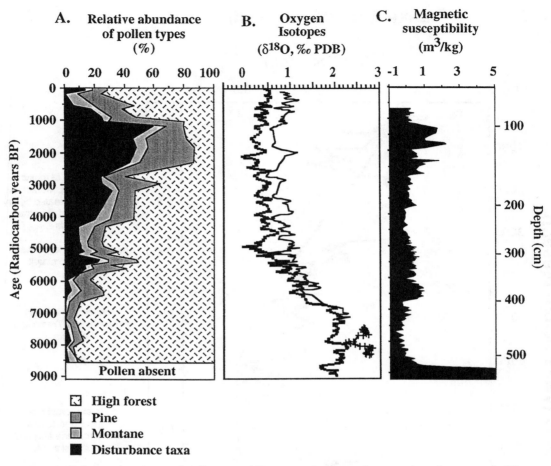

A. Relative abundance of pollen types (%)

B. Oxygen Isotopes (δ¹⁸O, ‰ PDB)

C. Magnetic susceptibility (m³/kg)

Pollen absent

- ⬚ High forest
- ▨ Pine
- ▢ Montane
- ■ Disturbance taxa

FIGURE 2.5. Relative abundance of pollen types (A), oxygen isotopes of ostracods and gastropods (B), and volume magnetic susceptibility (C) of a sediment core from Lake Petén-Itza, Guatemala. Note the rapid increase in disturbance pollen taxa and increase in magnetic susceptibility beginning at ~3,000 radiocarbon years B.P., interpreted to represent deforestation and increased erosion, respectively, by the Maya civilization. No corresponding change is observed in oxygen isotopes, suggesting relative climatic stability. *Source:* After figures 4 and 6 of Curtis et al. 1998.

conditions on land coincided with variations in trade wind strength and upwelling intensity offshore (Curtis et al. 1999).

Most lacustrine records in the circum-Caribbean indicate that the wettest period throughout the northern American tropics occurred between ca. 7000 and 5000 B.P. (Bradbury et al. 1981; Deevey et al. 1983; Leyden 1985; Piperno et al. 1990). Hodell et al. (1991) suggested that this period of high precipitation was related to an increase in the intensity of the annual cycle caused by the earth's precessional cycle. Today, minimum earth–sun distance (perihelion) occurs during Northern Hemisphere winter (January 3) resulting in slightly warmer winters and cooler summers. During the early Holocene, however, perihelion coincided with Northern Hemisphere summer, thereby increasing the seasonal contrasts in incoming solar radiation (figure 2.6). On the basis of modern meteorological studies of Caribbean climate, increases in the intensity of the annual cycle correlate with years of high precipitation during the twentieth century (Hastenrath 1984). Likewise, an enhancement of the annual cycle during the early Holocene, caused by the earth's precessional cycle, may have resulted in wetter summers (rainy season) and drier winters, i.e., increased seasonality and an overall increase in annual precipitation.

Increased seasonality may explain the widespread occurrence of laminated sediments and evidence for increased fire frequency in the early Holocene, as indicated by increased charcoal abundance in lake sediments (Piperno et al. 1990; Hodell et al. 1991; Leyden et al. 1994; Emiliani et al. 1991). Alternatively, the appearance of carbon and burnt weedy material in the early Holocene may reflect the early use of fire by prehistoric humans (Piperno et al. 1990). Emiliani et al. (1991) have even suggested that the onset of widespread forest burning by humans may have contributed to the transition from the last ice age to the interglacial conditions of the Holocene.

LATE HOLOCENE

A trend toward distinctly drier climate in the circum-Caribbean was initiated beginning about 3000 B.P. In Lake Miragoane, Haiti, a two-step increase in $\delta^{18}O$ values between 3200 and 2400 B.P. indicates increasing E/P and lower lake level (figure 2.4; Hodell et al. 1991). Pollen assemblages indicate a shift near 2500 B.P. from mature mesic forests dominated by Moraceae to dry forest taxa with abundant weeds (Higuera-Gundy et al. 1999). In Lake Chichancanab, Mexico, oxygen isotopic values and sulfur (gypsum) content of the sediment began to increase at 3000 B.P., indicating the inception of a late Holocene drying trend (Hodell et al. 1995). In Lake Valencia, Venezuela, proxy indicators suggest a trend toward higher salinity conditions beginning at 2800 B.P. (Bradbury et al. 1981; Binford 1982; Leyden 1985). Dry conditions beginning at 3200 B.P. have also been inferred from palynological evidence from the Bahamas (Kjellmark 1996). Although there is considerable variability in the timing and magnitude of increased aridity in the circum-Caribbean during the late Holocene, pollen and geochemical evidence indicate that the late Holocene was drier than the early-to-mid Holocene.

The paleoclimatic record from Lake Chichancanab suggests that the drying trend that began at ca. 3000 B.P. culminated in very arid conditions between 1300 and 1100 B.P. (equivalent to approximately A.D. 800 to 1000; Hodell et al. 1995). Oxygen isotope results from nearby Lake Punta Laguna, Mexico, further resolved distinct drought events that lasted on the order of decades to centuries (figure 2.7; Curtis et al.

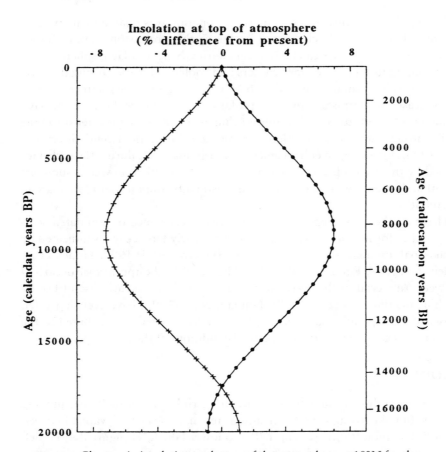

FIGURE 2.6. Changes in insolation at the top of the atmosphere at 10° N for the months of February (crosses) and August (filled circles) from 20,000 years to present. The distance between the two curves represents insolation forcing that affects the intensity of the annual cycle, which has been linked to interannual variability in rainfall in the Caribbean (Hastenrath 1984). Note that the seasonal insolation difference was high during the early and mid Holocene and diminishes throughout the late Holocene.

1996). Exceptionally dry periods were centered at ca. 1510, 1171, 1019, 943 B.P. (equivalent to A.D. 585, 862, 986, and 1051 ± 50). This period of increased drought frequency ended abruptly at about A.D. 1000 when increased moisture returned to the Yucatán Peninsula. Evidence of drying from other areas in Mexico and Central America suggests that severe droughts between ca. 1300 and 1100 B.P. may have been common and widespread. For instance, Sr/Ca ratios of bulk sediments from Lago de Pátzcuaro, Michoacán, Mexico, and desiccation in La Piscina de Yuriria, Mexico, between ca. 1500 and 900 B.P. indicate drought (Metcalfe 1995; Metcalfe et al. 1994). Increased charcoal abundance between 1180 and 1110 B.P. in records from Costa Rica suggests increased frequency of forest fires during this period (Horn and Sanford

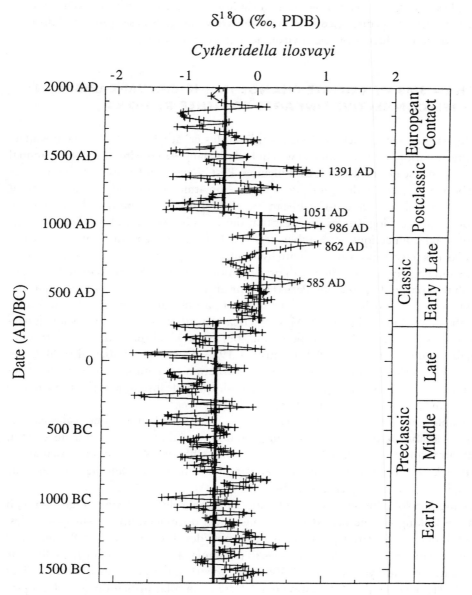

$$\delta^{18}O \ (\permil, \ PDB)$$

Cytheridella ilosvayi

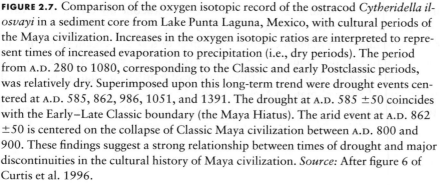

FIGURE 2.7. Comparison of the oxygen isotopic record of the ostracod *Cytheridella ilosvayi* in a sediment core from Lake Punta Laguna, Mexico, with cultural periods of the Maya civilization. Increases in the oxygen isotopic ratios are interpreted to represent times of increased evaporation to precipitation (i.e., dry periods). The period from A.D. 280 to 1080, corresponding to the Classic and early Postclassic periods, was relatively dry. Superimposed upon this long-term trend were drought events centered at A.D. 585, 862, 986, 1051, and 1391. The drought at A.D. 585 ±50 coincides with the Early–Late Classic boundary (the Maya Hiatus). The arid event at A.D. 862 ±50 is centered on the collapse of Classic Maya civilization between A.D. 800 and 900. These findings suggest a strong relationship between times of drought and major discontinuities in the cultural history of Maya civilization. *Source:* After figure 6 of Curtis et al. 1996.

1992). Drought has also been reported in California and Patagonia at about the same time (Stine 1994). As discussed later in this essay, we suggest these episodes of drought had a profound influence on the environment and culture of Mesoamerica.

MECHANISMS OF CLIMATE CHANGE IN THE CIRCUM-CARIBBEAN AND TELECONNECTIVE LINKAGES TO OTHER REGIONS

On the basis of recent meteorological studies (Hastenrath 1976, 1984), we postulate that precipitation in the Caribbean was linked to changes in the intensity of the annual cycle on time scales of decades to millennia. Past changes in Caribbean climate may have been associated with large-scale departures in atmospheric and oceanic fields that affected not only the circum-Caribbean but also left their imprint on a broad region. Historical meteorological data suggest that the Caribbean–Central America area has complex teleconnections with other regions. For example, rainfall in the circum-Caribbean is positively correlated with hydrometeorological conditions in sub-Saharan Africa and negatively correlated with precipitation in the U.S. Central Great Plains and SST off the coast of Ecuador and Peru (Hastenrath 1976; Hastenrath and Kaczmarczyk 1981). Prior to the paleolimnological studies of past climate from the Yucatán Peninsula, the paleoclimatic history of Mesoamerica was inferred largely by assuming teleconnections with areas beyond the region, but from which paleoclimatic data were available (Gunn and Adams 1981; Folan et al. 1983; Messenger 1990; Gill 1994; Gunn et al. 1995). Although such methods of reconstruction are tenuous, comparison of paleoclimatic records among regions can potentially provide insight into the atmospheric and oceanic processes responsible for past climate change.

Possible causes of past climate change on decadal-to-millennial time scales include solar variability and insolation forcing, random variability in the climate system, ocean-atmosphere dynamics (El Niño–southern oscillation, North Atlantic oscillation, North Atlantic deep water production, etc.), and injection of volcanic aerosols into the atmosphere (Rind and Overpeck 1994).

Changes in solar luminosity may play an important role in climate change, although data are lacking for the period prior to historical documentation of sunspot cycles. Spectral analysis of variation in cosmogenic nuclides, such as ^{14}C or ^{10}Be, indicates periods of solar variability from decades to millennia: 11- and 22-year cycles, 88-year cycle; ca. 200-year cycle, and ca. 2,500-year cycle (Damon and Linick 1986; Stuiver et al. 1991). Lean et al. (1995) estimated a decrease in solar insolation of ca. 0.25 percent during the Maunder Minimum (A.D. 1660–1710) and concluded that solar variability has affected global climate since A.D. 1610. Rind and Overpeck (1994, 1995) used a general circulation model to estimate the effect of a 0.25 percent reduction in the solar constant and found about a 0.5°C cooling of global mean temperature. Low-latitude cooling resulted in a decrease in the intensity of Hadley circulation by 5 to 10 percent, which could affect climate in Mesoamerica.

Long-term variations in the earth's orbital geometry create changes in the seasonal distribution of radiation (see figure 2.6). The most important parameter affecting insolation at low latitudes is the earth's precessional cycle that varies with a characteristic period of 19,000 and 23,000 years, with harmonic periods of 11,000 and 9,500 years (Short et al. 1991; Hagelberg et al. 1994). Hodell et al. (1991) suggested that long-term changes in Caribbean climate could be explained by orbitally induced variations in seasonal insolation that modified the intensity of the annual cycle. For example, compared with the present, early Holocene insolation at 10°N was almost 8 percent higher in August and ca. 8 percent lower in February (see figure 2.6), resulting in an increase in the intensity of the annual cycle. We suggest these early Holocene conditions caused increased rainfall in the circum-Caribbean region (Hastenrath 1976, 1984). In support of this hypothesis, long-term changes in interseason insolation differences during the Holocene appear to match the long-term pattern of circum-Caribbean rainfall inferred from oxygen isotope data (see figure 2.4). Climate was dry but becoming wetter during the earliest Holocene (10,500 to 8500 B.P.), the wettest conditions occurred in the early to mid-Holocene (8500 to 3000 B.P.), and dry conditions returned in the late Holocene beginning at 3000 B.P.

McIntyre and Molfino (1996) predicted that the modality of the tropical Atlantic climate system changed at ca. 3000 B.P. in response to precessional forcing. Their mechanism involves a teleconnective linkage between the equatorial and high-latitude North Atlantic. Insolation forcing at low latitudes affects winds that control the advection of heat to the high-latitude North Atlantic via the strength of the western boundary current (i.e., the Gulf Stream). When the tropical easterlies diminish, warm waters are released from the Caribbean and Gulf of Mexico via the Gulf Stream to the North Atlantic. At ca. 3000 B.P., precipitation in the subpolar latitudes of the North Atlantic increased, whereas precipitation decreased in the circum-Caribbean and West Africa (McIntyre and Molfino 1996).

Although orbital (Milankovitch) forcing provides a plausible explanation for millennial-scale climate variability during the Holocene, other mechanisms must be responsible for rapid climate changes on submillennial time scales. Natural variability in climate can result from unforced oscillations that arise from atmospheric-oceanic dynamics or thermodynamics (Rind and Overpeck 1995). If parts of the climate system with long time constants or high heat capacities are involved, oscillations on multi-decadal scales are possible (Rind and Overpeck 1995). Such random changes in the climate system lack external forcing and are therefore unpredictable.

The El Niño–southern oscillation is perhaps the best example of climate change resulting from variability in the ocean-atmosphere system. Because of the dependence of rainfall in the Central America–Caribbean region on the position of the North Atlantic subtropical high, the North Atlantic oscillation, as measured by the inverse pressure gradient between the Azores high and the Icelandic low, may influence Caribbean climate. Changes in the production of North Atlantic deep water play an important role in the climate of the North Atlantic region and may introduce climate variability on a variety of time scales. Strong teleconnective linkages have been suggested between Central America and the high-latitude North Atlantic (Gill 1994). For example, protracted droughts in Central America may correlate with a more southerly position of the North Atlantic high, strong trade winds, a low and strong trade wind inversion, and cold SST in the North Atlantic. Keigwin (1996) documented millennial-scale changes of SST in the Sargasso Sea, including a 1°C warming of SST during the medieval climate anomaly (MCA) about 1000 B.P. and a 1°C cooling of SST during the Little Ice Age (LIA) about 400 B.P. The MCA appears to correlate with an abrupt return of abundant rainfall to Yucatán at about A.D. 1000 (Hodell et al. 1995; Curtis et al. 1996), but there is no existing evidence for circum-Caribbean climate change associated with the LIA.

Lastly, volcanic eruptions can influence climate by injecting large amounts of ash and gases high into the atmosphere. Large volcanic eruptions are generally followed by tropospheric cooling because of increased reflection of solar radiation by sulfate aerosols in the troposphere and stratosphere. The Intra-Americas Sea is rimmed to the east and west by active plate boundaries that contain many volcanoes capable of providing large quantities of aerosols. In Middle America, for example, the eruption of El Chichón was correlated with unusual atmospheric phenomena around A.D. 1300 and possibly A.D. 623 (Tilling et al. 1984; Gill 1994). Unfortunately, volcanic ash does not preserve well in tropical, lacustrine sediments, and the cores we studied are some distance from likely source areas. Volcanic activity can also have cultural implications, including eruptions as a natural hazard (Sheets 1979, 1981), improved soil fertility by volcanic ash, and the use of volcanogenic materials in ceramics, obsidian for cutting implements or jewelry, and extrusive rocks for building stone (Ford and Rose 1995).

IMPACT OF CLASSIC MAYA CULTURE AND CLIMATE CHANGE ON THE ENVIRONMENT OF THE YUCATÁN PENINSULA

As an example of the complex relationships among culture, climate, and environment, we review the natural and human history of the Yucatán Peninsula during the time of occupation by the Maya civilization. Beginning about 3000 to 4000 B.P., the Maya began to transform the landscape of the southern Yucatán lowlands significantly by clearing forests for construction and agriculture. Paleoenvironmental evidence suggests that the Department of Petén, Guatemala, was subjected to a protracted period of human disturbance (vegetation removal and soil erosion) between ca. 3000 and 1000 B.P. The pollen record from Lake Petén-Itza illustrates that lowland forest taxa declined and disturbance taxa increased rapidly beginning about 2800 B.P., presumably as a consequence of forest clearance by the Maya (figure 2.5; Islebe et al. 1996). This episode of dramatic deforestation around Lake Petén-Itza coincided with documented forest removal around other Petén lakes (see figure 2.1b), including Petenxil (Cowgill et al. 1966), Quexil (Deevey 1978, Wiseman 1978, Brenner 1983b; Vaughan et al. 1985), Salpeten (Leyden 1987), Sacnab (Deevey et al. 1979) and Macanche (Vaughan et al. 1985), and with deforestation and disturbance in Belize (Hansen 1990) and northern Yucatán (Leyden et al. 1998). The Petén was especially densely populated during the Late Classic period (A.D. 600 to 800), with regional estimates varying from 3 million to 14 million people (Rice and Rice 1990; Rice and Puleston 1981; Thompson 1954; Adams et al. 1981). Extensive forest removal led to rapid erosion of watershed soils and siltation of local water bodies, giving rise to a thick deposit of inorganic colluvium that underlies many Petén lakes, referred to as the "Maya clay" (Deevey and Rice 1980; Deevey et al. 1980; Brenner 1983a, 1994; Rice et al. 1985; Binford et al. 1987). Erosion led to rapid loss of soil nutrients from upland sites in the karst terrain and may have had a detrimental impact on Maya agriculture. For instance, it has been suggested that the Classic Maya collapse was, in part, precipitated by human-induced deforestation and soil erosion (Deevey et al. 1979; Binford et al. 1987), but these human-mediated disturbances of terrestrial and aquatic ecosystems in the Petén must be considered in the context of ecologic change caused by natural climate variability.

Evidence from Lake Chichancanab, Mexico, indicates that central Yucatán climate began to dry beginning at ca. 3000 B.P. (Hodell et al. 1995). Was the palynologically documented decline in high forest taxa at this time a consequence of climate change, human disturbance, or both? Is it possible that drier climate may have induced cultural change and contributed to the development of Maya agricultural practices during the Preclassic period? The emergence of *Zea* (maize) in pollen spectra at the time of decline in high forest taxa argues for the role of humans in causing forest disturbance. In a core from Lake Petén-Itza, no significant changes occurred in the oxygen isotope record after 4800 B.P. (Curtis et al. 1998), yet the pollen record documents a progressive increase in disturbance taxa and decrease in lowland forest taxa (figure 2.5; Islebe et al. 1996). This suggests that human disturbance was the primary agent for vegetation change.

The highest-resolution record of Precolumbian climate change on the Yucatán Peninsula comes from Lake Punta Laguna, Mexico, because of its high sediment accumulation rate (figure 2.7; Curtis et al. 1996). The oxygen isotope record indicates alternating wet and dry conditions lasting for multiple decades or centuries. The Classic and earliest Postclassic periods from A.D. 280 to 1080 were generally dry relative to climate before or after this time interval. This raises the question whether Classic Maya culture evolved because of a shift toward drier climate, or whether climatic drying was a response to the environmental impact of expanding Classic Maya population. Superimposed upon the relatively drier conditions of the Classic and earliest Postclassic periods were four distinct drought events centered at A.D. 585, 862, 986, and 1051 (see figure 2.7). The first drought episode (A.D. 585 ± 50) coincides with the Early Classic–Late Classic boundary and the Maya Hiatus (Curtis et al. 1998). The Maya Hiatus lasted from A.D. 525/536 to 590/681, depending upon location, and was marked by a sharp drop in stelae and monument erection, abandonment in some areas, and social upheaval (Gill 1994). The drought centered at A.D. 862 ± 50 coincided with the collapse of Classic Maya civilization between A.D. 800 and 900 and is the same event documented in Lake Chichancanab between A.D. 800 and 1000 (Hodell et al. 1995). The terminal Classic drought has yet to be documented in the southern Yucatán lowlands, but only one lake has been extensively studied, Petén-Itza, and it may be relatively insensitive to climate change owing to its large volume (Curtis et al. 1998).

The role of environmental change in the collapse of Classic Maya civilization during the ninth century A.D. serves to illustrate the complex interplay among climate, culture, and environment. Paleoclimatic evidence for drought in central and northern Yucatán is strong (Hodell et al. 1995; Curtis et al. 1996), and corroborative evidence of dry conditions at about the same time is emerging from elsewhere in Central America, including the central highlands of Mexico (Metcalfe et al. 1994; Metcalfe 1995) and Costa Rica (Horn and Sanford 1992). In addition to paleoclimate change, paleolimnological evidence from the southern lowlands documents widespread human-induced deforestation during the Late Classic period. The denudation of landscapes by the Maya altered natural biogeochemical cycles as watershed soils were subject to severe erosion. The coupled climatic and human-induced changes on the environment of the southern lowlands support the claim that Late Classic Maya populations, with high demands for protein, may have been confronted with declining crop yields due to both soil nutrient depletion and decreased rainfall. As a consequence, cultural stability faltered, at least in part, because of resource depletion and climate change (Leyden et al. 1998).

Following the Classic Maya collapse, forests recovered and soils stabilized after human and climatic pressures on regional vegetation were reduced. Detailed pollen analysis can provide an estimate of the time required for forest regeneration following the amelioration of environmental stresses. Human pressures were strongly curtailed after A.D. 900, whereas climatic pressure was reduced at about A.D. 1000 by an abrupt increase in rainfall (Hodell et al. 1995; Curtis et al. 1996). The pollen record from Lake Petén-Itza suggests that forest recovery commenced at ca. 1000 B.P. (about A.D. 1025), shortly after environmental pressures were reduced (figure 2.5; Islebe et al. 1996). However, a pollen sequence from Lake Chilonche, in the savanna of Petén, suggests that reforestation didn't begin until after about A.D. 1600 (ca. 350 B.P.), following the arrival of Europeans in Petén (Brenner et al. 1990). Additional studies of cores from six Petén lakes are under way to resolve this discrepancy by determining rates of deforestation and reforestation surrounding the Classic Maya collapse ca. A.D. 850.

The role of human and natural climatic disturbance in the Maya region during the late Holocene is relevant to modern society because human-induced deforestation is a recurring phenomenon in Petén. Today, the human population in Petén is expanding in response to increasing land pressures in the Guatemalan highlands, and this growth is already having environmental consequences. The population of Petén is estimated to have been 3,027 in the early 1700s, 25,207 by the end of the 1960s, and 300,000 by the mid-1980s. Schwartz (1990) suggested that 40–50 percent of the Petén landscape was deforested or degraded by the mid-1980s, and as much as 60 percent of the vegetation may have been disturbed by 1989. A better understanding of the history of human impact and climate change on the ecology of the Petén may be useful for guiding future management decisions regarding land and water resources.

LESSONS FROM THE PAST AND IMPLICATIONS FOR THE FUTURE

The example of the Classic Maya clearly demonstrates that mesoamerican ecosystems were impacted by human activities long before the arrival of Europeans in the New

World. Maya occupation of the karsted lowland environment of the Yucatán Peninsula resulted in widespread deforestation, accelerated rates of soil erosion, and nutrient sequestering in lake sediments (Brenner 1983a,b). In addition, oxygen isotope analysis of shell carbonate in lake sediment cores suggests that climate, specifically precipitation, has been highly variable in the circum-Caribbean region during the Holocene (Hodell et al. 1995; Curtis et al. 1996). Periods of prolonged drought lasting for decades to centuries have recurred during the late Holocene, altering conditions for the natural vegetation and affecting agriculture. Human perturbation of the environment and protracted drought in Mesoamerica culminated during the terminal Classic period (ca A.D. 800 to 900), and may have contributed to the collapse of Classic Maya civilization. This natural historical experiment provides an important opportunity to study the response of social and economic systems to human disturbance and climate change. In addition, the period immediately following the collapse and climatic amelioration provides a measure of the resilience and recovery time of ecosystems once human and natural climatic pressures were removed. Investigation of these questions in the Precolumbian past is especially germane to contemporary problems of tropical deforestation and climate change.

ACKNOWLEDGMENTS

This research was supported by NSF award EAR-9709314 and a grant from the National Geographic Society. We thank two anonymous reviewers for their constructive and helpful criticism of the manuscript.

REFERENCES

Adams, R. E. W., W. E. Brown Jr., and T. P. Culbert. 1981. Radar mapping, archeology, and ancient Maya land use. *Science* 213: 1457–1463.

Binford, M. W. 1982. Ecological history of Lake Valencia, Venezuela: Interpretation of animal microfossils and some chemical, physical, and geological features. *Ecological Monographs* 52: 307–333.

Binford, M. W., M. Brenner, T. J. Whitmore, A. Higuera-Gundy, E. S. Deevey, and B. W. Leyden. 1987. Ecosystems, paleoecology, and human disturbance in subtropical and tropical America. *Quaternary Science Reviews* 6: 115–128.

Bond, G., W. Broecker, S. Johnsen, L. Labeyrie, J. McManus, J. Andrews, S. Huon, R. Jantsschik, S. Clasen, C. Simet, K. Tedesco, M. Kias, G. Bonani, and S. Ivy. 1992. Evidence for massive discharges of icebergs into the North Atlantic ocean during the last glacial period. *Nature* 360: 245–249.

Bradbury, J. P. 1997. Sources of glacial moisture in Mesoamerica. *Quaternary International* 43/44: 97–110.

Bradbury, J. P., B. W. Leyden, M. L. Salgado-Labouriau, W. M. Lewis Jr., C. Schubert, M. W. Binford, D. G. Frey, D. R. Whitehead, and F. H. Weibezahn. 1981. Late Quaternary environmental history of Lake Valencia, Venezuela. *Science* 214: 1299–1305.

Bradbury, J. P., R. M. Forester, W. A. Bryant, and A. P. Covich. 1990. Paleolimnology of Laguna de Cocos, Albion Island, Rio Hondo, Belize. In M. D. Pohl, ed., *Ancient Maya Wetland Agriculture: Excavations on*

Albion Island, Northern Belize, pp. 119–154. Boulder, Colo.: Westview Press.

Brenner, M. 1983a. Paleolimnology of the Maya region. Ph.D. dissertation, 249 pp., University of Florida, Gainesville.

———. 1983b. Paleolimnology of the Petén Lake District, Guatemala, 2: Mayan population density and sediment and nutrient loading of Lake Quexil. *Hydrobiologia* 103: 205–210.

———. 1994. Lakes Salpeten and Quexil, Petén, Guatemala, Central America. In E. Gierlowski-Kordesch and K. Kelts, eds., *Global Geological Record of Lake Basins*, vol. 1, pp. 377–380. Cambridge: Cambridge University Press.

Brenner, M., B. W. Leyden, and M. W. Binford. 1990. Recent sedimentary histories of shallow lakes in the Guatemalan savannas. *J. Paleolimnology* 4: 239–252.

Broccoli, A. J., and S. Manabe. 1987. The influence of continental ice, atmospheric CO_2, and land albedo on the climate of the last glacial maximum. *Climate Dynamics* 1: 87–99.

Bush, M. B., and P. A. Colinvaux. 1990. A pollen record of a complete glacial cycle from lowland Panama. *J. Vegetation Science* 1: 105–118.

Colinvaux, P. A. 1972. Climate and the Galapagos Islands. *Science* 240: 17–20.

Cowgill, U. M., G. E. Hutchinson, A. A. Racek, C. E. Goulden, R. Patrick, and M. Tsukada. 1966. The history of Laguna de Petenxil, a small lake in northern Guatemala. *Memoirs of the Connecticut Academy of Arts and Sciences* 17: 1–126.

Curtis, J. H. 1997. Climatic variation in the circum-Caribbean during the Holocene. Ph.D. dissertation, 148 pp., University of Florida, Gainesville.

Curtis, J. H., D. A. Hodell, and M. Brenner. 1996. Climate variability on the Yucatán Peninsula (Mexico) during the past 3,500 years and implications for Maya cultural evolution. *Quaternary Research* 46: 37–47.

Curtis, J. H., M. Brenner, D. A. Hodell, R. A. Balser, G. A. Islebe, and H. Hooghiemstra. 1998. A multi-proxy study of Holocene environmental change in the Maya Lowlands of Petén, Guatemala. *J. Paleolimnology* 19: 139–159.

Curtis, J. H., M. Brenner, D. A. Hodell. 1999. Changes in moisture availability in the Lake Valencia Basin, Venezuela (late Pleistocene to present). *Holocene* 9(5): 609–619.

Damon, P. E., and T. W. Linick. 1986. Geomagnetic-heliomagnetic modulation of atmospheric radiocarbon production. *Radiocarbon* 208: 266–278.

Deevey, E. S. 1978. Holocene forests and Maya disturbance near Quexil Lake, Petén, Guatemala. *Polish Archives of Hydrobiology* 25 (1/2): 117–129.

Deevey, E. S., and D. S. Rice. 1980. Coluviación y retención de nutrientes en el distrito lacustre del Petén central, Guatemala. *Biótica* 5: 129–144.

Deevey, E. S., D. S. Rice, P. M. Rice, H. H. Vaughan, M. Brenner, and M. S. Flannery. 1979. Mayan urbanism: Impact on a tropical karst environment. *Science* 206: 298–306.

Deevey, E. S., M. Brenner, M. S. Flannery, and G. H. Yezdani. 1980. Lakes Yaxha and Sacnab, Petén, Guatemala: Limnology and hydrology. *Archiv für Hydrobiologie*, Supplement 57: 419–460.

Deevey, E. S., M. Brenner, and M. W. Binford. 1983. Paleolimnology of the Petén Lake District, Guatemala, 3: Late Pleistocene and Gamblian environments of the Maya area. *Hydrobiologia* 103: 211–216.

Emiliani, C., and D. B. Ericson. 1991. The glacial/interglacial temperature range of the surface water of the oceans at low latitudes. In H. P. Taylor Jr., J. R. O'Neil, and I. R. Kaplan, eds., *Stable Isotope Geochemistry: A Tribute to Samuel Epstein*, pp. 223–228. San Antonio, Tex.: Lancaster Press.

Emiliani, C., D. A. Price, and J. Seipp. 1991. Is the Postglacial artificial? In H. P. Taylor Jr., J. R. O'Neil, and I. R. Kaplan, eds., *Stable Isotope Geochemistry: A Tribute to Samuel Epstein*, pp. 229–234. San Antonio, Tex.: Lancaster Press.

Fairbanks, R. G. 1989. A 17,000-year glacio-eustatic sea level record: Influence of glacial melting rates in the Younger Dryas event and deep-ocean circulation. *Nature* 342: 637–642.

Flower, B. P., and J. P. Kennett. 1990. The Younger Dryas cool episode in the Gulf of Mexico. *Paleoceanography* 5: 949–961.

Folan, W. J., J. Gunn, J. D. Eaton, and R. W. Patch. 1983. Paleoclimatological patterning in southern Mesoamerica. *J. Field Archaeology* 10: 454–468.

Fontes, J. C., and R. Gonfiantini. 1967. Comportement isotopique au cours de l'évaporation de deux bassins sahariens. *Earth and Planetary Science Letters* 3: 258–266.

Ford, A., and W. I. Rose. 1995. Volcanic ash in ancient Maya ceramics of the limestone lowlands: Implications for prehistoric volcanic activity in the Guatemalan highlands. *J. Volcanology and Geothermal Research* 66: 149–162.

Gill, R. B. 1994. *The Great Maya Droughts.* Ph.D. dissertation, University of Texas, Austin.

Gray, C. R. 1987. *History of Tropical Cyclones in the Caribbean, 1886 to 1986.* Jamaica: National Meteorological Service.

———. 1993. Regional meteorology and hurricanes. In G. A. Maul, ed., *Climatic Change in the Intra-Americas Sea,* pp. 87–99. London: Edward Arnold.

Grimm, E. C., G. L. Jacobson Jr., W. A. Watts, B. C. S. Hansen, and K. A. Maasch. 1993. A 50,000-year record of climate oscillations from Florida and its temporal correlation with the Heinrich Events. *Science* 261: 198–200.

Guilderson, T. P., R. G. Fairbanks, and J. L. Rubenstone. 1994. Application of the Sr and Ca method to paleothermometry of Barbados corals spanning the period of deglaciation. *Science* 263: 663–665.

Gunn, J., and R. E. W. Adams. 1981. Climatic change, culture, and civilization in North America. *World Archaeology* 13: 87–100.

Gunn, J. D., W. J. Folan, and H. R. Robichaux. 1995. A landscape analysis of the Candelaria watershed in Mexico: Insights into paleoclimates affecting upland horticulture in the southern Yucatán Peninsula semi-karst. *Geoarchaeology* 10: 3–42.

Hagelberg, T. K., G. Bond, and P. deMenocal. 1994. Milankovitch band forcing of sub-Milankovitch climate variability during the Pleistocene. *Paleoceanography* 9: 545–558.

Hansen, B. C. S. 1990. Pollen stratigraphy of Laguna de Cocos. In M. D. Pohl, ed., *Ancient Maya Wetland Agriculture: Excavations on Albion Island, Northern Belize,* pp. 155–186. Boulder, Colo.: Westview Press.

Hastenrath, S. 1966. On general circulation and energy budget in the area of the Central American seas. *J. Atmospheric Sciences* 23: 694–711.

———. 1967. Rainfall distribution and regime in Central America. *Archiv Meteor. Geophys. Bioklim.* Ser. B 15: 201–241.

———. 1976. Variations in low-latitude circulation and extreme climatic events in the tropical Americas. *J. Atmospheric Sciences* 33: 202–215.

———. 1984. Interannual variability and the annual cycle: Mechanisms of circulation and climate in the tropical Atlantic sector. *Monthly Weather Review* 112: 1097–1107.

———. 1991. *Climate Dynamics of the Tropics.* Boston, Mass.: Kluwer Academic Publishers.

Hastenrath, S., and E. B. Kaczmarczyk. 1981. On spectra and coherence of tropical climate anomalies. *Tellus* 33: 453–462.

Higuera-Gundy, A., M. Brenner, D. A. Hodell, J. H. Curtis, B. W. Leyden, and M. W. Binford. 1999. Climate and vegetation of southwest Haiti: Late Pleistocene to present. *Quaternary Research* 52(2): 159–170.

Hodell, D. A., J. H. Curtis, G. A. Jones, A. Higuera-Gundy, M. Brenner, M. W. Binford, and K. T. Dorsey. 1991. Reconstruction of Caribbean climate change over the past 10,500 years. *Nature* 352: 790–793.

Hodell, D. A., J. H. Curtis, and M. Brenner. 1995. Possible role of climate in the collapse of Classic Maya civilization. *Nature* 375: 391–394.

Holmes, J. A. 1996. Trace-element and stable-isotope geochemistry of non-marine ostracod shells in Quaternary palaeoenvironmental reconstruction. *J. Paleolimnology* 15: 223–235.

———. 1998. A late Quaternary ostracod record from Wallywash Great Pond, a Jamaican marl lake. *J. Paleolimnology* 19: 115–128.

Horn, S. P., and R. L. Sanford Jr. 1992. Holocene fires in Costa Rica. *Biotropica* 24: 354–361.

Hughen, K. A., J. T. Overpeck, L. C. Peterson, and S. Trumbore. 1996. Rapid climate change in the tropical Atlantic region during the last deglaciation. *Nature* 380: 51–54.

Islebe, G. A., H. Hooghiemstra, and K. van der Borg. 1995. A cooling event during the Younger Dryas Chron in Costa Rica. *Palaeogeography, Palaeoclimatology, Palaeoecology* 117: 73–80.

Islebe, G. A., H. Hooghiemstra, M. Brenner, J. H. Curtis, and D. A. Hodell. 1996. A Holocene vegetation history from lowland Guatemala. *Holocene* 6: 265–271.

Keigwin, L. D. 1996. The Little Ice Age and Medieval Warm Period in the Sargasso Sea. *Nature* 274: 1504–1508.

Kjellmark, E. 1996. Late Holocene climate change and human disturbance on Andros Island, Bahamas. *J. Paleolimnology* 13: 133–145.

Kutzbach, J. E., and T. Webb III. 1993. Conceptual basis for understanding Late-Quaternary climates. In H. E. Wright Jr., J. E. Kutzbach, T. Webb III, W. F. Ruddiman, F. A. Street-Perrott, and P. J. Bartlein, eds., *Global Climates since the Last Glacial Maximum*, pp. 5–11. Minneapolis: University of Minnesota Press.

Lean, J., J. Beer, and R. Bradley. 1995. Reconstruction of solar irradiance since 1610: Implications for climate change. *Geophysical Research Letters* 22: 3195–3198.

Lean, J., and D. A. Warrilow. 1989. Simulation of the regional climatic impact of Amazon deforestation. *Nature* 342: 411–413.

Leventer, A., D. F. Williams, and J. P. Kennett. 1982. Dynamics of the Laurentide Ice Sheet during the last deglaciation: Evidence from the Gulf of Mexico. *Earth and Planetary Science Letters* 59: 11–17.

Leyden, B. W. 1984. Guatemalan forest synthesis after Pleistocene aridity. *Proceedings of the National Academy of Sciences USA* 81: 4856–4859.

———. 1985. Late Quaternary aridity and Holocene moisture fluctuations in the Lake Valencia Basin, Venezuela. *Ecology* 66(4): 1279–1295.

———. 1987. Man and climate in the Maya Lowlands. *Quaternary Research* 28: 407–414.

Leyden, B. W., M. Brenner, D. A. Hodell, and J. H. Curtis. 1993. Late Pleistocene climate in the Central American lowlands. In P. K.

Swart, K. C. Lohmann, J. McKenzie, and S. Savin, eds., *Climate Change in Continental Isotopic Records*, pp. 165–178. Washington, D.C.: American Geophysical Union.

———. 1994. Orbital and internal forcing of climate on the Yucatán Peninsula for the past ca. 36 ka. *Palaeogeography, Palaeoclimatology, Palaeoecology* 109: 193–210.

Leyden, B. W., M. Brenner, and B. H. Dahlin. 1998. Cultural and climatic history of Cobá, a lowland Maya city in Quintana Roo, Mexico. *Quaternary Research* 49: 111–122.

Maasch, K. A., and R. J. Oglesby. 1990. Meltwater cooling of the Gulf of Mexico: A GCM simulation of climatic conditions at 12 ka. *Paleoceanography* 5: 977–996.

Markgraf, V. 1989. Climatic history of Central and South America since 18,000 BP: Comparison of pollen records and model simulations. In H. E. Wright Jr., J. E. Kutzbach, T. Webb III, W. F. Ruddiman, F. A. Street-Perrott, and P. J. Bartlein, eds., *Global Climates since the Last Glacial Maximum*, pp. 357–385. Minneapolis: University of Minnesota Press.

McIntyre, A., and B. Molfino. 1996. Forcing of Atlantic equatorial and subpolar millennial cycles by precession. *Science* 274: 1867–1870.

Messenger, L. C. Jr. 1990. Ancient winds of change: Climatic settings and prehistoric social complexity in Mesoamerica. *Ancient Mesoamerica* 1: 21–40.

Metcalfe, S. E. 1995. Holocene environmental change in the Zacapu Basin, Mexico: A diatom-based record. *Holocene* 5: 196–208.

Metcalfe, S. E., F. A. Street-Perrott, R. B. Brown, P. E. Hales, R. A. Perrott, and F. M. Steininger. 1989. Late Holocene human impact on lake basins in central Mexico. *Geoarchaeology* 4: 119–141.

Metcalfe, S. E., F. A. Street-Perrott, S. L. O'Hara, P. E. Hales, and R. A. Perrott. 1994. The palaeolimnological record of environmental change: Examples from the arid frontier of Mesoamerica. In A. C. Millington and K. Pye, eds., *Environmental Change in Drylands: Biogeographical and Geomorphological Perspectives*, pp. 131–145. Chichester: John Wiley & Sons.

Metcalfe, S. E., A. Bimpson, A. J. Courtice, S. L. O'Hara, and D. M. Taylor. 1997. Climate change at the monsoon/westerly boundary in northern Mexico. *J. Paleolimnology* 17: 155–171.

Newell, R. E. 1973. Climate and the Galapagos Islands. *Nature* 245: 91–92.

Nieuwolt, S. 1977. *Tropical Climatology: An Introduction to the Climates of the Low Latitudes.* London: John Wiley & Sons.

Overpeck, J. T., L. C. Peterson, N. Kipp, J. Imbrie, and D. Rind. 1989. Climate change in the circum-North Atlantic region during the last deglaciation. *Nature* 338: 553–557.

Piperno, D. R., M. B. Bush, and P. A. Colinvaux. 1990. Paleoenvironments and human settlements in late-glacial Panama. *Quaternary Research* 33: 108–116.

Prell, W. L., and J. D. Hays. 1976. Late Pleistocene faunal and temperature patterns of the Colombia Basin, Caribbean Sea. In R. M. Cline and J. D. Hays, eds., *Investigations of Late Quaternary Paleoceanography and Paleoclimatology,* pp. 201–220. Boulder, Colo.: Geological Society of America.

Rice, D. S., and D. E. Puleston. 1981. Ancient Maya settlement patterns in the Petén, Guatemala. In W. Ashmore, ed., *Lowland Maya Settlement Patterns,* pp. 121–156. Albuquerque: University of New Mexico Press.

Rice, D. S., and P. M. Rice. 1990. Population size and population change in the Central Petén Lake Region, Guatemala. In T. P. Culbert and D. S. Rice, eds., *Precolumbian Population History in the Maya Lowlands,* pp. 123–148. Albuquerque: University of New Mexico Press.

Rice, D. S., P. M. Rice, and E. S. Deevey. 1985. Paradise lost: Classic Maya impact on a lacustrine environment. In M. Pohl, ed., *Prehistoric Lowland Maya Environment and Subsistence Economy,* pp. 91–105, Peabody Museum Papers 77. Cambridge, Mass.: Harvard University Press.

Rind, D., and J. Overpeck. 1994. Hypothesized causes of decade-to-century-scale variability: Climate model results. *Quaternary Science Reviews* 12: 357–374.

———. 1995. Modeling the possible causes of decadal-to-millennial-scale variability. In Douglas G. Martinson et al., eds., *Natural Climate Variability on Decade-to-Century Time Scales,* pp. 187–198. Washington, D.C.: National Academy Press.

Rosenmeier, M. F., D. A. Hodell, J. H. Curtis, J. B. Martin, and M. Brenner. Submitted. Impact of climate and human-induced vegetation change on watershed hydrology: Implications for the interpretation of Holocene $\delta^{18}O$ records in the lowland Neotropics, Department of Petén, Guatemala. *J. Paleolimnology.*

Schwartz, N. B. 1990. *Forest Society: A Social History of Petén, Guatemala.* Philadelphia: University of Pennsylvania Press.

Sheets, P. D. 1979. Maya recovery from volcanic disasters: Ilopango and Cerén. *Archaeology* 32(3): 32–42.

———. 1981. Volcanoes and the Maya. *Natural History* 90(8): 32–41.

Short, D. A., J. G. Mengel, T. J. Crowley, W. T. Hyde, and G. R. North. 1991. Filtering of Milankovitch cycles by Earth's geography. *Quaternary Research* 35: 157–173.

Spero, H. J., and D. F. Williams. 1990. Evidence for seasonal low-salinity surface waters in the Gulf of Mexico over the last 16,000 years. *Paleoceanography* 5: 963–975.

Stine, S. 1994. Extreme and persistent drought in California and Patagonia during mediaeval time. *Nature* 369: 546–549.

Street-Perrott, F. A., P. E. Hales, R. A. Perrott, J. Ch. Fontes, V. R. Switsur, and A. Pearson. 1993. Limnology and palaeolimnology of a tropical marl lake: Wallywash Great Pond, Jamaica. *J. Paleolimnology* 9: 3–22.

Stuiver, M., T. F. Braziunas, B. Becker, and B. Kromer. 1991. Climatic, solar, oceanic, and geomagnetic influences on late-Glacial and Holocene atmospheric $^{14}C/^{12}C$ change. *Quaternary Research* 35: 1–24.

Thompson, J. E. S. 1954. *The Rise and Fall of Maya Civilization.* Norman: University of Oklahoma Press.

Thompson, R. S., C. Whitlock, P. J. Bartlein, S. P. Harrison, and W. G. Spaulding. 1993. Climatic changes in the western United States since 18,000 yr B.P. In H. E. Wright Jr., J. E. Kutzbach, T. Webb III, W. F. Ruddiman, F. A. Street-Perrott, and P. J. Bartlein, eds., *Global Climates since the Last Glacial Maximum,* pp. 468–513. Minneapolis: University of Minnesota Press.

Tilling, R. I., M. Rubin, H. Sigurdsson, S. Carey, W. A. Duffield, and W. I. Rose. 1984. Holocene eruptive activity of El Chichon volcano, Chiapas, Mexico. *Science* 224: 747–749.

Vaughan, H. H., E. S. Deevey, and S. E. Garrett-Jones. 1985. Pollen stratigraphy of two cores from the Petén Lake District. In M. Pohl, ed., *Prehistoric Lowland Maya Environment and Subsistence Economy,* pp. 73–89, Peabody Museum Papers 77. Cambridge, Mass.: Harvard University Press.

Watts, W. A. 1975. A late Quaternary record of vegetation from Lake Annie, south-central Florida. *Geology* 2: 344–346.

Watts, W. A., and B. C. S. Hansen. 1994. Environments of Florida in the Late Wisconsin and Holocene. In B. A. Purdy, ed., *Wet Site Archaeology,* pp. 307–323. Caldwell, N.J.: Telford Press.

Watts, W. A., and M. Stuiver. 1980. Late Wisconsin climate of northern Florida and the origin of species-rich deciduous forest. *Science* 210: 325–327.

Watts, W. A., B. C. S. Hansen, and E. C. Grimm. 1992. Camel Lake: A 40,000-yr record of vegetational and forest history from northwest Florida. *Ecology* 73: 1056–1066.

Wilson, E. M. 1984. Physical geography of the Yucatán Peninsula. In E. H. Moseley and E. D. Terry, eds., *Yucatán: A World Apart,* 2nd ed., pp. 5–40. Tuscaloosa: University of Alabama Press.

Wiseman, F. M. 1978. Agricultural and historical ecology of the Maya Lowlands. In P. D. Harrison and B. L. Turner II, eds., *Prehispanic Maya Agriculture,* pp. 63–115. Albuquerque: University of New Mexico Press.

ANDREW M. GRELLER

3 | VEGETATION IN THE FLORISTIC REGIONS OF NORTH AND CENTRAL AMERICA

In treating so vast and complex an area as North and Central America, it was necessary to make decisions about the depth of detail that any one region would receive. Those decisions were made from the perspective of an anglophone North American. The treatment of Mexico relies mostly on Rzedowski (1981), whose monumental study of the Mexican flora is a model for all vegetation reviews. New information from Central America has come mainly from Costa Rica, which country marks the end of extensive influence of the Holarctic flora, being covered by tropical vegetation except for the highest peaks. The West Indies long ago received treatments in English comparable to that of Mexico, in an overview by Beard (1955), as well as in comprehensive treatments of individual islands: Puerto Rico and the Virgin Islands, Cuba, Jamaica, and the Windward and Leeward Islands. Overviews of North American vegetation have been readily available since the mid twentieth century. Detailed regional treatments and one general treatment have become available since the 1950s for the Great Plains (e.g., J. E. Weaver 1954) and from the 1970s for Canada (Rowe 1972), the Intermountain Region (Cronquist et al. 1972), the Pacific Northwest (Franklin and Dyrness 1973), California (Barbour and Major 1988), Florida (Platt and Schwartz 1990; Myers and Ewel 1990), all of eastern North America (Miyawaki et al. 1994), and North and Central America (Knapp 1965; Barbour and Billings 1988). "Ecoregions" of North America are recently treated by Bailey (1996, 1997). These units are approximately coincidental with major vegetation types in Barbour and Christensen (1993). Bailey's (1997) "Ecoregion Divisions" are approximately

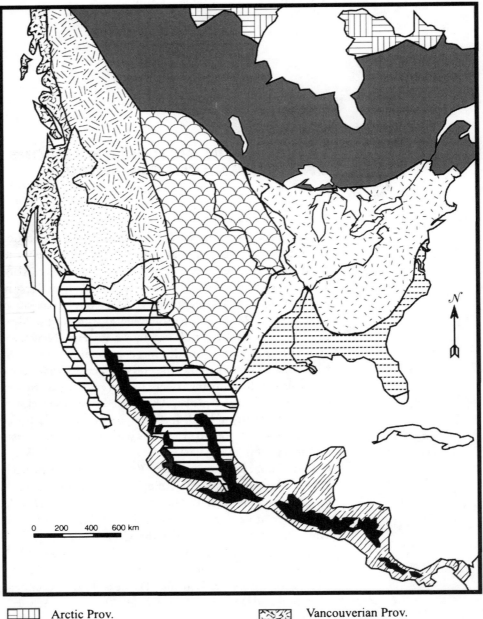

Arctic Prov.

Canadian Prov.

Appalachian Prov.

North American Prairies Prov.

Central American Prov.

West Indian Prov.

Mexican Xerophytic Prov.

Vancouverian Prov.

Rocky Mountain Prov.

Great Basin Prov.

Atlantic & Gulf Coastal Plain Prov.

Mexican Highlands Prov.

Sierra Madre Occidental Prov.

Californian Prov.

FIGURE 3.1. Floristic regions of North and Central America.

coincidental with floristic provinces, and with many floristic regions, in figure 3.1 (present study).

Because the emphasis in this book is on environments related to the rise of indigenous civilizations, it was decided to retain in the text detailed treatments of the vegetation of regions where indigenous peoples created the most complex societies, namely Central America, Mexico, southwestern United States, and southeastern North America.

It was decided to organize the vegetation by floristic region, following the treatment of North American floristic regions and provinces prepared by Cronquist in Takhtajan (1986). Thorne (1993) also follows Cronquist-Takhtajan, with a minor modification. A floristic approach permits a more comprehensive treatment of major vegetation, because the emphasis is on the region rather than on a particular vegetation type. Any one vegetation type may be found throughout a number of floristic regions. The map of floristic regions is given in figure 3.1; it incorporates information from Cronquist's (1982) map with some significant differences. For Mexico and Central America it follows Rzedowski (1981), Knapp (1965), and Atlas Censal de El Salvador (1955). A map (1:15,000,000) including Mexican and Central American ecoregions was developed for Dinerstein et al. (1995) using digital databases from satellite imagery. That map is detailed in its depiction of existing vegetation units but broad in its system of classification (physiognomical and geographical), so it cannot easily be used to compare with the previous ones. Nevertheless, large areas of broad-leaved tropical forests, pine-oak forests, grassland, desert scrub, etc., are seen to be generally congruent. Similar forest-type and ecoregion maps from digital satellite images are available for the whole or portions of the United States (e.g., "Forest Type Groups of the United States," 1: 7,500,000, U.S. Department of Agriculture, Forest Service, 1993). Tables 5-3, 5-4, 5-5, and 7-1 in Dinerstein et al. list the conservation status of forests among the 178 "ecoregions" in Latin America (including South America) and the Caribbean. A number of regional or local examples of Mexican, Central American, and Caribbean forests qualify as "Critical, Endangered or Vulnerable" and "Highest Priority." The reader is directed to that publication for information on specific vegetation types. Because the environmental conditions in the Americas prior to the first human occupation reveal so much about climate variation and consequent plant migrations and compositional changes in vegetation, paleobotany has much to offer students of contemporary flora and vegetation. North American plant life and environments from end-Cretaceous through the Tertiary and during the Late Quaternary are admirably treated by Graham (1993) and Delcourt and Delcourt (1993), respectively. The reader is referred to those treatments of paleobotany, which is otherwise beyond the scope of the present study. In the following discussion, *vegetation* refers to the plant cover over the landscape, *flora* refers to the taxonomic enumeration of the individual plants in a given region, and *phytochorion* is an unspecified taxonomic unit of flora. Formal units of flora, such as kingdom, region, etc., follow Takhtajan (1986). *MAP* is mean annual precipitation. *MAT* is mean annual temperature. Classification of the floristic regions and provinces discussed here is outlined in table 3.1.

TABLE 3.1. CLASSIFICATION OF FLORISTIC REGIONS AND SOME PROVINCES OF NORTH AND CENTRAL AMERICA

Holarctic Floristic Kingdom
Boreal Subkingdom
  Circumboreal Region
    Arctic Province
    Canadian Province
  Atlantic North American Region
    Appalachian Province
    North American Prairie Province
  Rocky Mountain (Cordilleran) Region
    Rocky Mountain Province
    Vancouverian Province
    Great Basin Province
Madrean Subkingdom
  Madrean Region
    Californian Province
    Sierra Madre Occidental Province
  Southeastern North American Region (Proposed)
    Atlantic and Gulf Coastal Plains Province
    Mexican Highlands (Sierra Madre Oriental–Serranías Meridionales–Serranías Transístmicas) Province

Neotropical Floristic Kingdom
  Mexican Xerophytic (Sonoran) Region
    Peninsular (Baja California) Province
    Northwestern Coastal Plain (Sonoran) Province
    Central Plateau (Altiplanicie or Chihuahuan) Province
    Coastal Plain Province of the Northeast of Mexico (Tamaulipan Province)
    Tehuacán-Cuicatlán Valley Province
  Central American Region
    Central American Province
    West Indian (Antillean) Province

HOLARCTIC FLORISTIC KINGDOM

This is the largest of all floristic kingdoms and occupies more than half of the terrestrial world—all of Europe, northern Africa, eastern Asia north of the tropics, and nearly all of North America. Takhtajan estimates there are 60 endemic or nearly endemic families of vascular plants restricted to the Holarctic flora, but none of them is large and most have only one genus. In addition, some of them (jojoba, ocotillo, and crossosoma families) are endemic to the southwestern deserts of North America. In the present treatment, following Rzedowski (1981), these are considered with the Neotropical Floristic Kingdom. Also, Takhtajan tends to split families so that the Ruscaceae, Hemerocallidaceae, and Trilliaceae are distinguished from the Liliaceae. A number of families have many genera and species in the Holarctic flora: the ferns (Pteridaceae, Polypodiaceae, Aspleniaceae, and Aspidiaceae); conifers (pine and cypress); many families of monocot flowering plants (orchid, grass, sedge, rush, and iris), woody dicot flowering plants (magnolia, laurel, witch hazel, beech, birch, walnut, tea, willow, heath, mallow, spurge, rose, dogwood, aralia, and buckthorn), and herbaceous dicots (buttercup, bean, chenopod, pink, smartweed, mustard, carrot, mint, bellflower, and aster) (Takhtajan 1986). D. E. Brown et al. (1980) provide black-and-white photographs of the major types of vegetation in North America.

BOREAL SUBKINGDOM

The most extensive of all the Holarctic subkingdoms is the Boreal. It has four regions: Circumboreal (European-Canadian), Eastern Asiatic, North American Atlantic, and Rocky Mountain. Three of these are represented in North America.

CIRCUMBOREAL REGION

This is the largest floristic region on earth and includes the whole northern part of the Northern Hemisphere landmasses. In North America it includes most of Canada and most of Alaska. No endemic families are found here, and even endemic genera are few. Nevertheless, many genera are shared only between the Circumboreal and East Asiatic regions; most of these are in the well-studied mountains of Europe. The Alaska-Yukon area is one of the richest in endemic species. In this area the characteristic conifers are species of pine, spruce, fir, and larch. In Canada, hemlock and northern white cedar are also found. Among the numerous woody, broad-leaved genera present are species of: oak, beech, birch, alder, maple, hornbeam, poplar, willow, ash, elm, linden, walnut, hackberry, hop hornbeam, apple, mountain ash, spiraea, bladdernut, rhododendron, honeysuckle, viburnum, elder, buckthorn, and blueberry. Scattered widely in the boreal area are wet meadows. In central Canada prairies cover a large area. Both of these vegetation types are characterized by herbaceous plants. Northern areas are characterized by vast bogs and by tundra. In the mountains, forest vegetation forms altitudinal belts dominated by coniferous trees. Above the forest belts is a zone of high mountain herbaceous vegetation (alpine tundra or subalpine meadow), which is very rich in endemic taxa. Two provinces of the Circumboreal Floristic Region occur in North America: the Arctic Province and the Canadian Province.

ARCTIC PROVINCE

Tundra, a treeless expanse of low vegetation, is the most characteristic type in the province. In the south it is dominated by shrubby communities, mostly willow (*Salix*) and birch (*Betula*). Heath shrubs are also very typical, especially species of blueberry (*Vaccinium*) and bearberry (*Arctostaphylos*), as well as herbaceous plants of the sedge family, such as cottonsedge (*Eriophorum*) and *Carex*. In the more northern area vegetation is in a single layer. Dominant plants are species of *Dryas,* several prostrate willows (*Salix*), and perennial herbs in the grass, mustard, and pink families. Also characteristic are species of saxifrage, lousewort (*Pedicularis*), cinquefoil (*Potentilla*), and *Oxyria digyna*. Annual herbs are rare or absent. This province will not be treated further because of its harsh growing conditions, which historically have not permitted agriculture. Human life in the area is sustained by hunting caribou and marine mammals and by animal husbandry (reindeer, in the Old World).

CANADIAN PROVINCE

Coniferous forest is the major vegetation type in the region. Trees are mostly small, less than 20 m tall and with a trunk diameter less than 60 cm dbh (diameter at breast height). Nevertheless, the forest can be dense, and the understory sparse, with only mosses, lichens, and mycorrhizal herbs abundant. Elliott-Fisk (1988) describes three latitudinal zones of forest south of the Arctic tundra in the Canadian Province: forest-tundra ecotone, the farthest north, is present as patches of woodland, tree clones, and individual trees in a tundra matrix, north of the treeline; lichen woodland where black spruce is the usual dominant; and closed forest, generally described as spruce-feathermoss, where the spruces are black and white spruce and the feathermosses are *Hylocomium splendens* and *Pleurozium schreberi*.

Elliott-Fisk (1988) also describes nonforested boreal vegetation, specifically shrublands on better drained sites, and bogs on wet sites. The reader is referred to that author for details and a bibliography. More detailed data on composition of vegetation in eastern Canada can be obtained from Miyawaki et al. (1994).

ATLANTIC NORTH AMERICAN REGION

This region stretches across North America from the Atlantic Ocean to the Rocky Mountains, north to the Canadian Province of the Circumboreal Region and, in its original concept (Cronquist 1982), south to the Gulf of Mexico. At its core is the southern Appalachian Mountains. In the present treatment the region comprises only two provinces, the Appalachian and the North American Prairies provinces. Northward, around the Great Lakes, there is a broad transition between the Canadian Province and the Appalachian Province; here, conifers often become dominants. Southward, on the Piedmont, there is a transition to broad-leaved evergreen-conifer vegetation. Westward, in the east-central United States, there is a transition to the prairie in the oak savanna area (formerly "Prairie Peninsula").

APPALACHIAN PROVINCE

For the purposes of this treatment, the Appalachian Province can be divided into three sections: the deciduous forest, the oak savanna (transition to the Prairie Province), and the conifer-hardwood (Great Lakes–St. Laurent transition to the Canadian Province). On the south is the Atlantic and Gulf Coastal Plain Province, which shows an increasing dominance of forests by broad-leaved angiosperms southward to southern Florida and eastward to the Atlantic coast.

The dominant zonal vegetation type is deciduous forest. This is a vegetation type in which cold-deciduous trees dominate in one to three layers. The tallest layer can exceed 35 m and is closed. The subcanopy tree layer is usually present and can reach 12 m. One or two shrub layers are present, and in the well-developed herbaceous layer, spring-flowering herbs and ferns are especially conspicuous on mesic sites. Bryoids may be patchy, scattered, or absent. Lichens and mosses grow as epiphytes on the main trunks. Evergreen dicotyledonous taxa are present as small trees, shrubs, and herbs. Evergreen dicotyledonous small trees and shrubs occur at the southern transition, in the Piedmont, and also in some mountain habitats. A number of evergreen herb taxa and a very few of evergreen shrubs can be found on some nutrient-poor soils. Greller (1988, with modifications) recognizes the following subzones of deciduous forest: mixed mesophytic subzone (mixed mesophytic, western mesophytic, and oak-chestnut associations), sugar maple subzone (beech-maple, maple-basswood, maple-beech-birch-buckeye associations), oak-hickory subzone (oak-hickory association, oak savanna), oak-pine subzone (oak-pine association). Deciduous forest is best developed in the valleys and lower slopes of the southern Appalachian Mountains (mixed mesophytic association). More than two dozen tree species are abundant in the mixed mesophytic forest community, with species of *Acer, Aesculus, Betula, Carya, Fagus, Liriodendron, Quercus,* and *Tilia.* The number of dominants decreases in all directions from the southern Appalachian center. To the north, around the Great Lakes–St. Laurent the dominants are *Tsuga canadensis, Pinus strobus, Picea rubens, Acer saccharum, Fagus grandifolia;* to the far northwest, *Acer saccharum* and *Tilia americana;* to the west, *Carya* spp. and *Quercus* spp.; to the south and southeast, in the Piedmont, *Quercus* and *Pinus* with a mixture of more southern trees, especially *Nyssa, Cornus, Oxydendrum, Liquidambar,* and *Carya.* On the high mountains of the Appalachians and in the New York–New England mountains the following conifers dominate a coniferous subalpine belt: *Picea rubens, Abies balsamea, A. fraseri.* Other plants that range deeply into the south along the Appalachian chain are *Betula lutea, Acer pensylvanicum,* and *Sorbus americana.* Shale barrens, granite flatrocks, limestone glades, all have endemic species or species that share a distribution with the Prairie Province (Takhtajan 1986). More detailed data on vegetation composition can be obtained from Braun (1950); Miyawaki et al. (1994); and Greller (1988).

In the western portion of the Appalachian Province, from Indiana to Missouri-Iowa–southern Wisconsin (the former Prairie Peninsula), a transition occurs from dense, closed, relatively tall deciduous forest in the east, through woodland and savanna in the midwest, to prairie west of the Mississippi River. Closed forests persist on the plains only along river valleys. Locally, the transition is often expressed as a mosaic of forest, woodland-savanna, and prairie patch. Heikens and Robertson (1995), for example, describe a series of depauperate communities related to rocky, silty, shallow, and nutrient-poor soils in southern Illinois: xeric, oak (*Quercus*)-dominated forests; scattered trees in a prairie matrix (barrens); eastern redcedar (*Juniperus virginiana*) stands with an open shrub layer and low herbaceous cover (sandstone or limestone glades); and loess hill prairie communities. In addition, limestone substrates in the northern South permit the development of a blue ash-oak savanna-woodland in the bluegrass region of Kentucky (Baskin et al. 1987); cedar glades (open areas on rock, gravel, or shallow soils within redcedar forests) in the central basin of Tennessee; and barrens, or prairie patches in east Tennessee (Chester 1989). McPherson (1997) notes that "midwestern oak savannas are discontinuously distributed from northern Minnesota to southeastern Texas." Originally, these are estimated to have covered "20 million hectares in a continuous band along the eastern edge of the Great Plains." European settlers plowed the land for agriculture in the nineteenth century, and it is still in cultivation. Currently they encompass only 12,000 ha in a fragmented distribution. "Dominant overstory plants are winter-deciduous oaks." In southern savannas, post oak (*Quercus stellata*) and blackjack oak (*Q. marilandica*) are characteristic; black hickory (*Carya texana*), black oak (*Q. velutina*), and white oak (*Q. alba*) are locally abundant. Minor species are eastern redcedar (*Juniperus virginiana*), honey mesquite (*Prosopis glandulosa*), cedar elm (*Ulmus crassifolia*), common persimmon (*Diospyros virginiana*), and sugar hackberry (*Celtis laevigata*). In Missouri this community gives way to a more northern group of trees: bur oak (*Q. macrocarpa*) and northern pin oak (*Q. ellipsoidalis*) on moist sites, and black oak and white oak as codominants on other sites, with a large number of subdominant trees (McPherson 1997). Farther north, in Nebraska, Iowa, South Dakota, and Illinois, the following trees increase in importance: white oak, black oak, red oak (*Q. rubra*), and chinquapin oak (*Q. muhlenbergii*), especially on limestone; and bitternut hickory (*Carya cordiformis*), shagbark hickory (*C. ovata*), elms (*Ulmus* spp.), on moist sites. A number of small trees, shrubs, and woody vines occur in the understory. Still farther north, in North Dakota, Minnesota, and Wisconsin, the important trees are bur oak, red oak, northern pin oak, black oak, sugar maple (*Acer saccharum*), on the general uplands; and on wet sites, white elm (*Ulmus americana*) and hackberry (*Celtis occidentalis*). Understory plants are primarily grasses and forbs that are common in the Prairie Province. The most widespread is little bluestem (*Andropogon scoparius*). In the southern part, subdominant species are warm-season grasses such as Indiangrass (*Sorghastrum nutans*), big bluestem (*Andropogon gerardii*), brownseed paspalum (*Paspalum plicatulum*), as well as some introduced grasses. Farther north these are gradually augmented by cool-season graminoids, including the exotic Kentucky bluegrass (*Poa pratensis*), sedges (*Carex* spp. and *Cyperus* spp.), and needlegrass (*Stipa* spp.). Summer-flowering, herbaceous dicots are common throughout, including hog peanut (*Amphicarpaea*), false toadflax (*Comandra*), spurge (*Euphorbia*), and bee balm (*Monarda*).

In the Great Lakes–St. Laurent region and south in the Appalachian chain at increasingly higher elevations from New York State and New England, a mosaic of pure conifers, mixed conifers and northern hardwoods, and pure northern hardwoods occurs. Two types of mixed conifer-hardwood associations can be recognized: a southern, lower elevation forest with Canadian hemlock (*Tsuga canadensis*), eastern white pine (*Pinus strobus*) and hardwoods such as sugar maple (*Acer saccharum*), red maple (*Acer rubrum*), beech (*Fagus grandifolia*), yellow birch (*Betula lutea*), and some others; a northern, upper elevation forest with red spruce (*Picea rubens*) and some hardwoods, including birches, beech, and maples. On sandy sites pines (*Pinus strobus, P. resinosa, P. banksiana*) predominate; in low wetlands black spruce (*Picea mariana*), northern white cedar (*Thuja occidentalis*), and larch (*Larix laricina*) predominate; in burned areas aspen (*Populus tremuloides*), balsam poplar (*Populus balsamifera*), paper birch (*Betula papyrifera*), and gray birch (*B. populifolia*) combine or form pure stands. West of the Great Lakes and north of the Great Plains prairie, in Manitoba, Saskatchewan, and Alberta, Canada, is the Aspen-Fescue Parkland (*Populus tremuloides* and *Festuca*), a mosaic of fescue prairie with groves of aspen in low spots. It forms a transition between the conifer-hardwoods forest region and the prairies.

NORTH AMERICAN PRAIRIE PROVINCE

According to Takhtajan (1986), a zonation of grassland vegetation occurs along a gradient of increasing moisture from west to east: shortgrass prairie ($\leq$ 3 dm tall), midgrass prairie ($\leq$ 1 m tall), and tallgrass prairie (1–3 m tall). Sims (1988) notes, further, that a north-south gradient of temperature is a fundamentally important factor. The vegetation varies with edaphic and physiographic conditions, even with annual variations in seasonal distribution of precipitation and warmth. Fires play an important role in establishing and maintaining the forest-prairie boundary. Absence of fire promotes woody plant growth. Gallery forests extend deep into the province. Grasslands are the largest of the North American vegetation formations, originally covering nearly 300 million ha of the 770 million ha in the United States and about 50 million in Canada and 20 million in Mexico. In the United States, 125 million ha remain in grassland, and grassland continues to constitute the largest type of natural vegetation (Sims 1988). *Helianthus annuus* may have originated in the south of this province, or in grasslands of the Sierra Madre Occidental.

ROCKY MOUNTAIN (CORDILLERAN) REGION

As recognized here, this region includes the Rocky Mountain Floristic Province, extending from the border between the Yukon and British Columbia south through Idaho and Montana to New Mexico and Texas. Also included in this region are the following phytochoria of McLaughlin (1989): Southern Rocky Mountains–Mogollon, Colorado Plateau, Apachian, Great Basin, and Columbia Plateau. The Columbia Plateau is a transitional region between the Great Basin, Vancouverian, and Rocky Mountain provinces. The Colorado Plateau is a transitional region between the Rocky Mountain and the Great Basin provinces, as are the many mountain ranges in southeastern Utah and northern Arizona.

The dominant vegetation cover is coniferous forest on the higher portions of the mountains and sagebrush shrublands, palouse grassland, and desert grassland on the lowlands, with pygmy conifer woodland, deciduous oak, or mountain mahogany shrubland on the foothills. Among the dominant conifers are species of *Abies, Picea, Pinus, Pseudotsuga, Thuja,* and *Tsuga;* more restricted or subdominant, rarely dominant, are the following: *Chamaecyparis, Cupressus, Juniperus, Larix, Calocedrus, Sequoia, Sequoiadendron,* and *Taxus.* Only *Pinus contorta* and *Pseudotsuga menziesii* are widespread and common in both the Rocky Mountain and Vancouverian provinces. *Juniperus* and *Pinus* sect. *cembroides* form pygmy woodlands throughout the Great Basin Province but also in the southern portion of the Rocky Mountain Province. *Pinus ponderosa* is common, as well, on the Colorado Plateau. "At middle and upper elevations in the Great Basin Province, the plant communities look more and more like those of the Southern Rocky Mountain Province: *Pseudotsuga menziesii, Picea engel-*

mannii or *Picea pungens*, and *Abies bifolia* form patches of forest in favorable sites, with *Picea* and *Abies* at higher elevations than *Pseudotsuga*. *Populus tremuloides* forms large, clonal groves as in the southern Rocky Mountains. No other region of the New World holds so great a diversity of conifers. Alpine tundra is present on the highest peaks; its development diminishes southward. Above timberline in the Great Basin Province, the common species also occur in other provinces of the Rocky Mountain Region" (Takhtajan 1986).

ROCKY MOUNTAIN PROVINCE

Peet (1988) notes that altitudinal life zones have long been used to classify vegetation in the Rocky Mountains. These are generally recognized as foothill, montane, subalpine, and alpine. Within these belts, dominant species are used to designate vegetation types. He notes that regionally certain altitudinal belts may be lacking, while in other areas more or different vegetation types are present. Also, vegetation varies continuously along environmental gradients. Peet (1988: figure 7) provides "gradient mosaic" diagrams for five mountain ranges at latitudes varying from 32°N (Madrean) to 53°N (Far Northern). Alexander (1985) lists in outline form the forest types of the Rocky Mountains (including also the Great Basin and Black Hills) in the United States. Komarkova et al. (1988) provide a detailed classification and description of woodland and forest types in southwestern Colorado, in the Elk, San Juan, and Sawatch mountains.

VANCOUVERIAN PROVINCE

This region has received reviews by Franklin and Dyrness (1973) and Franklin (1988). The western slopes of the Olympic Mountains in Washington support a temperate rain forest, dominated by the giant conifers sitka spruce (*Picea sitchensis*), western redcedar (*Thuja plicata*), and western hemlock (*Tsuga heterophylla*); these are draped with epiphytic bryophytes and *Usnea* (old man's beard lichen). A spruce-hemlock forest continues north to Kodiak Island, where mountain hemlock (*Tsuga mertensiana*) gradually becomes a dominant, with *T. heterophylla* and *Picea sitchensis*. Alder (*Alnus rubra*) is the only deciduous tree that extends to the Gulf of Alaska. *Thuja* goes no farther north than the base of the Alaska Panhandle. Southward and inland from the Olympic Peninsula drought becomes an increasing factor, so that fire frequency increases; Douglas fir (*Pseudotsuga menziesii*), which has a thick bark, becomes a forest dominant. On the east side of the Cascade Range and southward in the Sierra Nevada, the forests are also drier, and ponderosa pine (*Pinus ponderosa*) becomes a dominant, as it is in the Rocky Mountains. In southern Oregon and northern California the Puget Trough is interrupted by a series of low mountains connecting the Coast and Cascade ranges; this is the Klamath region. Many taxa of the Californian flora occur here. The Klamath region also has a large number of local endemics, associated with the serpentine and peridotite rocks that occupy large areas within the granite rock matrix.

Among the widely distributed species in the Sierra Nevada are the trees *Pinus contorta, P. flexilis,* and *P. ponderosa. Pinus monticola* is widespread but reaches its best development in northern Idaho and adjacent areas of the Rocky Mountain Province.

Herbaceous taxa link the Sierra Nevada with the southern Cascade Range and with such mountains of the Great Basin Province as the Inyo Mountains, Toiyabe and Toquima mountains, and Steens Mountain, and with the Wallowa Mountains of the Rocky Mountain Province. In addition there are herbs of cosmopolitan distribution in North America and circumboreal distribution at high latitudes (Takhtajan 1986).

The Vancouverian Province extends south into California also along the northern Coast Ranges. On the seaward side they are immersed in fog during the summer as well as the winter, although the actual precipitation is not very great. Here is the home of *Sequoia sempervirens,* the coast redwood, arguably the tallest tree in the world. *Pseudotsuga* intermingles with the *Sequoia* or replaces it locally. Sword fern (*Polystichum munitum*), Pacific rhododendron (*Rhododendron macrophyllum*), salal (*Gaultheria shallon*), evergreen huckleberry (*Vaccinium ovatum*), redwood sorrel (*Oxalis oregana*), inside-out flower (*Vancouveria parviflora*), and whipplevine (*Whipplea modesta*) are prominent understory plants that extend here from farther north. From the south, broad-leaved evergreens such as California laurel (*Umbellularia californica*), *Myrica californica,* and tanbark (*Lithocarpus densiflora*) represent the Californian Province (Takhtajan 1986).

GREAT BASIN PROVINCE

Big sagebrush desert, shrub steppe, and pygmy conifer woodland are the major vegetation types in the Great Basin Province. As in the Rocky Mountain Province, altitudinal zonation of vegetation is prominent. With increasing elevation one finds desert shrubland, open woodland or chaparral brushland, open or patchy forest, and alpine tundra. North-facing slopes are forested or wooded, adjacent south-facing slopes are more open. Cold air drainage in canyons causes the reversal of zonation, with trees extending to lower elevation in canyons than on ridges. Barbour and Christensen (1993) give the following account of forest zonation: between 1,500 and 2,500 m is a pinyon-juniper woodland that covers 170,000 km²; it is 3–7 m tall, with rounded crowns, and has an understory of cold-desert shrubs, C_3 bunchgrasses and forbs, with a varying cover. Above this is a zone of *Abies concolor* montane forest, followed by a subalpine woodland (at 3,000 m) dominated by *Pinus flexilis* and/or *P. longaeva;* the latter, at nearly 5,000 years, are the oldest living things on earth. On the driest ranges the pinyon pines and sagebrush extend to the summits. On the more mesic ranges the montane zone includes *Pinus ponderosa* and *Pseudotsuga menziesii,* and the subalpine zone includes *Abies lasiocarpa* and *Picea engelmannii.*

Sagebrush (*Artemisia tridentata*) is the most widespread plant in the province. It is a shrub < 2 m tall. It dominates valleys, foothills, and plains, extending high up on desert mountains, even above timberline. Grasses intermingle with the shrubs and can be dominant locally (*Agropyron, Poa, Festuca, Stipa,* etc.). On calcareous soils the dominant shrub is blackbrush (*Coleogyne ramosissima*) and the grasses are needlegrass (*Stipa arida, S. speciosa*) and blue grama (*Bouteloua gracilis*). The pronghorn antelope (*Antilocapra americana*) was apparently the only large native herbivore on the sagebrush when Europeans first arrived. West (1988) speculates that aboriginal people arriving 13,000 B.P. hunted the large mammals to extinction.

Immediately above the sagebrush zone, where precipitation is sufficient, there is an open juniper or juniper-pinyon woodland, with sagebrush as an important member of the community. The juniper is *Juniperus osteosperma;* the pinyon is *Pinus monophylla* in the Great Basin and *P. edulis* on the Colorado Plateau. *Pinus ponderosa* and *Quercus gambelii* enter into the pygmy woodland east and south of the Great Basin, especially on the Colorado Plateau. *Quercus gambelii* forms a dense chaparral. *Pinus ponderosa* forms an open parkland or a forest on the Kaibab Plateau. On the higher mountains of the Great Basin the conifer forests and aspen parkland occur. These are similar in altitudinal zonation and species dominance to those in the southern Rocky Mountains. At timberline, spruce, fir, and Douglas fir may form the krummholz (a transition zone between alpine tundra and subalpine coniferous forest). Often there is a more open timberline forest of *Pinus flexilis* (limber pine), *P. longaeva* (intermountain bristlecone pine), or *Pinus albicaulis* (whitebark pine). On the alpine tundra are many species of herbaceous plants that also occur in the Rocky Mountain Province. The southern border of the Great Basin Province is marked by the replacement of sagebrush by *Larrea* and *Ambrosia* (creosote bush and ragweed), where *Yucca brevifolia* (Joshua tree, as in the Mohave Desert) and *Coleogyne ramosissima* (blackbrush) are most abundant (Takhtajan 1986).

MADREAN SUBKINGDOM

In this phytochorion evergreen taxa of the following families are prominent members of the vegetation of most regions: Pinaceae, Fagaceae, Lauraceae, Arecaceae, Ericaceae, Rosaceae. Some deciduous taxa are Aceraceae, Platanaceae, Hippocastanaceae (*Aesculus*), Caesalpiniaceae (*Bauhinia, Cercis*). Traditionally, this phytochorion has comprised flora of subhumid to arid regions with low freeze frequencies. In the present conception of the subkingdom, the floras of arid regions are excluded and treated with either the Boreal Subkingdom (Great Basin Province) or with the Neotropical Kingdom (Mexican Xerophytic [Sonoran] Region). Instead, floras are included of humid and perhumid climates with low freeze frequencies, characterized by Holarctic evergreen and some deciduous taxa, accompanied by some cold-tolerant Neotropical taxa. Vegetation ranges from oak-dominated montane rain forest to open evergreen oak woodland and savanna and edaphic grasslands.

MADREAN REGION

For this study the Madrean Region comprises the Californian Province and the Sierra Madre Occidental Subprovince of Takhtajan (1986). It includes the Madrean Woodlands and related broad-leaved evergreen-dominated vegetation in the Sky Islands of Arizona and in the mountains of southeastern New Mexico and western Texas. It is characterized by a subhumid to semiarid climate at lower elevations, gradually changing to marginally humid at the highest elevations. Light freezes in winter characterize the climate throughout, although mean annual temperatures decrease with increasing elevation. The typical sequence of vegetation is: broad sclerophyll woodland (California) or oak woodland, often mixed with pine, cypress, and/or juniper, or oak savanna, at lowest elevations; grassland at the bases of hills above floodplains; oak shrubland (chaparral) on thin, rocky soils; mixed conifer montane forest; and fir forest, with spruce, pine, or Douglas fir, in the subalpine zone.

CALIFORNIAN PROVINCE

Californian Province vegetation varies from grasslands and shrublands to savannas, woodlands, and forests. Excluding grasslands, woody angiosperms dominate most of the lowland vegetation; conifers dominate in the mountains and on extreme soils. Evergreen angiosperms dominate woodlands in southern and central California; deciduous angiosperms become more common northward and dominate in wooded wetlands.

Barbour (1988) reviews the major vegetation types of the Californian Province. He recognizes the following arborescent types: blue oak woodland, southern oak woodland, mixed evergreen forest, and a number of montane and subalpine forests. Minnich (1976) maps and describes a "remarkable diversity of plant communities" in the vegetation of the San Bernardino Mountains, east of Los Angeles. The diversity is due to the great elevational (1,220 m–3,506 m above sea level) and topographical variation. Vegetation is arranged in altitudinal belts. MAP ranges from 380 mm (San Bernardino) to 1,020 mm (mountain crest) to 200 mm (edge of Mohave and Colorado deserts). Minnich lists the following vegetation types: California sage scrub, chaparral (chamise, mixed, oak, and desert), oak woodland, conifer forests in chaparral, marginal forests, western coniferous forest, timberland (sic) chaparral, lodgepole pine forest, pinyon woodland, Great Basin sage scrub, desert oak scrub, juniper woodland, Joshua tree woodland, desert vegetation, and barren. Some of these are more characteristic of the Great Basin, Sonoran, or Mohave deserts and are described in other sections of the present work.

SIERRA MADRE OCCIDENTAL PROVINCE

D. E. Brown (1994) includes a series of mountain vegetation belts within his Madrean evergreen woodland type; in altitudinal sequence from the lower slopes they are: encinal (oak woodland), oak-pine woodland, and montane conifer (pine) forest. On the south slope of Santa Catalina Mountain and higher elevations of the Pinaleno Mountains of southern Arizona, Niering and Lowe (1984) illustrate and describe the altitudinal sequence of vegetation types (see also Fitzhugh et al. 1987; Muldavin et al. 1990). From base to top these are: desert (< 1,200 m); grassland (ca. 1,200–1,300 m); open oak woodland (ca. 1400–1700 m, dominated by *Quercus oblongifolia* and *Q. emoryi*, with understory graminoids and rosette monocots); pine-oak woodland (ca. 1,600–2,150 m, dominated by *Quercus arizonica*, *Q. emoryi*, *Q. rugosa*, *Pinus cembroides*, *P. leiophylla* var. *chihuahuana*, and *Juniperus deppeana*); canyon forests (1,200–1,800 m, dominated by *Cupressus arizonica*, with *Fraxinus* and *Alnus*); pine forest (ca. 2,100–2,500 m, dominated by *Pinus ponderosa*, with *Quercus hypoleucoides* and graminoids); fir forest (ca. 2,500–2,750 m, dominated by *Abies concolor*,

Pseudotsuga menziesii, Pinus strobiformis, and *Pinus ponderosa*); and spruce forest (ca. 2,750–3,050 m, dominated by *Picea engelmannii* and *Abies lasiocarpa*). Similar altitudinal sequences are found in the Rincon, Chiricahua and Huachuca mountains of Arizona and mountains in southern New Mexico.

This region has a long history of human habitation, probably due to the presence of so many different types of plant communities in close altitudinal proximity, with a great variety of herbs, vines, and grasses that thrive in open vegetation. Indeed, the relatives of many domesticated plants, e.g., maize (*Zea mays*), squash (*Cucurbita pepo*), a domesticated chenopod (*Chenopodium berlandieri*), sunflower (*Helianthus annuus*), and sumpweed (*Iva annua*), are native to this province.

The following is a detailed treatment of the vegetation from the southwestern United States south through the length of the Sierra Madre Occidental.

EVERGREEN OAK WOODLAND. It is difficult to separate oak woodlands from oak shrublands. The principal characters that one uses to distinguish trees from shrubs is structure and form of branching. Oaks tend to average 4–5 m tall but lack a single well-defined trunk. Other oaks reach heights of only 2–2.5 m tall but have a single trunk. Some species of oaks can grow as either shrubs or trees. Various authors can refer to the same stand as matorral (shrubland) and bosque (woods). In northern Mexico, open oak woods are the common structural type, referred to in English as woodland, oak-grassland, and oak savanna. Most southern Mexican oak communities are dense and with closed formations, but some stands are open, with wide spaces between trees so that shrub or herb layers can develop; this is especially so in the edges between oak forest and shrubland.

Oak woodland occupies a middle position in the Sierra Madre Occidental, where it forms zonal vegetation between lowland tropical deciduous forest, xerophytic shrublands and pastures, and upland pine-oak and pine forests. In the Sierra Surotato, Sinaloa, oak woods occur between 600 and 1,400 m. In the northeast of Sonora and adjacent portions of Chihuahua, oak woods occur between 1,000 m and 2,800 m; farther south in the Río Mayo, Sonora, drainage they occur between 900 and 1,500 m. The most widespread oak species is *Quercus chihuahuensis.* It occurs in the driest, hottest sites and reaches 4–8 m tall. Trunks are often thick with a wide crown. Foliage is grayish due to heavy pubescence. Other common oaks are *Q. santaclarensis, Q. emoryi, Q. arizonica, Q. oblongifolia, Q. chuchipensis, Q. durangensis,* and *Q. hypoleuca.* Leaves are generally small, thick, and stiff. *Pinus cembroides* and various species of *Juniperus* are often found with the oaks. At higher elevations or in protected, moist sites the following oaks are prominent: *Quercus albocincta, Q. tuberculata, Q. mcvaughii, Q. penninerva, Q. sipuraca,* and *Q. epileuca.* These species appear at higher elevations in pine-oak and pine forests. Goldberg (1982) describes evergreen oak woodlands of low stature on north-facing sites at 600–1,100 m, in a matrix of tropical deciduous woodland, in central Sonora, Mexico. Soils under oak woodland are more acidic and nutrient-deficient than under tropical trees, at this lower limit of the oak woodland. Dominant oaks are *Quercus chihuahuensis, Q. albocincta,* and *Q. tuberculata.* Trees are widely spaced and grassland occurs between them, leading

Gentry (1946) to describe some stands as "pasture with oak-juniper." The high mountains at the tip of the Baja California peninsula appear to be related floristically to the Sierra Madre Occidental. There are two zones of vegetation in which oaks play a prominent role, a midelevation oak woodland type in which *Quercus tuberculata* (1,000–1,500 m) is a dominant, and a high-elevation type (1,300–2,200 m) of oak woods in which *Quercus devia* is the dominant in pure stands on moist, low sites; more often it occurs with *Pinus cembroides* var. *lagunae*.

In the south of Nayarit, lower elevation transitions to tropical deciduous forest occur at 1,000 m. The dominant oaks are *Quercus aristata, Q. elliptica,* and *Q. planipocula;* they form stands 10–15 m tall. Diversified oak woods occur in the Transverse Volcanic Belt. In the western part the oaks are xerophytic and low; *Quercus resinosa,* with large leaves, occurs with *Juniperus flaccida.*

In the Sierra Madre Oriental, near Monterrey, Nuevo León, at 430 m, is an evergreen woodland of *Quercus fusiformis.* To the southeast of Piedras Negras, Coahuila, is an oak woodland of *Quercus mohriana.* Both are some 10 m tall and surrounded by more xerophytic vegetation. They may be maintained by close proximity to springs.

INTERIOR CHAPARRAL (D. E. BROWN 1994). This shrubland type is important in Arizona (1.4 million ha) and parts of northern Mexico. In Arizona it occurs at midelevations (1050–2,000 m) on foothills, mountain slopes, and in canyons, from the Virgin Mountains on the border with Nevada, southeastward to the central sub-Mogollon regions of the state, where it is "of major occurrence and importance." Southeastward it occurs as scattered stands in the drier mountains of southeast Arizona, southern New Mexico, southwest Texas, and northeastern Chihuahua to the limestone mountains of Coahuila and Nuevo León; in eastern Chihuahua and western Coahuila it is also an important vegetation type, occurring at 1,700–2,450 m. It often borders (and integrates with) ponderosa pine, or encinal, or pinyon-juniper woodlands at upper elevations, and desert shrubland or semidesert grassland at lower elevations. Precipitation is bimodal, with steady winter rains and sudden, high-intensity summer thundershowers. It varies from 380 to 635 mm/year, usually falling as rain but also as snow at higher elevations. Most chaparral shrubs have dense, compact crowns (1–2.5 m tall) and small, evergreen sclerophyllous leaves. After fires, most species readily sprout from root crowns, or they regenerate from seed germination (desert ceanothus and deerbrush, pointleaf and Pringle manzanitas). Many species are disjunct in the California chaparral, e.g., shrub live oak, sugar sumac, hollyleaf buckthorn, manzanitas, birchleaf mountain-mahogany, hairy mountain-mahogany, yellowleaf silktassel, brickellbush, ceanothus, and California fremontia.

Brown recognizes two variants of interior chaparral. He describes the Arizona chaparral as being characterized by shrub live oak (*Quercus turbinella*), either occurring in pure stands or accompanied by shrubs such as birchleaf mountain-mahogany (*Cercocarpus montanus* var. *glaber*), skunkbush sumac (*Rhus trilobata*), silktassels (*Garrya wrightii, G. flavescens*), and desert ceanothus (*Ceanothus greggii*). Also present are the following shrubs: hollyleaf buckthorn (*Rhamnus crocea*), cliffrose (*Purshia stansburiana*), desert olive (*Forestiera neomexicana*), sophoras (*Sophora* spp.), Arizona rosewood (*Vauquelinia californica*), Lowell ash (*Fraxinus anomala* var. *lowellii*), barberry (*Berberis fremontii*), and, at higher elevations, manzanita (*Arctostaphylos*).

Brown describes the Coahuila chaparral as having an important role played by Coahuila scrub oak (*Quercus intricata*); other oaks are Vasey oak (*Q. pungens*), Pringle oak (*Q. pringlei*), as well as some others (*Q. invaginata, Q. tinkhamii, Q. laceyi, Q. organensis, Q. grisea,* and *Q. toumeyi*). Taxa shared with the Arizona chaparral are *Garrya wrightii, Cercocarpus montanus* var. *paucidentatus, Ceanothus greggii, Fallugia paradoxa, Rhus trilobata,* and *Arctostaphylos pungens.* There are a number of sister species of Arizona chaparral genera represented in the Coahuila flora: *Rhus virens* ssp. *choriophylla, Garrya ovata, Cowania plicata, Berberis trifoliata, Vauquelinia angustifolia, Fraxinus greggii,* and *Fendlera linearis.* Also present in Coahuila are madrones (*Arbutus texana, A. arizonica*), sages (*Salvia ramosissima, S. roemeriana,* and *S. regla*), as well as "nonchaparral associates" such as *Mimosa biuncifera, Acacia greggii, Ptelea trifoliata, Nolina erumpens,* and *Cupressus arizonica.* Brown says the Interior Chaparral is floristically continuous from southern Nevada through the Sierra Madre Oriental.

PINE-OAK, PINE-OAK-JUNIPER MIXED FORESTS, AND DECIDUOUS OAK FORESTS. Mixed communities of oak and pine occur widespread over the landscape, often accounting for as much cover as either pure oak or pure pine forests. In any stand the proportion of pine and oak strongly influences the cover and composition of the understory layers. As in the oak forests, physiognomy of the oak species is often deciduous. But the leafless period of the oaks is often brief and does not coincide among the various species that compose the forest, so that some green is present throughout the year. White (1948) describes the floristic composition of pine and oak dominated woodlands, between 1,300 and 1,800 m, at Río Bravispe, northeastern Sonora. The oaks are listed as *Quercus arizonica, Q. emoryi, Q. hypoleucoides,* and *Q. toumeyi*; pines are *Pinus arizonica, P. strobiformis, P. cembroides, P. leiophylla* var. *chihuahuana,* and *P. ponderosa*; and other species include *Arbutus arizonica* and *Garrya wrightii.* A similar type of mixed woods is described as montane low forest by Muller (1947), in Coahuila, Mexico; oaks include *Quercus gravesii, Q. hypoleucoides, Q. laceyi, Q. arizonica, Q. sinuata, Q. mohriana,* and the pine is *Pinus cembroides.* Also *Juniperus pachyphloea* and *J. flaccida, Arbutus xalapensis,* and *Fraxinus cuspidata* are listed, and in the understory: *Salvia, Garrya, Rhus trilobata, Ptelea trifoliata, Bumelia lanuginosa, Cercocarpus,* and *Vitis arizonica.* Juniper woodlands are described for Chihuahua and for the northwestern spur of the Sierra Madre Oriental. There, juniper (*Juniperus mexicana*) is often associated with *Pinus cembroides* on limestones slopes, or surrounded by desert shrublands or gypsophilic grassland. *Juniperus deppeana* forms low forests in the Sierra Madre Occidental and in other ranges throughout Mexico.

Rzedowski (1981) brings together numerous records of pine-oak woods in Durango, where the type forms zonal forests at 1,500–3,150 m. Dominant oaks are *Quercus arizonica, Q. sideroxyla, Q. rugosa, Q. urbanii, Q. coccolobifolia, Q. laxa, Q. durifolia, Q. magnolifolia, Q. viminea, Q. crassifolia, Q. microphylla,* and *Q. striatula.* Dominant pines are *Pinus engelmannii, P. leiophylla* var. *chihuahuana, P. arizonica, P. strobiformis, P. reflexa, P. lumholtzii,* and *P. rugosa.* Gentry (1946, 1957) describes oak-pine forests between 1,370 m and 2,130 m on Sierra Surotato, near the border with Chihuahua State, in which (at 1,830 m) *Quercus pallescens* and *Q. epileuca* are the dominants; other oaks are *Q. durifolia, Q. gentryi, Q. candicans, Q. coccolobifolia,* and *Q. penninerva;* the pines are *Pinus macrophylla, P. ayacahuite, P. oocarpa, P. lumholtzii,* and *P. pseudostrobus.* In mesic canyons on Sierra Surotato pines and oaks are replaced by more mesophytic trees: *Styrax argenteus, Ostrya virginiana,* and *Cornus disciflora;* along streams: *Cornus excelsa, Garrya laurifolia,* and *Magnolia schiedeana;* shrubs are *Rhus allophylloides* and *Rhamnus* spp.; herbs are *Crotalaria, Erythrina montana, Triumfetta, Ipomoea, Aristida,* and *Muhlenbergia* (Gentry 1946, 1957). At 1,830 m on Sierra Surotato is a deciduous oak forest with the epiphytes *Tillandsia inflata* and various mosses; the understory is open (Gentry 1946: figure 3). In Jalisco, at higher elevations (1,800–2,600 m), the dominant oak is deciduous, *Q. resinosa;* it has a large leaf and is whitish on the back. This oak forms savannalike stands and can reach heights of 6–10 m, or it can form matorrales only 3 m tall. *Quercus potosina,* with *Q. eduardii* and *Q. coccolobifolia,* occur on the driest slopes. *Quercus depressipes, Q. grisea,* and *Q. oblongifolia* form similar woodlands. These communities resemble some deciduous oak and mixed forests of the northern Sierra Madre Oriental (see Muller 1939).

In Jalisco the oak-pine forests reach 18 m tall; the dominant oaks are *Quercus obtusata, Q. viminea, Q. gentryi, Q. urbanii,* and *Q. rugosa.* Dominant pines are (lower elevations) *Pinus michoacana, P. oocarpa;* (higher elevations) *Pinus lumholtzii, P. leiophylla, P. teocote, P. engelmannii.* On the western slopes of the Sierra Madre Oriental, Rzedowski (1981) describes woodlands ("bosque seco de encino-pino") in which the pines (*Pinus montezumae* and *P. patula*) are quantitatively more important than the evergreen oaks (*Quercus affinis, Q. diversifolia,* and *Q. rugosa*). Forests can be of low stature, with a one-layered canopy, such as one that occurs near Jacala, Hidalgo. Here the dominant oaks are *Quercus mexicana* (evergreen) and *Q. obtusata* (deciduous); other trees are *Pinus teocote, Juniperus flaccida,* and *Arbutus xalapensis;* understory plants are *Rhus virens, R. trilobata, Senecio aschenbornianus, Amicia zygomeris,* and *Eupatorium berlandieri.* The herb layer varies with the openness of the stand; grasses are often important. In the same area, at higher elevations (900–2,100 m), are open pine-oak forests ("bosque humido de pino-encino") 20 m tall, where oaks have rounded crowns and deciduous or evergreen leaves. Orchids, ferns, and bromeliads festoon the branches. The most common oaks are *Quercus clivicola, Q. polymorpha, Q. grisea,* and *Q. canbyi;* common pines are *Pinus montezumae* and *P. teocote;* also present are *Juniperus flaccida, Arbutus,* and *Juglans.* In the Sierra de la Laguna, at the tip of Baja California, there is an extensive oak-pine woods, at 1,300–2,200 m, in which *Quercus devia* is the dominant, occurring with the emergent *Pinus cembroides* var. *lagunae.* Understory plants include *Arbutus, Heteromeles, Prunus,* and *Salix* and *Populus* along watercourses. These woodland types are described in floristic detail by León de la Luz and Domínguez-Cadena (1989).

JUNIPER SHRUBLANDS. These are evergreen stands that vary from matorrales 50 cm tall to forests of 15 m in height; mainly they grow 2–6 m tall. These are open stands, with a lower shrub layer and a herbaceous layer. Often they occur within other more extensive forest, shrubland, or grassland communities. Branches are usually infested with the vascular plant parasite *Phoradendron* and the fungus *Gymnosporangium*. Vines and epiphytes are rare. This type is widespread throughout Mexico, from Baja California and Tamaulipas to Chiapas. It is usually confined to a narrow belt between forests of oak and pine above and pasturelands, "matorral xerófilo" or "bosque tropical caducifolio," below. They appear largely to be secondary communities, at elevations rarely below 1,500 m. They thrive on a great variety of substrates but usually on thin soils above bedrock. Species of *Juniperus* are integrated into the drier forests and woodlands of *Pinus* and *Quercus*. They also form open treelike stands at high elevations.

PINE WOODLAND. Pinelands are characteristic vegetation of Mexico and occupy large areas of the country. There are at least 35 species of *Pinus*, 37 percent of the world total (Rzedowski 1981). Pines are morphologically rather similar but are ecologically diverse. They are particularly well adapted to colder climates in Mexico, as well as to subhumid moisture regimes and to acidic and overly well-drained substrates. Pines often occur throughout the oak forests, with junipers, and also in subalpine fir forests. Thus pines form a part of a complex landscape mosaic, which extends to temporal (successional) relationships.

Pine woodlands occur throughout the length of the Sierra Madre Occidental. Such types occur as well in the Sierra Madre Oriental in Tamaulipas; on mountains in the Central Plateau, especially in Coahuila; north and south in mountains on the Baja Peninsula, in the Transverse Volcanic Belt, Sierra Madre del Sur, and the mountains of Chiapas. The best development of pineland occurs between 1,500 and 3,000 m in the mountains of Mexico. At this level, mean annual temperatures (MAT) range from 10 to 20°C. All pine areas are affected by freezes. Mean annual precipitation (MAP) ranges from 600 to 1,000 mm, concentrated in six or seven months of the year. Within those climatic parameters, pines occur on igneous rocks, gneisses, sands, and limestones, usually at pH ratings of 5–7 (acidic). Pines have podzolic soils, either brown or red, and the soil is generally covered with needles, then a humus layer 10–30 cm thick; lower horizons are nutrient-deficient. Mycorrhizal relationships are well developed, which is attested to by the great variety of fungi associated with the pine woodlands. Often the limits of pines are associated with the contacts between igneous rocks and sediments of marine origin (Rzedowski 1981).

Pine woodlands are fire-prone, and humans manage pine stands for grazing by use of fire. If fires are too frequent, the pinelands are degraded into shrublands or grasslands. Pines are well known to form secondary wooded communities after disturbance of the primary, more mesic vegetation. Judging from the abundance of pine roots in areas now occupied by other, more mesic types of vegetation, pinelands must have been more extensive in historical times than today (Rzedowski 1981:288). Pinelands are a source of timber, fuelwood, resins, and nuts. These lands are put under agriculture after pines are removed; principal crops are corn, beans, wheat and fruits of temperate climates.

Pine woodlands also occur at higher elevations in northern Mexico, reaching 3,650 m, and form forests in the central and southern mountains to a maximum of

4,100 m, the highest limits of arborescent vegetation on earth. These are related to Rocky Mountain subalpine forests. Caribbean pinelands, in hot, humid climates, with an understory of tropical woody plants, occur on the eastern side of Central America; these are treated with tropical Central American vegetation.

Structurally, pines often form two-layered communities, termed savannas, in which a single species of pine is the sole dominant of the tree layer. The understory may be of two types: a single herbaceous layer dominated by grasses or a variety of herbaceous plants, especially Asteraceae. Fires favor the maintenance of grasses. In areas of high atmospheric humidity, epiphytes (mosses and lichens) and lianas may be well developed. Alternatively, the understory can be woody. Where oaks occur in the pine woodlands they often form an understory tree layer.

The widespread and continuous pine woodlands in the Sierra Madre Occidental appear to be climatically controlled, according to Rzedowski (1981). In northwestern Mexico, along the Río Mayo from Sonora to Chihuahua, are upper montane pine forests reaching from 1,830 to 2,750 m; at lower elevations of 1,500–1,830 m are lower montane pine forests; below that are the oak woods at 900–1,500 m; and at the lowest elevations are tropical deciduous forests (300–900 m) and spiny tropical woodlands (< 300 m). At the higher levels, the understory flora is strongly Holarctic; at lower elevations, Neotropical flora are of increasing importance.

SUBALPINE FIR AND PINE FORESTS. Subalpine fir (*Abies*) forests are confined to the high mountains, between 2,400 and 3,600 m. Elevations above 3,600 m are too dry for *Abies,* and pine (*Pinus*) forests replace them. Subalpine forests are found not only in the Sierra Madre Occidental but also in the other major Mexican ranges, including some in Chiapas and in Guatemala. Rzedowski (1981) estimates that they cover no more than 0.16 percent of Mexico. Although such subalpine conifer forests occur on a variety of well-drained bedrock substrates, black, organic podzol soils are very frequent above 3,000 m. Fir and pine forest environmental conditions on the high mountains of Mexico differ considerably from those of subalpine spruce-fir forests in the Rocky Mountains and from those of the Boreal (Taiga) Forest Region (Rzedowski 1981:303). Mexican subalpine forests live under conditions of relatively high mean annual temperatures (7–15°C), ca. 60 days with freezing temperatures but only rare snows, high equability, great diurnal temperature variation, high humidity, high MAP (> 1,000 mm/year), and fewer than four dry months.

SOUTHEASTERN NORTH AMERICAN REGION

This is a newly conceived phytochorion. It was created to unite the flora and vegetation of warm-temperate, humid regions in southeastern and southern North America. The region includes the southeastern coastal plains of the United States, as well as Bermuda, and the montane zones of the Sierra Madre Oriental, the Sierra Madre del Sur (Takhtajan 1986), and other mountains (Rzedowski 1981) in Chiapas and southward. Throughout the region the following families have representatives as forest dominants: Taxodiaceae (bald cypress), Pinaceae (pine), Fagaceae (beech), Magnoliaceae (magnolia), Lauraceae (laurel), Oleaceae (olive), Aquifoliaceae (holly), Theaceae (tea), Ulmaceae (elm), and Arecaceae (palm). Clearly, the Mexican upper montane broad-leaved forests are predominantly Holarctic (and predominantly deciduous in and near Veracruz State), while the lower montane forests show a mixture of Holarctic and Neo-tropical taxa. In the northern part of Mexico the lower montane forests are dominated by Holarctic taxa (evergreen *Quercus* spp. and deciduous *Chaetoptelea [Ulmus] mexicana*). From Chiapas to Central America, lower montane forests are dominated by mixed tropical families, or evergreen *Quercus* is a codominant (bosque mesófilo de montaña). Similarly, deciduous Holarctic taxa dominate broad-leaved forests of the Gulf and Atlantic coastal plains, with some evergreen Holarctic taxa as codominants or subdominants. Farther south, in the infrequent broad-leaved forests of peninsular Florida, evergreen oaks dominate with Neotropical taxa including the cabbage palm (*Sabal palmetto*) and figs (*Ficus* spp.).

ATLANTIC AND GULF COASTAL PLAINS PROVINCE

Taxa representing tropical, warm-temperate, and deciduous forest floras have been present in the region since the early Tertiary. Progressive cooling, culminating with the deep penetration of Pleistocene ice sheets into south-central North America, eliminated all but the deciduous flora and some broad-leaf evergreens. These taxa persisted in the Mississippi Valley and on a Florida peninsula that extended 200 km farther south than today and was 600 km across. Takhtajan (1986) classifies the flora of the region as Atlantic and Gulf Coastal Plain Province of the North American Atlantic Region of the Boreal Subkingdom of the Holarctic Kingdom. Most of the plants occur in all or parts of a **U**-shaped area coinciding with the Atlantic and Gulf coastal plains and extending northward in the Mississippi Valley to Indiana. Takhtajan includes all but the southern tip of Florida in the province. The Apalachicola River region and central Florida–outer coastal plains (Central Floridian) region can be treated as separate subdivisions.

Many species are endemic to both the Appalachian Province and the Gulf and Atlantic Coastal Plains Province, including *Liriodendron tulipifera* (tuliptree), *Carpinus caroliniana* (hornbeam), *Fagus grandifolia* (beech), some *Quercus* spp. (oaks), *Carya* spp. (hickories), *Nyssa sylvatica* (sourgum), *Cornus florida* (flowering dogwood), and *Chionanthus virginicus* (fringe-tree). Others are confined to just the southern part of the Appalachian Province and to this one, including *Castanea pumila* (chinquapin), *Betula nigra* (river birch), *Bumelia lycioides* (southern buckthorn), *Acer barbatum* (southern sugar maple), *A. leucoderme* (southern black maple), *Toxicodendron vernix* (poison sumac), and *Aralia spinosa* (Hercules' club) (Takhtajan 1986).

An important local area of endemism is along the lower Apalachicola River in western Florida and adjacent Georgia. James (1961) cites R. A. Howard as recognizing 39 endemic species in "northern peninsular Florida along the Georgia border." This is the only location for *Torreya taxifolia* and *Taxus floridana* (Florida yew). Several species in the Asteraceae are narrow endemics to the Apalachicola region, and such endemism occurs in other families as well. An Apalachicolan phytochorion is proposed for this region.

Central Florida is another area of endemism, especially in the near-endemic vegetation type called the Florida scrub. Myers (1990) lists a large number of endemic genera; Harper (1949) gives ranges for many endemic Florida plants. James cites R. A. Howard as recognizing 189 species endemic to the "lake district" of central Florida. Wilbur (1970) lists six species of *Asimina* as endemic to central Florida. In addition, C. L. Brown and Kirkman (1990) map the following woody plants as centered on peninsular Florida, but ranging north only on the coastal fringes of the Gulf and Atlantic coastal plains: *Taxodium ascendens* (pond cypress), *Sabal palmetto*, *Myrica cerifera* (wax myrtle), *M. inodora*, *Quercus chapmanii* (Chapman oak), *Q. geminata* (sand live oak), *Q. hemisphaerica* (laurel oak), *Q. myrtifolia*, *Q. virginiana* (live oak), *Lyonia ferruginea*, and *Viburnum obovatum*. I believe the high number of endemics in the entire area that includes central Florida and the Gulf and Atlantic coastal fringes justifies the recognition of a Central Floridian phytochorion as a distinct subunit of the Gulf and Atlantic Coastal Plain Province. The floristic connection between Florida and the Madrean flora was suggested by D. I. Axelrod, who contended that the antecedents of scrub probably appeared during the early Tertiary as part of the sclerophyllous and microphyllous Madro-Tertiary flora and spread from the southern Rocky Mountains and northern Mexico along the Gulf coast to Florida (Myers 1990). Since nearly all of peninsular Florida was inundated during the maximum sea level during the Pleistocene, it is likely that Florida's "morphologically highly distinctive species, well isolated from their closest relatives" (James 1961), are paleoendemics that represent a flora widespread during the Tertiary.

Pine forests are an important feature of the Coastal Plain Province (*Pinus palustris*, *P. elliottii* var. *elliotti*, *P. taeda*, and *P. glabra*) because of the poor, sandy soil and frequency of fires. In areas protected from fires, the pine forests give way to hardwoods (dicotyledonous angiosperms) such as oaks (*Quercus falcata*, *Q. michauxii*, and *Q. nigra*), hickories (*Carya glabra* and *C. tomentosa*) sweetgum (*Liquidambar styraciflua*), persimmon (*Diospyros virginiana*), and others. In moist lowlands above the swamps, beech (*Fagus grandifolia* var. *caroliniana*), magnolia (*Magnolia grandiflora*), laurel oak (*Quercus hemisphaerica*), and American holly (*Ilex opaca*) are important. Swamps have bald cypress (*Taxodium distichum*), gum (*Nyssa aquatica*), and ash (*Fraxinus caroliniana*). Taxa with mainly tropical relatives are the palms (*Sabal minor* and *Serenoa repens*), Spanish moss (*Tillandsia usneoides;* Bromeliaceae), and species of the genera *Chaptalia*, *Vernonia*, *Cyperus*, *Rhynchospora*, *Scleria*, *Panicum*, and *Paspalum*. Classification of forests is based on Christensen (1988). Ware et al. (1993) contend *Pinus palustris* (longleaf pine) was the most important pine in precolonial upland communities; today, after extensive lumbering, *Pinus taeda* (loblolly pine) dominates successional communities throughout the coastal plains. More detailed, semiquantitative data on vegetation composition can be obtained from Miyawaki et al. (1994).

MEXICAN HIGHLANDS (SIERRA MADRE ORIENTAL−SERRANÍAS MERIDIONALES−SERRANÍAS TRANSÍSTMICAS) PROVINCE

Greller (1990) briefly reviewed the floristic nature of this region. The most striking feature of the plant geography of eastern and southern Mexico (above the lowland tropical rain forest) is the dual nature of the flora and vegetation. This duality is manifested by the mixture of Neotropical and Holarctic flora and by the mixture of ever-green and deciduous trees. The latter is a consequence of the representation of both Boreal (mainly deciduous) and Madrean-Tethyan (mainly evergreen) taxa in the ele-vational belts of forest that occupy most of the area. Subalpine zone stands are domi-nated by Holarctic and endemic taxa and vary from stands dominated by evergreen angiosperms to those dominated by evergreen Holarctic conifers. The dominant vege-tation on good soils in the lower montane zone is broad-leaved forest dominated by Neotropical trees, a few species of evergreen oaks, and some other Holarctic trees, and in the upper montane zone by many evergreen oaks and other Holarctic genera, as well as some endemic (*Chiranthodendron*) and some austral (*Podocarpus, Weinmannia, Drimys*) genera. Plants of western U.S. affinities (pines, oaks, alder, madrone, syca-more) and of West Indian affinities also occur (Miranda and Sharp 1950). Overly well-drained soils are dominated by pine (*Pinus*) forests with oak (*Quercus*) understories. According to Rzedowski (1993), the upper montane (cloud) forests of the entire prov-ince have a group of genera in common with the eastern United States, eastern Asia, and the mountains of South America. These are discussed below. The Sierra Madre Occidental is surrounded by desert and dry tropical woodland and has a semiarid cli-mate at lower elevations, while the higher elevations have a subhumid climate that supports conifer-dominated forests (*Pinus, Abies, Picea,* and *Pseudotsuga*). I treat it as a separate province.

According to Rzedowski (1981) the Sierra Madre Oriental is a separate province that includes most of or patches of Coahuila, Nuevo León, Tamaulipas, San Luis Po-tosí, Querétaro, Hidalgo, Veracruz, and Puebla states. The substrate is limestone rock. Oak-dominated forests are widespread, but pine-dominated communities and some others are present as well. Takhtajan (1986) notes that a considerable number of spe-cies characteristic of the Appalachian Floristic Province occur in the upper montane (cloud) forests of the Sierra Madre Oriental, the number diminishing southward, in-cluding eastern white pine (*Pinus strobus*), witch hazel (*Hamamelis virginiana*), sweet-gum (*Liquidambar styraciflua*), sycamore (*Platanus occidentalis*), wood-nettle (*La-portea canadensis*), beech (*Fagus grandifolia*), hornbeam (*Carpinus caroliniana*), hop hornbeam (*Ostrya virginiana*), pecan (*Carya illinoensis*), shagbark hickory (*C. ovata*), wild black cherry (*Prunus serotina*), redbud (*Cercis canadensis*), sourgum (*Nyssa syl-vatica*), flowering dogwood (*Cornus florida*), buckthorn (*Rhamnus caroliniana*), par-tridge berry (*Mitchella repens*), beech drops (*Epifagus virginiana*), and several species of oak (*Quercus*), alder (*Alnus*), and other genera. The following Mexican endemics are "closely related vicariants" of Appalachian species: yew (*Taxus globosa*), magno-lias (*Magnolia dealbata* and *M. schiedeana*), *Fagus mexicana, Carya mexicana,* bay-berry (*Myrica pringlei*), linden (*Tilia longipes*), maple (*Acer skutchii*), and many oth-ers. There are a number of genera common to Mexico, Central America, and eastern Asia. Other genera are widespread with broad Laurasian connections (Rzedowski 1981).

The deciduous taxa become less important in forests farther south, so that the mon-tane forests of Central America are evergreen, although some of the dominant families have Madrean Holarctic affinities (Magnoliaceae, Fagaceae, Aquifoliaceae, Theaceae, Lauraceae, Rosaceae). Lauer (1968) presents a schematic profile of the altitudinal zonation of vegetation on mountains from Mexico to Colombia. On wind-swept slopes but also on protected interior slopes there are great expanses of oak forest; these constitute the most characteristic type of vegetation in the Sierra Madre Oriental. On the western slopes of the Sierra Madre Oriental, where climate is subhumid, oak and pine-oak woodland ("bosque seco de encino-pino" of Martin) is extensive and greatly enriched by Holarctic (*Arbutus, Arctostaphylos, Carya, Cupressus, Juglans, Pinus, Pseudotsuga*) and some tropical (*Bauhinia*) taxa. Dense oak and pine-oak forests occur on extreme substrates and at high elevations ("bosque humido de pino-encino" of Martin, in Rzedowski 1981). In his province of the Serranías Meridionales, Rzedowski (1981) includes the Transverse Volcanic Belt that extends from Jalisco and Colima to Veracruz, the Sierra Madre del Sur (Michoacán and Oaxaca), and the mountainous complex north of Oaxaca. Pine forests and oak forests are the most common type of vegetation and have an extent comparable to the Sierra Madre Oriental. He also recognizes a separate province, the Serranías Transístmicas, which includes the mountains of Chiapas and extends south into the mountainous regions of Central America. Here also forests of pine and oak predominate. At the Isthmus of Tehuantepec many Holarctic taxa reach their southern limits, so that the mountain forests from Chiapas to Costa Rica–Panama show greater floristic affinities with mountainous South America than with northern Mexico (Kappelle et al. 1992). Descriptions of Central American mountain vegetation follow.

Knapp (1965) recognizes three life zones in the mountains of Mexico–Central America: tierra templada, tierra fria, and tierra helada, in order of increasing elevation. In the tierra templada he distinguishes mountain rain forests with oaks from oak and pine forests with a strong dry period. He also recognizes swamp forests dominated by Holarctic taxa (*Alnus, Fraxinus, Juglans, Populus, Platanus, Salix,* and *Chaetoptelea*) in Mexico and Guatemala.

At elevations of 1,500–3,350 m in Guatemala are forests dominated by conifers, including cypress (*Cupressus*), juniper (*Juniperus*), fir (*Abies*), Montezuma cypress (*Taxodium*), which reach their southern limits in Guatemala. Pine forests and pine with juniper are characteristic of the lower portion of this belt; common species are *Pinus oocarpa, P. strobiliformis, P. ayacahuite, P. montezumae.* At higher elevations forests are composed of *Abies guatemalensis*, mixed with *Cupressus* or *Pinus*. Alpine vegetation occurs in open areas above 3,500 m; it is dominated by herbs, grasses, and some shrubs. The flora includes endemics, Mexican–Central America taxa, Rocky Mountain taxa, and Andean taxa.

In El Salvador, in the temperate climates of the northern mountains around Chalatenango and north of Metapan, on the slopes of Cerro de Monte Cristo, are pine and oak forests with epiphytes of bromeliads and orchids and accompanied by *Inga* sp. (Fabales), *Cedrela mexicana* (Meliaceae), *Perymenium* spp. (Asteraceae), *Clethra* spp. (Clethraceae), and *Nectandra sinuata* (Lauraceae); there are also forests of *Chaetoptelea mexicana* (Ulmaceae) and of *Liquidambar*. At higher elevations, 1,800–2,000 m, are cloud forests dominated by giant oaks that are covered with mosses, ferns, selaginellas, cacti, lycopods, bromeliads, and fungi. On the cones of the Santa Ana and San Miguel volcanoes, and on the highest peaks, Monte Cristo and El Pital, which are exposed to coastal storms, are high savannas (páramo) of ericaceous plants, with *Myrica mexicana* and agaves.

On Cerro Chirripo (3,819 m), in the Cordillera de Talamanca, Costa Rica, Kappelle (1991) describes a number of vegetation types above 2,000 m: lower montane forest, upper montane forest, subalpine forest, and páramo. These are illustrated by profiles in Kappelle et al. (1995). Lower montane forest can be treated in two sections, the lower belt and the upper belt. The lower belt (1,500–2,000 m) is dominated by many tropical and some Holarctic families. The upper belt (2,000–2,400 m) is characterized by a mixture of canopy trees that reach 35 m in height. The oaks (*Quercus* spp.) form the upper canopy (ibid.: figure 8), while the understory layers have the laurels (*Ocotea* spp., *Nectandra,* and *Persea*) as well as some Holarctic, austral, and tropical evergreen taxa. In the understory is the palm *Geonoma hoffmanniana,* occasionally accompanied by the bamboos *Chusquea tomentosa* and *Aulonemia viscosa.*

Upper montane forest can be treated in two sections, the upper belt and the lower belt. The lower belt (2,400–2,800 m) comprises a tall (45 m) oak forest dominated by *Quercus copeyensis;* it is always found with *Weinmannia pinnata.* In the subcanopy are a mixture of Holarctic, austral, and some tropical evergreens. The understory is dominated by *Chusquea tomentosa* (to 6 m tall); *C. talamancensis* only occurs at the upper limit of this subzone. The upper belt (2,800–3,200 m) is characterized by an evergreen oak forest that reaches heights of 35 m. *Quercus costaricensis* is the exclusive canopy dominant, and *Chusquea talamancensis* with *C. tomentosa* dominates the understory. At the upper limit of this subzone one finds *Schefflera, Drimys, Weinmannia,* and *Myrsine,* taxa that are typical of the subalpine forest. The presence here of trees and treelets of *Quercus copeyensis, Cleyera,* holly (*Ilex*), *Podocarpus, Saurauia,* avocado (*Persea*), marlberry (*Ardisia*), and *Styrax* manifest the relationship with the lower belt. At nearby Cerro is the Villa Mills study site of Holdridge, where the dominant oak is *Quercus costaricensis* (61 percent of basal area), with a melastome (*Miconia bipulifera,* 9.9), *Weinmannia pinnata* (9.8), an aralioid (*Didymopanax pittierii,* 4.8), a blueberry (*Vaccinium consanguinium,* 3.9), and fifteen other species (Hartshorn in Janzen 1983).

Subalpine forest (3,100–3,400 m) is composed of trees generally no taller than 12 m, twisted and with limbs covered by epiphytic cryptogams such as mosses, liverworts, and lichens. The canopy on the Pacific slope is dominated by *Comarostaphylis arbutoides* (35–45 percent cover); on the Atlantic slope the dominant is *Vaccinium consanguinium* (5–20 percent cover). *Weinmannia trianae* var. *sulcata* is found on both sides as an accompanying tree (to 40 percent cover). Small trees and shrubs are present in the gaps of the canopy. The oak, *Quercus costaricensis,* is present as a subcanopy tree at the lower elevational fringe of this forest. *Chusquea talamancensis,* a bamboo that reaches 5 m tall, provides between 50 and 100 percent ground cover.

Páramo (3,300–3,819 m) is a tropical alpine vegetation type characterized at

3,300 m by bamboos and shrubs. The dominant species of this rare vegetation type is *Chusquea subtessellata,* a bamboo that reaches 2 m tall and grows in dense clumps that cover 60 percent of the ground. It is accompanied by a few shrubs, none of which cover more than 10 percent of the ground.

NEOTROPICAL FLORISTIC KINGDOM

According to Takhtajan (1986), this kingdom extends from the southern, tropical portion of Florida south throughout the West Indies, parts of the lowlands and shores of Mexico, all of Central America, and the greater part of South America. In the present treatment it also includes the Mexican Xerophytic (Sonoran) Region (the Mexican Desert), but not the high mountain forests from Chiapas to Panama that are dominated by Holarctic evergreen angiosperms. Takhtajan lists 27 families as endemic, the best known of which is the Heliconiaceae. Many families and at least 450 genera have a pantropical distribution, so the ties to the Paleotropical Kingdom are strong. Nevertheless, Takhtajan gives estimates of 3,000 to 3,660 genera that are endemic to the Neotropical Kingdom.

MEXICAN XEROPHYTIC (SONORAN) REGION

Thorne (1993), following Cronquist in Takhtajan (1986), treats the hot deserts of North America in the Sonoran Province, for which he recognizes four subprovinces. He therefore considers the Sonoran Province as belonging to the Madrean Region of the Madrean Subkingdom of the Holarctic Kingdom. The present treatment follows Rzedowski (1981) in concept and nomenclature, and includes Takhtajan's Sonoran Province in the Neotropical Kingdom. Thus, the mainly tropical floras of the desert regions of Mexico are here united with the rest of the Neotropics, where most of the families and genera range. Rzedowski suggests intensive research could result in the recognition of the xerophytic area as a subkingdom. This region includes most of Baja California, all of Sonora, northern Mexico and the adjacent portions of the United States, the Chihuahuan Desert, and the arid and semiarid region from Texas, Nuevo León, and Tamaulipas south to Veracruz. It is present in small portions of Central America, as well. In area it comprises approximately half of Mexico. The following five provinces are included: (1) Peninsular (Baja California) Province, (2) Northwestern Coastal Plain (Sonoran) Province, (3) Central Plateau (Altiplanicie) Province, (4) Tamaulipan Province (Coastal Plain Province of the Northeast of Mexico), and (5) Tehuacán-Cuicatlán Valley Province. A brief description of each province follows.

PENINSULAR (BAJA CALIFORNIA) PROVINCE

Rzedowski states that because of the physical isolation of the Baja California peninsula it has many restricted taxa and so is the easiest to characterize. Climate is varied regionally but tends to be less arid in the southern part. The vegetation is termed "matorrales xerófilos (succulent desert shrubland)," with cacti, deciduous low trees (especially in the south), and Fouquieriaceae (boojum tree, *Fouquieria columnaris*, in the north), throughout. Endemic genera are *Alvordia, Burragea, Coulterella, Pachycormus,* and *Pelucha.*

NORTHWESTERN COASTAL PLAIN (SONORAN) PROVINCE

This province occupies the greater part of the state of Sonora and extends to Sinaloa in a narrow coastal strip; it extends northward to include southern Arizona and a small part of California. Its flora is similar to that of Baja California, but endemics are fewer. The climate is hot and arid or semiarid. Vegetation is mainly matorrales xerófilos and "bosque espinoso (spiny woodland)." In the southern part it is augmented by Central American taxa so that there is a large zone of codominance there. Among the endemic genera are *Agiabampoa, Canotia,* and *Carnegiea.* Many taxa are shared only with Baja California.

CENTRAL PLATEAU (ALTIPLANICIE OR CHIHUAHUAN) PROVINCE

Extending from Chihuahua and Coahuila to the arid and semiarid portions of Jalisco, Michoacán, Mexico State, Tlaxcala, and Puebla, this province also includes the northeast of Sonora, southern parts of New Mexico, and the Trans-Pecos region of Texas. Altitude of the desert varies from 1,000 m in the north to 2,000 m in the south, so that low temperatures, throughout, are influential in the structure and composition of the vegetation. Mean annual temperatures range from 14 to 23°C and average 18.6°C. Mean annual precipitation ranges from 150 to 400 mm and averages 235 mm. This is the largest of all the hot desert provinces. MacMahon (1988) recognizes three regions: the Trans-Pecos in the north, which includes 40 percent of the desert; Mapimian, in the middle, which includes eastern Chihuahua, Coahuila, and part of Durango, an area of basin and limestone range topography with many playas; and the southern region, the Saladan, in Zacatecas, San Luis Potosí, and adjacent states, where elevations range from 500 m in the valley floors to limestone and igneous mountains over 3,000 m. The number of endemic species is very high, and their diversity is favored by extreme geological substrates. Limestone is the preponderant substrate, although gypsum (hydrous calcium sulfate) and igneous rocks also occur. Dunes derived from gypsum deposits are a characteristic landform of the Chihuahuan Desert (MacMahon 1988). Gypsum-tolerant plants are especially well represented in the flora. At the eastern edge of the province, plants of the Tamaulipan region are important. Common vegetation is matorrales xerófilos, but grasslands are frequent and also "bosque espinoso (mezquitál)."

Takhtajan estimates 8–10 percent endemism for the native flora of the Chihuahuan Desert. Rzedowski (1981) notes that 16 genera are endemic to the province and lists the following as examples: *Ariocarpus, Eutetras, Grusonia, Lophophora, Sartwellia,* and *Sericodes.* He further states that the number of endemic species is very high and their diversity is favored by the geological substrata, especially the gypsum deposits. Takhtajan (1986:180–181) gives a list of gypsophilous endemics. MacMahon accepts Henrickson and Johnson's value of some 70 endemic species for Chihuahuan Desert gypsum soils, and the following endemic genera: *Dicranocarpus, Marshalljohnstonia,* and *Strotheria*. Some other genera occurring principally on gypsum soils are *Selinocarpus, Nerisyrenia, Sartwellia,* and *Pseudoclappia*. Only 40 halophytes occur in the Chihuahuan Desert, 25 of which are endemic, including three genera: *Meiomeria, Reederochloa,* and *Pseudoclappia* (MacMahon 1988).

Vegetation is mainly matorrales xerófilos with *Larrea* as the dominant shrub (often occurring with *Flourensia cernua*), and with *Yucca* and large shrubs or even arborescent cacti. Grasslands and mesquite-dominated bosque espinoso are also frequent on alluvial plains, valleys, and riverbeds. Ocotillo (*Fouquieria* spp.) and lechuguilla (*Agave lecheguilla*) occur over most of the Chihuahuan Desert and are considered among the best indicators by MacMahon (1988). That author recognizes the following subdivisions of desert scrub and woodland, following Henrickson and Johnson: Chihuahuan Desert scrub (70 percent of all vegetation), lechuguilla scrub, yucca woodland, *Prosopis-Atriplex* scrub, alkali scrub, gypsophilous scrub, cactus scrub, and riparian woodland. They further divide the Chihuahuan Desert scrub into the following phases: larrea scrub (40 percent of Chihuahuan Desert scrub), mixed desert scrub, sandy arroyo scrub, canyon scrub, and sand dune scrub. Mixed dominants such as *Larrea-Agave* and *Fouquieria-Agave* appear to be the rule in Chihuahuan Desert vegetation. In areas where desert scrub communities occur on limestone, grassland occurs on igneous rock. On limestone slopes the following plants are common constituents of the vegetation: *Agave lecheguilla, Fouquieria, Hechtia scariosa, Leucophyllum frutescens, Euphorbia antisyphilitica, Dyssodia* spp. (dogweeds), *Condalia* spp., *Viguiera stenoloba, Perezia nana,* and *Parthenium argentatum* (guayule). On igneous rock slopes the vegetation is commonly arborescent cactus scrub, and the following taxa are characteristic: *Opuntia* spp., *Myrtillocactus geometrizans,* and *Yucca carnerosana*. Vegetation on gypsum soils is floristically depauperate and characterized by perennial herbs or dwarf shrubs. Widespread taxa that occur on gypsum dune substrates are *Yucca elata*, Indian rice grass (*Oryzopsis hymenoides*), *Atriplex canescens*, tobosa (*Hilaria mutica*), and *Rhus trilobata*.

COASTAL PLAIN PROVINCE OF THE NORTHEAST OF MEXICO (TAMAULIPAN PROVINCE)

Almost completely coincidental with the physiographic feature of the same name, this province also includes a portion of the adjacent state of Texas (Texas Plains). In Mexico it occupies nearly the entire state of Tamaulipas, the northeast two-thirds of Nuevo León, and small parts of Coahuila, San Luis Potosí, and the extreme north of Veracruz. Its northwest limit is difficult to define because of a broad floristic transition with the Altiplanicie Province. Climate is generally semiarid and hot, with great extremes of temperatures. The vegetation is mainly bosque espinoso and matorrales xerófilos. Central American taxa are important in the southern part of the province. Endemism is not as pronounced as in the Altiplanicie, but the number of endemics apparently justifies recognition as a separate province. Among the endemic taxa are *Clappia, Nephropetalum, Pterocaulon,* and *Runyonia*.

This province is not so dry as the others, being characterized by a semiarid, hot-warm climate. Severe frosts occur over several consecutive days in winters several times per century and cause the complete killing of the upper portions of susceptible plants. For Linares, Nuevo León, mean annual precipitation is 749 mm, mean annual temperature is 22.3°C. D. E. Brown et al. (1980) argue that this province is distinctive on both physiognomic and biotic criteria, which include (1) the abundance of shrubs and short microphyllous trees which make up 20–90 percent of cover, where intervening areas are covered by forbs and grasses, (2) a large number of plants and animals that have their centers of distribution here, including *Fouquieria macdougalii* (tree ocotillo) and *Lophortyx douglasii* (elegant quail), (3) the absence or poor representation of numerous characteristic Sonoran Desert species: *Larrea tridentata* (creosote bush), *Simmondsia chinense* (jojoba), *Cercidium microphyllum* (palo verde), and (4) the appearance and heavy representation of numerous southern tropical forms not found in the Sonoran Desert to the north, e.g., *Acacia cymbispina* and *Felis pardalis* (jaguar). According to Rzedowski (1981), the vegetation is mainly "Tamaulipan thornscrub" (which he classifies as a type of "matorral espinoso") and is generally dominated by *Acacia* spp., *Prosopis glandulosa* (honey mesquite), *P. reptans* var. *cinerascens,* and *Cercidium* spp.; others include *Aloysia gratissima, Castela texana, Celtis pallida, Karwinskia humboldtiana,* and *Ziziphus obtusifolia.* In the vicinity of Linares, Nuevo León, in the piedmont of the Sierra Madre Oriental, Reid et al. (1990) describe floristic and structural variation in the Tamaulipan thornscrub. The most distinctive community occurs on dry, calcareous rock sites (caliche); the major species are *Malpighia glabra, Calliandra conferta, Zanthoxylum fagara, Condalia hookeri, Forestiera angustifolia, Amyris texana, Diospyros palmeri, Chamaecrista greggii, Krameria ramosissima,* and *Jatropha dioica.* Slightly less distinctive is the floristic assemblage on "dry, deep soils;" here the common species are *Diospyros palmeri, Acacia wrightii, Helietta parvifolia, Viguiera stenoloba, Acacia farnesiana, Eysenhardtia polystachya, Citharexylum berlandieri, Neopringlea integrifolia, Fraxinus greggii,* and *Xylosma flexuosa.* On "humid, deep soils" and "humid, skeletal soils" the common taxa are *Gochnatia hypoleuca, Prosopis laevigata, Celtis pallida, Chamaecrista greggii, Fraxinus greggii, Schaefferia cuneifolia, Neopringlea integrifolia, Ehretia anacua, Croton torreyanus,* and *Ziziphus obtusifolia.* Those authors note that *Acacia farnesiana* forms a secondary shrubland vegetation and that *Karwinskia humboldtiana* and *Opuntia leptocaulis* are abundant in heavily grazed sites. Also in the vicinity of Linares, Rzedowski (1981) notes that Muller had found extensive shrublands of the cycad *Dioon edule* on lutitas (clayey, impermeable soil derived from sedimentary rock). Mesquite-grassland (savanna parkland) is also widespread and includes as dominants the following grasses: *Bouteloua hirsuta, Aristida roemeriana, Trichachne hitchcockii,* and *Tridens texanus* (Takhtajan 1986). Van Auken and Bush (1997), under experimental conditions in central Texas, find that belowground competition with *Bouteloua curtipendula* significantly reduces the dry mass of *Prosopis glandulosa,* suggesting that spread of honey mesquite is related to denudation of associated grasslands. For southern Texas, Archer (1990) describes the savanna parkland as clusters of woody shrubs beneath arborescent honey mesquite (*Prosopis glandulosa* var. *glandulosa*), in a matrix of grasses (*Paspalum setaceum, Bouteloua rigidiseta, Chloris cucullata, Aristida* spp., *Bouteloua tri-*

fida, and *Cenchrus incertus*) and herbs (*Evolvulus* spp., *Eupatorium* spp., *Verbesina* spp., and *Zexmenia hispida*). The shrubs include *Zanthoxylum fagara* (a coriaceous broad-leaved evergreen); *Condalia hookeri, Diospyros texana, Schaefferia cuneifolia, Ziziphus obtusifolia* (all drought-deciduous); and *Berberis trifoliata* (a sclerophyllous evergreen). With time, closed canopy woodland can form by the lateral growth and coalescence of the shrub understories.

TEHUACÁN-CUICATLÁN VALLEY PROVINCE

In a small area (10,000 km²) of the southeastern section of Puebla State and adjacent portions of Oaxaca and part of Veracruz is an isolated area of arid, hot climate, between elevations of 545 and 2,458 m. Its flora comprises 123 families, 630 genera, and 1,460 species of phanerogams and is largely dominated by tropical taxa (58 percent of genera), with Mexican endemics or near-endemics second in importance (13 percent of genera, 30 percent of species). Some endemic genera are *Oaxacania, Pringleochloa,* and *Solisia.* This province appears to be related to South American floras (with *Aloysia, Castela, Condalia, Gochnatia, Grabawkia, Maytenus, Nicotiana,* and *Prosopis*) more than to those of the east of North America (taxa such as *Celtis, Cercidium, Coldenia, Cryptantha, Dalea, Flourensia, Grindelia, Hoffmanseggia, Hymenoxys, Nama, Sanvitalia, Schkuhria,* and *Zaluzania*). The closest floristic relations are with the Valle de Mezquitál, Hidalgo (89.9 percent similarity), Cuenca del Río Estorax, Querétaro (85.5 percent), and Cuenca de Río Balsas (83.0 percent). The vegetation types are tropical deciduous forest, gallery forest, spiny forest, and matorral xerófilo. The latter is the most widespread and variable vegetation type, being dominated variously by taxonomic life-forms such as *Neobuxbaumia tetetzo* (tetecheras), *Yucca periculosa* (izotales), *Lemaireocereus weberi* (cardonales), and *Escontria chiolla* (quiotillales) (Villaseñor et al. 1992).

In the arid and semiarid parts of Puebla, extending into adjacent Oaxaca, *Yucca periculosa* is common and accompanied by *Nolina* or by *Beaucarnea.* Groups (manchones) of *Nolina parviflora* dominate locally, near Perote, Veracruz, and Alchichica and Libres, Puebla, on volcanic slopes. These plants reach 2–4 m tall, forming above a shrub layer of *Agave obscura* and species of *Salvia, Chrysactinia,* and *Dalea.* Manchones of *Nolina parviflora* occur in other places in the center of Mexico, although they may well be anthropogenic and pyrogenic types of vegetation.

In the Balsas River drainage, in the Cuicatlán depression, in the Papaloapan Basin, as well as in the drainage of the Río Tehuantepec, Rzedowski (1981) recognizes the following communities dominated by Cactaceae: (1) cardonales of *Lemaireocereus,* (2) quiotillales of the monotypic *Escontria chiolla,* (3) tetecheras of *Neobuxbaumia tetetzo,* and (4) a community of *Cephalocereus hoppenstedtii.* He notes that the latter two are particularly scenic, for great numbers of these columnar cacti, few branching at all, grow in such density that they do not permit the co-occurrence of other tall plants. At greater elevations (2,400–2,700 m), where MAP is 500 mm, in the Papaloapan River basin, on igneous rock substrates, shrubland dominants are the following cacti: *Opuntia macdougaliana, O. huajuapensis,* and *Lemaireocereus chichipe.* Other genera are *Mimosa, Senecio, Tecoma, Parthium, Eysenhardtia, Bursera, Ipomoea,* and *Dasylirion*. This community also occurs farther north in the Río Balsas drainage, in Guerrero, and terminates in Michoacán.

CENTRAL AMERICAN REGION (CARIBBEAN REGION OF RZEDOWSKI 1981; INCLUDING THE CARIBBEAN REGION 23 OF TAKHTAJAN 1986)

This region includes the southern Florida Peninsula, the Florida Keys, the Bahamas, the Greater and Lesser Antilles, all of lowland Central America from Mexico to Panama, the shores of Ecuador, Colombia, and western Venezuela, the Revillagigedo Islands, the Galapagos Islands, and Cocos Island. It excludes the highlands of Central America and the island of Bermuda, all of which are here treated in the Holarctic Madrean Subkingdom. Takhtajan's Caribbean Region in Mexico includes only the tropical lowland plains and coasts. Takhtajan recognizes only one endemic family (Plocospermataceae, often treated as part of Loganiaceae) but a large number of endemic genera and species. Howard (1973) lists a large number of genera that are found on the mainland of Central America and northern South America and are also present in the Antilles. His table 1 lists pan-Caribbean plants; tables 4–7 list Caribbean plants variously distributed in the Central American Region but missing in the Lesser Antilles, due to the impoverished flora of the latter islands.

CENTRAL AMERICAN PROVINCE

The following vegetation types may be present from Veracruz and Chiapas south into Panama. They are given detailed descriptions by Rzedowski (1981), which form the basis for the following treatment.

Tropical lowland evergreen rain forest (TEF; "bosque tropical perennifolio" of Rzedowski 1981) occupies a nearly continuous extension in the east and southeast of Mexico; there is a narrow fringe on the Pacific side of the Sierra Madre de Chiapas, isolated from the east by the Isthmus of Tehuantepec. TEF comprised, originally, 11 percent of the territory of Mexico, although it is now much reduced and fragmented (Rzedowski 1981). In Mexico, it is often found on deep, alluvial, well-drained limestone-derived soils, with near-neutral pH. Climate is very warm-hot, and humid to perhumid, with a short dry period. Shifting agriculture has been practiced here since the Formative period. Presently these lands are often cleared for pasturage, sown with grasses, and maintained as a grassland by fire during the dry season, with reseeding after burning. On deep soils of river floodplains, sugarcane, maize, citrus, plantain, mango, and some other fruits are grown. Coffee is grown at the higher elevations of TEF. Mahogany (*Swietenia macrophylla*) and cedar (*Cedrela mexicana*) are the two major lumber trees of commercial value. At the extreme north of its distribution, in a floodplain in southeastern San Luis Potosí, TEF is dominated by one or two canopy tree species. The numbers of species and dominants increase to the south. Canopy trees average 30 m tall or greater; they have straight trunks that do not branch for half or two-thirds their height. Scattered emergent giants exceed 45 m. Buttressed bases are common and well developed, e.g., in fig (*Ficus*) and *Sterculia*. Crowns are often pyramidal or spherical in outline. The crowns are often bound together by the tangled growth of liana branches. Strangler figs are present. Two other arborescent layers are present between 5 and 20 m above the ground; their crowns tend to be pyramidal to elongate, and their leaves darker green than trees of the canopy. Orchids, bromeliads, and lichens are common epiphytes. Bromeliads can be massive and form cisterns or hold water in their leaf axils. The forest floor is dark and populated by palms (*Chamaedorea* spp.), ferns with divided leaves, and some broad-leaved graminoids.

Tropical semideciduous forest (SDF; "bosque tropical subcaducifolio" of Rzedowski 1981) is a tall forest where some trees lose all their leaves during the dry period, others remain green, and still others lose their leaves only for a few weeks. Intensity of defoliation often reflects the severity of the drought in any year. The flowering period of the trees coincides with a drought period and defoliation; it is often spectacular because the trees are then covered with attractive blossoms. Distribution of this type in Mexico is most extensive, but discontinuous, on the Pacific side, and it also occurs in the Central Depression and in the Yucatán Peninsula. Ecologically, it forms a mosaic with other vegetation types: tropical deciduous forest, palmares, savannas, etc. It occupies about 4 percent of land cover in Mexico, 0–1,300 m, perhaps higher in Guerrero and Oaxaca. At its upper limit it contacts encinares, pinares, and bosque mesófilo de montaña (tropical evergreen montane forest). Climate is torrid-hot, subtemperate, and subhumid, with precipitation of 1,000–1,600 mm. The dry period can last five to seven months, but atmospheric humidity remains high throughout the year. Understory plants flower in the dry period. Freezes are negligible. Distribution is extended to ravines and floodplains in drier climates. SDF occurs on deep soils, with good drainage, acidic-neutral pH, on a great variety of bedrock types. Flora is Neotropical with a few Holarctic taxa, and many endemics. Although not the preferred vegetation type for agriculture, it is deforested for local uses and planted with maize, plantain, coffee, sugar, truck crops, and tobacco, where soils are deep and accessible. Dominant trees are *Enterolobium cyclocarpum* (Fabales), *Cedrela mexicana* (Meliaceae), *Roseodendron* (= *Tabebuia*) *donnellsmithii* (Bignoniaceae), *Dalbergia granadillo* (Fabales), *Astronium graveolens* (Anacardiaceae), *Hymenaea courbaril* (Fabales), *Platymiscium dimorphandrum* (Fabales). Dominance is shared by up to five taxa. Height of the canopy varies from 15 to 40 m and is frequently 20–30 m. The canopy is continuous except where *Enterolobium* occurs as an emergent. Trunks are straight and slender; diameters average 30–80 cm and rarely surpass 1 m. Exceptions are *Enterolobium* and *Ficus*, which often develop thicknesses to 2 and 3 m at the base. Most trees have relatively small crowns, exceptions being, again, *Enterolobium* and *Ficus*, which have large, spreading crowns that branch from low on the trunk. Leaves are usually dark green, often compound and deciduous, with leaflets or leaves of medium size or smaller, and with entire margins. Palms are rare in the canopy, although *Attalea cohune* is common in the littoral of the Pacific coast. Among the most colorful trees in flower are *Andira inermis* (Fabales), *Belotia* (= *Trichspermum*) *mexicana* (Tiliaceae), *Bernoullia flammea* (Bombacaceae), *Calycophyllum candissimum* (Rubiaceae), *Cochlospermum vitifolium* (Bixaceae or Cochlospermaceae), *Cordia alliodora, C. elaeagnoides* (Boraginaceae), *Luehea candida* (Tiliaceae), *Plumeria rubra* (Apocynaceae), *Poeppigia procera* (Fabales), *Tabebuia donnellsmithii, T. palmeri,* and *T. rosea* (Bignoniaceae). An understory of smaller, mainly evergreen trees is usually present, 8–15 m tall and reaching 50 percent cover. The shrub layer is variable with available light, occasionally absent. Frequently palms are present and almost always members of the Rubiaceae. A ground layer is sometimes present in the poor light but often absent on undisturbed sites, in flat or gently sloping terrain. Woody vines, commonly members of the Bignoniaceae, are often abundant, especially in protected sites and ravines. Epiphytes of the Bromeliaceae and Orchidaceae are present, but not as abundant as in evergreen lowland forests. Xeromorphic taxa are especially common. Crustose lichens are especially common on the bark of trees and can cover the trunks completely. Strangler figs are not common. Bryophytes and pteridophytes are scarce. Cycads occur sporadically. Fabales are the most common group of angiosperms, but they are not as dominant here as in some of the drier forest types.

Tropical semideciduous woodland (SDW) is the name I give to a woodland that is characterized by trees whose trunks are relatively short, with diameters rarely exceeding 50 cm, and with wide crowns so that under optimal conditions the crowns meet to form a nearly closed canopy, and the forest appears dense. Tree height ranges from 5 to 15 m, most often 8–12 m. Trunks are twisted and branch close to the ground. Bark is often colorful and shiny, continuously exfoliating. Tree crowns are convex or plane, forming a uniform canopy. Width of the crown is often equal to the height of the tree. Foliage of deciduous trees is usually light green, compound, and of nanophyllous size.

There are two phases of this type, one in which the majority of trees are characterized by drought-deciduous leaves, the other in which the majority of trees are evergreen. Cacti are often present in the deciduous phase of the woodland. The two types of semideciduous woodland are often treated under separate categories, but their general structure and composition and climatic requirements allow a unified classification. Rzedowski (1981) discusses the two separately, naming the more deciduous phase "bosque tropical caducifolio." He gives a long list of synonyms for that phase, including the "deciduous seasonal forest" of Beard. In Asia forests of this type are commonly included under the category of monsoonal, but that pattern of rain-bearing winds is not characteristic of Mexico and Central America. Based on a Bailey analysis of stations listed for this type by Rzedowski, climate is very warm-hot, subtemperate, and subhumid to semiarid (see Greller 1990). Extreme minimum temperatures go no lower than 0°C. Mean annual temperatures range from 20 to 29°C (or higher). Rainfall occurs in one season following a variable dry period of five to eight months (December–May). Mean annual precipitation ranges from 300 to 1,800 mm and is usually 600–1,200 mm. At the middle or near the end of the dry season, when temperatures reach their annual maxima, many woody species are covered with flowers. Many of the plants never possess leaves and flowers simultaneously. Rzedowski assigns this arborescent type an intermediate position in a sequence of increasing drought, from bosque tropical subcaducifolio to bosque tropical caducifolio to bosque espinoso.

Semideciduous woodland is particularly well developed on the Pacific side of Mexico and Central America. In Mexico it covered the land uninterruptedly from Sonora and the southwest of Chihuahua to Chiapas and then south into Central America. It occurs between 0 and 1,900 m and mainly below 1,500 m; in the Central Depression of Chiapas it does not occur above 800 m, because of low temperatures. In the northwest it is restricted to the lower slopes of the Sierra Madre Occidental. Farther south the woodland comes into contact with the littoral zone and extends onto the Serranías and into the valleys of the Santiago and Balsas rivers. In the extreme south of Baja California there is a small patch on the lower slopes of the Sierra de la Laguna and Sierra de la Giganta, where it receives less than 500 mm of precipitation. In the Isthmus of Tehuantepec the semideciduous woodland traverses the continental divide and occupies a large part of the Central Depression of Chiapas. On the Atlantic side in Mexico it exists in only three patches: in the south of Tamaulipas and to adjacent states; in the center of Veracruz between Nautla, Alvarado, Jalapa, and Tierra Blanca, and including the port of Veracruz; and the northern part of the Yucatán Peninsula, occupying the largest part of Yucatán State and a bit of Campeche. In all it accounts for 8 percent of land cover in Mexico. This type occurs on a wide variety of landscape forms and soil types, generally well drained and young, derived from parent material. It is replaced by bosque espinoso on deep alluvial soils that Rzedowski claims are nutrient poor, compared to well-drained slopes. Commercial lumbering has little impact on this type of woodland because the trees are suitable only for local uses. Cattle-raising is the principal land use over much of the extent of semideciduous woodland. In Morelos and Guerrero open woodland and grassland are artificially maintained for pasturage; in other locations exotic grasses are introduced. Certain legumes are used for tanning and extraction of essential oils; henequen (*Agave fourcroydes*) is grown in this type, yielding sisal and sisal hemp. Maize, beans, chickpeas, and garlic are grown, as well as warm-climate fruits.

The flora of the tropical semideciduous woodland is mainly Neotropical with a large number of endemics above the species level; these are concentrated particularly in the Balsas Depression, in the Yucatán Peninsula, and the northeast of Mexico. Families of the Fabales dominate the landscape with many species, numerous individuals, and extensive cover. The genus *Bursera* (Burseraceae) is extremely important, principally on the Pacific side, especially at the higher elevations. *Bursera* species are the absolute dominants in the Balsas Depression, Guerrero. Vegetation is usually in one tree layer, although rarely two are present. Emergent trees are also rarely seen. A shrub layer is variable with tree crown density. Herbs are poorly developed and can be absent completely in undisturbed flatlands, but partially developed on slopes. Vines and epiphytes are rare and best developed in ravines or on favorable exposures. *Tillandsia* species and crustose lichens are the common epiphytes. Columnar and candelabra-form cacti are often present, as well as rosette-form succulent Agavaceae, especially in the drier phases of the deciduous type. Cycads are rarely present.

The evergreen phase of the semideciduous woodland is represented by the *Quercus oleoides* (Fagaceae; Virentes section of *Quercus*) woodland. *Quercus oleoides* forms open stands from sea level to an altitude of 800 m (Tamaulipas, Veracruz, and Chiapas), mainly on the Atlantic coasts of Mexico, Belize, and Honduras, from the Tropic of Cancer south, to the north of Costa Rica on the Pacific Ocean side, where it reaches its southern limit (Montoya Maquin 1966). It grows in a wide range of generally hot temperatures (mean annual temperatures of 14.2°C in Jalapa, Veracruz, Mexico, but also 27.4°C, in Liberia, Costa Rica). Precipitation ranges from 705 to 3,185 mm. It occurs in a variety of suboptimal habitats that include alluvial quartz gravel soils, red clays from volcanic cinders, black clays over basalt rock, sandy soils near the sea, degraded sites, and acidic swamps. It is the dominant in moderately dense to open evergreen woodland that reaches 15–30 m tall. But *Quercus oleoides* can also assume shrub form. Montoya Maquin considers the woodland to be an "edaphic climax" community, in the tropical lowland evergreen rain forest zone. In Oaxaca and Puebla it is accompanied by *Quercus glaucescens, Q. sororia, Q. peduncularis,* and *Q. perseafolia,* as well as by *Byrsonima crassifolia,* on a variety of suboptimal sites, at an altitude lower than the associations of pine woodland. In Tabasco and Chiapas, *Quercus oleoides* forms part of the shrub savannas to altitudes above 1,300 m. In Guatemala it occurs in Petén (scarcely, in wooded swamps), in Alta Verapaz, Izabál, Zacapa, and Chiquimula, on flatlands or rugged slopes frequently with pines, below 300 m. It occurs as tree islands in pine woods and even in adjacent acid swamps. In Belize, *Quercus oleoides* is present in forests mixed with pines (*Pinus oocarpa* or *P. caribaea*), often forming tree islands of considerable extent. It is rare under similar conditions in Honduras. *Quercus oleoides* apparently is extensively distributed in Nicaragua, but information is incomplete. In the northeast of Nicaragua it is present in open pinelands. On the most degraded sites it is present with *Byrsonima crassifolia* and *Curatella americana.* In this area it also forms shrublands with *Miconia albicans* and *Henriettella seemannii* that represent an early stage of succession to leafy forests. In northwestern Nicaragua it is reported to be associated with *Pinus pseudostrobus* and other species of oak such as *Q. peduncularis* var. *sublanosa, Q. oocarpa, Q. eugeniaefolia,* and *Q. sapotaefolia.* Here the oaks are generally present in the understory of the pine woods that grow on thin, nutrient-deficient, yellow, brown, or black acidic soils on bedrock. In Costa Rica *Quercus oleoides* has been recorded as present in Guanacaste, Heredia, and San José provinces. At present it is confirmed only for Guanacaste, where it occurs in a 15 km radius around Liberia, in habitats associated with streams and ravines. Shrublands are the typical secondary vegetation on sites from which *Q. oleoides* has been removed.

Thorn woodland (TW; "bosque espinoso" of Rzedowski 1981) is a heterogenous formation of low woods dominated by spiny trees. They are able to tolerate climates drier than those of the semideciduous woodland, usually in hot to torrid temperature regimes. Their water need is greater than that of the spiny shrubland that characterizes the deserts of Mexico. Nevertheless, the two formations often occur adjacent to one another in different habitats. In the Yucatán Peninsula, thorn woodland occurs in the lowlands with deep soils that are poorly drained after heavy rains but dry out during the rest of the year. Leaf function is deciduous or semievergreen. Thorn woodland occupies a continuous expanse on the northwest coastal plain of Mexico, from Sonora to the southern part of Sinaloa and continues along the Pacific coast as isolated patches to the Balsas Depression and the Isthmus of Tehuantepec. From the northwest it is distributed obliquely to the Gulf of Mexico, including areas of San Luis Potosí and northern Veracruz. In the Altiplano it is known as "bajos" and occupies a large area of Guanajuato, as well as adjacent areas of Michoacán and Querétaro. Many isolated patches occur farther to the north, in San Luis Potosí, Zacatecas, Coahuila, Nuevo León, and Chihuahua. It is discontinuous and sporadic in Chiapas and the Yucatán Peninsula. Locally, it forms a mosaic with other types of vegetation and is therefore difficult to map in small scale. Rzedowski (1981) estimates that it covers ca. 5 percent of the landscape of Mexico. Altitudinal limits are 0–2,200 m, in climates from hot to mild and from subhumid to arid. Mean annual temperatures range from 20 to 29°C, in an annual range of 4–18°C; precipitation ranges from 350 to 1,200 mm/year, with five to nine dry months. Tall "mezquitales" of *Prosopis* spp. are included in this type, but shrublands of *Prosopis* are treated under the category of spiny shrubland of warm climates. Thorn woodland is most characteristic of flat to slightly sloping terrains, but locally it may occur on hills. Soils are often deep, rich in organic matter, and good for agriculture. An exception occurs in the northeast coastal plain and on the Yucatán Peninsula, where it occurs, respectively, on shallow, clayey, alkaline soils of shallow depth or on periodically inundated sites where it is edaphic vegetation in a region of evergreen or semideciduous forest. Throughout Mexico, sites occupied by TW have long been cleared for agriculture that required little irrigation. Many of the regions originally covered by TW were irrigated in the 1940s and 1950s to create grazing lands or for the cultivation of sugarcane, bananas, wheat, rice, cotton, and locally coconuts, peanuts, sweet potatoes, yuca (*Manihot esculenta*), and diverse tree fruits. Trees of the TW are locally exploited for charcoal manufacture or for fine craft work, for their fruits, or for dye (*Haematoxylon campechianum*). Floristic relations are with the Neotropical Kingdom, and there are many taxa shared with desert shrublands; endemic taxa are common especially to the north. This type of vegetation grows 4–15 m tall and is often dense in the canopy. Mostly the trees are deciduous (*Cercidium, Prosopis, Escontria, Ziziphus, Acacia, Podopterus*), but one association, dominated by *Pithecellobium dulce,* is evergreen, and a few species in many communities are evergreen or semievergreen. Leaf size is leptophyllous or nanophyllous. Trunks branch from the base but do not diverge until 2 m or higher; crowns are rhomboidal, ellipsoidal, or spherical and are relatively small. There is one arborescent stratum, with occasional emergents. Tree branches are often covered with small, xerophytic bromeliads of the genus *Tillandsia*. The shrub layer is generally well developed and rich in taxa with spines. Dense stands lack an herb layer, although in some places extensive patches of spiny bromeliads make transit impossible. Open communities have a rich herb layer, especially annuals, which appear after rains. Pteridophytes and bryophytes are rare. The Fabales are the most prominent taxonomic group. Gymnosperms are absent.

Savannas are grass-dominated vegetation types in which trees, 3–6 m tall, occur as scattered individuals or clumps. Tree trunks are twisted and leaves are sclerophyllous. Vines are uncommon, but there are epiphytes of the Orchidaceae and Bromeliaceae and hemiparasites of the Loranthaceae. Common tree genera are *Byrsonima, Curatella, Crescentia* spp., *Coccoloba, Acoelorraphe, Quercus oleoides,* and Melastomataceae. Common grasses are tall, 80–100 cm, and rough; they often grow in dense bunches. Their hemicryptophyte life-form and the abundant culms protect the growing points from fires. Common grass genera are *Paspalum, Andropogon, Aristida, Imperata, Trichachne, Leptocoryphum, Axonopus,* and *Digitaria,* accompanied by Cyperaceae, Fabales, and Asteraceae, but only the Cyperaceae are important in primary production. This type of vegetation is similar from Mexico to South America and to the West Indies. In Mexico savannas are best developed in the southeast, in Campeche, Tabasco, Chiapas, and Veracruz. They also occur in much reduced form on the Pacific coast from Chiapas to Sinaloa. Savannas occur in hot climates with rainfall between 1,000 mm and 2,500 mm and a dry period of zero to six months. They develop on poor soils, usually with clay in the substrate, where relief is low and drainage is deficient. Soils are acidic and rich in organic matter. There is a pronounced alternation of wet and dry conditions. Disturbed soils are often associated with savanna vegetation. Flora is tropical, and endemism is rare. Rzedowski (1981) shows savannas near Huimanguillo, Tabasco, where the common trees are *Curatella americana,* and near Escarcega, Campeche, where *Byrsonima crassifolia, Crescentia cujete,* and *Acoelorraphe wrightii* are common trees. *Acoelorraphe* forms palmares within the savanna; *Quercus oleoides* forms encinares. Forests of *Byrsonima, Curatella,* and *Crescentia* are geographically, floristically, and ecologically related to the savannas, but they are closed communities of 5 m tall trees, with a poorly developed herbaceous layer. They often occur as large patches within the savanna, covering a considerable area. On the Pacific side similar low forests occur on clayey, poorly drained, black soils but also on upper slopes, especially on metamorphic rock. Gallery forests and tropical evergreen forests occur in and around savannas. According to Rzedowski, Sarukhan considers the low forests to be ecotonal between savanna and tall tropical evergreen forests. Savannas are important cattle-raising habitats, even though the grasses are tough in the dry season and fires are started to stimulate new growth. Fire is frequent, and the species have long since adapted to it (Rzedowski 1981).

Palm communities ("palmares" of Rzedowski 1981) are dominated by members of the Arecaceae. They are graceful and distinctive communities of hot and humid to subhumid climates, south of 23°N latitude. Although they can be found on the Pacific coast, Gulf coast, or Caribbean coast, they occur sporadically and do not cover more than 1 percent of the surface of Mexico. They generally occur below 300 m but have been recorded in Mexico above 2,000 m. Their occurrence is usually attributed to disturbance or suboptimal edaphic conditions. Soils are diverse, often deep, and frequently inundated, either by a perched water table or by a high groundwater table, as is the case near the coast. Nevertheless, they have been recorded on limestone slopes with shallow, rocky soil (near springs?). Their presence is often associated with human activity. Many are secondary communities replacing lowland evergreen rain forest, semideciduous forest, or semideciduous woodland. Palm products have been used by humans since ancient times: fruits and seeds are eaten or fats and soap obtained from the seeds; trunks are used in construction; fronds are used for thatching, weaving, bags, etc.; and hearts of palm can also be consumed. One of these palm communities may have been the site of the domestication of *Acrocomia aculeata* (coyol palm) (Lentz 1990). Palms can form forests up to 40 m tall or shrublands 50–30 cm tall. Often tall trees and low shrubs of the same species form a two-layered community. Some stands are dense and shade the soil; other stands are open and an understory develops. The palms form monocultures, with an occasional strangler of *Ficus*. Palms are often categorized by the type of leaf dissection: palmate or pinnate. Fan-palm communities of *Sabal mexicana,* up to 15 m tall, occur on the Gulf coast and slopes as well as on the Pacific side. On the Atlantic side it ranges from Tamaulipas to Chiapas, but recent conversion to pastureland has eliminated it in the area where Tamaulipas, San Luis Potosí, and Veracruz meet. On the coast it occurs in sand; in the Central Depression of Chiapas it occurs on rain-flooded plains. *Sabal* palms occur also as dominants in mixed communities of other palms and tropical evergreen rain forest species in Chiapas. In Yucatán, mixed *Sabal* palm stands occur in the transition between low wetlands and upland rain forests. Mixed *Sabal* palm stands in central Michoacán are thought to be influenced by cutting the primary vegetation and by frequent fires thereafter. *Brahea dulcis* forms extensive stands, at 1,200–2,200 m, in the Balsas drainage area, extending to Oaxaca, and also on the higher parts of the Papaloapan Depression and along the length of the Sierra Madre Oriental south from Tamaulipas. It appears to be associated with rocks rich in calcium carbonate. They occur in the transition between semideciduous woodland and oak forest and can withstand a regular occurrence of freezes. In Sonora and in Baja California, *Washingtonia* palms occur in canyons near the shoreline, in a discontinuous distribution. These trees are commonly 10–15 m tall but can reach 20 m. Palm stands of *Acoelorraphe wrightii* (paurotis palm, tasiste) in the Yucatán Peninsula, Tabasco, and southeast Veracruz are isolated in patches on the shores of lagoons, flooded lands, and ravines, where they endure inundation by brackish water. In Tabasco they occur as part of the savannas, often forming islets of trees 2–5 m tall. The low palm stand of *Crysophila nana* is a rarer coastal community in Tonala, Chiapas, where the palms grow on the shallow soil of slopes.

Among the communities of palms with pinnately dissected leaves, the most impressive are stands of *Attalea cohune* that reach 15–30 m tall and are so dense that they cast deep shade on the ground. *Attalea* forests occupy a narrow belt, rarely more than 10 km long and 5 km wide, along the littoral Pacific, from Nayarit to Oaxaca, where the substrate is deep sand with a high water table. Often they are upland from beaches bordering bays. In this community, *Attalea* is the absolute dominant; the other trees present are of little importance: *Ficus* spp. (figs, Moraceae), *Brosimum alicastrum* (breadnut, Moraceae), *Dendropanax* (Araliaceae), *Enterolobium* (Fabales), and *Bursera* (Burseraceae). An understory is present only after disturbance of the canopy. *Orbignya cohune* is present in the semideciduous forest at its littoral border.

Palm stands of *Scheelea* spp. and *Attalea cohune* occur on deep, moist soils that are susceptible to frequent inundation, at the base of the Yucatán Peninsula. On the Gulf coastal plain of Mexico, at elevations mostly below 200 m, from Veracruz and adjacent Oaxaca to the northeast of Campeche and northern Chiapas, stands of *Scheelea liebmannii* are characteristic. Although *Scheelea* stands usually represent secondary vegetation as a result of human disturbance, that palm appears to represent primary vegetation on inundated riverine soils in the Bajo (Basin) de Papaloapan, along the Río Usumacinta and its tributaries in the north and northeast of Chiapas and adjacent areas of Tabasco. These palms reach 15–18 m tall. The understory varies from dominance by the grass *Andropogon glomeratus* to a dense impenetrable shrub layer. At the other end of the vegetation complex in which it occurs, *Scheelea liebmannii* dominates a lowland evergreen rain forest with *Sweetia panamensis*. *Scheelea preussii* dominates palm stands under similar conditions on the Pacific side of Chiapas.

In a hot and humid to subhumid climate, *Roystonea* (royal palm) forms pure stands along the littoral from southern Veracruz and Tabasco to the northeastern Yucatán Peninsula. *Roystonea* also occurs as a constituent in the tropical lowland evergreen rain forest, on poorly drained soils. Also on the north of the Yucatán Peninsula is an association dominated by *Pseudophoenix sargentii,* in which the plants are of low stature. Another feather palm stand in the littoral belt of the Yucatán Peninsula is *Thrinax parviflora,* which occurs in immediate contact with mangrove forests. Readers are referred to Rzedowski (1981) for mangrove forests and freshwater wetlands of Mexico.

GENERAL OVERVIEW OF CENTRAL AMERICA

The following section is an overview of the flora and vegetation of Central America. Standley and Steyermark (1945) estimate there are 8,000 species of vascular plants in all of Guatemala, many of which are endemic. Orchidaceae, Fabales, and Asteraceae are especially prominent. They recognize eleven floristic regions: limestone plains of Petén, mangrove swamps along both coasts, rain forests of the Atlantic coast, low savannas of Izabál and Petén, mixed forests of the Pacific plains, arid desert plains-chaparral of the Oriente plateaus and valleys of the Río Motagua and Río Blanco, wet mountain forests of Alta Verapaz, mixed mountain forests of the Pacific bocacosta, upland mixed forests of temperate and cold regions, coniferous forests, and alpine regions on the fourteen major volcanoes.

Knapp (1965) gives an overview of the vegetation of Central America and adjacent Mexico. He recognizes three major types of lowland tropical forest (with canopy height): rain forest (35–50 m), moist forest (25–35 m), and dry forest (15–25 m). In the present study these are, respectively, lowland evergreen rain forest, semideciduous forest, and semideciduous woodland (mainly represented by deciduous trees). Knapp's figure 138 provides a map of the nine major vegetation types he recognizes for the area from Chiapas through Nicaragua. The five countries are briefly reviewed in the following paragraphs.

Guatemala adjoins Chiapas on the south. Caribbean slash pine (*Pinus caribaea*) savanna occurs along the Atlantic coast, which is located in the southeast of the state. Belize, which occupies most of that coast between Yucatán and Honduras, is largely mapped as pine savanna. Most of Guatemala is evergreen rain forest, as is central Belize. In western Guatemala are high mountain ranges bearing altitudinal bands of forest that are increasingly dominated by Holarctic taxa with elevation (see Mexican Highlands Province). West of the mountains, along the Pacific coast, are tropical semideciduous forests. The Pacific slope forests extend into adjacent El Salvador and south into Nicaragua. El Salvador has scattered mountain ranges covered by montane forests with oak and pine; these also continue through western Nicaragua. In the interior of El Salvador are extensive semideciduous woodlands which extend, as well, into Nicaragua, where they become an important feature of the western interior. Most of Honduras is mountainous in a continuous chain from Chiapas and Guatemala. High rainfall areas of the north are covered by rain forest. Much of central Honduras is mapped as forest with oak and pine. Thin soils over mafic bedrock (rich in magnesium, aluminum, and iron) in the west-central part have extensive pine forests. High elevations (1,500–1,800 m) throughout Honduras have warm and mild, equable climates where dominance by Holarctic taxa increases with elevation. Tropical evergreen rain forest occurs on the northern, Atlantic coast. Eastern Honduras is mapped largely as evergreen rain forest and slash pine savanna. Eastern Nicaragua is largely slash pine savanna. Interior eastern and central lowlands are mapped as evergreen rain forest. Central highlands are mapped as lower montane rain forest with oaks and some upper montane (cloud) forest. Two large areas of drier forest with oak and pine occur in the west-central mountains.

Knapp treats Costa Rica and Panama in a separate map (1965: figure 139). Most of Costa Rica is mountainous and wet, with drier forests on the northwest lowlands and rain forests in the southwest, east, and on the central mountain chain. Here are mountains that can reach above the level of forest. Holarctic taxa are present only at the higher elevations. Mountain vegetation of Costa Rica is treated under Mexican Highlands Province. Hartshorn (in Janzen 1983) reviews the extensive observations of Holdridge on the lowland vegetation of Costa Rica (Hartshorn 1988). A particularly well-studied rain forest is La Selva, Puerto Viejo de Sarapiqui, Heredia, Costa Rica. Hartshorn (in Janzen 1983) summarizes data on composition, noting that the following trees are major dominants (percentage of basal area in parentheses) in the "virgin forest": *Pentaclethra macroloba* (Fabales, 29.2), *Carapa guianensis* (Meliaceae, 19.4), *Pterocarpus officinalis* (Fabales, 16.6). *Pentaclethra* is the only major dominant in a number of the most mesic sites. Understory palms are varied and abundant: *Welfia, Socratea, Iriartia, Geonoma, Synecanthus, Asterogyne,* and *Calyptogyne*. The dominance by compound-leaved trees suggests that occasional droughts may influence forest composition.

Panama is the southernmost country of Central America. It is oriented east-west in an **S**-curve. It has a maximum width of 193 km and comprises 81,840 km². Nine-tenths of the area is below 366 m. The Atlantic slope has a moist climate (ca. 330 cm/year precipitation). The Pacific coast has a drier climate, with about half the rainfall of the Atlantic slope. Dense forests dominate lowlands, such as Barro Colorado Island (see Knight 1975), and uplands on the Atlantic slope. These forests appear similar to those along the whole eastern coast of Central America north to Mexico. Savanna and more open forests in which drought-deciduous trees are widespread dominants are characteristic of the Pacific slope; these are used for grazing cattle. Vegetation on the Pacific slope is similar to that of Costa Rica, El Salvador, Guatemala, and Mexico to the north, according to Standley and Steyermark (1945). In addition there are extensive mangrove swamps, coastal beaches, inland lakes, rivers and ponds, highlands (900–2,500 m) comprising lower and upper montane forests, and subalpine regions (above 2,500 m). The subalpine region is especially well developed on Volcán de Chiriqui (ca. 4,000 m), near the Costa Rican border, and comprises plants and communities as described by Kappelle (1991) for the same elevations in Costa Rica (Schery 1945).

GEOGRAPHICAL VARIATION IN FOREST TYPES OF THE GULF SEMIDECIDUOUS FOREST ZONE

Yucatán Peninsula phytochorion of Rzedowski (1981) occupies the entire territory of that peninsula, but its southern limits are not well defined. It includes at least a part of Belize and the Department of Petén in Guatemala. The climate is hot and humid at the base of the peninsula, and a gradient of dryness exists in a southeast to northwest direction. The substrate is limestone, and the physiography is flat plains. The relationships to the West Indian phytochorion are more accentuated than in any other part of Mexico. Floristic richness decreases toward the northwest. The vegetation consists historically of tropical deciduous forest, semideciduous forest, and evergreen forest. The legume families are well represented and dominate the zonal vegetation. Lundell (1945) also emphasizes that the predominant floristic relationship of the northern plain of Belize is to the Yucatán Peninsula. Standley and Steyermark (1945) note that the flora of the limestone plains of Petén in Guatemala is identical to the Yucatán Peninsula and includes many endemic species. No more recent works are available, so that these studies remain the most comprehensive accounts of that vegetation. Extensive treatment of Yucatán vegetation is included in this section because of close affinities to the Petén region of Guatemala.

The optimal forests of the Yucatán Peninsula, usually treated as tropical lowland evergreen (LEF), are dominated by zapota (*Manilkara zapota*), *Swietenia macrophylla, Brosimum alicastrum, Rheedia edulis, Lucuma campechiana, Calophyllum brasiliense, Crysophila argentea, Ficus, Piper,* and *Psychotria* spp. Miranda recognized a number of associations of "zapotal."

According to Rzedowski (1981), the semideciduous forest of the Yucatán Peninsula is dominated by *Vitex gaumeri* (Verbenaceae) and a number of codominants. *Brosimum alicastrum* (Moraceae) is codominant in the northeast of Yucatán State and adjacent Quintana Roo, where the following trees are present: *Coccoloba* (Polygonaceae), *Guettarda* (Rubiaceae), *Simarouba glauca* (Simaroubaceae), etc. In the south and east of the peninsula *Vitex* dominates with *Sideroxylon gaumeri* (Sapotaceae) and *Caesalpinia gaumeri* (Caesalpiniaceae). Some of the trunks of *Enterolobium, Ceiba,* and *Cedrela* reach diameters of 1–2 m and have very large crowns. In the coastal fringe of Campeche, *Vitex* is associated with *Cedrela mexicana* (Meliaceae). Also here are *Aspidosperma* (Apocynaceae), *Bursera, Ficus, Gyrocarpus* (Hernandiaceae), *Metopium* (Anacardiaceae), and *Pileus* (*Jacaratia,* Caricaceae). In deeper soils on the Yucatán Peninsula, *Enterolobium cyclocarpum* and *Ceiba pentandra* dominate; they are accompanied by *Astronium, Brosimum, Cedrela, Ficus* spp., *Spondias mombin,* and *Vitex gaumeri.*

An association of the semideciduous woodland has been reconstructed by Miranda from fragments and secondary woodlands in the north of Yucatán (Rzedowski 1981). It is a totally deciduous type, 15–20 m tall, with the dominants *Lysiloma bahamensis*

(Mimosaceae) and *Piscidia piscipula* (Fabaceae); other trees are *Alvaradoa, Bursera, Cedrela, Chlorophora, Cordia, Ehretia, Gyrocarpus, Lochocarpus, Neomillspaughia, Simarouba,* and *Trichilia.* On the coastal fringe of Yucatán a type of SDW low in stature (6–15 m tall) and of xerophytic aspect has been described for thin soils on flatlands. Common trees are *Bursera simaruba, Caesalpinia vesicaria, Ceiba aesculifolia, Chlorophora tinctoria, Diospyros cuneata, Guaiacum sanctum, Hauya trilobata, Metopium brownei, Parmentiera aculeata, Piscidia piscipula,* and candelabra-form cacti. Secondary vegetation in the SDW region of Yucatán is the result of shifting agriculture, in which disturbance occurs every 15 years, whereas SDW requires 50 years to regenerate (Rzedowski 1981). SDW is replaced by spiny woods of *Acacia, Cassia, Gymnopodium, Mimosa, Pithecellobium. Gymnocarpium* or *Mimosa* form pure stands of secondary forest following the weed stage after abandonment of cultivation. In some parts of Quintana Roo the palm *Pseudophoenix* sp. is abundant. In the high watershed of Papaloapan a simple type of SDW is found dominated by *Cyrtocarpa,* with *Bursera, Amphipterygium, Ceiba, Cassia, Euphorbia, Pseudosmodingium, Gyrocarpus, Leucaena.* Around the port of Veracruz, the SDW has been reconstructed from surviving patches. Trees listed in Rzedowski (1981) are *Cordia, Piscidia,* and *Pithecellobium,* accompanied by *Parmentiera, Tabebuia, Ehretia, Lysiloma, Crescentia,* and *Enterolobium.* The patch of SDW in southern Tamaulipas, southeastern San Luis Potosí, and adjacent Querétaro and Veracruz occurs on soils derived from marine sedimentary rocks, between 50 and 800 m. Dominants are *Bursera simaruba, Lysiloma divaricata,* and *Phoebe tampicensis.* Other frequent trees are *Acacia, Beaucarnea (Nolina), Cedrela, Lysiloma, Zuelania,* and *Piscidia.* In the isolated Sierra de Tamaulipas, two variants of SDW have been described, one dominated by *Bursera simaruba* and *Lysiloma divaricata* and the other by *Phoebe tampicensis* and *Pithecellobium flexicaule.* SDW also occurs as edaphic vegetation in desert canyons in Hidalgo, Querétaro, Guanajuato, and San Luis Potosí, where *Bursera morelensis* dominates. Secondary vegetation in SDW areas includes *Ipomoea* woodlands, *Dodonaea-Tecoma* shrublands, spiny shrublands of *Acacia* and *Stenocereus* or *Opuntia,* pastures of Asteraceae and Poaceae, and weed fields. In the SDW area of San Luis Potosí, the secondary shrublands are dominated by *Acacia, Croton,* and *Karwinskia;* also present are palmares of *Sabal mexicana,* "aquichales" of *Guazuma ulmifolia,* and secondary woods of *Piscidia piscipula.*

Thorn woodland occurs in the Yucatán Peninsula, extending to Tabasco and northern Chiapas; it is called tinto. Miranda (in Rzedowski 1981) describes TW in the lowlands and gulleys with deep and inundated soils. It is a woodland 4–12 m tall, relatively rich in epiphytes and vines, dominated by *Haematoxylon campechianum.* TW is also present on hillsides in Oaxaca, to 900 m, where it forms stands less than 8 m tall, and in the southeast of San Luis Potosí and in the Sierra de Tamaulipas. In the latter location TW forms dense, impenetrable stands less than 12 m tall, dominated by *Pithecellobium, Eysenbeckia,* and *Phyllostylon;* also present are *Bumelia, Capparis, Cercidium,* and *Prosopis.* Near Matamoros, Tamaulipas, TW passes into a mezquitál of *Prosopis glandulosa, Pithecellobium flexicaule,* and *Cercidium macrum,* 6–8 m tall, with a shrub layer of spiny plants. It also occurs in adjacent Nuevo León, where *Cordia* is a dominant as well. In the understory is a shrub layer 3–5 m high, with *Acacia, Celtis, Porlieria, Ptelea,* and *Yucca filifera.* Low savannas of Izabál and central Petén develop in a climate with a marked seasonal distribution of rainfall in which the dry season lasts from November to April. Many species lose their leaves at this time, and fires sweep the country. Vegetation is characterized by opens stands of *Pinus caribaea, Curatella,* and *Byrsonima crassifolia* in a matrix of Poaceae, Cyperaceae, Fabales, and Asteraceae.

Lundell (1937) believes the savannas to be of human origin and that "flatland forest" once occupied the entire area. That flatland forest has the following widespread trees: *Matayba oppositifolia* (Sapindaceae), *Metopium brownei* (Anacardiaceae), *Spondias mombin* (Anacardiaceae), *Guettarda combsii* (Rubiaceae), *Calyptranthes* sp. (Myrtaceae), *Hirtella* sp. (Chrysobalanaceae), *Alibertia edulis* (Rubiaceae), *Cnestidium rufescens* (Connaraceae), *Nectandra* spp. (Lauraceae), *Vitex gaumeri* (Verbenaceae), *Simarouba glauca* (Simaroubaceae), *Zanthoxylum* spp. (Rutaceae), and *Ficus* spp. (Moraceae). There is a three-tiered canopy, with the upper canopy reaching to 30 m and dominated by *Swietenia macrophylla, Termnalia excelsa, Cassia* sp., and *Matayba* sp.; a middle layer, 18–25 m, dominated by *Matayba oppositifolia;* and a lower tree layer. Below this there are three layers: tall shrubs, low shrubs (especially *Psychotria homputalis*), and forest floor plants, mainly ferns. Canopy trees have leaves of reduced size, glossy surface, pubescence, thick cuticle, all characteristic of "xerophytes," according to Lundell (1937). Many have compound leaves, as well. On the deep soils of limestone valleys in central Petén are luxuriant forests of mixed deciduous and evergreen trees that are subject to severe fires. Lundell calls this forest semideciduous. It appears to correspond to the same type in the Yucatán Peninsula described above and to include large leguminous trees, e.g., *Enterolobium cyclocarpum, Schizolobium parahybum,* and *Tipuana lundellii,* as well as *Terminalia excelsa* and *Aspidosperma* sp. Second-story trees are *Orbignya cohune* (Arecaceae), *Pouteria* sp. (Sapotaceae), *Alseis* sp. (Rubiaceae), and *Sickingia* sp. (Rubiaceae). There is a lower tree layer, a shrub layer, lianas, scrambling graminoids, epiphytes, and herbs (Lundell 1937).

In Belize, in climates with a pronounced dry season, where MAP is 1,500–3,000 mm, all along the coastal plain but also on 323.7 km² of granite hills that rise to 1,000 m, extensive forests of *Pinus caribaea* develop on poor, acidic soils. The understory is of tropical hardwoods. Fires occur periodically (Lamb 1950).

WEST INDIAN (ANTILLEAN) PROVINCE

In structure and generic composition West Indian plant communities show strong resemblances to vegetation types in Central America. This province includes the Greater Antilles (Cuba, Hispaniola, Jamaica, Puerto Rico), Lesser Antilles (Windward and Leeward Islands), southern Florida, and the Bahamas (Takhtajan 1986). The 1,000 named islands of the West Indies form an archipelago of small landmasses 1,930 km long. In the southern part of their distribution their north-south orientation separates the Caribbean Sea from the Atlantic Ocean. Floristic and vegetational diversity is enhanced by elevation (2,955 m maximum, on Hispaniola), temperature range, and edaphic conditions (sands, salt-gypsum, volcanic deposits, laterite, and serpentinite). At low elevations, mean annual temperatures range from 24.9°C in Havana to 26.1°C in

Dominica; mean annual precipitation ranges generally from 1,157 mm in Havana to 1,979 mm in Dominica. Many areas receive less than 1,016 mm annually, with six or seven months of reduced rainfall occurring as two dry periods. Some areas, such as the lowland station on Dominica and the mountain stations in Puerto Rico and Martinique, show no months of deficiency less than 102 mm (Howard 1973).

Takhtajan accepts Good's treatment of the flora as containing about 200 endemic genera, with about 50 percent of all the species as endemic. Cuba, the largest island, has about 6,000 species, 47 percent endemic, including the cycad *Microcycas*. Each of the four islands of the Greater Antilles has its own endemic genera or genera endemic to various combinations of two or more islands (Howard 1973). Shreve (1945) notes that Jamaica, although connected to the coast of Honduras by a shallow marine bank 380 miles long, has "less relation to the flora of Honduras and Central America than the floras of any of the Greater Antilles." Jamaica has a weak relation to Central America and in general has little relation to the flora of any other Caribbean island. It has the highest number of endemics of the Greater Antilles. Howard lists only one family as endemic to the Antilles, the Picrodendraceae (often treated as Euphorbiaceae), and only two large genera (20+ species): *Wallenia* (Myrsinaceae) and *Calycogonium* (Melastomataceae). The families with the largest number of genera are (in order of decreasing size): Asteraceae, Poaceae, Leguminosae, Rubiaceae, Orchidaceae, Euphorbiaceae, and Melastomataceae. The families with the largest number of species are Poaceae, Leguminosae, Rubiaceae, Asteraceae, Euphorbiaceae, Orchidaceae, and Myrtaceae. Although most genera are small, with 5 or fewer species, there are ten genera with 60 or more species. Most of the species are present in the Greater Antilles: *Croton* (Euphorbiaceae), *Eugenia* (Myrtaceae), *Miconia* (Melastomataceae), *Panicum* and *Paspalum* (Poaceae), *Peperomia* (Piperaceae), *Phyllanthus* (Euphorbiaceae), *Pilea* (Urticaceae), *Psychotria* (Rubiaceae), and *Rondeletia* (Rubiaceae).

Vegetation of the West Indian Province was reviewed and described by Beard (1949) and Howard (1973). Zonal vegetation comprises lowland semideciduous forests, semideciduous (deciduous and dry evergreen) woodlands; lowland thornscrub (evergreen bushland). Seasonal and dry-evergreen forests are the typical, zonal vegetation of the lowlands of the Lesser Antilles (Beard 1949). The sequence of forests with increasing elevation on well-drained soils in the lowlands of the Lesser Antilles is given by Beard as deciduous seasonal forest, semievergreen, evergreen seasonal, and rain forest. On shallow, compact, poorly drained soils, the sequence of dry evergreen formations is as follows: evergreen bushland, dry evergreen forest, lower montane rain forest. Ewel and Whitmore (1973), using the Holdridge system, classify the "ecological life zones" for a vegetation map of Puerto Rico and the U.S. Virgin Islands as follows: subtropical rain forest, subtropical wet forest, subtropical moist forest, subtropical dry forest, and in the montane zone, subtropical lower montane rain forest and subtropical lower montane wet forest. They estimate that over 60 percent of Puerto Rican vegetation could be classified as subtropical moist forest, 24 percent as subtropcal wet forest, and 13.8 percent as subtropical dry forest.

Montane-zone forest vegetation consists of lower and upper montane evergreen forest and thicket. Whitmore (1984) gives the general forest zonation for tropical mountains as lowland, lower montane, and upper montane. For the Luquillo Mountains of

Puerto Rico, P. L. Weaver and Murphy (1990) list the following altitudinal forest types (my estimates of the equivalent zones from Whitmore 1984 are in parentheses): subtropical wet forest (tabonuco, the lower montane forest), lower montane wet forest (palm, part of the lower montane forest), another lower montane wet forest (Colorado, the upper montane forest), and lower montane rain forest (dwarf, part of the upper montane forest). In table 6 of Weaver and Murphy, the authors give parameters of stand structure for the four types. In their table 7, they give summaries of stand dynamics for the forests. P. L. Weaver (1989) documents changes in dominance in Colorado (upper montane) forest after severe hurricanes.

CONCLUSION

Floristic and vegetational changes from the Arctic to the Equatorial Zone follow global latitudinal climate belts, with deflections related to continental position. The Holarctic flora covers all of North America. At the U.S.-Mexico border, this flora occupies the mountains, whereas the lowlands are the realm of the Neotropical flora. The Holarctic flora is confined to progressively greater elevations with decreasing latitude. Winter length and severity decreases with decreasing latitude in the Holarctic realm. Where winters are most pronounced, as in Canada and most of the United States, Boreal (Subkingdom) flora dominates. Here the plant cover is formed by deciduous trees and shrubs (moister climates, in the east) and by mainly evergreen conifers (drier climates, in the west); evergreen angiosperms and bryophytes are confined to wet habitats or occupy habitats close to the ground. In the rainshadow of the Rocky Mountains, the interior of midlatitudinal North America is occupied by cold deserts that are dominated by low shrubs and by grasslands, which show floristic changes of aspect with the seasons. The cool, wet climates of the Pacific Northwest, subject to occasional summer droughts and severe fires, are occupied by tall evergreen conifers. These forests are unique for the midlatitudes of earth and contain flora and vegetation persisting from the Tertiary period. In the extreme southern parts of the United States and throughout all but the highest elevations in the mountains of Mexico, the Madrean (Subkingdom) flora occupies the land. Here winters are mild and of short duration. Consequently, evergreen angiosperms, evergreen conifers, and deciduous angiosperms co-occur, with relative dominance of each largely determined by the frequency of low temperatures (which favor deciduous angiosperms) and soil nutrients (scarcity favoring conifers). The height and openness of the Madrean vegetation depends upon effective precipitation, so that southeastern North America, having abundant year-around precipitation, is (under natural conditions) largely in forest, whereas western North America, with dry summers or low annual precipitation, is in evergreen woodland, grassland, or desert. In Mexico and southward, the Madrean landscapes are confined to the large mountain masses. South of the Sierra Madre Occidental, the zonal mountain vegetation below treeline is wholly of forests, which range from pine and pine-oak domination in the drier climates or on overly well-drained sites, to rain forests teeming with a rich biota of diverse life-forms. Following Rzedowski, I treat the hot deserts of the

Southwest as representative of the Neotropical Kingdom. The dominant life-forms are intricately adapted to survival under hot, arid conditions: shrubs with ephemeral leaves, stem-succulents that range in size and structure from pincushionlike through sprawling shrubs to columnar branched and unbranched trees, leaf-succulents in a great range of sizes and of branching and flowering patterns, and evergreen shrubs with a variety of root-system adaptations. From Sonora State southward (and to a lesser extent in southern Tamaulipas and northern Veracruz) thorn woodlands fringe the hot deserts, then gradually replace them on the landscape where precipitation is greater. Drought-deciduous tropical woodland in turn replaces the thorn woodland farther south and along the bases of mountain ranges. From Chiapas State, Mexico, southward, the landscape of the lowlands is mainly of Neotropical forests, ranging from drought-deciduous woodlands through semideciduous forests to evergreen rain forests. Throughout much of Mexico, the Isthmus of Tehuantepec, and in southern Central America, rainfall is often greatly variable (10–40 percent) from year to year, so that families of compound-leaved trees (e.g., Mimosaceae, Caesalpiniaceae, and Fabaceae) are often dominants. The Isthmus of Tehuantepec marks the southern limit of regular cold air masses, so that Holarctic taxa are confined to the mountain forests southward. However, Caribbean slash pine, a taxon with Holarctic family affinities, dominates a tropical understory on extensive areas of Belize and eastern Honduras, south to Nicaragua. On the highest parts of the mountains in Costa Rica, taxa of South American distribution occur; they are present in vegetation types such as páramo (mountain perennial tall grassland). In summary, this is a region of great plant diversity due to an immense variability in climatic patterns, physiographic conditions, and floral history.

ACKNOWLEDGMENTS

This contribution is dedicated to Dr. Pierre Dansereau, then Curator of Ecology, New York Botanical Garden, and Adjunct Professor, Department of Botany, Columbia University, who inspired and instructed my lifelong interest in vegetation geography. It is also dedicated to the following persons, whose care, guidance, and concern enabled me to undertake and complete this work: Stephen Green, M.D.; Frederick Fein, M.D.; Jeffrey Scavron, M.D.; Tracy Stopler Kasdan, M.S., R.D.; Virginia Novak, M.S.

The author thanks Bill Evans and Steve McLaughlin for extensive written comments and helpful conversations on sections of the manuscript. He is grateful to Dennis Wm. Stevenson for reviewing the manuscript and making useful suggestions.

REFERENCES

Alexander, R. R. 1985. *Major Habitat Types, Community Types, and Plant Communities in the Rocky Mountains.* General Technical Report RM-123. Fort Collins, Colo.: USDA Forest Service, Rocky Mountain Forest and Range Experiment Station.

Archer, S. 1990. Development and stability of grass/woody mosaics in a subtropical savanna parkland, Texas, U.S.A. *J. Biogeography* 17: 453–462.

Atlas Censal de El Salvador. 1955. Ministerio de Economia, Dirección General de Estadistica y Censos.

Bailey, R. G. 1996. *Ecosystem Geography.* New York: Springer-Verlag.

——. 1997. Ecoregions Map of North America (1:15,000,000): Explanatory Note. USDA Forest Service, Misc. Publ. 1548.

Barbour, M. G. 1988. California upland forests and woodlands. In M. G. Barbour and D. W. Billings, eds., *North American Terrestrial Vegetation,* pp. 131–164. Cambridge: Cambridge University Press.

Barbour, M. G., and D. W. Billings, eds. 1988. *North American Terrestrial Vegetation.* Cambridge: Cambridge University Press.

Barbour, M. G., and N. L. Christensen. 1993. Vegetation. In Flora of North America Editorial Committee, ed., *Flora of North America,* vol. 1, pp. 97–131. New York: Oxford University Press.

Barbour, M. G., and J. Major, ed. 1988. *Terrestrial Vegetation of California.* California Native Plant Society, Spec. Publ. 9.

Baskin, J. M., C. C. Baskin, and R. L. Jones. 1987. *The Vegetation and Flora of Kentucky.* Richmond: Kentucky Native Plant Society, Eastern Kentucky University.

Beard, J. S. 1949. *Natural Vegetation of the Windward and Leeward Islands.* Oxford Forest Memoirs. Oxford: Clarendon Press.

——. 1955. The classification of tropical American vegetation-types. *Ecology* 36: 89–100.

Braun, E. L. 1950. *Deciduous Forests of Eastern North America.* Philadelphia: Blakiston.

Brown, C. L., and L. K. Kirkman. 1990. *Trees of Georgia and Adjacent States.* Portland, Ore.: Timber Press.

Brown, D. E., ed. 1994. *Biotic Communities: Southwestern United States and Northwestern Mexico.* Salt Lake City: University of Utah Press.

Brown, D. E., C. H. Lowe, and C. P. Pase. 1980. A digitized systematic classification for ecosystems with an illustrated summary of the natural vegetation of North America. General Technical Report RM-73. Fort Collins, Colo.: USDA Forest Service, Rocky Mountain Forest and Range Experiment Station.

Chester, E. W., ed. 1989. The Vegetation and Flora of Tennessee. *J. Tennessee Academy of Science* 64(3): 57–207.

Christensen, N. L. 1988. Vegetation of the Southeastern Coastal Plain. In M. G. Barbour and D. W. Billings, eds., *North American Terrestrial Vegetation,* pp. 317–363. Cambridge: Cambridge University Press.

Cronquist, A. 1982. Map of the floristic provinces of North America. *Brittonia* 34: 144–145.

Cronquist, A., P. H. Holmgren, N. H. Holmgren, and J. L. Reveal. 1972. *Intermountain Flora,* vol. 1. New York: Hafner.

Delcourt, P. A., and H. R. Delcourt. 1993. Paleoclimates, paleovegetation, and paleofloras during the Late Quaternary. In Flora of North America Editorial Committee, ed., *Flora of North America,* vol. 1, pp. 71–94. New York: Oxford University Press.

Dinerstein, E., D. M. Olson, D. J. Graham, A. L. Webster, S. A. Primm, M. P. Bookbinder, and G. Ledec. 1995. *A Conservation Assessment of the Terrestrial Ecoregions of Latin America and the Caribbean.* Washington, D.C.: World Bank.

Elliott-Fisk, D. L. 1988. The Boreal Forest. In M. G. Barbour and D. W. Billings, eds., *North American Terrestrial Vegetation,* pp. 33–62. Cambridge: Cambridge University Press.

Ewel, J. J., and J. L. Whitmore. 1973. The ecological life zones of Puerto Rico and the U.S. Virgin Islands. Forest Service Research Paper ITF-18. Rio Piedras, P.R.: USDA Institute of Tropical Forestry.

Fitzhugh, E. L., W. H. Moir, J. A. Ludwig, and F. Ronco, Jr. 1987. Forest habitat types in the Apache, Gila, and part of the Cibola National Forests, Arizona and New Mexico. General Technical Report RM-145. Fort Collins, Colo.: USDA Forest Service, Rocky Mountain Forest and Range Experiment Station.

Franklin, J. F. 1988. Pacific Northwest forests. In M. G. Barbour and W. D. Billings, eds., *North American Terrestrial Vegetation,* pp. 103–130. Cambridge: Cambridge University Press.

Franklin, J. F., and C. T. Dyrness. 1973. *Natural Vegetation of Oregon and Washington.* Corvallis: Oregon State University Press.

Gentry, H. S. 1946. Notes on the vegetation of Sierra Surotato in northern Sinoloa. *Bulletin of the Torrey Botanical Club* 73: 451–462.

———. 1957. *Los Pastizales de Durango.* Mexico, D.F.: Edic. del Inst. Mex. de Recursos Naturales Renovables, A.C.

Goldberg, D. E. 1982. The distribution of evergreen and deciduous trees relative to soil types: An example from the Sierra Nevada, Mexico, and a general model. *Ecology* 63: 942–951.

Graham, A. 1993. History of the vegetation: Cretaceous (Maastrichian)-Tertiary. In Flora of North America Editorial Committee, ed., *Flora of North America,* vol. 1, pp. 57–70. New York: Oxford University Press.

Greller, A. M. 1988. Deciduous forest. In M. G. Barbour and W. D. Billings, eds., *North American Terrestrial Vegetation,* pp. 287–316. Cambridge: Cambridge University Press.

———. 1990. Comparison of humid forest zones in eastern Mexico and southeastern United States. *Bulletin of the Torrey Botanical Club* 117: 382–396.

Harper, R. M. 1949. A preliminary list of the endemic flowering plants of Florida. *Quarterly J. Florida Academy of Sciences* 12: 1–19.

Hartshorn, G. S. 1988. Tropical and subtropical vegetation of Meso-America. In M. G. Barbour and W. D. Billings, eds., *North American Terrestrial Vegetation,* pp. 365–390. Cambridge: Cambridge University Press.

Heikens, A. L., and P. A. Robertson. 1995. Classification of barrens and other natural xeric forest openings in southern Illinois. *Bulletin of the Torrey Botanical Club* 122: 203–214.

Howard, R. A. 1973. The vegetation of the Antilles. In A. Graham, ed., *Vegetation and Vegetational History of Northern Latin America,* pp. 1–38. Amsterdam: Elsevier.

James, C. W. 1961. Endemism in Florida. *Brittonia* 13: 225–244.

Janzen, D. H., ed. 1983. *Costa Rican Natural History.* Chicago: University of Chicago Press.

Kappelle, M. 1991. Distribución altitudinal de la vegetation del Parque Nacional Chirripo, Costa Rica. *Bresenia* 36: 1–14.

Kappelle, M., A. M. Cleef, and A. Chaveri. 1992. Phytogeography of Talamanca montane *Quercus* forests, Costa Rica. *J. Biogeography* 19: 299–315.

Kappelle, M., J.-G. Van Uffelen, and A. M. Cleef. 1995. Altitudinal zonation of montane *Quercus* forests along two transects in Chirripo National Park, Costa Rica. *Vegetatio* 119: 119–153.

Knapp, R. 1965. *Die Vegetation von Nord- und Mittelamerika.* Stuttgart: Gustav Fischer Verlag.

Knight, D. H. 1975. A phytosociological analysis of species-rich tropical forest on Barro Colorado Island, Panama. *Ecological Monographs* 45: 259–284.

Komarkova, V., R. R. Alexander, and B. C. Johnson. 1988. Forest vegetation of the Gunnison and parts of the Uncompahgre National Forests: A preliminary habitat type classification. General Technical Report RM-163. Fort Collins, Colo.: USDA Forest Service, Rocky Mountain Forest and Range Experiment Station.

Lamb, A. F. A. 1950. Pine forests of British Honduras. *Empire Forestry Review* 29: 219–226.

Lauer, W. 1968. Problemas de la división fitogeográfica en America Central. In C. Troll, ed., *Colloquium Geographicum, Band 9: Geoecology of the Mountainous Regions of the Tropical Americas,* pp. 139–156. UNESCO.

Lentz, D. 1990. *Acrocomia mexicana:* Palm of the Ancient Mesoamericans. *J. Ethnobiology* 10(2): 183–194.

León de la Luz, J. L., and R. Domingüez-Cadena. 1989. Flora of the Sierra de la Laguna, Baja California Sur, Mexico. *Madroño* 36: 61–83.

Lundell, C. L. 1937. *The Vegetation of Petén.* Publ. 478. Washington, D.C.: Carnegie Institution.

———. 1945. The vegetation and natural resources of British Honduras. In F. Verdoorn, ed., *Plants and Plant Science in*

Latin America, pp. 270–273. Waltham, Mass.: Chronica Botanica.

MacMahon, J. A. 1988. Warm deserts. In M. G. Barbour and D. W. Billings, eds., *North American Terrestrial Vegetation,* pp. 231–264. Cambridge: Cambridge University Press.

McLaughlin, S. P. 1989. Natural floristic areas of the western United States. *J. Biogeography* 16: 239–248.

McPherson, G. R. 1997. *Ecology and Management of North American Savannas.* Tucson: University of Arizona Press.

Minnich, R. A. 1976. Vegetation of the San Bernardino Mountains. In J. Latting, ed., *Plant Communities of Southern California,* pp. 99–124. Special Publ. 2. Berkeley: California Native Plant Society.

Miranda, F., and A. J. Sharp. 1950. Characteristics of the vegetation in certain temperate regions of eastern Mexico. *Ecology* 31: 313–333.

Miyawaki, A., K. Iwatsuki, and M. M. Grandtner. 1994. *Vegetation in Eastern North America.* Tokyo: University of Tokyo Press.

Montoya Maquin, J. M. 1966. Notas fitogeográficas sobre el *Quercus oleoides* Cham. y Schlect. *Turrialba* 16: 57–66.

Muldavin, E., F. Ronco, Jr., and E. F. Aldon. 1990. Consolidated stand tables and biodiversity data base for southwestern forest habitat types. General Technical Report RM-190. Fort Collins, Colo.: USDA Forest Service, Rocky Mountain Forest and Range Experiment Station.

Muller, C. H. 1939. Relations of vegetation and climatic types of Nuevo León, Mexico. *American Midland Naturalist* 21: 687–729.

———. 1947. Vegetation and climate in Coahuila, Mexico. *Madroño* 9: 33–57.

Myers, R. L. 1990. Scrub and high pine. In R. L. Myers and J. J. Ewel, eds., *Ecosystems of Florida,* pp. 150–193. Orlando: University of Central Florida Press.

Myers, R. L., and J. J. Ewel, eds. 1990. *Ecosystems of Florida.* Orlando: University of Central Florida Press.

Niering, W. A., and C. H. Lowe. 1984. Vegetation of the Santa Catalina Mountains:

Community types and dynamics. *Vegetatio* 58: 3–28.

Peet, R. K. 1988. Forests of the Rocky Mountains. In M. G. Barbour and D. W. Billings, eds., *North American Terrestrial Vegetation,* pp. 63–101. Cambridge: Cambridge University Press.

Platt, W. J., and M. W. Schwartz. 1990. Temperate hardwood forests. In R. L. Myers and J. J. Ewel, eds., *Ecosystems of Florida,* pp. 194–229. Orlando: University of Central Florida Press.

Reid, N., D. M. Stafford Smith, P. Beyer-Muenzel, and J. Marroquin. 1990. Floristic and structural variation in the Tamaulipan thornscrub, northeastern Mexico. *J. Vegetation Science* 1: 529–538.

Rowe, J. 1972. Forest regions of Canada. Canadian Forest Service Publ. 1300. Ottawa: Department of the Environment.

Rzedowski, J. 1981. *Vegetación de México.* Mexico, D.F.: Editorial Limusa.

———. 1993. Diversity and origins of the phanerogamic flora of Mexico. In T. P. Ramamoorthy, R. Bye, A. Lot, and J. Fa, eds., *Biological Diversity of Mexico: Origins and Distribution,* pp. 129–144. New York: Oxford University Press.

Schery, R. W. 1945. A few facts concerning the flora of Panama. In F. Verdoorn, ed., *Plants and Plant Science in Latin America,* pp. 284–287. Waltham, Mass.: Chronica Botanica.

Shreve, F. 1945. The vegetation of Jamaica. In F. Verdoorn, ed., *Plants and Plant Science in Latin America,* pp. 287–289. Waltham, Mass.: Chronica Botanica.

Sims, P. L. 1988. Grasslands. In M. G. Barbour and D. W. Billings, eds., *North American Terrestrial Vegetation,* pp. 265–286. Cambridge: Cambridge University Press.

Standley, P. C., and J. A. Steyermark. 1945. The vegetation of Guatemala: A brief review. In F. Verdoorn, ed., *Plants and Plant Science in Latin America,* pp. 275–278. Waltham, Mass.: Chronica Botanica.

Takhtajan, A. 1986. *Floristic Regions of the World.* Berkeley: University of California Press.

Thorne, R. F. 1993. Biogeography. In Flora of North America Editorial Committee, ed.,

Flora of North America, vol. 1, pp. 132–153. New York: Oxford University Press.

Van Auken, O. W., and J. K. Bush. 1997. Growth of *Prosopis glandulosa* in response to changes in aboveground and belowground interference. *Ecology* 78: 1222–1229.

Villaseñor, J. L., P. Dávila, and F. Chiang, 1992. Fitogeográfia del Valle de Tehuacán-Cuicatlán. In S. P. Darwin and A. L. Welden, eds., *Biogeography of Mesoamerica,* pp. 293–301. New Orleans: Tulane University.

Ware, S. A., C. Frost, and P. D. Doerr. 1993. Southern mixed hardwood forest: The former longleaf pine forest. In W. H. Martin, S. G. Boyce, and A. C. Echternacht, eds., *Biodiversity of the Southeastern United States, vol. 1: Lowland Terrestrial Communities,* pp. 447–493. New York: Wiley.

Weaver, J. E. 1954. *North American Prairie.* Lincoln, Neb.: Johnsen.

Weaver, P. L. 1989. Forest changes after hurricanes in Puerto Rico's Luquillo Mountains. *Intersciencia* 14: 181–192.

Weaver, P. L., and P. G. Murphy, 1990. Forest structure and productivity in Puerto Rico's Luquillo Mountains. *Biotropica* 22: 69–82.

West, N. E. 1988. Intermountain deserts, shrub steppes, and woodlands. In M. G. Barbour and D. W. Billings, eds., *North American Terrestrial Vegetation,* pp. 209–230. Cambridge: Cambridge University Press.

White, S. S. 1948. The vegetation and flora of the region of the Río Bravispe in northeastern Sonora, Mexico. *Lloydia* 11: 229–302.

Whitmore, T. C. 1984. *Tropical Rain Forests of the Far East.* 2nd ed. Oxford: Clarendon Press.

Wilbur, R. L. 1970. Taxonomic and nomenclatural observations on the eastern North American genus *Asimina* (Annonaceae). *J. Elisha Mitchell Scientific Society* 86: 88–96.

DAVID L. LENTZ

4 | ANTHROPOCENTRIC FOOD WEBS IN THE PRECOLUMBIAN AMERICAS

The greatest environmental impact brought about by ancient Americans revolved around their interaction with domesticated organisms, mostly plants, that were incorporated into highly interconnected trophic webs with humans as primary consumers. A trophic web is a set of populations whose interactions are intensely linked and act as a subunit of a larger community, with only loose connections to other subunits (Putnam 1994:40). The evolution of anthropocentric trophic webs had five profound effects on the Precolumbian landscape: (1) they greatly extended the range and enlarged populations of organisms within the human-centered trophic webs; (2) species, such as ruderals and vermin, that could adapt to environmental disturbances caused by the expansion of human settlements proliferated; (3) habitat for native species outside the webs was reduced; (4) plants and animals sought as prey species were affected, most commonly reduced in numbers; and (5) the efficient and flexible nature of cultivated organisms within the trophic webs allowed large-scale expansion of human populations into aggregated settlements in several different regions. Because of differences in available resources and variability in human groups, not all trophic webs throughout the Americas were the same. While some Native Americans remained totally dependent on wild foods until European contact, others incorporated domesticates into their food procurement strategies. Before attempting to discuss the overall impact of these developments, let us examine the effects of climate and how relationships, initially predator-prey interactions, between humans and other members of their trophic webs began.

CLIMATE

The climate in Central America at the end of the Pleistocene was cooler than it is today with less pronounced seasonality cycles (Byrne 1987). Modern or post-Pleistocene vegetation became established sometime between 11,000 and 6,000 years ago (Hunter et al. 1988; Pearsall 1995). However, there have been many prolonged wet periods and droughts since the end of the Pleistocene (see Hodell et al., this volume), and this caused a shifting in the ranges of most species, both north and south and up and down in elevation, since the beginning of the Holocene (Webb 1987).

As the Pleistocene drew to a close and the Pleistocene megafauna disappeared with it, perhaps as a result of human influences (Martin and Wright 1967; MacPhee and Marx 1997), an ameliorated climate and a different set of predator options faced the remaining and proliferating human populations at the outset of the Holocene. Gone were the large game animals that formed at least part of Paleoindian food intake, so there must have been some alteration of dietary habits. Fortunately for the early Holocene occupants, many of the plants adapted to the new climate were annuals, "*r* strategists" with a high intrinsic rate of increase (Jones 1997:158). This meant they would set seed at the end of a short growing season and produce copious numbers of seeds to insure reproductive success in an unpredictable environment. Geophytes, or plants with underground storage organs (e.g., tubers), also proliferated at the end of the Pleistocene and throughout the Holocene. They are the kind of plants that do well in a variable climate and thrive in disturbances caused by human occupation. Generally, reports of late Pleistocene–early Holocene sites do not include information about plant remains, but the ones that have been published tell us something of that early transition to a diversified hunting and gathering adaptation (table 4.1).

Archaeological sites from late Pleistocene–early Holocene times provide a set of randomly connected snapshots of the transition from hunting and gathering to early agriculture. What we can see from this filtered mosaic is a portent of subsequent patterning: exploitation of tree fruits began early and continued through the contact period, weedy annuals and geophytes were gathered (often from campsites and other areas of human disturbance), and some became the ancestors of domesticates. All the while, symbiotic microorganisms were developing vital interactions with many useful plants. One of the few large-seeded New World grasses, teosinte, was gathered as a wild food. Later, through a series of fortunate mutations, it was transformed into a diminutive form of maize, which was adopted and exploited by incipient mesoamerican horticulturalists. All told, the picture of the New World foragers that preceded agriculturalists is not a sharp one, but what does seem clear is that the transition was gradual, probably taking more than six or seven millennia. During that time, humans gained knowledge about useful organisms and developed much tighter connections with all components of their trophic webs.

TABLE 4.1. ARCHAEOLOGICAL EVIDENCE FOR EARLY PLANT USE

Site	Date	Plant remains	Author
Monte Verde, Chile	11,000 B.C.	wild potatoes (*Solanum* sp.)	Dillehay 1989 Ugent et al. 1987
Pedra Pintada, Brazil	9–8000 B.C.	palms (*Attalea* spp. and *Astrocaryum vulgare*), other trees (*Hymenaea* spp., *Sacoglottis guianensis, Bertholletia excelsa* and *Byrsonima crispa*)	Roosevelt et al. 1996
La Yeguada, Panama	9–8000 B.C.	burnt tree species	Piperno et al. 1990, 1991a, b
Hester, U.S.	9–6000 B.C.	hickory (*Carya* spp.), acorns (*Quercus* spp.), black walnuts (*Juglans nigra*), hackberry (*Celtis* sp.), and wild plums (*Prunus americana*)	Lentz 1986
Gainesville, U.S.	8500–8000 B.C.	acorns, hickory nuts, persimmon (*Diospyros virginiana*)	B. D. Smith 1986
Rodgers Shelter, U.S.	8500–8000 B.C.	hickory nut, black walnut, hackberry (*Celtis occidentalis*)	Wood and MacMillan 1976
San Isidro, Colombia	8000 B.C.	palms (*Acrocomia aculeata*), wild avocados (*Persea* sp.) and roots	Gnecco Valencia 1994 Piperno and Holst 1997
Peña Roja, Colombia	7200 B.C.	palm (*Oenocarpus* sp.)	Cavelier et al. 1995
Guilá Naquitz, Mexico	8–6000 B.C.	acorns, hackberry, mesquite (*Prosopis juliflora*), *Dalea*, prickly pear (*Opuntia* sp.), squash (*Cucurbita pepo*) in early levels. Bottle gourds (*Lagenaria* sp.), avocados (*Persea americana*), and common beans (*Phaseolus vulgaris*) in more recent levels	Flannery 1986 C. E. Smith 1986 B. D. Smith 1997a Whitaker and Cutler 1986
Zohapilco, Mexico	6000–2200 B.C.	goosefoot (*Chenopodium* sp.), *Amaranthus* sp., tomatillo (*Physalis* sp.), and teosinte (*Zea mays* ssp. *mexicana*)	Niederberger 1979
Koster, U.S.	5000–2900 B.C.	squash, marshelder (*Iva annua*), hickory, acorns, black walnuts, pecan (*Carya illinoensis*), goosefoot (*Chenopodium* spp.), grape (*Vitis* sp.), smartweed (*Polygonum* spp.), wild bean (*Strophostyles* sp.), hackberry, wild plum, pokeweed (*Phytolacca americana*), hawthorn (*Crataegus* sp.), green briar (*Smilax* sp.), *Viburnum* sp., Solomon's seal (*Polygonatum* sp.)	Conard et al. 1984 Asch et al. 1972
Ocampo Caves, Mexico	4500–1100 B.C.	bottle gourd, squash (*Cucurbita pepo*), avocado, acorns, common beans, runner beans, foxtail grass, peppers, sunflower (*Helianthus annuus*), amaranth, and maize	B. D. Smith 1997a MacNeish 1992
Tehuacán Valley, Mexico	3500 B.C.–A.D. 1500	foxtail grass (*Setaria* cf. *macrostachya*), peppers (*Capsicum annuum*), amaranthus, mesquite, avocado, ciruela (*Spondias mombin*), prickly pear, peanuts (*Arachis hypogaea*), jack beans (*Canavalia* sp.), runner beans (*Phaseolus coccineus*), sieva beans (*P. lunatus*), common beans, guava (*Psidium guajava*), cotton (*Gossypium hirsutum*), squash (*Cucurbita pepo, C. mixta, C. moschata*), and maize (*Zea mays*)	C. E. Smith 1967 Long et al. 1989

TROPHIC WEBS IN THE AMERICAS

As with most aspects of nature, humans are parts of networks of groups of organisms that are intertwined through multiple facets of their life cycles. Most of these networks, or trophic webs, have anthropogenic orientations as a result of several thousand years of both conscious and unconscious human manipulation. Humans have organized the funneling of major portions of the food web output for themselves. This is similar to what some authors have referred to as "agroecology," yet is more comprehensive in scope. Rindos (1984:122) defines agroecology as a system that includes humans, their domesticates, and associated ruderals. Defining the interaction in this way reflects Western cultural bias; humans grow food (keep out the weeds), eat it, and then get rid of the waste. Or in a more urban context, humans go to the supermarket, buy food (herbicides have already taken care of the weeds), consume it, throw out the trash, and flush the waste. This is not a cycle, but a unidirectional flow from autotrophs to heterotrophs. The cycle of a trophic web is more intricate because it includes large groups of other organisms that are closely and essentially involved with agroecological systems in addition to the autotrophs and heterotrophs that are generally the sole topics of discussion.

Figure 4.1 outlines some of the key elements in human-centered trophic webs. At the center are humans and their domesticated animals feeding off domesticated plants (autotrophs) as the core interaction of the agroecological system. Their remains and waste products are consumed by scavengers and detritivores (mostly bacteria, fungi, and protistans). These organisms are largely responsible for the breakdown and fragmentation of macromolecules (carbohydrates, proteins, fats, etc.) that ultimately release minerals and other nutrients into the soil. Lacking the dung of large domesticated animals (although the Andean peoples with their camelids were exceptions), many Native American groups were careful to recycle human waste. For example, the Aztecs at the time of the Conquest were observed saving their excrement for later use (Díaz 1963:233). Today, the Yucatec Maya intentionally defecate in their dooryard gardens (Hanks 1990:335) to fertilize them. Even though they were probably unaware of the details, Native Americans were aiding the cyclical flow of nutrients within their trophic webs.

Also in figure 4.1 are wild plants and animals that were always part of the Precolumbian resource base. Note that in close association with the autotrophs are mycorrhizal fungi and bacteria that have coevolved with their host plants. These organisms are essential components of the trophic web in all regions of Precolumbian America (and all other agricultural systems, for that matter). They live in close proximity and form a symbiotic relationship with the root systems of many of the most important crops, promoting the absorption of minerals and in some cases even manufacturing vital nutrients.

The *Rhizobium* spp., *Azorhizobium* spp., and *Bradyrhizobium* spp. symbionts (Giller and Wilson 1991:32) of beans and other legumes are well known for their ability to convert gaseous nitrogen into solid nitrates that can be absorbed by the host

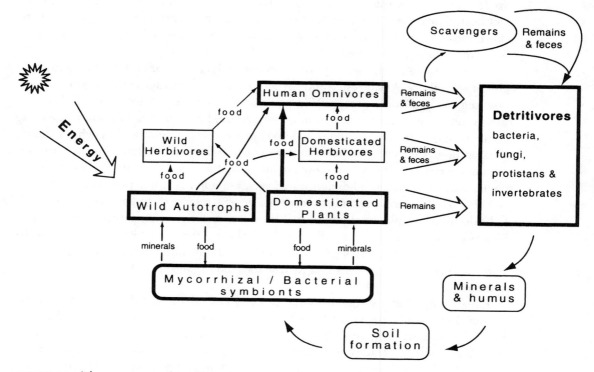

FIGURE 4.1. A human-centered trophic web.

plant (Harley and Smith 1983:369). Nitrogen-fixing bacteria have two dramatically important effects on the heterotrophic components of the web: they help to enrich the surrounding soil with nitrogen compounds, and they make nitrogen available to the leguminous host plants (which in turn convert it to compounds rich in nitrogen, e.g., amino acids and proteins). When consumed, legumes are highly nutritious dietary components and undoubtedly helped to sustain many bean-eating Precolumbians when no meat was available. As a component of the agricultural system, the nitrogen-fixing bacteria in leguminous root nodules helped to restore soils impoverished by crops that absorb nitrates but have no direct mechanism to return what they have consumed.

Mycorrhizal fungi work in a different way; they do not manufacture usable minerals, but through their extended hyphae and close proximity to living root tissues they greatly enhance the host plant's ability to absorb minerals that are in the soil, especially phosphorus (Harley and Smith 1983:38). Many domesticates form mycorrhizal associations (e.g., maize, tobacco, beans, potatoes, manioc, cacao, tomatoes, avocados, and cotton) and have stunted growth in the absence of appropriate symbionts (S. E. Smith and Read 1997). Modern reforestation efforts, plantation projects, and agricultural plantings are often accompanied by fungal inoculations to insure proper mycorrhizal development (Harley and Smith 1983:404). Even though mycorrhizal species are less host specific than nitrogen-fixing bacteria, and a range of mycorrhiza may infect host plants, there is a marked difference in symbiotic effectiveness (ibid.:378). In short, most crops that form symbiotic relationships with ground-dwelling microorganisms need the right symbionts to grow and mature effectively.

Understanding the agroecology of trophic webs is critical for understanding the origins of agriculture. As suggested above, there is much more to agricultural systems than just humans and domesticates. Wild plants, for example, did not simply fall out of use as food, even after the adoption of agriculture, but remained important resource components of all webs. In addition to asking how, where, why, and by whom plants became domesticated, it is important to consider how wild plants adapted to the human-centered trophic webs. How are humans, domesticates, and wild plants interconnected? What input did domesticated verses nondomesticated plants have on food webs? Did wild plants continue to serve as a genetic reserve through introgression and hybridization? As plant tending turned from a singular process, gathering, to a complex process, agriculture, how did this affect the question of who was involved with what aspects of plant tending? All of these questions address important aspects of human-centered trophic webs and their associated organisms, yet none can be easily answered.

Equally important as organisms that can be seen with the unaided eye are ones that cannot. Root symbionts might have had two effects on early agriculture: (1) they enhanced its productivity, and (2) they may well have influenced its rate of expansion, especially in terms of trade from one region to another. Many Native American groups, such as the Maya (Vogt 1993:56), traditionally intercropped their fields with maize and beans. Studies have shown that maize, beans, and squash grown in polyculture produce higher yields than monoculture stands of the same crops (Tuxill and Nabhan 1998). The nitrogen-fixing bacteria in the root nodules of beans improve usable nitrogen content of the soil, and mycorrhizal fungi in the roots of maize and beans enhance their ability to absorb phosphorus. Conversely, if a crop was transported to a region without its symbionts, as might have occurred when seeds were exchanged through long-distance trade, its productivity in the new region would have been reduced. This lowered productivity may have acted as a disincentive for the crop's adoption for general use. Perhaps this is why maize was not popular for a long time after its first introduction into eastern North America; possibly its symbionts were not available, and the newly introduced crop did not do well without them. In the case of down-the-line trade of cultigens to neighboring lands, it is more likely that the right symbionts would be in the area of introduction. Also, the shorter time span involved might have made it more likely that the spores of symbionts would be transported along with the seeds, thus insuring that the correct symbionts would be present in the new area.

INTEGRATING DATA AND THEORY: THE MESOAMERICAN EXAMPLE

Archaeological and paleoenvironmental studies are beginning to reveal some of the conditions that existed in the early Archaic that may have precipitated the incorporation of domesticates into the New World trophic webs. These include the availability of domesticable species with coevolved symbionts, technocultural preadaptations for agriculture, e.g., storage and food processing capability, and a climatic regime characterized by strong seasonality. As described above, many different wild plants were exploited for their high yield of nutrients and, in some cases, were processed using a variety of specialty tools (e.g., grinding and chopping stones). Surpluses may have been stored for later consumption in storage pits and through other kinds of storage technology. These are common features found at the earliest Tehuacán Valley sites (Byers 1967), Guilá Naquitz (Flannery 1986), and Tamaulipas (MacNeish 1958). The ability to store surplus food was an important step in the growing hegemony of the energy flow in trophic webs as it gave humans the ability to monopolize available nutrients and exclude competing organisms. Also, the innovation of storage facilities with sequestered food supplies created a buffer against the vagaries of an unpredictable yet periodically bountiful environment. Ford (1968) observed that the reduction of risk from periodic drought was a strong motivating factor in community planning among the Tewa in the North American Southwest. This reduced risk may have been accompanied by concomitant reduced mobility because quantities of stored food and grinding stones are not readily portable. Archaic hunters and gatherers need not have been completely sedentary, i.e., inhabiting one location throughout the year, to make the investment in storage facilities and heavy grinding stones worthwhile. The preponderance of archaeological data from the Tehuacán and Oaxaca valleys indicate the occupants were seasonally transhumant. Nevertheless, the presence of storage units clearly implies the intent to occupy sites for extended periods or at least to reoccupy them.

Increased sedentism most likely affected the population structure by increasing fecundity and reducing infant and child mortality. Kelly notes (1995:259) that in hunting and gathering groups, a highly variable diet as a result of seasonality as well as frequent aerobic activity and prolonged breastfeeding results in lowered fecundity. Also, infant and child mortality is high in hunting and gathering groups because mobility exposes children to higher risks of accidents, disease, and parasites. As such, mobility and group size are inversely correlated through a complex array of social, biological, and psychological mechanisms. All of these factors combine so that once storage and specialty tool innovations have been adopted, they increase the cost of residential mobility, making it more difficult to abandon whatever food production and storage capabilities have been developed. Neo-Darwinists (Bettinger 1991) might comment that these adaptations would lead to increased fitness, so the reproductive units adopting sedentary and stable food-obtaining mechanisms would have an advantage over competing groups without the adaptations. In this way, hunters and gatherers can become enmeshed in what systems theorists (e.g., Bertalanffy 1962; Miller 1965) refer to as deviation amplifying behavior. Postprocessualists (Hodder 1995) would argue that these innovations led to economic and cultural attachments to particular places. Indeed, neither perspective need be opposed; storage and sedentism certainly involved a host of physical, economic, and cultural changes, all of which were interrelated in varying ways for different groups of peoples.

The food storage habits of Archaic Americans probably helped them endure the uncertain nature of the climatic pattern, and these innovations may have been coupled with other options for coping with uncertainty. One option was to develop techniques that would enhance the productivity of desirable plants. This could have been done in many ways, short of a headlong plunge into agriculture. Steward (1934) observed Paiutes in Nevada, essentially hunters and gatherers, irrigating open fields supporting wild sunflowers (*Helianthus bolanderi*), goosefoot (*Chenopodium* spp.), sage (*Salvia columbariae*), wild rice (*Oryzopsis hymenoides*), and *Eleocharis* sp. that had not been planted or even cultivated. (Cultivation here means to turn the soil to promote growth of a specific plant, whether wild or domesticated.) Other groups, e.g., Native Californians, burned areas to promote growth of certain weedy species that would gain a competitive advantage over less useful plants following a fire (Bean and Lawton 1976: 30). Many ethnographic instances of the broadcasting of wild seeds have been recorded (Harlan 1975:23). This enhancement of desirable wild plants would have been another predomestication mechanism to expand the food supply. When coupled with burning, dissemination of wild seeds would have been an effective way of enhancing growth of useful food plants. Plants, especially ones with a plastic nature like weedy annuals, would soon have responded to this kind of activity so that characteristics (such as uniform seed set, larger seed size, tough rachises, etc.) more compatible with human activities would evolve rapidly and quite possibly without the knowing intervention of human manipulators.

Directed change began to occur when humans started disseminating plant propagules and, ultimately, selected plants with desired phenotypes, just as is practiced among the Tzotzil Maya, who set aside the largest and best ears of maize from each harvest for future planting (Vogt 1993:55). Accordingly, human decision-making would have played a huge role in this process, not only in terms of developing and selecting improved cultivars but also in terms of adopting previously developed cultigens from other areas. Some of the New World crops may have been domesticated only once and then disseminated from a single source. This is an example of the way human agency played a pivotal role: creativity was an essential component of the domestication process. Crop development would have led to improved productivity, would impact a sedentary group's ability to sustain larger numbers, and, ultimately, led to inclusion of domesticated plants and animals in the human-centered trophic web.

Even as Precolumbian American societies became more dependent on agriculture, they continued to exploit wild plants (mostly ruderals) and especially wild animals. For example, the Maya, among the most dependent on agriculture of New World peoples, relied upon wild plants as a component of their diet throughout the prehistoric period (Vogt 1993:51; Lentz 1991). Yucatec Maya commonly plant a field for two or three years then allow it to fallow for five to fifteen years (Hanks 1990:358–361). While land is left in fallow, it is not useless; weedy herbaceous species, pioneer shrubs, and trees move in and become established, making excellent sites for gathering wild foods and medicinal plants and hunting game that would come to forage on the verdant growth in a fallowed field. Many pioneer species are legumes (e.g., *Acacia* spp., *Gliricidia sepium,* and *Inga* spp.), and these have *Rhizobium* spp. symbionts in their root nodules, thus improving the nitrate content of the soil. Nitrogen fixation undoubtedly is a significant factor in the rejuvenation of fallow fields. Most pioneer species have mycorrhizal symbionts that help to incorporate minerals into the biomass of the fields, which, after burning, is available for use by crop plants. This pattern probably was repeated in various agricultural regions of the New World, often with different species involved, forming unique webs in different environmental settings.

MAJOR NEW WORLD DOMESTICATES

The best estimates indicate that the domestication process began in the New World at least by 8000 B.C. The development of crops did not happen in the same place at the same time. It is possible that domestication was an idea that developed in certain areas first and diffused into other areas, thus encouraging neighboring groups to experiment with local plants of apparent utility. Table 4.2 outlines a list of plants domesticated in the Americas and their probable wild ancestors. The approximate distribution of New World domesticates at the time of European contact and the natural distribution of their wild progenitors also is included in the table. Following is a discussion of a few of the more important American crops, all derived from weedy annuals or geophytes, beginning with the first domesticated.

SQUASHES

In what may have been the earliest New World domesticate, *Cucurbita pepo* appears to have been under domestication in the Oaxaca Valley at least by 8000 B.C. (B. Smith 1997a,b). Probably the plant was grown initially for its seeds rather than its flesh, which, in wild cucurbits, is quite bitter (Whitaker and Bemis 1975). *C. fraterna*, a wild gourd of northeastern Mexico, has been proposed as the wild progenitor of *C. pepo* (Nee 1990), but molecular studies have questioned this relationship (Decker-Walters et al. 1993). Other early domesticated Cucurbitaceae from archaeological contexts include *Lagenaria siceraria* (gourd) (4490 B.C.) and *C. argyrosperma* (3085 B.C.) from the Tamaulipas caves (B. Smith 1997a,b).

BEANS

Beans, both common and lima, were important components of the trophic web in most parts of the Precolumbian Americas where agriculture was practiced. Data generated by electrophoretic analysis of a seed storage protein, phaseolin, suggest beans may have originated in both Mesoamerica and South America. Apparently, the "S" phaseolin protein, common in most bean cultivars, seems to have originated in Mesoamerica, while "B," "A," "C," "H," and "T" phaseolin types have South American origins (Gepts 1990). The chronology of early bean remains from several archaeological sites has been reevaluated using accelerator mass spectrometry (AMS) technology (Kaplan and Lynch 1999) and the results give us fresh insights. The earliest common bean from the Andean Highlands at Guitarrero Cave, a site excavated in the 1970s (Lynch 1980), was dated at 4337 ± 55 B.P. The earliest bean remains, pods from Chilca Cave on the

TABLE 4.2. CROPS OF THE AMERICAS, THEIR WILD PROGENITORS, AND DISTRIBUTION AT TIME OF CONTACT

Domesticated Plants	Distribution at Contact	Wild Progenitors	Wild Distribution	Reference
Dicots				
Amaranthaceae				
Amaranthus cruentus L. amaranth	Mexico, Central America, tropical South America	*A. powellii* S. Wats?	Mexico, eastern and western United States	Towle 1961, Harlan 1995, Cole 1979, Pickersgill and Heiser 1977
A. caudatus L. amaranth	Andes: Peru–northwest Argentina	*A. quitensis* HBK.?	north Argentina, south-central Andes	Harlan 1995, Cole 1979
A. hypochondriacus L. amaranth	Mexico, southwest and eastern United States, Central America	*A. hybridus* L.?	temperate and tropical New World (Mexico?)	Harlan 1995, Cole 1979, Pickersgill and Heiser 1977
Anacardiaceae				
Anacardium occidentale L. cashew	Central America, lowland South America	wild *A. occidentale*	south Brazil	Mitchell and Mori 1987
Spondias purpurea L. jocote	Mexico, Central America, West Indies, North and South America	wild *S. purpurea*	Mexico, Central America	Williams 1981
Annonaceae				
Annona cherimola Mill. cherimoya	Peru	wild *A. cherimola*	Andean valleys of Peru, Ecuador	Purseglove 1974
A. muricata L. guanabana	Central America to coastal Peru	*A. montana* Macfad.	Costa Rica to northern South America, West Indies	Safford 1914, Purseglove 1974, Schatz 1987
Apiaceae				
Arracacia xanthorrhiza Bancr. arracacha	Andean Bolivia to Venezuela	?	Andean Peru? Bolivia?	Hermann 1992, Purseglove 1974
Asteraceae				
Helianthus annuus L. sunflower	southwest to eastern United States, Mexico, Central America	*H. annuus* ssp. *lenticalaris* (Dougl.) C. Kll. *H. annuus* ssp. *jaegeri* Heiser *H. annuus* ssp. *annuus* L.	western United States, Mexico	Heiser 1985, Rieseberg and Seiler 1990
Iva annua L. marshelder	southwestern and central United States	wild *I. annua*	central United States	Smith 1995, Gleason and Cronquist 1991
Basellaceae				
Ullucus tuberosus Calda. ssp. *tuberosus* Sperling ulluco	Andean Argentina to Venezuela	*U. tuberosus* ssp. *aborigineus* Sperling	Andean Peru to Argentina	King 1998
Bignoniaceae				
Crescentia cujete L. calabash	Mexico to northern South America	wild *C. cujete*	Mexico, Central America	Williams 1981

Taxon	Wild relative	Distribution	Wild distribution	Reference
Bixaceae				
Bixa orellana L. achiote	*B. excelsa* Gleason & Krukoff	Mexico to Amazonia	southwestern Amazon	Schultes 1984
Brassicaceae				
Lepidium meyenii Walp. maca	wild *L. meyenii*	central Peruvian Highlands	Peru, Bolivia, Argentina	King 1988
Cactaceae				
Opuntia spp. prickly pear	wild *Opuntia* spp.	Mexico, southwestern United States, Central America	Mexico, southwestern United States	Williams 1981, Ebeling 1986
Caricaceae				
Carica papaya L. papaya	*C. microcarpa* Jacq. *C. pubescens* Lenne & Koch. *C. goudotiana* (Tr. & Planch.) Solms	Central America to Amazonia, coastal Ecuador	northwestern South America	Prance 1984
Chenopodiaceae				
Chenopodium berlandieri ssp. *nuttaliae* (Safford) H. D. Wilson & C. B. Heiser goosefoot	*C. berlandieri* ssp. *berlandieri* Moq.	central Mexico, central United States	eastern and western United States to Mexico	Wilson 1990
C. quinoa Willd. quinoa	*C. hircinum* Schrad?	Andes, Colombia to northern Chile, northern Argentina	northern Argentina, Uruguay, southern Brazil	Wilson 1990
Convolvulaceae				
Ipomoea batatas (L.) Lam. sweet potato	*I. trifida* (HBK.) G. Don.	lowland Neotropics	Central America, northern South America ??	Austin 1978, Piperno and Pearsall 1998: 199, Nee 1990, Purseglove 1974
Cucurbitaceae				
Cucurbita argyrosperma Huber squash	*C. sororia* Bailey	southwestern United States to Guatemala	Pacific coast of Mexico to Nicaragua	Nee 1990
C. ficifolia Bouché squash	?	Andean South America, Mexico	Bolivia?	Nee 1990
C. maxima Duch. squash	*C. andreana* Naud	western slopes of the Andes	Uruguay and northern Argentina	Nee 1990
C. moschata (Lam.) Poir butternut	?	southwestern United States to northern South America	southern Central America? to Colombia?	Piperno and Pearsall 1998: 199, Nee 1990
C. pepo L. squash, pumpkin	*C. fraterna* Bailey	eastern United States, southwestern United States, Mexico	northeastern Mexico to southeastern United States?	Nee 1990
Lagenaria siceraria bottle gourd	?	North America, South America, Central America, Mexico, West Indies	southeastern Africa	J. D. Sauer 1993, Purseglove 1974
Sechium edule (Jacq.) Sw. chayote	*S. compositum* (Donn. Sm.) C. Jeffrey	Mexico, Guatemala, Honduras	Mexico, Guatemala	Newstrom 1991, Lentz 1991

(*continued*)

TABLE 4.2. (Continued)

Domesticated Plants	Distribution at Contact	Wild Progenitors	Wild Distribution	Reference
Erythroxylaceae				
Erythroxylum coca Lam. var. *ipadu* T. Plowman coca	Andean Ecuador to northern Argentina	*E. coca* Lam.	eastern Peru	Plowman 1984
E. novogranatense (Morris) Hieron. coca	coastal Peru to Costa Rica?	*E. coca* Lam.	eastern Peru	Plowman 1984, Schultes 1984
Euphorbiaceae				
Manihot esculenta Crantz. manioc	Central America, Mexico, Amazon Basin, Pacific coast South America	*M. esculenta* ssp. *flabellifolia* *M. tristis* Muell-Arg. *M. irwinii* Rogers	Guianas or Central Brazil	C. O. Sauer 1952, Stone 1984, Fregene et al. 1994
Lauraceae				
Persea americana Mill. avocado	Mexico to Peru	wild *P. americana*	Guatemala to central Mexico	B. D. Smith 1992, Schieber and Zentmeyer 1978
Leguminosae				
Arachis hypogaea L. peanut	South America lowlands, West Indies, midelevation Andes	*A. monticola* Krop. & Rio	northwestern Argentina	Stalker 1990, J. D. Sauer 1993, Piperno and Pearsall 1998
Canavalia ensiformis (L.) DC sword bean	Mexico, West Indies, Central America, coastal Peru and Ecuador	*C. brasiliensis* (Benth.) Mart.	Central America and West Indies	Purseglove 1974, Kaplan and Kaplan 1988
C. plagiosperma Piper. jack bean	coastal Ecuador and Peru	*C. piperi* Killip & Macbride	central Brazil, northeastern Bolivia, northwestern Argentina	Purseglove 1974, Piperno and Pearsall 1998, J. D. Sauer 1964
Indigofera suffruticosa Mill. añil	Mexico to South America, West Indies	wild *I. suffruticosa*?	Mexico, Central America	Dering 1895, Williams 1981
Pachyrhizus erosus (L.) Urban jícama	Mexico, tropical South America, Central America			Harlan 1995
Phaseolus acutifolius A. Gray tepary bean	southwestern United States, Mexico, Central America to northern Nicaragua	*P. acutifolius* var. *acutifolius* (Wild.)	northwestern Mexico, south-western United States	Pratt and Nabhan 1988
P. coccineus L. scarlet runner bean	Mexico	*P. coccineus* ssp. *formosus* (Kunth.) Maré, Masch. & Stain.	Mexico, Guatemala	Delgado et al. 1988
P. lunatus L. lima bean	Mexico to Andes and west coast South America	*P. lunatus* var. *silvester* Baudet	southern Mexico to central Argentina below 1,000 m	Baudoin 1988, Kaplan and Kaplan 1988

Species	Distribution	Wild ancestor/relative	Range	References
P. vulgaris L. common bean	Andes (Chile–Ecuador), Central America, Mexico, southwest and eastern United States	*P. vulgaris* var. *mexicanus* A. Delgado *P. aborigineus* Burk.	west-central Mexico to southern Andes	Gepts 1990, Kaplan and Kaplan 1988, Delgado et al. 1988, Brücher 1968
Malphigiaceae				
Byrsonima crassifolia (L.) DC nance	Mexico to northern South America	wild *B. crassifolia*	Mexico (to northern South America?)	Williams 1981
Malvaceae				
Gossypium barbadense L. cotton	Central America, West Indies, Andes, western South America, Amazonia	wild *G. barbadense*?	western South America	Percy and Wendell 1990, Purseglove 1974, Fryxell 1979
G. hirsutum L. cotton	Mexico, southwestern United States (?), Central America, West Indies, northeastern Brazil	wild *G. hirsutum* *G. thurberi* Tod.	Central America and South America	Purseglove 1974
Moraceae				
Brosimum alicastrum Swartz. breadnut, ramón	Mexico, Central America	wild *B. alicastrum*	Mexico	Purseglove 1974
Myrtaceae				
Psidium guajava L. guava	Mexico to Peru, West Indies	wild *P. guajava*	Panama, Neotropical	Popenoe 1948, Purseglove 1974
Oxalidaceae				
Oxalis tuberosa Mol. oca	Andean Chile to Venezuela	?	Andes above 2,500 m	King 1988
Rosaceae				
Prunus capuli Cav. Mexican cherry	Mexico, Central America	wild *P. capuli*	Mexico, western Guatemala?	Standley and Steyermark 1946
Sapotaceae				
Lucuma mammosa Gaertn. f. zapote	Mexico, Central America, West Indies, tropical South America	wild *L. mammosa*	southern Mexico to northern Nicaragua	Morton 1987
L. obovata Kuntze caimito	South America, West Indies	wild *L. obovata*	Colombia to Chile	Piperno and Pearsall 1998; Stanley and Steyermark 1946
Manilkara zapota (L.) van Royen sapodilla	Mexico, Central America, West Indies, lowland tropical South America	wild *M. zapota*?	southern Mexico, Central America	Morton 1987
Solanaceae				
Capsicum annuum L. chile, bell pepper	southern Mexico to Colombia, West Indies	*C. annuum* var. *glabrisculum* (Dun.) Heiser & Pickersgill	southern United States to northern South America	Andrews 1995, Pickersgill 1971, Eshbaugh et al. 1983, Heiser 1976

(*continued*)

TABLE 4.2. (Continued)

Domesticated Plants	Distribution at Contact	Wild Progenitors	Wild Distribution	Reference
C. baccatum L. ají pepper	southern Amazon to Argentina, South America Pacific coast	C. praetermissum Heiser & Smith	southern Brazil	Andrews 1995, Eshbaugh et al. 1983, McLeod et al. 1982, Pickersgill 1984
C. chinense Jacq. habañero pepper	Amazonia, West Indies, Central America, coast of South America: Ecuador to Chile	C. chacoense var. tomentosum A. T. Hunziker	Bolivia	McLeod et al. 1982, Andrews 1995, Eshbaugh et al. 1983
C. frutescens L. tabasco pepper	West Indies, Central America, Amazon, Pacific coast Ecuador to Chile	wild C. frutescens	western Amazon of Colombia and Peru	Andrews 1995, McLeod et al. 1982
C. pubescens Ruiz & Pav. rocoto pepper	midelevation (2,000–2,500 m) Andes	C. cardenasii Heiser & Smith	Bolivia, upper Amazon	Andrews 1995, Eshbaugh et al. 1983
Lycopersicon esculentum Mill. tomato	Mexico to South America	L. esculentum var. cerasiflora (Dun.) A. Gray	northwestern Peru (foothills of Andes)	Rick and Holle 1990
Nicotiana tabacum L. tobacco	eastern South America, Colombia, Central America, Mexico, West Indies	N. sylvestris Spregazzini & Comes N. tomentosiformis	northwestern Argentina, Bolivia	Heiser 1992, Goodspeed 1954, Pickersgill 1977
N. rustica L. tobacco	western–eastern United States, Mexico, eastern Canada, South America	N. rustica var. pavoni (Dunal) Goodsp.?	Andean South America	Goodspeed 1954, Hill 1952, Purseglove 1974
Solanum tuberosum L. potato	pan Andean	S. brevicaule Bitter	Andean Peru, Bolivia	Grun 1990, Van den Berg et al. 1998
Sterculiaceae				
Theobroma cacao L. cacao, chocolate	southern Mexico to coast of Ecuador, Amazonia	T. cacao ssp. sphaerocarpum (A. Chevalier) Schultes	upper Amazon, northeastern South America	Young 1994, Schultes 1984
Tropaeolaceae				
Tropaeolum tuberosum R. & P. mashua	Andean Argentina to Venezuela	T. tuberosum ssp. silvestre Sparre	Colombia to Argentina above 2,400 m	King 1988
Monocots				
Agavaceae				
Agave fourcroydes Lem. henequin	Mexico, Central America	wild A. fourcroydes	Mexico	Purseglove 1972
A. sisalana Perr. sisal	Mexico, Central America	wild A. sisalana	Mexico, Central America	Purseglove 1972
Araceae				
Xanthosoma sagittifolium (L.) Schott malanga	Central America, tropical South America, West Indies	?	Central America	Purseglove 1972, Ghosh et al. 1988

Species	Distribution	Wild ancestor	Distribution	References
Arecaceae				
Acrocomia aculeata (Jacq.) Lodd. ex Mart. coyol	Mexico to Colombia, Amazon	wild *A. aculeata*	Central America	Lentz 1990, Henderson et al. 1995
Attalea cohune Mart. cohune	Mexico to Colombia low elevations	*A. cohune?*	Mexico to Nicaragua	Henderson et al. 1995
A. speciosa Mart. ex Spreng. babassu	Panama to Bolivia below 1000 m, Brazil	*A. speciosa?*	southern Central America to northern South America	Henderson and Balick 1991, Henderson et al. 1995, Piperno and Pearsall 1998
Bactris gasipaes HBK. pejibaye palm	coastal Ecuador to Central America, Amazonia	*B. microcarpa* Huber	western Amazon	Prance 1984, Clement 1986
Chamaedorea tepejilote Liebm. pacaya	Mexico to Colombia	wild *C. tepejilote?*	Mexico to Colombia	Henderson et al. 1995, Williams 1981
Elaeis oleifera (Kunth) Cortés corozo	Honduras to Colombia, Amazon	*E. oleifera?*	Honduras to northern Colombia	Henderson et al. 1995, Balée 1989
Bromeliaceae				
Ananas comosus (L.) Merrill pineapple	central Mexico to Amazon, coastal Ecuador and Peru	*A. microstachys* Lindman	western Amazonia	Schultes 1984, Ducke 1946
Cannaceae				
Canna edulis Kerr. achira	Neotropics, subtropics	?	northern Andes	Gade 1966, C. O. Sauer 1952, Piperno and Pearsall 1998
Dioscoreaceae				
Dioscorea trifida L. yam	Central America, northern South America, West Indies	?	northern South America	Piperno and Pearsall 1998, Coursey 1967, Purseglove 1972
Marantaceae				
Calathea allouia Lindl. leren	Central America, northern South America, West Indies	?	Brazil, Venezuela and/or Guianas	Piperno and Pearsall 1998, Purseglove 1972
Maranta arundinacea L. arrow root	West Indies, Central America and South America	?	West Indies, northern South America, Central America?	Piperno and Pearsall 1998, Purseglove 1972, Sturdevant 1969
Orchidaceae				
Vanilla planifolia Andrews vanilla	Mexico, Central America	wild *V. planifolia*	Mexico, Central America, West Indies, northern South America	Dering 1895, Rain 1996, Masefield et al. 1969
Poaceae				
Setaria macrostachya HBK. foxtail millet	Mexico, West Indies, southwestern United States	wild *S. macrostachya*	Mexico, southwestern United States	Rominger 1962, Ebeling 1986
Zea mays L. maize	southwest to eastern United States, Mexico, Central America, South America	*Z. mays* ssp. *parviglumis* Iltis and Doebley	Rio Balsas Valley, Mexico	Doebley 1990

Pacific coast of Peru, were dated to 5616±57 B.P. (Kaplan and Lynch 1999). A tepary bean (*Phaseolus acutifolius*) and a common bean pod from the Tehuacán Valley date to approximately 2300 B.P. Based on these dates, the oldest archaeological bean remains from South America are far older than the earliest material from Mesoamerica. At face value, these archaeological data suggest that the first beans were domesticated in South America and then introduced to Mesoamerica sometime prior to 2,500 years ago. However, many questions remain unanswered about the distribution of phaseolins in wild Central and South American *Phaseolus* populations. Clearly, much more work needs to be done in the bean paleoethnobotanical arena.

As mentioned above, beans are key dietary and ecological components of most human-centered trophic webs. Enhancing their vegetative success, common beans form symbiotic relationships with *Rhizobium leguminosarum* and several species of vesicular-arbuscular mycorrhizae in the genus *Glomus* (Harley and Smith 1983:38). Without the input of legumes and their symbionts, American agroecological systems would have been far less productive.

MAIZE

Zea mays was the staple crop for most agricultural Precolumbian Americans from North America to temperate areas of South America. Studies based on cytological and molecular data reveal that maize was derived from wild populations of *Zea mays* ssp. *parviglumis* in south-central and western Mexico (Doebley et al. 1987, 1984; Iltis and Doebley 1984; Dorweiler et al. 1993). A central question is how this seemingly dissimilar, albeit large-seeded, wild grass evolved into the crop that became the mainstay of many New World agriculturalists. A complete answer to this question is still lacking, partly because we have so little paleoethnobotanical evidence relating to teosinte as a wild food or maize as an early cultigen. Evidence from the Zohapilco and Tehuacán Valley sites offers a few pieces to a puzzle that remains largely a blank slate. At Zohapilco, remains of amaranth, tomatillo, and teosinte were recovered from a setting suggesting that hunters and gatherers were collecting wild plant foods (Niederberger 1979). The inhabitants of the site were experimenting with at least two plants, amaranth and teosinte, that later developed into important domesticates.

Let us focus on teosinte for a moment; it was marginally useful but desirable as a wild food mostly because it has larger seeds than most other New World grasses. Each seed is contained within a hard fruitcase attached to a fragile spikelet, which releases the ripened fruitcase readily. The seed can be removed from the fruitcase by toasting and grinding or popping. It is a good bit of work for a small reward, but Archaic Mexicans found it worthwhile, probably as a minor dietary component. Tehuacán Valley sites, occupied somewhat later, provide evidence from earliest levels that inhabitants were essentially hunters and gatherers who relied on collected plant foods such as *Setaria*, amaranth, prickly pear (*Opuntia* spp.), avocado (*Persea* sp., probably wild), and mesquite (*Prosopis juliflora*) (Byers 1967). In subsequent levels, caches of *Setaria* seeds suggested that the inhabitants were cultivating this plant as well as planting domesticated peppers (*Capsicum* sp.), cotton (*Gossypium hirsutum*), squash (*Cucurbita* spp.), and agave (*Agave* sp.) (C. E. Smith 1967:232). No teosinte was recovered from the Tehuacán Valley sites. However, in a higher and more recent stratum, ca. 3500 B.C.

(Long et al. 1989), maize appears. The first maize at Tehuacán represents a fully domesticated cultigen, although it seems primitive in many aspects, i.e., the cobs are small, possess tiny kernels, and have long glumes (Benz and Iltis 1990). The major mutations giving rise to polystichous spikes and naked edible grains without hard fruitcases had already occurred by the time maize was introduced. It is likely that maize was domesticated elsewhere and introduced into the Tehuacán Valley sometime before or around 3500 B.C. By the time maize was introduced into the valley, the inhabitants were already involved in incipient horticulture. Maize seems to have been adopted quickly and became an important part of the subsistence strategy in subsequent periods. As this happened, Tehuacán farmers had larger and better maize cultivars, and as maize became more important in the diet through time, foods like *Setaria* and mesquite became less popular.

How did maize develop from the weedy grass with hard fruitcases to a cultigen that fed most Americans by the time Columbus arrived? There is only one way that maize came to be the cultigen that it is; the Archaic Mexicans, who were already familiar with agriculture, observed favorable phenotypes in teosinte that arose spontaneously through mutation, seized upon them and planted those seeds. The modifications they selected, larger spikes and naked kernels, would not have survived in the wild otherwise. To aid in its productivity, maize forms mycorrhizal associations with at least two species of fungi, *Glomus mosseae* and *Gigaspora margarita* (S. E. Smith and Read 1997:30).

QUINOA

Chenopodium quinoa was a major component of the Andean subsistence assemblage, especially at high elevations where it thrives in areas less frost-tolerant cultigens cannot. The earliest quinoa was reported from the Zaña site (6400–4000 B.C.) in northern Peru (Piperno and Pearsall 1998), although the antiquity of these seeds has been questioned (Rossen et al. 1996). Also, lines of botanical information indicate the cultigen originated in the southern Andes, then dispersed to the northern Andes and ultimately to coastal areas (Wilson 1990).

Other Archaic period evidence of Andean quinoa came from Lake Junin around 2500 B.C. (Hastorf 1993:114) and Pachamachay (Pearsall 1980). Holden (1994) described stomach contents from burials at the El Morro site in northern Chile that date from 2000 to 500 B.C. Numerous other prehistoric remains of this important cultigen have been found in Peru, Chile, and northern Argentina (Towle 1961). Possibly the association of this plant with humans was brought about during domestication of alpacas and llamas. Quinoa, one of the favorite foods of camelids, quite possibly began its association with humans when it volunteered in corrals from seeds transported in stomachs of llamas or alpacas (Kuznar 1993). Two subspecies of *Chenopodium berlandieri* were cultivated in Mexico (*nuttalliae*) and what is now the United States (*jonesianum*) and may have been domesticated independently (Wilson 1990).

POTATOES

Solanum tuberosum was, and still is, the principal staple crop of the mountainous regions and altiplano of the southern Andes. Evidence for the use of wild potatoes dates back to late Paleoindian–early Archaic times at Monte Verde (Ugent et al. 1987). The earliest domesticated potato remains, dated to 2000–1200 B.C., seem to come from the Casma Valley on the Pacific coast of Peru, a curiosity because they are highland plants (Ugent et al. 1982). Reports of potatoes from Tres Ventanas Cave, also on the coast of Peru, claim an early date of 8000 B.C., but these data have been challenged (B. Smith 1995).

Although the earliest archaeological potatoes come from the coast of Peru, botanical information suggests that the origin of the domesticate was in the mountain valleys of the Andes. The plant thrives above 3,350 m and proliferates exceptionally well in the puna zone of the Lake Titicaca region. Kitchen middens of early Andean hunter-gatherers could have provided the staging ground for the domestication of the potato (Ugent 1970), because the cold-hardy crop is ideally suited for this region of frequent frosts and short growing seasons. It was grown intensively during Tiwanaku times and even earlier through the use of irrigation and terracing. Potatoes are stored as *chuño*, or freeze-dried tubers, that are prepared by trampling moisture out of them and letting them dry in the sun followed by overnight freezing. Dehydrated potatoes store well and are an essential part of the Andean diet. They not only provide sustenance in lean times but also can be easily transported and have been traded for maize, pottery, fruits, and textiles with coastal inhabitants for centuries (Correll 1962). One reason for the success of potatoes are mycorrhizal symbionts (*Glomus* spp.) in the roots (S. E. Smith and Read 1997:30). But the potato, with its many varieties, is only one, albeit the most prominent, of Andean tuber crops. Other Andean root crops include oca (*Oxalis tuberosa*), ullucu (*Ullucus tuberosus*), and añu (*Tropaeolum tuberosum*).

MANIOC

Manihot esculenta, which probably originated in northeastern South America, was undoubtedly one of the most important crops in the Neotropical lowlands, yet it has been a paleoethnobotanically elusive cultigen. Literature regarding ancient manioc is clouded with many reports of poor archaeological documentation or questionable botanical identification. Problematic reporting of manioc has been decried by many authors (e.g., Heiser 1965; Cutler 1968; C. E. Smith 1968; DeBoer 1975; Ugent et al. 1986). One of the earliest reports of domesticated manioc comes from the Zaña Valley in northern Peru, where Piperno and Pearsall (1998) claim a preceramic date for manioc roots retrieved from below a house floor. As with the quinoa, however, AMS dates of associated materials indicate a historic or even modern origin of those plant remains (Rossen et al. 1996). Piperno and Holst (1997) also report "manioc-like" starch grains from grinding stones recovered from the Aguadulce rock shelter in a stratum that dates to 5000–4000 B.C. Another discovery of archaeological manioc, dating to as early as 1800 B.C., came from manioc starch granules found in the coastal Casma Valley of northwestern Peru (Ugent et al. 1986; Ugent 1994). Small stone chips, possibly from a manioc grater (2100–1600 B.C.) were found at Parmana in Venezuela (Roosevelt 1984). Similar concentrations of stone chips, also likely from a manioc grater, were found in middle Formative deposits (ca. 900 B.C.) at the Yarumela site in central Honduras (Lentz et al. 1997). Elsewhere in Central America, casts of manioc tubers were identified at the Cerén site (ca. 590 A.D.) in El Salvador (Lentz et al. 1996). Callen (1967) reports manioc remains from Santa Maria times (ca. 900–200 B.C.) at Coxcatlan Cave, Tehuacán Valley, Mexico. Towle (1961) cites many examples of dried manioc roots from Early Horizon coastal Peru (900–200 B.C.) to more recent times, as well as references to ancient pottery and textiles with clear depictions of manioc roots from the Nazca, Moche, Chimu, and other early Peruvian cultures.

Much discussion has been generated, both archaeologically and ethnographically, about the distinction between sweet and bitter manioc with reference to cyanide-containing properties of roots, but studies have shown cyanide contents are actually quite variable, do not correspond to any morphological characteristics, and probably are more a function of culturally associated belief systems than actual toxicity (Nye 1991). Furthermore, elaborate processing procedures are more related to the desired end product (manioc flour) than are required for purposes of detoxification (Nye 1991). Future archaeological discussions regarding ancient manioc should follow in light of this understanding.

TREE CROPS

Many tree fruits were of great importance to the Precolumbian inhabitants of both North and South America. In what is now the midwestern and eastern United States, naturally occurring nut and fruit trees, such as hickories (*Carya* spp.), chestnuts (*Castanea* spp.), persimmons (*Diospyros* spp.), beech (*Fagus americana*), wild plum (*Prunus americana*), hackberries (*Celtis* spp.), and oaks (*Quercus* spp.), were important food sources from early Holocene times until historic times, even after the introduction of maize and the adoption of other cultigens (C. E. Smith 1986; Meltzer and Smith 1986). Although these trees and other representatives of eastern forests may have been encouraged by not cutting them down or by clearing away competing species less useful to humans, none were really domesticated.

Notwithstanding considerable obstacles, e.g., seeds that do not breed true and characteristically long generation times, ancient Americans were successful at domesticating a number of excellent fruit-bearing trees. One of the most important, at least economically if not nutritionally, was the cacao or chocolate tree (*Theobroma cacao*). Although the plant originated in northern South America (Young 1994; Schultes 1984), some believe it was domesticated by the ancient Maya (e.g., Krickenberg 1946). Yet, seeds lose their germinating power after two weeks (León 1959) and do not travel well, so it seems hard to imagine that long-distance trade was the best explanation for the introduction of cacao in Mesoamerica. Alternatively, it has been proposed that cacao was transported as small seedlings from South America to Central America (Stone 1984). This explanation seems less parsimonious than a down-the-line movement of seeds from a plant initially cultivated in South America, probably for the delicious aril or pulp surrounding the seeds. In addition, with a down-the-line exchange of germ plasm, the crop would have been less likely to outdistance its mycorrhizal symbionts. However cacao arrived in Mesoamerica, the Maya certainly developed and improved the tree crop from South American seed stock and created new culinary dimensions, such as the exquisite beverage derived from the seeds.

Another domesticated tree, cashew, originated in southern Brazil (Mitchell and Mori 1987) yet somehow got transported up to Central America by middle Formative (900–400 B.C.) times (Miksicek 1991; Lentz et al. 1997). Other domesticated trees in Neotropical lowland areas included ciruelas (*Spondias* spp.), anonas (*Annona* spp.), papaya (*Carica papaya*), avocado (*Persea americana*), sapodilla (*Manilkara achras*), zapote (*Lucuma mammosa*), and guava (*Psidium guajava*). Palms were enormously important as well, especially coyol (*Acrocomia aculeata*), cohune (*Attalea cohune*), cocoyol (*Bactris major*), peach palm (*B. gasipaes*), pacaya (*Chamaedorea tepejilote*), corozo (*Elaeis oleifera*), and babassu (*Attalea speciosa*). Of these, probably only peach palm and pacaya were truly domesticated. The others were encouraged and cultivated, and many of them, such as coyol, had their seeds moved and thus their range enlarged by humans, but they remained essentially wild plants, fully capable of surviving on their own. Other trees in the wild-but-encouraged category include black zapote (*Diospyros ebenaster*), inga (*Inga* spp.), mesquite (*Prosopis* spp.), ciruela de fraile (*Bunchosia armeniaca*), nance (*Byrsonima crassifolia*), acerola (*Malpighia punicifolia*), and ramón (*Brosimum alicastrum*).

From the list of plants in table 4.2, we can observe a number of interesting aspects of plant-human interactions. Most obvious is the widespread dissemination of plant species useful to humans. Many species had small ranges initially that became greatly expanded by human action. The examples of cacao and cashew described above help illustrate an important point about Precolumbian trade interactions. Many archaeologists, who like to think of the groups they work on as arising sui generis without outside influences, tend to downplay the diffusion of artifacts, trade goods, and especially ideas in Precolumbian America. Dissemination of plant germ plasm, however, clearly shows humans were transporting seeds and knowledge of plants from one region to another and in some cases from very early times. This undoubtedly occurred in conjunction with the trade of other commodities; there is no evidence to suggest that Precolumbian traders specialized in unusual crop seeds only. Maize is an obvious example of a well-traveled crop that from its limited original range in western Mexico made its way as far north as New England and Canada and as far south as temperate South America in Argentina long before the first European ever set foot in the New World.

Another observation from table 4.2 is that there is much we do not know about a number of important domesticates. For some plants we can only guess as to the wild progenitors, and for others we have no idea. This lack of knowledge can be solved by focusing intensive collection in areas of suspected origins. Newly developed molecular techniques will help to sort out answers, if adequate plant material can be obtained from regions whose flora is poorly known. Moreover, archaeologists interested in the origins of agriculture should investigate sites in the home range of the ancestral populations to find early evidence of plant domestication.

WEEDS AND PIONEERS

If domesticated plants were so widely disseminated, what were the plants that were replaced by the spreading wave of human-centered trophic webs? For an indication of the geographic areas and vegetation types that were most heavily impacted, the reader should look to the essays on vegetation in this volume. Although there were many plants whose numbers were diminished by human influence, there were others that proliferated in the wake of human-centered trophic webs, and these are generally referred to in agronomy circles as weeds. There are a number of weedy genera that spread throughout the hemisphere, ready to move into a trash heap or abandoned agricultural plot when the opportunity arose. Most of these weeds share an *r* selection strategy with many of the first domesticates, and in fact, some are congeners with early domesticates. Native panamerican weedy genera include the numerous taxa found in table 4.3. For the most part these fast-growing herbaceous plants act as early colonizers in successional sequences, soon followed by shrubs and pioneer trees that may eventually shade out herbaceous invaders.

Pioneer trees and shrubs tend to grow quickly and reach maturity within a couple of decades. Examples of the woody pioneers in southeastern North America are listed in the first part of table 4.4 (Harper 1944; Radford et al. 1968), while woody pioneers

TABLE 4.3.   NATIVE PANAMERICAN WEEDY GENERA

Family	Common Name	Genera
Amaranthaceae	amaranth	*Amaranthus*
Asteraceae	sunflower	*Ambrosia*
		Aster
		Bidens
		Eupatorium
		Mikania
Chenopodiaceae	pigweed	*Chenopodium*
Commelinaceae	spiderwort	*Commelina*
Convolvulaceae	morning glory	*Ipomoea*
Cyperaceae	sedge	*Carex*
		Cyperus
		Eleocharis
		Fimbristylis
Euphorbiaceae	spurge	*Phylanthus*
Leguminosae	bean	*Cassia*
		Desmodium
		Crotalaria
Malvaceae	mallow	*Sida*
Onagraceae	evening primrose	*Ludwigia*
Passifloraceae	passion flower	*Passiflora*
Poaceae	grasses	*Andropogon*
		Cenchrus
		Panicum
		Paspalum
		Tripsacum
Phytolaccaceae	pokeweed	*Phytolacca*
Smilacaceae	catbrier	*Smilax*
Solanaceae	nightshade	*Physalis*
		Solanum

Source: This was prepared by cross-checking lists of weedy plants from North America (Harper 1944; Radford et al. 1968), Central America (Garcia et al. 1975; Lentz 1989), and South America (Myint 1994).

TABLE 4.4.   WOODY PIONEERS

Southeastern North America	Central America	Lowland South America
Aesculus pavia	Acacia cookii	Alchornea triplinervia
Alnus serrulata	A. pennatula	Aparisthmium cordatum
Catalpa bignonioides	Acrocomia aculeata	Astrocaryum huicungo
Celtis spp.	Ardisia revoluta	Bixa orellana
Cephalanthus occidentalis	Bactris major	Casearia macrophylla
Diospyros virginiana	Calliandra emarginata	C. ulmifolia
Ilex vomitoria	Cassia fistula	Cecropia sp.
Liquidambar styraciflua	C. grandis	Coccoloba sp.
Morus rubra	Cecropia peltata	Cordia sp.
Pinus echinata	Diospyros cuneata	Doliocarpus dentatus
P. taeda	Genipa caruto	Geonoma sp.
Platanus occidentalis	Gliricidia sepium	Inga quaternata
Populus deltoides	Guazuma ulmifolia	Isertia hypoleuca
Prunus angustifolia	Inga spp.	Jacaranda copaia
Rhamnus caroliniana	Iresine arbuscula	Miconia tomentosa
Rhexia spp.	Leucaena brachycarpa	Pharus virescens
Rhus glabra	Pinus caribaea	Pollalesta discolor
Sambucus canadensis	Saurauia villosa	Rinorea racemosa
Sassafras albidum	Solanum atitlanum	Sorocea hirtella
	Spondias mombin	Vismia angusta
	Thevetia ovata	
	Trema micrantha	

in Central America are listed in the second section (Lentz 1989), and pioneer woody plants from the Amazon region described in a study of forest regrowth (Denevan and Treacy 1987) can be found in the third section. A quick glance at each of the three sets of plants reveals some overlap in genera among the three regions, but overall the lists reflect the division between Holarctic and Neotropical floras (see Greller, this volume). Central America is the meeting ground between the two floras and so has representatives from both. One common genus in all the Americas is *Pinus,* part of the Holarctic flora, whose range extends from southeastern North America across Central America and even into the highlands of South America. It seems likely that fire climax pine savannas developed as a result of repeated anthropogenic burning in several areas (Perry 1991). All of the plants listed in this section are pioneer species that are adapted to the kinds of disturbances found in nature and created by humans. Although some may be viewed as intrusive weeds, most, as important trophic web components, are valuable to humans as sources of food, medicines, and construction materials; as hosts for nitrogen fixers; and as part of the gene pool for domesticates through periodic hybridization and introgression.

SUMMARY AND IMPLICATIONS

The process of creating human-centered trophic webs in the Americas, which included domesticated plants, animals, weedy species that adapted to human disturbance, and associated symbionts, began some time prior to 8000 B.C. To date, the earliest cultigen (*Cucurbita pepo*) positively identified as such was unearthed in the Oaxaca Valley. Other plants were domesticated in distant regions, but whether these were independent events or the result of idea diffusion is hard to say. Sometime before 500 B.C. in southern Mexico, domesticated maize, beans, and squash came together, probably along with roots and other crops, as an agriculturally and nutritionally complementary system that formed the economic foundation for mesoamerican civilizations. Trade networks exchanging bird feathers and other exotic commodities for turquoise in western Mexico carried mesoamerican germ plasm to southwestern North America sometime before 1500 B.C. (Riley et al. 1990), where vibrant cultural traditions developed along with crop varieties adapted to arid lands.

Maize did not arrive in the Eastern Woodlands of North America until some time around A.D. 175 (Chapman and Crites 1987). It might have entered the region from Mexico via the Southwest, or the Gulf coast, while others suggest a Central or South American introduction via the Caribbean (Riley et al. 1990). However it arrived in North America, it was not used extensively for many centuries. Before maize, eastern North American cultures were dependent on a set of locally domesticated or quasi-domesticated plants, i.e., goosefoot, knotweed, maygrass, little barley, marsh elder, sunflower, and squashes, along with other wild plant and animal foods. It has been asserted that locally domesticated goosefoot, marsh elder, squash, and sunflower represent original domestication events that apparently took place within a few centuries (2500–2000 B.C.) of each other (B. Smith 1995). In a recent discovery, early sunflower evidence was unearthed from pre-Olmec deposits in Tabasco, Mexico, a find that brings into question the North American domestication of sunflower. Maize cultivation, along with beans and squash, became important around A.D. 800, and Mississippian societies flourished.

In the Neotropical lowlands of Central and South America, domestication of manioc and other root crops began at least by 2000 B.C. and probably much earlier. Although some have described the Neotropical lowlands, particularly the Amazon region, as lacking in resource potential (Meggers 1971; Gross 1975; Ross 1978; Harris 1984), others (Lathrap 1973; Roosevelt 1998) have pointed to archaeological evidence that demonstrates the presence of substantial prehistoric settlements along the floodplains of the Amazon, which survived on manioc, maize, squashes, palms, fruit trees, and other plant food sources as well as the bounty of the forest and riverine fauna. Balée (1989, 1992, 1993) suggests the settlement pattern we now observe in the Amazon basin is actually a cultural artifact, and in the Precolumbian past, human populations were large enough to create "anthropogenic forests." Balée's assessment seems plausible for certain parts of Amazonia, such as the eastern várzea areas, where relatively large populations settled. In any case, the importance of the Neotropical lowlands in the development of domestication and cultural evolution in the Americas is gaining wider recognition (Piperno and Pearsall 1998).

In the Andes, agricultural economies, based largely on plant foods derived from quinoa, potatoes, and other root crops, as well as domesticated animals, viz., the llama, alpaca, and guinea pig, began around 3000–2000 B.C. (B. Smith 1995). Maize, beans, squash, and peppers, which grow less well at high altitudes, were more marginal contributors in the mountainous regions. However, Hastorf and Johannessen (1993) point out that even though maize may have been introduced to the region around 2000 B.C., it remained in a role of secondary crop until about A.D. 500, when the use of maize beer, or chicha, took on a role in ritualized feasting. From then until historic times the use of maize in ceremonial activities accounted for its major use in the region.

In summary, arrays of wild and domesticated plants with their microbiological symbionts and multicellular animals were incorporated into trophic webs throughout much of the temperate and tropical Americas. These webs supported stratified societies that modified their surroundings in a variety of ways by favoring useful species, through vegetation replacement, through introduction of domesticated animals, and in many cases through earth-moving activities that ultimately transformed the landscape of the Americas in distinctly human ways.

REFERENCES

Andrews, J. 1995. *Peppers*. Austin: University of Texas Press.

Asch, N. B., R. I. Ford, and D. L. Asch. 1972. *Paleoethnobotany of the Koster Site: The Archaic Horizons*. Reports of Investigations 24. Springfield: Illinois State Museum.

Austin, D. F. 1978. The *Ipomoea batatas* complex, 1: Taxonomy. *Bulletin of the Torrey Botanical Club* 105: 114–129.

Balée, W. 1989. The culture of Amazonian forests. In D. A. Posey and W. Balée, eds., *Resource Management in Amazonia*, pp. 1–21. Bronx: New York Botanical Garden.

———. 1992. People of the fallow: A historical ecology of foraging in lowland South America. In K. Redford and C. Padoch, eds., *Conservation of Neotropical Forests: Working from Traditional Resource Use*, pp. 35–57. New York: Columbia University Press.

———. 1993. Indigenous transformation of Amazonian forests: An example from Maranhâo, Brazil. *L'Homme* 33(2–4): 231–254.

Balick, M. 1979. Amazonian oil palms of promise: A survey. *Economic Botany* 33: 11–28.

Baudoin, J. P. 1988. Genetic resources, domestication, and evolution of lima bean, *Phaseolus lunatus*. In P. Gepts, ed., *Genetic Resources of Phaseolus Beans: Their Maintenance, Domestication, Evolution, and Utilization*, pp. 393–407. Dordrecht, Neth.: Kluwer.

Bean, L. J., and H. Lawton. 1976. Some explanations for the rise of cultural complexity in native California with comments on proto-agriculture and agriculture. In L. J. Bean and T. C. Blackburn, eds., *Native Californians: A Theoretical Retrospective*, pp. 19–48. Socorro, N.Mex.: Ballena Press.

Benz, B. F., and H. H. Iltis. 1990. Studies in archaeological maize, 1: The "wild" maize from San Marcos Cave reexamined. *American Antiquity* 55: 500–511.

Bertalanffy, L. von. 1962. General system theory: A critical review. *General Systems, Yearbook of the Society for General Systems Research* 7: 1–20.

Bettinger, R. L. 1991. *Hunter-Gatherers: Archeological and Evolutionary Theory*. New York: Plenum Press.

Brücher, H. 1968. Die Evolution der Gartenbohne *Phaseolus vulgaris* L. aus der sudamerikanische Wildbohne *Ph. aborigineus* Buch. *Angewandte Botanik* 42: 119–124.

Byers, D., ed. 1967. *The Prehistory of the Tehuacan Valley, vol. 1: Environment and*

Subsistence. Austin: University of Texas Press.

Byrne, R. 1987. Climatic change and the origins of agriculture. In L. Manzanilla, ed., *Studies in the Neolithic and Urban Revolutions: The V. Gordon Childe Colloquium, Mexico, 1986,* pp. 21–34. British Archaeological Reports International Series 349.

Callen, E. O. 1967. The first New World cereal. *American Antiquity* 32: 535–538.

Cavelier, I., C. Rodriguez, L. F. Herrera, G. Morcote, and S. Mora. 1995. No solo de caza vive el hombre: Ocupación del bosque Amazónico, holoceno temprano. In I. Cavelier and S. Mora, eds., *Ambito y Ocupaciones Tempranas de la América Tropical,* pp. 27–44. Santa Fé de Bogota: Instituto Colombiano de Antropología Fundación Erigaie.

Chapman, J., and G. Crites. 1987. Evidence for early maize (*Zea mays*) from the Icehouse Bottom Site, Tennessee. *American Antiquity* 52: 352–354.

Clement, A. B. 1986. Domestication of the Pejibaye palm (*Bactris gasipaes*): Past and present. In M. Balick, ed., *The Palm-Tree of Life: Biology, Utilization, and Conservation,* vol. 6 of *Advances in Economic Botany,* pp. 155–174. Bronx: New York Botanical Garden.

Cole, J. 1979. *Amaranth from the Past for the Future.* Emmaus, Penn.: Rodale.

Conard, N., D. L. Asch, N. B. Asch, D. Elmore, H. Gove, M. Rubin, J. A. Brown, M. D. Wiant, K. B. Farnsworth, and T. G. Cook. 1984. Accelerator radiocarbon dating evidence for prehistoric horticulture in Illinois. *Nature* 308: 443–446.

Correll, D. S. 1962. *The Potato and Its Wild Relatives.* Renner: Texas Research Foundation.

Coursey, D. G. 1967. *Yams.* London: Longman.

Cutler, H. 1968. Origins of agriculture in the Americas. *Latin American Research Review* 3: 3–22.

DeBoer, Warren. 1975. The archaeological evidence for manioc cultivation: A cautionary note. *American Antiquity* 40(4): 419–33.

Decker-Walters, D. S., T. W. Walter, C. W. Cowan, and B. D. Smith. 1993. Isozymic characterization of wild populations of *Cucurbita pepo. J. Ethnobiology* 13: 55–72.

Delgado, S., A. A. Bonet, and P. Gepts. 1988. The wild relative of *Phaseolus vulgaris* in Middle America. In P. Gepts, ed., *Genetic Resources of Phaseolus Beans: Their Maintenance, Domestication, Evolution, and Utilization,* pp. 153–184. Dordrecht, Neth.: Kluwer.

Denevan, W., and J. Treacy. 1987. Young managed fallows at Brillo Nuevo. In W. Denevan and C. Padoch, eds., *Swidden-Fallow Agroforestry in the Peruvian Amazon,* pp. 8–46. Bronx: New York Botanical Garden.

Dering, H. N. 1895. *Report on the Cultivation of Cacao, Vanilla, India-Rubber, Indigo, and Bananas in Mexico.* London: Harrison & Sons.

Díaz, B. 1963. *The Conquest of New Spain.* London: Penguin.

Dillehay, T. 1989. *Monteverde: A Late Pleistocene Settlement in Chile, 1.* Washington, D.C.: Smithsonian Institution Press.

Doebley, J. W. 1990. Molecular evidence and the evolution of maize. *Economic Botany* 44(3): 6–27.

Doebley, J. W., M. M. Goodman, and C. W. Stuber. 1984. Isoenzymatic variation in *Zea* (Gramineae). *Systematic Botany* 9: 203–218.

Doebley, J. W., W. Renfroe, and A. Blanton. 1987. Restriction site variation in the *Zea* chloroplast genome. *Genetics* 117: 139–147.

Dorweiler, J., A. Stec, J. Kermicle, and J. Doebley. 1993. Teosinte glume architecture 1: A genetic locus controlling a key step in maize evolution. *Science* 262: 233–235.

Ducke, A. 1946. Plantas de cultura precolombiana na Amazônia Brasileira: Notas sôbre as especies ou formas espontaneas que supostamente hes teriam dado origem. *Boletim Técnico do Instituto Agronômico de Norte* 8.

Ebeling, W. 1986. *Handbook of Indian Foods and Fibers of Arid America.* Berkeley: University of California Press.

Eshbaugh, W. H., S. L. Guttman, and M. J.

McLeod. 1983. The origin and evolution of domesticated *Capsicum* species. *J. Ethnobiology* 3: 49–5.

Flannery, K. V. 1986. *Guilá Naquitz: Archaic Foraging and Early Agriculture in Oaxaca, Mexico.* New York: Academic Press.

Ford, R. I. 1968. *An Ecological Analysis Involving the Population of San Juan Pueblo, New Mexico.* Ph.D. dissertation, Department of Anthropology, University of Michigan. Ann Arbor: University Microfilms.

Fregene, M. A., J. Vargas, J. Ikea, F. Angel, J. Tohme, R. A. Asiedu, M. O. Akoroda, and W. M. Roca. 1994. Variability of chloroplast DNA and nuclear ribosomal DNA in cassava (*Manihot esculenta* Crantz) and its wild relatives. *Theoretical and Applied Genetics* 89: 719–727.

Fryxell, P. A. 1979. *The Natural History of the Cotton Tribe.* College Station: Texas A&M University Press.

Gade, D. W. 1966. Achira, the edible canna: Its cultivation and use in the Peruvian Andes. *Economic Botany* 20: 407–415.

Garcia, J. G. L., B. MacBryde, A. R. Molina, and O. Herrera-MacBryde. 1975. *Prevalent Weeds of Central America.* Corvallis, Ore.: International Plant Protection Center.

Gepts, P. 1990. Biochemical evidence bearing on the domestication of *Phaseolus* (Fabaceae) beans. *Economic Botany* 44(3): 28–38.

Ghosh, S. P., T. Ramanujam, J. S. Jos, S. N. Moorthy, and R. G. Nair. 1988. *Tuber Crops.* New Delhi: Oxford and IBH Publishing.

Giller, K. E., and K. J. Wilson. 1991. *Nitrogen Fixation in Tropical Cropping Systems.* Wallingford, U.K.: CAB International.

Gleason, H., and W. Cronquist. 1991. *Manual of Vascular Plants of Northeastern United States and Adjacent Canada.* 2nd ed. Bronx: New York Botanical Garden.

Gnecco Valencia, C. 1994. *The Pleistocene-Holocene Boundary in the Northern Andes: An Archaeological Perspective.* Ph.D. dissertation, Department of Anthropology, Washington University, St. Louis, Mo. Ann Arbor, Mich.: University Microfilms.

Goodspeed, T. H. 1954. *The Genus Nicotiana.* Waltham, Mass.: Chronica Botanica.

Gross, D. 1975. Protein capture and cultural development in the Amazon Basin. *American Anthropologist* 77(3): 526–549.

Grun, P. 1990. The evolution of cultivated potatoes. *Economic Botany* 44(3): 39–55.

Hanks, W. F. 1990. *Referential Practice: Language and Lived Space among the Maya.* Chicago: University of Chicago Press.

Harlan, J. R. 1975. *Crops and Man.* Madison, Wisc.: American Society of Agronomy.

———. 1995. *The Living Fields.* Cambridge: Cambridge University Press.

Harley, J. L., and S. E. Smith. 1983. *Mycorrhizal Symbiosis.* New York: Academic Press.

Harper, R. 1944. *Preliminary Report on the Weeds of Alabama.* Wetumpka, Ala.: Wetumpka Printing.

Harris, M. 1984. Animal capture and Yanomamo warfare: Retrospect and new evidence. *J. Anthropological Research* 40(1): 183–201.

Hastorf, C. 1993. *Agriculture and the Onset of Political Inequality before the Inka.* Cambridge: Cambridge University Press.

Hastorf, C., and S. Johannessen. 1993. Pre-Hispanic political change and the role of maize in the Central Andes of Peru. *American Anthropologist* 95(1): 115–138.

Heiser, C. B. Jr. 1965. Cultivated plants and cultural diffusion in nuclear America. *American Anthropologist* 67: 930–949.

———. 1976. *The Sunflower.* Norman: University of Oklahoma Press.

———. 1985. Some botanical considerations of the early domesticated plants north of Mexico. In R. I. Ford, ed., *Prehistoric Food Production in North America,* pp. 57–72. Anthropological Papers 75. Ann Arbor: Museum of Anthropology, University of Michigan.

———. 1992. On possible sources of tobacco of prehistoric eastern North America. *Current Anthropology* 33: 54–56.

Henderson, A., and M. Balick. 1991. *Attalea crassispatha,* a rare and endemic Haitian palm. *Brittonia* 43: 189–194.

Henderson, A., G. Galeano, and R. Bernal. 1995. *Field Guide to the Palms of the Americas.* Princeton: Princeton University Press.

Hermann, M. 1992. *Andean Roots and Tu-*

bers: Research Priorities for a Neglected Food Resource. Lima: International Potato Center.

Hill, B. D. 1952. A new chronology of the Valdivia ceramic complex from the coastal zone of Guayas Province, Educador. Ñawpa Pacha 10(12): 1–32.

Hodder, I. 1995. Theory and Practice in Archaeology. London: Routledge.

Holden, T. G. 1994. Dietary evidence from the intestinal contents of ancient humans with particular reference to desiccated remains from northern Chile. In Jon Hather, ed., Tropical Archaeobotany, pp. 65–85. London: Routledge.

Hunter, M. L., G. L. Jacobson, and T. Webb III. 1988. Paleoecology and the course-filter approach to maintaining biological diversity. Conservation Biology 2: 375–385.

Iltis, H. H., and J. F. Doebley. 1984. Zea: A biosystematical odyssey. In W. F. Grant, ed., Plant Biosystematics, pp. 587–616. New York: Academic Press.

Jones, A. M. 1997. Environmental Biology. London: Routledge.

Kaplan, L., and L. Kaplan. 1988. Phaseolus in archaeology. In P. Gepts, ed., Genetic Resources of Phaseolus Beans, pp. 125–142. Dordrecht, Neth.: Kluwer.

Kaplan, L., and T. F. Lynch. 1999. Phaseolus (Fabaceae) in archaeology: AMS radiocarbon dates and their significance for pre-Columbian agriculture. Economic Botany 53: 261–272.

Kelly, R. L. 1995. The Foraging Spectrum. Washington, D.C.: Smithsonian Institution Press.

King, S. 1988. Economic Botany of the Andean Tuber Crop Complex: Lepidium meyenii, Oxalis tuberosa, Tropaeolum tuberosum, and Ullucus tuberosus. Doctoral dissertation, City University of New York.

Krickenberg, W. 1946. Etnología de América. Mexico City: Fondo de Cultura Económica.

Kuznar, L. 1993. Mutualism between Chenopodium, herd animals, and herders in the south central Andes. Mountain Research and Development 13(3): 257–265.

Lathrap, D. 1973. The "hunting" economy of the tropical forest zone of South America: An attempt at historical perspective. In D. R. Gross, ed., People and Cultures of Native South America, pp. 349–352. New York: Doubleday.

Lentz, D. L. 1986. Archaeobotanical remains from the Hester site: The late Paleo-Indian and early archaic horizons. Midcontinental J. Archaeology 11(2): 270–279.

———. 1989. Contemporary plant communities in the El Cajon region. In K. G. Hirth, G. Lara Pinto, and G. Hasemann, eds., Archaeological Research in the El Cajon Region 1: Prehistoric Cultural Ecology, pp. 59–94. University of Pittsburgh Memoirs in Latin American Archaeology 1.

———. 1990. Acrocomia mexicana: Palm of the ancient Mesoamericans. J. Ethnobiology 10: 183–194.

———. 1991. Maya diets of the rich and poor: Paleoethnobotanical evidence from Copan. Latin American Antiquity 2(3): 269–287.

Lentz, D. L., M. P. Beaudry-Corbett, M. L. Reyna de Aguilar, and L. Kaplan. 1996. Foodstuff, forests, fields, and shelter: A paleoethnobotanical analysis of vessel contents from the Cerén site, El Salvador. Latin American Antiquity 7(3): 247–262.

Lentz, D. L., C. R. Ramirez, and B. W. Griscom. 1997. Formative period subsistence and forest product extraction at the Yarumela site, Honduras. Ancient Mesoamerica 8: 63–74.

León, J. 1959. Origen del cultivo del cacao. Actas del 33 Congresso International de Americanistas 1: 251–258.

Long, A., B. F. Benz, D. J. Donahue, A. J. T. Jull, L. J. Toolin. 1989. First direct AMS dates on early maize from Tehuacán, Mexico. Radiocarbon 31(3): 1035–1040.

Lynch, T., ed. 1980. Guitarrero Cave: Early Man in the Andes. New York: Academic Press.

MacNeish, R. S. 1958. Preliminary Archaeological Investigations in the Sierra de Tamaulipas, Mexico. Transactions of the American Philosophical Society 48(6). Philadelphia.

———. 1992. The Origins of Agriculture and Settled Life. Norman: University of Oklahoma Press.

MacPhee, R. D. E., and P. A. Marx. 1997.

The 40,000-year plague: Humans, hyperdisease, and first-contact extinctions. In S. Goodman and B. Patterson, eds., Natural Change and Human Impact in Madagascar, pp. 169–217. Washington, D.C.: Smithsonian Institution Press.

Martin, P. S., and H. E. Wright, eds. 1967. Pleistocene Extinctions: The Search for a Cause. New Haven, Conn.: Yale University Press.

Masefield, G. B., M. Wallis, S. G. Harrison, and B. E. Nicholson. 1969. Oxford Book of Food Plants. London: Oxford University Press.

McLeod, M. J., S. I. Guttman, and W. H. Eshbaugh. 1982. Early evolution of the chili peppers (Capsicum). Economic Botany 36: 361–368.

Meggers, B. 1971. Amazonia: Man and Culture in a Counterfeit Paradise. Chicago: Aldine.

Meltzer D., and B. Smith. 1986. Paleoindian and early archaic subsistence strategies in eastern North America. In S. W. Neusius, ed., Foraging, Collection, and Harvesting: Archaic Period Subsistence and Settlement in the Eastern Woodlands, pp. 3–31. Carbondale: Southern Illinois University.

Miksicek, C. 1991. The ecology and economy of Cuello. In N. Hammond, ed., An Early Maya Community in Belize, pp. 70–84. Cambridge: Harvard University Press.

Miller, J. G. 1965. Living systems: Basic concepts. Behavioral Science 10: 193–237.

Mitchell, J., and S. Mori. 1987. The Cashew and Its Relatives (Anacardium: Anacardiaceae). Bronx: New York Botanical Garden.

Morton, J. F. 1987. Fruits of Warm Climates. Winterville, N.C.: Creative Systems.

Myint, A. 1994. Common Weeds of Guyana. Georgetown, Guyana: National Agricultural Research Institute.

Nee, M. 1990. The domestication of Cucurbita (Cucurbitaceae). Economic Botany 44(3): 56–68.

Newstrom, L. 1991. Evidence for the origin of chayote, Sechium edule (Cucurbitaceae). Economic Botany 45(3): 410–428.

Niederberger, C. 1979. Early sedentary economy in the Basin of Mexico. Science 203: 131–142.

Nye, M. 1991. The mis-measure of manioc (Manihot esculenta, Euphorbiaceae). Economic Botany 45(1): 47–57.

Pearsall, D. M. 1980. Pachamachay ethnobotanical report: Plant utilization of a hunting base camp. In J. Rick, ed., Prehistoric Hunters of the High Andes, pp. 191–232. New York: Academic Press.

———. 1995. Domestication and agriculture in the New World tropics. In T. D. Price and A. B. Gebauer, eds., Last Hunters, First Farmers, pp. 157–192. Santa Fe, N. Mex.: School of American Research Press.

Percy, R. G., and J. F. Wendell. 1990. Allozyme evidence for the origin and diversification of Gossypium barbadense L. Theoretical and Applied Genetics 79: 529–542.

Perry, J. P. Jr. 1991. The Pines of Mexico and Central America. Portland, Ore.: Timber Press.

Pickersgill, B. 1971. Relationships between weedy and cultivated forms in some species of chili peppers (genus Capsicum). Evolution 25: 683–691.

———. 1977. Taxonomy and the origin and evolution of cultivated plants of the New World. Nature 268: 591–595.

Pickersgill, B., and C. B. Heiser. 1977. Origins and distribution of plants domesticated in the New World tropics. In C. A. Reed, ed., Origins of Agriculture, pp. 803–835. The Hague: Mouton.

———. 1984. Migrations of chili peppers, Capsicum spp., in the Americas. In D. Stone, ed., Pre-Columbian Plant Migration, pp. 105–123. Papers of the Peabody Museum of Archaeology and Ethnology. Cambridge: Harvard University Press.

Piperno, D., M. B. Bush, and P. A. Colinvaux. 1990. Paleoenvironments and human occupation in late-glacial Panama. Quaternary Research 33: 108–116.

———. 1991a. Paleoecological perspectives on human adaptation in Central Panama, 1: The Pleistocene. Geoarchaeology 6: 210–226.

———. 1991b. Paleoecological perspectives on human adaptation in Central Panama, 2: The Holocene. Geoarchaeology 6: 227–250.

Piperno, D., and I. Holst. 1997. The presence of starch grains on prehistoric stone tools from the humid tropics: Indications of early tuber use and agriculture in Panama. *J. Archaeological Science* 25: 765–776.

Piperno, D., and D. M. Pearsall. 1998. *The Origins of Agriculture in the Lowland Neotropics.* San Diego, Calif.: Academic Press.

Plowman, T. 1984. The origin, evolution, and diffusion of coca, *Erythroxylum* spp., in South and Central America. In D. Stone, ed., *Pre-Columbian Plant Migration,* pp. 125–164. Papers of the Peabody Museum of Archaeology and Ethnology. Cambridge: Harvard University Press.

Popenoe, W. 1948. *Manual of Tropical and Subtropical Fruits.* New York: Hafner Press.

Prance, G. T. 1984. The pejibaye, *Guilielma gasipaes* (HBK.) Bailey, and the papaya, *Carica papaya* L. In D. Stone, ed., *Pre-Columbian Plant Migration,* pp. 85–104. Papers of the Peabody Museum of Archaeology and Ethnology. Cambridge: Harvard University Press.

Pratt, R., and G. P. Nabhan. 1988. Evolution and diversity of *Phaseolus acutifolius* genetic resource. In P. Gepts, ed., *Genetic Resources of Phaseolus Beans,* pp. 409–440. Dordrecht, Neth.: Kluwer.

Purseglove, J. W. 1972. *Tropical Crops: Monocotyledons,* vols. 1 and 2. London: Longman.

———. 1974. *Tropical Crops: Dicotyledons.* New York: Wiley.

Putnam, R. J. 1994. *Community Ecology.* New York: Chapman & Hall.

Radford, A. E., H. E. Ahles, and C. R. Bell. 1968. *Manual of the Vascular Flora of the Carolinas.* Chapel Hill: University of North Carolina Press.

Rain, P. 1996. Vanilla: Nectar of the gods. In N. Foster and L. S. Cordell, eds., *Chiles to Chocolate: Food the Americas Gave the World.* Tucson: University of Arizona Press.

Rick, C. M., and M. Holle. 1990. Andean *Lycopersicon esculentum* var. *cerasiforme*: Genetic variation and its evolutionary significance. *Economic Botany* 44(3): 69–78.

Rieseberg, L. H., and G. J. Seiler. 1990. Molecular evidence and the origin and development of the domesticated sunflower (*Helianthus annuus,* Asteraceae). *Economic Botany* 44(3): 79–91.

Riley, T., R. Edging, and J. Rossen. 1990. Cultigens in prehistoric eastern North America. *Current Anthropology* 31(5): 525–535.

Rindos, D. 1984. *The Origins of Agriculture: An Evolutionary Perspective.* New York: Academic Press.

Rominger, J. M. 1962. *Taxonomy of Setaria (Gramineae) in North America.* Urbana: University of Illinois Press.

Roosevelt, A. C. 1984. Problems interpreting the diffusion of cultivated plants. In D. Stone, ed., *Pre-Columbian Plant Migration,* pp. 1–18. Papers of the Peabody Museum of Archaeology and Ethnology. Cambridge: Harvard University Press.

———. 1998. Ancient and modern hunter-gatherers of lowland South-America: An evolutionary problem. In W. Balée, ed., *Advances in Historical Ecology,* pp. 190–212. New York: Columbia University Press.

Roosevelt, A. C., M. L. da Costa, C. L. Machado, M. Michab, N. Mercier, H. Valladas, J. Feathers, W. Barnett, M. I. da Silveira, A. Henderson, J. Silva, B. Chernoff, D. S. Reese, J. A. Holman, N. Toth, and K. Schick. 1996. Paleoindian cave dwellers in the Amazon: The peopling of the Americas. *Science* 272: 373–384.

Ross, E. 1978. Food taboos, diet, and hunting strategy: The adaptation to animals in Amazonian cultural ecology. *Current Anthropology* 19(1): 1–36.

Rossen, J., T. D. Dillehay, and D. Ugent. 1996. Ancient cultigens or modern intrusions?: Evaluating plant remains in an Andean case study. *J. Archaeological Science* 23: 391–407.

Safford, W. E. 1914. *Classification of the Genus Annona with Descriptions of New and Imperfectly Known Species.* Washington, D.C.: GPO.

Sauer, C. O. 1952. *Agricultural Origins and Dispersals.* Bowman Memorial Lecture 5, Series 2. New York: American Geographical Society.

Sauer, J. D. 1964. Revision of *Canavalia. Brittonia* 16: 108–181.

———. 1993. *Historical Geography of Crop Plants: A Select Roster.* London: CRC Press.

Schatz, G. E. 1987. *Systematic and Ecological Studies of Central American Annonaceae.* Ann Arbor: University Microfilms.

Schieber, E., and G. A. Zentmeyer. 1978. Exploring for *Persea* in Latin America. *California Avocado Society Yearbook* 62.

Schultes, R. E. 1984. Amazonian cultigens and their northward and westward migrations in pre-Columbian times. In D. Stone, ed., *Pre-Columbian Plant Migration,* pp. 19–38. Papers of the Peabody Museum of Archaeology and Ethnology. Cambridge: Harvard University Press.

Smith, B. D. 1986. The archaeology of the southeastern United States: From Dalton to De Soto. In F. Wendorf and A. E. Close, eds., *Advances in World Archaeology 5,* pp. 1–92. New York: Academic Press.

———. 1992. *Rivers of Change.* Washington, D.C.: Smithsonian Institution Press.

———. 1995. The origins of agriculture in the Americas. *Evolutionary Anthropology* 3: 174–184.

———. 1997a. The initial domestication of *Cucurbita pepo* in the Americas 10,000 years ago. *Science* 276: 932–934.

———. 1997b. Reconsidering the Ocampo caves and the era of incipient cultivation in Mesoamerica. *Latin American Antiquity* 8(4): 342–383.

Smith, C. E. Jr. 1967. Plant remains. In D. S. Byers, ed., *Prehistory of the Tehuacán Valley, 1: Environment and Subsistence,* pp. 220–555. Austin: University of Texas Press.

———. 1968. The New World centers of origin of cultivated plants and the archaeological evidence. *Economic Botany* 22: 253–266.

———. 1986. Preceramic plant remains from Guilá Naquitz. In K. Flannery, ed., *Guilá Naquitz: Archaic Foraging and Early Agriculture in Archaic Mexico,* pp. 265–274. Orlando, Fla.: Academic Press.

Smith, S. E., and D. J. Read. 1997. *Mycorrhizal Symbiosis.* San Diego, Calif.: Academic Press.

Stalker, H. T. 1990. A morphological appraisal of wild species in section *Arachis* of peanuts. *Peanut Science* 17: 117–122.

Standley, P. C., and J. A. Steyermark. 1946. Annonaceae, flora of Guatemala. *Fieldiana, Botany* 24(4): 270–294.

Steward, J. 1934. Ethnography of the Owens Valley Paiute. *University of California Publ. Amer. Archaeol. Ethnol.* 33: 233–340.

Stone, D., ed. 1984. *Pre-Columbian Plant Migration.* Papers of the Peabody Museum of Archaeology and Ethnology. Cambridge: Harvard University Press.

Sturdevant, W. C. 1969. History and ethnography of some West Indian starches. In P. J. Ucko and G. W. Dimbleby, eds., *The Domestication and Exploitation of Plants and Animals,* pp. 177–199. Chicago: Aldine.

Towle, M. A. 1961. *The Ethnobotany of Pre-Columbian Peru.* Viking Fund Publication 30. Chicago: Aldine.

Tuxill, J., and G. P. Nabhan. 1998. *Plants and Protected Areas: A Guide to In Situ Management.* Kew, U.K.: Royal Botanic Gardens.

Ugent, D. 1970. The potato. *Science* 170: 1161–1166.

———. 1994. Chemosystematics in archaeology: A preliminary study of the use of chromatography and spectrophotometry in the identification of four prehistoric root crop species from the desert coast of Peru. In J. Hather, ed., *Tropical Archaeobotany,* pp. 215–226. London: Routledge.

Ugent, D., S. Pozorski, and T. Pozorski. 1982. Archaeological potato tuber remains from the Casma Valley of Peru. *Economic Botany* 36: 182–192.

———. 1986. Archaeological manioc (*manihot*) from coastal Peru. *Economic Botany* 40(1): 78–102.

Ugent, D., Thomas Dillehay, and C. Ramirez. 1987. Potato remains from a Late Pleistocene settlement in south-central Chile. *Economic Botany* 41: 17–27.

Van den Berg, R. G., J. T. Miller, M. L. Ugarte, J. P. Kardolus, J. Villand, J. Nienhuis, and D. M. Spooner. 1998. Collapse of morphological species in the wild potato *Solanum brevicaule* complex (Solanaceae: sect. *Petota*). *American J. Botany* 85(1): 92–109.

Vogt, E. Z. 1993. *Tortillas for the Gods: A Symbolic Analysis of Zinacanteco Rituals.* Norman: University of Oklahoma Press.

Webb, T. III. 1987. The appearance and disappearance of major vegetational assemblages: Long-term vegetational dynamics in eastern North America. *Vegetatio* 69: 177–187.

Whitaker, T. W., and W. P. Bemis. 1975. Origin and evolution of the cultivated *Cucurbita*. *Bulletin of the Torrey Botanical Club* 102: 362–368.

Whitaker, T. W., and H. C. Cutler. 1986. Cucurbits from preceramic levels at Guilá Naquitz. In K. Flannery, ed., *Guilá Naquitz: Archaic Foraging and Early Agriculture in Oaxaca, Mexico*, pp. 275–280. New York: Academic Press.

Williams, L. O. 1981. *The Useful Plants of Central America*. Tegucigalpa, Honduras: Escuela Agrícola Panamericana.

Wilson, H. D. 1990. *Quinoa* and relatives (*Chenopodium* sect. *Chenopodium* subsect. *Cellulata*). *Economic Botany* 44(3): 92–110.

Wood, W. R., and R. B. MacMillan. 1976. *Prehistoric Man and His Environments*. New York: Academic Press.

Young, A. M. 1994. *The Chocolate Tree: A Natural History of Cacao*. Washington, D.C.: Smithsonian Institution Press.

EMILY MCCLUNG DE TAPIA

5 | PREHISPANIC AGRICULTURAL SYSTEMS IN THE BASIN OF MEXICO

Agriculture was the fundamental component of the support systems that sustained prehistoric urban communities in the Basin of Mexico, from the Late Formative center of Cuicuilco through to the end of Late Postclassic Tenochtitlán (table 5.1). However, most of the available evidence regarding specific agricultural practices comes from sixteenth-century ethnohistorical and historical documents and ethnographic descriptions of traditional practices in different regions of Mesoamerica. Although some continuity between traditional practices and early Colonial period practices—roughly four centuries ago—may exist, it cannot simply be assumed that such continuity stretches back as far as would be necessary to develop a vision of practices in effect at the beginning of our era or even during the last millennium. On the contrary, it must be demonstrated on the basis of sound evidence, which is, unfortunately, a difficult task in a region that has been subjected to 3,000 years of constant human activity—particularly intense during the past century.

There would seem to be little hope for detecting direct evidence of early agricultural activities given the destructive expansion—in archaeological and landscape terms—of the urban zone of modern Mexico City and adjacent suburbs. However, the incorporation of techniques developed in other areas of scientific research together with rigorous archaeological method has enabled investigators to uncover significant evidence for early land use and alteration, thus enabling us to begin to provide answers to some of the questions about food production and distribution that have been asked for many years. Although recent

TABLE 5.1. ARCHAEOLOGICAL
CHRONOLOGY FOR THE BASIN
OF MEXICO

Date	Period
A.D. 1520	
	Late Postclassic
A.D. 1350	
	Middle Postclassic
A.D. 1150	
	Early Postclassic
A.D. 950	
	Epiclassic
A.D. 750	
	Classic
A.D. 150	
	Terminal Formative
250 B.C.	
	Late Formative
600 B.C.	
	Middle Formative
900 B.C.	
	Early Formative
1200 B.C.	

Source: Modified from Parsons 1991:26.

evidence is by no means complete, it suggests important directions for immediate and future research.

The specific aim of this paper is to summarize previous assumptions and debates about prehistoric agriculture in the Basin of Mexico and to discuss some of the more recent evidence that contributes to a better understanding of the problems involved. Rather than emphasize indirect approaches to reconstructing Prehispanic practices such as the use of ethnographic models and the extent to which they have fostered meaningful research, attention is turned to the direct evidence available for agricultural systems. Most of what is known during Prehispanic times comes from the results of archaeological investigation in the Teotihuacán Valley in the semiarid northeastern sector, although other regions such as Cuauhtitlan (Nichols 1980) and, recently, Cuicuilco (Palerm 1961; Pastrana and Fournier 1997) have contributed some information as well.

Recent evidence pertaining to the Aztec period provides some interesting examples because several very different areas insofar as environmental conditions are represented: Xaltocan, in the north-central sector; Otumba and other areas in the Teotihuacán Valley, in the northeastern sector; Cuauhtitlan, in the west-central sector; and Chalco, in the southeastern sector of the Basin of Mexico.

Finally, a brief mention will be given to the potential role of interdisciplinary approaches to these questions and the systematic incorporation of nonarchaeological data into the research design.

THE BASIN OF MEXICO

The Basin of Mexico in the central highlands incorporates the modern political entity known as the Federal District, as well as parts of the states of Mexico, Hidalgo, and Tlaxcala, and covers an area of approximately 7,500 km² (figure 5.1; Rzedowski and Rzedowski 1979).

The basin, situated in the transvolcanic belt that stretches northwest to southeast

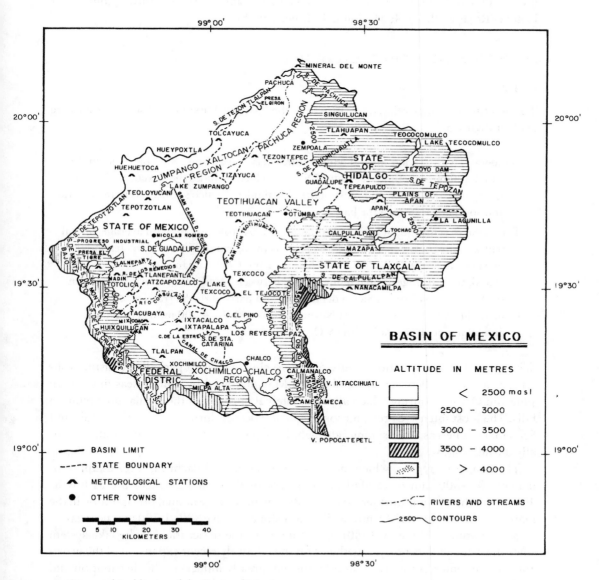

FIGURE 5.1. Geographical limits of the Basin of Mexico.

across central Mexico, became a closed hydrographic unit as a result of recent geological processes until major drainage projects were initiated during the nineteenth century. The region is delimited to the north by the Sierra de Pachuca, to the east by the Sierra Nevada (including the volcanoes Popocatepetl and Iztaccihuatl) and the Sierra de Río Frio, to the south by the Sierra del Ajusco, and to the west by the Sierra de las Cruces. The central plain is situated at an altitude of approximately 2,240 m above sea level, although several important volcanic structures interrupt the landscape, thereby forming a number of subregions. The lake system that characterizes the region has been greatly altered, beginning in the Prehispanic period and continuing throughout recent history (Rojas et al. 1974; McClung de Tapia 1990b).

ENVIRONMENTAL CHARACTERISTICS

CLIMATE

Temperature is closely related to altitude in the Basin of Mexico, and several zones are evident (figure 5.2):

> Temperate, with average annual temperatures between 12 and 18°C, between 2,000 and 2,800 m, sometimes reaching 2,900 m, which includes the Texcoco-Mexico, Zumpango-Xaltocan, and most of the Xochimilco-Chalco subregions in the central and southern sectors of the basin.

> Transition to semicold, in higher altitude zones close to 3,000 m, with average annual temperatures between 12 and 14°C, such as the Sierra de Tepotzotlan, Tezontlalpan, most of the Sierra de Pachuca, Chichicuatla, and Calpulalpan.

> Semicold, with average annual temperatures between 5 and 12°C, above 3,000 m, in the Sierra de las Cruces, Sierra del Ajusco, and parts of the Sierra Nevada.

> Cold, including parts of the Sierra Nevada that rise above 4,000 m, where average annual temperatures fall below 5°C.

With respect to precipitation (figure 5.3), two clearly defined periods are distinguishable: a dry season between November and May, followed by a rainy season between June and October, during which approximately 95 percent of the annual precipitation falls. The most intensive precipitation frequently occurs during the months of July and September, whereas a midsummer drought lasting several weeks is also common, usually during August.

The temperate region, which includes Teotihuacán and Pachuca, receives approximately 500–800 mm of rainfall per year, although annual averages are as low as 400 mm in the extreme northern sector of the basin. Average annual precipitation in the higher slopes is above 1,200 mm, while lower slopes receive 800–1,200 mm per year.

Some authors (Maderey 1980) have pointed to the desiccation of the lake system since the eighteenth century, together with the relatively recent expansion of the urban zone in the center of the basin, as a factor that greatly influences the distribution and

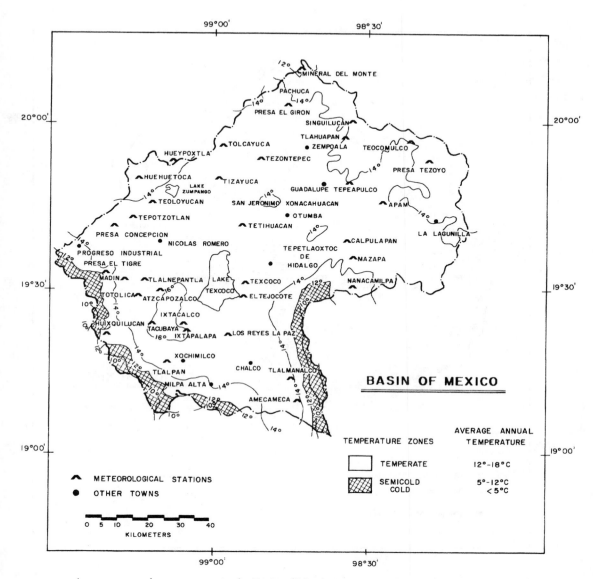

FIGURE 5.2. Average annual temperature in the Basin of Mexico.

intensity of temperature as well as precipitation. Similar effects, though of lower impact, on local microclimate were undoubtedly associated with the development of the urban centers of Teotihuacán in the northern basin and Tenochtitlán in the center. In fact it is probable that climatic conditions in the subregions of the basin were much more variable than is suggested by a summary of general tendencies. For example, Sanders et al. (1979:82) mention a greater than 50 percent probability that precipitation in any given year will be below average or begin late (mid-June). The risk of

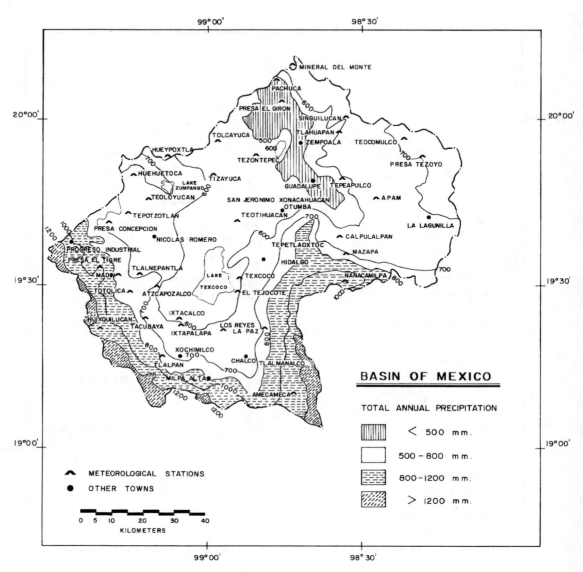

FIGURE 5.3. Average annual precipitation in the Basin of Mexico.

early and late frosts is also a significant factor (Sanders et al. 1979; Nichols 1987, 1988a), particularly in the upper piedmont (2,500–2700 m) and the lower alluvial plains (2,245–2,260 m) and, above all, in the northern sector of the region where precipitation is generally lower and least predictable (Sanders et al. 1979:83).

ECOLOGICAL ZONES

Sanders et al. (1979), based on Parsons (1974), distinguished a series of nine major ecological zones within the basin (figure 5.4), defined on the basis of characteristics

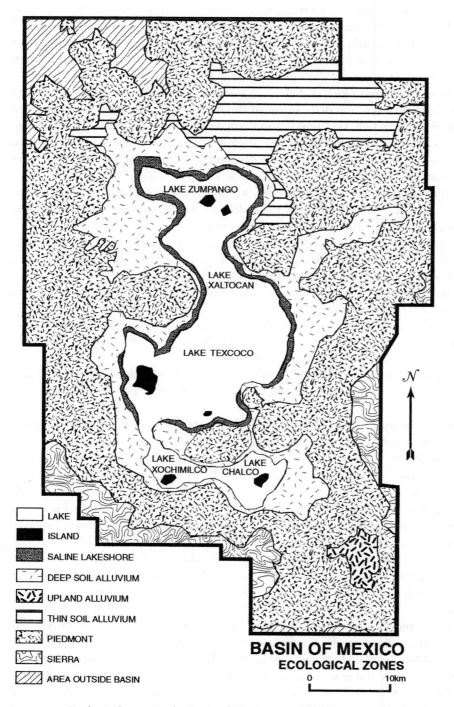

FIGURE 5.4. Ecological zones in the Basin of Mexico (modified from map 1 in Sanders et al. 1979).

such as temperature and precipitation, variation in soil depth and texture, slope gradient, subsurface moisture, and elevation above sea level. These zones are summarized here because of the significance they are believed to have had insofar as the dynamics of Prehispanic settlement patterns, population trends, and associated agricultural systems are concerned.

Lake system (approximately 2,235 m). A system composed of five lakes dominated the center of the Basin of Mexico: Xaltocan and Zumpango in the north; Texcoco, the largest, in the center; and Xochimilco and Chalco to the south. The freshwater spring-fed lakes of Xochimilco and Chalco were slightly higher and drained perennially into Lake Texcoco. The saline lakes of Xaltocan and Zumpango were also higher but drained seasonally into Lake Texcoco.

Saline lakeshore (2,328–2,245 m). This is a poorly drained band of saline soil surrounding Lakes Texcoco, Zumpango, and Xaltocan. It is unsuitable for agriculture and was utilized in Prehispanic times for salt production.

Deep soil alluvium (2,240–2,300 m). A broad strip of soil surrounding the lake system, characterized by low precipitation and high risk of frosts, thus generally unsuitable for agriculture without irrigation.

Thin soil alluvium (2,250–2,300 m). A zone of shallow soils in the northwestern basin, marginal for agricultural production.

Upland alluvium (2,450–2,600 m). Situated in the southeastern corner of the basin near Amecameca, this fertile zone of higher rainfall and lower risk of frost was probably covered largely by oak-pine forest in prehispanic times.

Lower piedmont (2,250–2,300/2,350 m). An area of variable width roughly parallel to the alluvial zones, with soil depths generally ranging between 5 and 50 cm today.

Middle piedmont (2,350–2,500 m). This belt parallels the lower piedmont and is characterized by similar characteristics, although slopes are slightly steeper.

Upper piedmont (2,500–2,700 m). This zone is characterized today by shallow soils and heavy erosion, although it is thought originally to have been largely covered with oak-pine forest.

Sierra (2,700–5,800 m). This zone is dominated by pine-oak forests. It is generally assumed to be marginal for agriculture in the basin, although it was undoubtedly an important source of forest products.

Although it has been postulated that the distribution of Prehispanic communities in these ecological zones during different periods may be associated with access to certain resources (Sanders 1965; Sanders et al. 1979) it should be stressed that archaeological research has rarely if ever been undertaken in order to demonstrate that site location is a function of specific resource distribution and availability in the Basin of Mexico. Some excavations have uncovered evidence suggestive of specific adaptations through time (e.g., Zohapilco, Niederberger 1975, 1987; Terremote-Tlaltenco, Serra 1988). In fact, little is known about the degree of intraregional organization, even during the height of Teotihuacán and Tula, prior to the dominance of Tenochtitlán in the Basin of Mexico.

BACKGROUND

Archaeological survey of the region was undertaken under the direction of W. T. Sanders and colleagues and students over a period of approximately two decades, beginning in 1960. Much of the data recovered is synthesized to some extent in the publication *The Basin of Mexico: Ecological Processes in the Evolution of a Civilization* (Sanders et al. 1979). A great deal, although not all, of the subsequent research in which evidence for agriculture has been recovered has been carried out by investigators associated with different stages of this project, essentially continuing the cultural ecological orientation of the original research and the kinds of questions it generated (Sanders and Price 1968).

A major objective of the cultural ecological approach as spelled out by Steward (1955:40–42) consisted of three aspects: to detect the interrelations between a society's exploitative or productive technology and the environment, to observe the behavioral patterns related to the exploitation of a particular area by means of a particular technology, and to determine the degree to which behavior patterns related to the exploitation of the environment affect other aspects of culture.

The Basin of Mexico Survey covered such a broad geographic area that only recently have excavations of some of the recorded sites been undertaken. Although many of the sites were originally characterized in terms of their inclusion in or proximity to specific ecological zones and presumed access to particular resources, actual excavations have focused mainly on other issues such as household structure, craft production, articulation with other centers, and ceremonial activities. Sustaining activities such as agriculture were generally assumed to be much like historical and ethnographic descriptions, and these were put forward as hypotheses, yet without a specific effort to recover archaeological evidence to support them. Important exceptions include excavations in Chalco (Parsons et al. 1981), Xaltocan (Nichols and Frederick 1993), Otumba (Charlton 1977, 1978, 1990), and elsewhere in the Teotihuacán Valley (González Quintero and Sánchez 1991; Nichols 1988b; Nichols et al. 1991).

From a somewhat different perspective and in spite of the overall risks to Prehispanic remains, the modern expansion of urban centers and communications networks has on occasion contributed to our knowledge of prehistoric agricultural activities, as in the case where irrigation canals or ancient field surfaces are exposed in excavations for roadbeds or other constructions. For example, Charlton (1977, 1990) was able to recognize and eventually excavate canals pertaining to a floodwater irrigation system in the Otumba area in the Teotihuacán Valley. Parsons et al. (1981) recovered evidence of chinampas, or raised fields, visible in the profiles of modern canals in Chalco and Xochimilco as well as in excavations. Avila López (1992) explored chinampas from the Iztapalapa peninsula in relation to salvage operations carried out during the construction of a market distribution center to serve the metropolitan area of Mexico City. Santley and Nichols excavated Formative period irrigation canals in Santa Clara Coatitlan, visible in the walls of a large extraction pit (Sanders and Santley 1977; Nichols 1982). The recent development of a shopping center and entertainment complex on the site of a nineteenth-century paper factory immediately to the south of the circular pyramid of Cuicuilco provided the opportunity for the discovery of a section of a small desiccated lake with evidence of a canal emptying into it, partially covered by the lava flow (Pastrana and Fournier 1997).

CLASSIFICATION OF AGRICULTURAL SYSTEMS

In this paper two broad categories of agricultural systems are discussed that developed in the Basin of Mexico from the time of the earliest permanent settlement in the southeastern sector by agriculturists, up through the Aztec period and the initial Colonial period. Each was characterized by several variants with varying degrees of sophistication.

Temporal (rainfall-dependent plant cultivation), including:
    Sloped piedmont cultivation
    Terraces
    Floodplain cultivation
Humidity control, including:
    Floodwater irrigation (piedmont and terraces)
    Drained fields
    Permanent irrigation
    Chinampas

While the archaeological evidence for most of these systems is slight, there is documented historical information indicating that all were functioning in different sectors of the basin at the time of the Spanish Conquest (Rojas 1988).

The general characteristics of each system permit a broad range of variation, adapted to specific local conditions. While no evolutionary sequences are postulated here, in some cases a logical progression from one type to another, occasionally cross-cutting the basic categories, can be hypothesized.

Some authors have placed emphasis on the evolution of increasing complexity of agricultural systems as a consequence of the increased demands placed on their productivity, resulting from increased population, based on the model proposed by Boserup (1965; cf. Sanders 1985; Rojas 1988). Thus certain manifestations are considered in terms of the increased expense of energy required to produce higher agricultural yields. Within this framework, Boserup suggests that extensive production gives way to intensive systems, characterized by a higher energy input per surface area of cultivation as a growing population places greater demands on productivity.

Other authors have emphasized the supposed quality of agricultural lands, in terms of their potential productivity under particular cultivation regimes, and have indirectly evaluated agricultural systems in terms of the yields necessary to sustain populations within certain ranges, based, however, on modern or recent productivity in the same area with some sort of adjustment made to take less advanced technology and lower-yield crops into consideration (Evans 1980; Nichols 1980).

The role of risk compensation has also been considered, insofar as the techniques associated with intensive cultivation can be considered as means to guarantee food production in high-risk circumstances such as unpredictable frosts, late rains, or torrential precipitation events that may occur at a critical moment of the crop's development (Nichols 1987). This hypothesis has little archaeological evidence to support it outside the obvious use of irrigation that lends itself to different interpretations.

Finally, still other authors have stressed the role of soil alteration, classifying agricultural systems in terms of the degree to which the soil is worked in preparation for and during cultivation. Rojas (1988:28) for example, distinguishes two broad categories: extensive systems with minimum soil alteration and intensive systems that involve working the soil. The degree to which the soil is worked is closely related to the intensiveness of agricultural production, ecological conditions, and the specific crops in question (ibid.: 32). However, from an archaeological perspective, investigators seldom encounter direct evidence adequate to measure such a variable and, rather, are obliged to rely on descriptions of historical or modern applications of traditional techniques.

The distinction between extensive and intensive agricultural systems will be retained here, but rather than emphasize them as consequences of population size and concomitant demand for subsistence resources, they will be examined in terms of cultural adaptations to particular environmental conditions and the range of variation in agricultural systems in relation to what the environment can sustain (McClung de Tapia 1990b, 1997).

The definitions of extensive versus intensive agricultural systems are relatively straightforward. Extensive agriculture ordinarily refers to systems that require large areas, only a portion of which is under cultivation at one time. Once a suitable plot is cleared, usually by cutting and later burning off dried vegetation, several years of annual planting are followed by a longer period of fallowing to allow regeneration of the soil nutrients to take place. The ratio of planted years to the number of years during which the plots are left to fallow varies, depending upon local ecological conditions. In tropical areas, regeneration of vegetation is rapid, but several stages of succession are necessary for soil nutrients to be sufficiently restored. In temperate zones, the period of cultivation may be reduced further, with a shorter period of fallowing in comparison to tropical variants. In both cases, large extensions of territory are necessary for farmers to maintain the system and allow sufficient time for the regeneration of fertility to take place in specific plots.

In an intensive system, an effort is made to increase the productivity of a plot in terms of greater crop yield rather than to increase the amount of land under cultivation. This definition contrasts somewhat with Rojas's point of view, mentioned further on. In general, it is assumed that the labor investment in an extensive system is lower in overall energetic terms with respect to intensive systems. Intensification is generally understood to involve enhancing the yield of a plot through more frequent planting, which may include yearly harvests with no fallowing or an increase in the number of harvests per year, of the same or different plants (rotation). Additional features of intensification may include planting more densely, increased attention to plants through fertilization, more frequent and careful weeding, and, often, humidity control in the form of conservation measures as well as direct applications such as irrigation (see Rojas 1988:64).

Various different agricultural systems undoubtedly coexisted in different subregions during the prehistoric occupation of the Basin of Mexico, resulting in an extensive use of available land resources and simultaneous intensification of production to the extent that additional suitable cultivable land was utilized as needs dictated. However, on occasion these needs may have been politically determined rather than being strictly economic, and although intensification in some zones may have been a response to the demands of an increasing population, in other instances it may have provided an opportunity to establish or maintain control over a particular area.

CHARACTERISTICS OF AGRICULTURAL SYSTEMS

The following section presents a description of the principal traditional agricultural systems thought to be relevant for the study of Prehispanic practices. To the extent that archaeological data permit, an attempt will be made to place these systems in regional and temporal perspective in the Basin of Mexico.

TEMPORAL (RAINFALL-BASED CULTIVATION)

Traditional rainfall-dependent agriculture is generally centered around maize (*Zea mays* L.) as the principal crop, although some interplanting with frijol, or beans (*Phaseolus vulgaris* or *P. coccineus*), and squash (*Cucurbita* spp.) occurs. Other important plants that are grown under this regime include nopal, or prickly pear (*Opuntia* spp.), and maguey, or century plant (*Agave* spp.). Fruit-bearing trees such as tejocote, or hawthorn (*Crataegus mexicana*), and capulin, or Mexican cherry (*Prunus serotina*), are also adapted to such conditions although they are normally cultivated in houselot gardens or as field borders. In the basin temporal cultivation is carried out on gentle to steep piedmont slopes, often on terraces involving varying degrees of construction and on alluvial surfaces where drainage is adequate. Maize plots are commonly referred to as milpa, whether the reference is to annual plots (barbecho) or to itinerant systems such as roza in tropical ecosystems and tlacolol in temperate zones.

Generally in the Basin of Mexico a single harvest of a principal crop is obtained from plots dependent upon seasonal rains. There may be little distinction between piedmont fields with some sort of ridge (bancal) to control erosion from runoff and simple terraces with ridges planted in rows of magueys to facilitate soil buildup behind the plants, thus combining efforts to limit erosion, increase soil depth, and retain humidity. More sophisticated terrace systems may include the construction of earth, adobe, or stone retaining walls (Donkin 1990). Rojas (1988:111) argues that this is in fact a form of intensification to the extent that the cultivable surface is increased and the period of time normally dedicated to fallowing is shortened. However, it is difficult to demonstrate for the Prehispanic period; we can hypothesize with respect to the overall area available for cultivation if certain systems are employed but can hardly speculate about fallowing times at this stage.

Some restricted areas of alluvial soils at lower elevations are also suitable for rainfall-dependent agriculture when additional sources of humidity are unavailable. Stretches of low-lying alluvial soils in the Basin of Mexico are often subject to greater risk of frosts, however. In addition, poor drainage is a factor that limits successful cultivation in some areas, resulting in inundation of crops during heavy rains.

HUMIDITY CONTROL

Humidity control takes many forms with varying degrees of technological sophistication. Irrigation is included here as a sophisticated form of humidity control, and chinampa cultivation is considered as a special case—an extremely intensive form of permanent irrigation, although other authors consider it to be a separate category (Rojas 1988). A logical step from rainfall-dependent piedmont and terrace cultivation is the incorporation of canal systems to divert excess runoff from seasonal storms and direct water systematically to the fields prior to planting or during the plants' growth. Not only is control exercised over the distribution of humidity with respect to the crops, but also the risks of sheet and gully erosion are greatly reduced. This type of application is clearly seasonal, but it provides control over the distribution of humidity to a broader area, depending upon the structure of the associated canal system and the size of the area to be irrigated.

Drained-field systems, on the other hand, develop in areas where a high water table, poor drainage, and, on occasion, associated springs provide too much humidity for successful cultivation. A canal system may be built to divert water elsewhere so those plots can be successfully cultivated without the risk of waterlogging plants or periodic inundation, especially during the rainy season. Often the same water can be diverted into a permanent irrigation system that services adjacent areas with lower natural humidity. Such plots generally permit at least two and occasionally three harvests per year, depending upon the crops.

Permanent irrigation refers to systems in which a perennial water source is available. Usually a complex system of canals with primary, secondary, and sometimes tertiary channels is developed, in which water is diverted to many plots on a rotating or otherwise systematic basis by means of a series of check dams and sluices that are controlled to assure an equitable distribution of humidity (Doolittle 1990). As in the case of drained fields, irrigated plots generally support at least two crops per year, although there is significant variability insofar as regional manifestations are concerned. Irrigation is permanent not in the sense that water is available to all plots at all times but rather in the sense that the source is perennial. In many parts of the Basin of Mexico today, irrigation is used to provide sufficient humidity for crops to get a head start in their growth cycle, after which they depend upon seasonal rains during the rest of their development.

Chinampa cultivation is a highly specialized form of intensive agriculture that developed largely in the southern basin, at the edges of Lakes Chalco and Xochimilco (Coe 1964; Armillas 1971; Parsons 1976, 1991; Parsons et al. 1981). Archaeological evidence for Prehispanic chinampas has also been detected at Xaltocan (Nichols and Frederick 1993), although the extension in the central sector of the Basin of Mexico is not known. This technique involved the reclamation of broad expanses of swamp at the edges of the lakeshores and consisted of the construction of artificial surfaces rising above the lake surface by alternating layers of mud and aquatic plants (sedges, bulrushes), secured by the roots of willow trees planted around the margins of the rectangular plots (Coe 1964; Armillas 1971). This construction technique may have originated as a means by which to stabilize island and lakeshore areas to support domestic and ceremonial construction (Tolstoy et al. 1977; McClung de Tapia et al. 1986; Serra 1988; Brumfiel 1992) prior to its application in agriculture.

ARCHAEOLOGICAL EVIDENCE FOR PREHISTORIC AGRICULTURAL SYSTEMS IN THE BASIN OF MEXICO

Rojas (1988) discusses in depth the available historical and ethnohistorical documentation for these systems as well as various ethnographic examples, which bear similarities to sixteenth-century and, in some cases, slightly later descriptions from various parts of Mesoamerica. Nonetheless, the author stresses the absence of clear indications that the techniques described were in fact utilized as such during earlier periods. However, some recent sources of archaeological evidence, together with paleoethnobotanical and paleoenvironmental data, suggest specific directions to answer questions about prehistoric agricultural systems.

As was mentioned previously, relatively little archaeological investigation has been undertaken in the Basin of Mexico with the specific purpose of recovering evidence for agricultural practices, with the notable exception of studies of irrigation systems. Much of the research during the fifties and sixties on the role of irrigation in primary state formation processes in the Basin of Mexico was a consequence of the impact of the hydraulic hypothesis (Sanders and Price 1968; Millon 1973). Wittfogel (1957) postulated a causal relationship between the development of large-scale hydraulic works including irrigation systems and the rise of centralized authority in regions characterized by semiarid climatic conditions. Almost no direct archaeological evidence was available at the time to support arguments favoring the position that the managerial requirements of irrigation systems provided the necessary conditions for increasing political complexity and eventual state formation. This situation was especially problematic in the case of Classic period Teotihuacán (approximately A.D. 150–650), early recognized as the first urban center of its dimension in Mesoamerica and the apparent center of a state system. Yet, available evidence for the use of irrigation in the surrounding region seemed to be later and could not be clearly associated to the period of Teotihuacán's development and expansion (Millon 1973:47)

On the other hand, various elements of indirect evidence suggested that the city of Teotihuacán and surrounding communities during the Classic period must have benefited from a number of different techniques of humidity control, ranging from floodwater irrigation systems in the piedmont, to intensive drainage of the alluvial plain in the area of permanent springs southwest of the city, to the canalization of this permanent water source to irrigate the lower alluvial plain. These included the presence of a high water table area that impeded successful cultivation in a large part of the region without some form of drainage, a permanent water source of great potential, clear evidence that the course of the San Juan River had been redirected through the center of the city in order to conform to its grid layout—clearly demonstrating the capacity to plan, direct, and carry out large-scale hydraulic projects—and the visible remains of a complex artificial floodwater system developed on the north slope of Cerro Gordo in evident association with Classic period sites. All of these indicators argued for the use of complex systems of humidity control at the time when the Classic period city was the most powerful civic-ceremonial, political, and economic focal point in central Mexico as well as much of Mesoamerica (Sanders et al. 1979:262–267).

Investigators were in agreement that irrigation was undoubtedly a significant component of Classic period agricultural production in the Teotihuacán Valley, in spite of the absence of direct evidence for its use. The main issue was one of scale. The absence of direct evidence associated with Teotihuacán occupation suggested a relatively small-scale endeavor, one that could hardly play a critical role in the dynamics of state formation.

The distribution of archaeological sites throughout the period of Prehispanic settlement in the basin indicates that communities were located in piedmont zones suitable for the development of agricultural terraces, but few excavations have been undertaken to examine the occupational history and the possible relation between settlement and agricultural technology in such contexts. Furthermore, while it is generally assumed that systems similar to those described in sixteenth-century and later documents probably dated to earlier periods, diverse kinds of archaeological evidence presently available support a more recent date—Middle and Late Postclassic periods—for most terrace and chinampa zones located in the region.

One of the problems for archaeological research is that Late Postclassic land-use patterns have obscured evidence for earlier agricultural systems for the most part. Lands that were known to be highly productive in earlier periods such as the Late Terminal Formative and Classic periods, were probably quite attractive during the Late Postclassic, by which time most of the region had been resettled, after a general displacement and reorganization of the population following the fall of Teotihuacán. Activities such as the refurbishment or reconstruction of previously established terrace systems, the development of floodwater irrigation systems in areas where rain-dependent terraces or sloping fields were present, and the expansion of cultivation on alluvial soils during later periods would obliterate much of the evidence for earlier land use.

Furthermore, it is not entirely clear that all terraces or chinampas were habitational either, because of the limited nature of archaeological excavations. Thus it is often difficult to document an occupational sequence of reuse of agricultural surfaces and associated house mounds based on artifactual materials.

Colonial and later occupations all but devastated many surface indications of remaining areas of Prehispanic agriculture and habitation. Recent urbanization in formerly rural sectors of the basin (Chalco, Teotihuacán, or Zumpango, to name just a few), together with the immense growth of the metropolitan zone of Mexico City and adjacent zones of the State of Mexico, has engulfed major portions of Prehispanic occupations as well.

The overall impact of human settlement in the region since the Colonial period is difficult to assess. Significant changes in agricultural practices occurred as European technology was introduced into central Mexico. Some population sectors in the basin were probably slower to respond to these shifts. The demographic response to the introduction of European diseases among the indigenous population took the form of drastic decreases, thus reducing the amount of land under cultivation, consequently affecting the maintenance of complex agricultural systems (Ruvalcaba 1985:22). Sheet and gully erosion in piedmont zones and severe flooding of alluvial and lakeshore areas were extreme ecological consequences of the social and demographic trends following the Spanish Conquest.

Fortunately, a large proportion of the region was surveyed during the sixties and seventies, and although the surface remains that were once visible—mainly sherd concentrations and occasional low mounds—may no longer be present, the possibility of excavation still remains in some instances. It should be noted that most of the archaeological evidence for Prehispanic agricultural practices prior to the Late Postclassic period has been detected through excavation rather than survey. More sophisticated techniques of remote sensing beyond Nichols's (1988b) application of infrared aerial photography, mainly in the Teotihuacán Valley, have not generally been utilized. Therefore, opportunities still exist to design appropriate research to test earlier hypotheses about agricultural systems that were based largely on the locations of sites detected during surface surveys in different ecological zones in various sectors of the basin (Parsons 1990). The availability of Universal Transverse Mercator (UTM) map coordinates for most of these sites (excepting the Teotihuacán region prior to the Late Postclassic period; see Evans 1980 for site coordinates pertaining to this period and Parsons 1983 for the rest) allow investigators to locate sites in specific zones where appropriate techniques can be applied.

At present, no specific archaeological data are available for the use of extensive cultivation techniques in the Basin of Mexico. During the earliest phases of Prehispanic settlement by agriculturalists, revealed by surface surveys, the location of small, presumably self-sufficient agricultural communities in piedmont zones throughout the region, together with the relative absence of sites in environmental zones other than the lakebed and shore, suggests that extensive techniques were utilized (Parsons 1991). The combination of slightly higher rainfall with a lower risk of frosts in the piedmont zone between 2,300 and 2,500 m contributed to making this a favorable zone for early settlement. Population growth and the expansion of larger communities in this region by the Late Formative are interpreted as a sign that floodwater systems were developed by this time, particularly in the southern sector of the basin (Parsons 1991:24).

TERRACES

Bancal-type terraces (in which ridges are built of packed earth, occasionally secured with rows of magueys) are extremely difficult to detect archaeologically, although Sanders et al. (1979:249) suggest that the underlying bedrock (*tepetate* in much of the basin) surface sometimes shows traces. Sanders et al. (1979:250–251) summarize evidence for Prehispanic stone-faced terraces presumed to be agricultural, including Middle-Late Formative structures at Zacatenco and Ticoman, a site in the Patlachique range, sites in the Tezoyuca hilltop complex, and Patlachique-phase sites in the Texcoco-Iztapalapa regions.

Classic period terraces are scarce. During this time the rural population appears to have resided in nucleated villages for the most part. Stone-faced terracing dating to this period is apparent on the slopes of Cerro Paula in the Temascalapa region north of Cerro Gordo. In the Texcoco region, terracing may have been practiced on the southern slopes of the Patlachique Range and in other dispersed piedmont areas.

Maguey cultivation and processing may have been a significant activity at Classic period communities such as Maquixco Bajo, situated in the north piedmont of the lower Teotihuacán Valley (Sanders et al. 1979:346–348). In sites such as this only indirect evidence for an agriculturally based economy is present, such as the absence of craft activities on the one hand and the presence of obsidian tools possibly employed as maguey scrapers on the other, in addition to the ecological characteristics of the site's location.

A similar situation holds for the Epiclassic period, following the demise of Teotihuacán up to the beginning of the Late Postclassic. Possible indications of Early Postclassic terraces are indefinite because of the overlying presence of more dense Late Postclassic occupations. Late Postclassic terraces detected at the piedmont sites of Cerro Gordo and Cihuatecpan (Evans 1989, 1992) in the Teotihuacán Valley indicated a consistent association of house mounds with these agricultural structures. Evans (1989:202–204) also cited Charlton's excavation of domestic structures at the Aztec site of TA-40 on the north slope of Cerro Gordo in association with terraces. Based on archaeological materials recovered from excavations at Cihuatecpan, including gray-black obsidian scrapers suitable for maguey sap production, basalt scrapers, large storage vessels, and large spindle whorls suitable for spinning maguey fibers, Evans (1992) postulated that maguey cultivation and processing was a significant economic activity in semiarid piedmont settlements in this region during the Late Postclassic. No paleoethnobotanical evidence was reported from this excavation. Benton (1991) and Nichols (1997) have detailed maguey fiber processing at the Late Postclassic site of Otumba as well; Brumfiel (1991) describes evidence of similar activities in Early and Late Postclassic contexts at Huexotla in the Texcoco region.

Most of the stone-faced terraces recorded in the Chalco-Xochimilco piedmont

region in the southern sector of the basin are assumed to be Late Postclassic constructions, given the relative absence of earlier settlements in this area from the Classic period up until resettlement during the Late Postclassic. In the Milpa Alta region, stone
terraces may even correspond to Posthispanic settlement, given the scanty evidence for
earlier occupation in this area.

IRRIGATION

Inconclusive archaeological evidence for the use of irrigation probably associated with
Late Formative settlement in the area of Cuicuilco in the southern basin was reported
by Palerm (1961). Although no excavations have clearly delineated irrigation systems
in this area, the presence of barrancas and springs flowing toward the alluvial plain in
the vicinity of Cuicuilco (Cordova et al. 1994) and the recent discovery of
a canal draining into the lake at the edge of the area recently excavated (Pastrana and
Fournier 1997) suggest their use.

In some cases, investigators have estimated the potential agricultural carrying capacity of certain regions within the basin in order to evaluate the size of population
that could be sustained under a particular agricultural system. For example, Sanders
and Santley (1977) argued that the potential productivity of agricultural land on the
alluvial plain in the Cuauhtitlan region was insufficient to sustain the Late Formative
period population, thus suggesting that permanent irrigation was likely.

Direct archaeological evidence is available that floodwater control to take advantage of abundant seasonal humidity was established in the basin by the Middle Formative period. A prehistoric system was identified in a small alluvial plain in the area
of Santa Clara Coatitlan southeast of the Guadalupe Range (Sanders and Santley 1977;
Sanders et al. 1979; Nichols 1982). Survey data suggested that the system had been
utilized during the Classic period, although the earliest canals were possibly Late Formative (Sanders and Santley 1977). Nichols's 1977 excavations indicated a Middle
Formative date for the beginning of the system, followed by its abandonment by the
Terminal Formative (Nichols 1982). The presence of coarse sediments in the canal fill
indicate a simple system with no regulation of water flow by means of reservoirs. Only
primary and secondary canals were detected (Nichols and Frederick 1993:126). The
prior use of extensive cultivation in the area is suggested as the probable cause of substantial sheet erosion, indicated by approximately 7 m of redeposited sediment on the
alluvial plain. The use of floodwater irrigation may have developed as a response to
erosive processes. In this case, the canal system was covered by a sterile layer of alluvial
sediment, over which lay a stratum of Tzacualli-phase ceramics, thus sealing the canal
deposits. However, archaeological sites in proximity to the irrigated area have not been
excavated, and no additional data such as paleoethnobotanical remains from the canals were recovered that could in this case provide evidence for possible crops and local
flora, given that the deposits were sealed.

The Teotihuacán Valley has been a continual focus of research related to the use of
irrigation for several decades. Archaeological evidence that floodwater systems were in
place by the Classic period comes mainly from this sector of the basin, where several
recent discoveries have contributed to our knowledge of agricultural systems in this
region. Millon (1957) excavated an abandoned floodwater irrigation system on the

piedmont of the region known as the Maravillas system. A Coyotlatelco phase date, following the demise of the Classic period urban center, was ultimately assigned for its construction based on associated ceramic materials. Sanders (1965; Sanders et al. 1979) described a complex floodwater system associated with Classic period settlements based on surface reconnaissance on the north slope of Cerro Gordo.

Charlton (1977, 1978, 1979, 1990) identified floodwater canals to the north of Otumba, a town of considerable importance during the Aztec period in the Teotihuacán Valley. Most of the canals appeared to be in use during the Postclassic period although the earliest canals were thought to have been constructed during the Terminal Formative as the city of Teotihuacán was expanding. Macrobotanical remains recovered from the canals included taxa adapted to semiarid climatic conditions and opportunistic plants such as *Amaranthus* and *Chenopodium* associated with cultivation, together with maize cupules (McClung de Tapia 1980, 1990a).

Floodwater canals were also discovered underlying a Classic period apartment compound at Tlailotlacan in the so-called Oaxaca Barrio at the western edge of the city of Teotihuacán (Nichols et al. 1991). The system involves primary and secondary canals as well as tertiary canals that distribute water on fields. The system is sealed by an overlying structure dating to the Early Tlamimilolpa phase (ibid.: 120), and ceramics from the canal fill suggest a Tzacualli-phase inception with abandonment occurring sometime during the subsequent Miccaotli phase (ibid.: 126). Pollen was poorly preserved in the samples recovered from one of the canals, but the dominance of Gramineae, Compositae, and cheno-ams together with a single grain of *Zea mays* suggests that opportunistic plants associated with maize cultivation are represented (ibid.: 126).

Nichols (1988b; Nichols et al. 1991; Nichols and Frederick 1993) excavated a floodwater canal system in the Tlajinga plain at the southern extreme of the Classic period city of Teotihuacán, visible as a linear feature on infrared aerial photographs. Ceramics in one of the canal systems indicated its construction during the Teotihuacán period (Nichols 1988b:23). Other canals were associated with the Late Aztec period or, possibly, the Mazapan phase (ibid.: 20, 23). Botanical remains were not recovered from these excavations because of their limited depth (generally between 0.35 and 1.6 m below the modern agricultural surface) and the likelihood of intrusive plant material.

CHINAMPAS

Although chinampa cultivation was characteristic of the Postclassic period, in the marshes of the central lakes in the basin, investigators speculated about the existence of possible Classic period chinampas in the area of permanent springs southwest of the urban zone of Teotihuacán (Millon 1973; Sanders 1965; Sanders et al. 1979). Excavations were undertaken in the high water table area south of the modern town of San Juan Teotihuacán in order to recover evidence for Teotihuacán period agricultural use (González Quintero and Sánchez 1991). Seventeen stratigraphic pits revealed clear evidence that the Prehispanic agricultural plots in this area were not chinampas in the traditional sense but drained fields, confirming Lorenzo's (1977) position. Teotihuacán ceramics indicated a Classic period development, however, confirming the suspicion of some authors that this zone may have provided subsistence products to the urban center. Although paleoethnobotanical remains were recovered from all strata, representing both potential cultivated plants as well as wild opportunists, frequencies were not reported. Samples for pollen analysis were apparently recovered but were not reported. Possible evidence of Teotihuacán construction in association with one of these plots is mentioned (González Quintero and Sánchez 1991:359). A potential direction for future research would be more extensive excavations in this zone to evaluate evidence for Teotihuacán occupation and associated construction, although the high water table will make excavations difficult.

Calnek (1972) and Parsons et al. (1981; Parsons 1991) refer to chinampas in the central and southern Basin of Mexico as residential in addition to their role in agricultural or horticultural production, but the limited nature of excavations in such contexts leaves room to question to what extent this was the norm. More extensive excavations are necessary in both of these cases, to understand the nature of household plant production.

Parsons' excavations in the Chalco region revealed clear evidence of the superposition of layers of vegetation and lake sediments characteristic of chinampa construction, although in some cases the compaction of layers made it difficult to recognize the overall structure (Parsons et al. 1981). These chinampas were consistently associated with Early Aztec (Middle Postclassic) occupations. Chinampas on the edge of Lake Xochimilco seem to have developed during the Late Aztec phase in association with population growth and the development of more sophisticated systems of hydraulic control (Parsons 1991:33). Evidence for Early and Late Postclassic chinampa construction was also recovered from a site west of Ayotzingo in the Chalco zone (Frederick 1997).

Recent excavation of a chinampa zone southeast of the island of Xaltocan suggests that this system was initially constructed in the central part of the lake system during the Early Aztec phase, based on associated ceramics and radiocarbon determinations (Nichols and Frederick 1993:141). Freshwater from a spring located near Ozumbilla at the base of Cerro Chiconauhtla, on the lakeshore to the east of Xaltocan, was channeled to the area of cultivation.

Paleoethnobotanical remains recovered from chinampa excavations in Xaltocan (McClung de Tapia and Martínez 1996) and Chalco (Popper 1988), in different stages of analysis or preparation for publication, contribute significant data to an understanding of prehistoric cultivation when available.

GARDENS

Another significant contribution to household subsistence would be the produce obtained from adjacent gardens suitable for the cultivation of fruit trees (capulin, tejocote), chile (*Capsicum* spp.), tomate (*Physalis* spp.), medicinal herbs and condiments (epazote), and ornamental plants. Unfortunately, sufficiently extensive excavations of houselots in the basin have not been undertaken with the objective of recovering evidence for this type of horticultural production, either in spatial terms or with respect to the kinds of plants that may have benefited from more intensive care as a result of their proximity to habitation areas. The probable role of chinampa plots as gardens in this sense should also be considered. To date there is not enough archaeological evidence available to evaluate the extent and localization of chinampas dedicated to intensive cultivation of staples, flowers and vegetables, or produce destined for household consumption. This would be a useful direction for future research.

Sanders et al. (1979:323) infer the presence of garden plots within household compounds and between residential clusters at the Late Formative village of Loma Torremote, situated in the lower piedmont in the western part of the Cuauhtitlan region. Some of the plant remains recovered from subsurface storage pits (Reyna Robles and González 1978) are certainly congruent with this type of cultivation, but no specific evidence is cited to bear out this hypothesis.

DISCUSSION

Recent geomorphological (Frederick 1997) and palynological (González Quintero 1986; Lozano et al. 1993; Lozano and Ortega 1994) studies in the Chalco region and geomorphological surveys in the Xaltocan region (Frederick 1992, 1997; Nichols and Frederick 1993) suggest that there is a great deal more to be learned about localized environmental variability in a region as broad as the Basin of Mexico. There is also evidence for climatic fluctuation through time. Variation in lake levels is one indicator of environmental changes, as are alterations in vegetation suggested by pollen sequences from radiocarbon-dated strata.

Preliminary analysis of phytolith frequencies in stratigraphic soil profiles may also suggest environmental variation. Research under way in the Teotihuacán Valley (McClung de Tapia et al. 1995; McClung de Tapia 1996) supports the hypothesis of some alternation of dry and humid periods, together with the possibility of cooler temperatures suggested by intemperization of soils (McClung de Tapia et al. 1998).

The preliminary results from a recent effort to recover macrobotanical remains as well as pollen and phytoliths from a series of stratigraphic soil profiles excavated in noncultural (offsite) areas of the lower piedmont and alluvial plains of the Teotihuacán Valley (McClung de Tapia 1996) illustrate some of the problems in interpretation of botanical remains that should be taken into consideration in the evaluation of macrobotanical materials recovered from the sites previously mentioned.

Pollen is generally poorly preserved in alluvial soils in the Teotihuacán region, even in superficial deposits. Phytoliths, on the other hand, appear to be quite well preserved, but their interpretation is somewhat complex due to the taphonomic processes involved in their deposition as well as their sensitivity to localized humidity conditions. Uncarbonized and a few carbonized macrobotanical remains are generally present in the surface strata to a depth of approximately 50 cm, at which point overall frequencies decline drastically. Conservation of seeds in strata below this point is minimal. However, carbonized materials continue to be present, together with uncarbonized specimens in low quantities. Of particular interest is the concentration of seeds from opportunistic plants, many of which represent edible genera, but all are components of the vegetation associated with cultivated plants and fallowed plots (Rzedowski et al. 1964; Castilla and Tejero 1983). From an archaeological perspective, it is important to study carefully the conditions of deposition in each context in order to determine which plant remains are prehistoric and which represent economically useful genera or pertain to the surrounding flora.

Botanical remains recovered from archaeological contexts in the Basin of Mexico are highly variable insofar as the information they provide. In addition, it is difficult to compare the results obtained from different studies, given diverse objectives and variable formats for reporting remains. Generally very little descriptive information about the plant remains themselves (particularly measurements and observations concerning preservation conditions) is given. The relevant frequencies of different genera associated in particular archaeological contexts are not usually emphasized and may contribute to their interpretation. Furthermore, the recovery of paleoethnobotanical evidence from prehistoric agricultural contexts is a recent component of archaeological research, and as such, little published information is as yet available (see table 5.2).

TABLE 5.2. ECONOMIC PLANTS RECOVERED FROM ARCHAEOLOGICAL CONTEXTS IN THE BASIN OF MEXICO

Agave spp. (century plant)
Alnus sp.
Amaranthus hybridus (quelite)
A. hypochondriacus (alegria)
Capsicum annuum (chile)
Casimiroa edulis (zapote blanco)
Chenopodium berlandieri (huauhtzontli)
Crataegus mexicana (tejocote)
Cucurbita sp. (squash)
Gossypium hirsutum (cotton)
Helianthus annuus (sunflower)
Mammillaria sp. (biznaga)
Myrtillocactus geometrizans (garambullo)
Opuntia spp. (prickly pear)
Persea americana (avocado)
Phaseolus coccineus (scarlet runner bean)
Physalis spp.
P. vulgaris (common bean)
Pinus sp.
Portulaca oleracea
Prunus serotina (capulin)
Quercus sp.
Salvia hispanica (chia)
Sechium edule (chayote)
Solanum spp.
Spondias sp. (ciruela)
Suaeda nigra (romerito)
Teloxys ambrosioides (epazote)
Zea mays (maize)
Zea mays ssp. *mexicana* (teosinte, Chalco type)

Problem-oriented archaeological research coupled with paleoethnobotanical studies would be a highly productive means by which to gain a better understanding of prehistoric agricultural systems, in terms of the crops produced and the conditions of cultivation. The study of prehistoric agricultural systems in the Basin of Mexico can benefit from the application of other techniques that contribute to an understanding of regional and local ecological conditions as well. On both levels, stratigraphic soil analyses in conjunction with systematic radiocarbon dating of bulk sediment and ^{13}C/^{12}C ratios can be judiciously employed to recover information related to prior environments. Finally, despite the many productive archaeological investigations that have already taken place, there is still much to be learned about ancient agricultural practices and early landscape modifications in the Basin of Mexico.

REFERENCES

Armillas, P. 1971. Gardens in swamps. *Science* 174: 653–661.

Avila López, R. 1992. Arqueología de chinampas en Iztapalapa. In C. J. González, ed., *Chinampas Prehispánicas*, pp. 81–154. Mexico City: Instituto Nacional de Antropología e Historia.

Benton, M. R. 1991. *Analysis of Maguey and Cotton Spindle Whorls from the Aztec City-State of Otumba*. Manuscript, Department of Anthropology, Dartmouth College, Hanover, N.H.

Boserup, E. 1965. *The Conditions of Agricultural Growth: The Economics of Agrarian Change under Population Pressure*. Chicago: Aldine.

Brumfiel, E. M. 1991. Agricultural development and class stratification in the southern Valley of Mexico. In H. R. Harvey, ed., *Land and Politics in the Valley of Mexico: A Two Thousand Year Perspective*, pp. 43–62. Albuquerque: University of New Mexico Press.

———, ed. 1992. *Xaltocan: Centro Regional de la Cuenca de México, Informe Anual de 1991*. Albion, Mich.: Albion College.

Calnek, E. 1972. Settlement pattern and chinampa agriculture at Tenochtitlan. *American Antiquity* 37(1): 104–115.

Castilla Hernández, M. E., and J. D. Tejero Diez. 1983. *Estudio florístico del Cerro Gordo (Próximo a San Juan Teotihuacán) y Regiones Aldeñas*. Undergraduate thesis in biology, Escuela Nacional de Estudios Profesionales Iztacala, Universidad Nacional Autónoma de México, Mexico City.

Charlton, T. H. 1977. *Report on a Prehispanic Canal System, Otumba, Edo: De Mexico Archaeological Investigations,* August 10–19, 1977. Report submitted to the Instituto Nacional de Antropología e Historia (manuscript), Department of Anthropology, University of Iowa, Iowa City.

———. 1978. *Investigaciones Arqueológicas en el Municipio de Otumba, Temporada de 1978, 1a Parte: Resultados Preliminares de los Trabajos de Campo*. Report submitted to the Instituto Nacional de Antropología e Historia (manuscript), Department of Anthropology, University of Iowa, Iowa City.

———. 1979. *Investigaciones Arqueológicas en el Municipio de Otumba, Temporada de 1978, 5a Parte: El Riego y el Intercambio: La Expansión de Tula*. Report submitted to the Instituto Nacional de Antropología e Historia (manuscript), Department of Anthropology, University of Iowa, Iowa City.

———. 1990. Operation 12, Field 20, Irrigation system excavations. In T. H. Charlton and D. L. Nichols, eds., *Preliminary Report on Recent Research in the Otumba City-State*, pp. 210–212. Mesoamerican Research Report 3. Iowa City: University of Iowa, Department of Anthropology.

Coe, M. D. 1964. The chinampas of Mexico. *Scientific American* 211: 90–98.

Cordova, C., A. L. Martín del Pozzo, and J. López Camacho. 1994. Palaeolandforms and volcanic impact on the environment of prehistoric Cuicuilco, Southern Mexico City. *J. Archaeological Science* 21: 585–596.

Donkin, R. A. 1990. *Agricultural Terracing in the Aboriginal New World*. Tucson: University of Arizona Press.

Doolittle, W. E. 1990. *Canal Irrigation in Prehistoric Mexico*. Austin: University of Texas Press.

Evans, S. T. 1980. *A Settlement System Analysis of the Teotihuacan Region, Mexico, A.D. 1350–1520*. Ph.D. thesis, Department of Anthropology, Pennsylvania State University, University Park.

———. 1989. El sitio Cerro Gordo: Un asentamiento rural del periodo azteca en la cuenca de México. *Estudios de Cultura Nahuatl* 19: 183–215.

———. 1992. The productivity of maguey terrace agriculture in central Mexico during the Aztec period. In T. W. Killion, ed., *Gardens of Prehistory: The Archaeology of Settlement Agriculture in Greater Mesoamerica*, pp. 92–115. Tuscaloosa: University of Alabama Press.

Frederick, C. D. 1992. El contexto geomorfológico de Xaltocan: Un reconocimiento inicial. In E. M. Brumfiel, ed., *Xaltocan: Centro Regional de la Cuenca de México, Informe Anual de 1991*, pp. 70–73. Albion, Mich.: Albion College.

———. 1997. *Landscape Change and Human Settlement in the Southeastern Basin of Mexico*. Manuscript, Department of Archaeology and Prehistory, University of Sheffield, U.K.

González Quintero, L. 1986. Análisis polínico de los sedimentos. In J. L. Lorenzo and L. Mirambell, coords., *Tlapacoya: 35000 Años de Historia del Lago de Chalco*, pp. 157–166. Mexico City: Instituto Nacional de Antropología e Historia.

González Quintero, L., and J. E. Sánchez Sánchez. 1991. Sobre la existencia de chinampas y el manejo del recurso agrícola-hidráulico. In R. Cabrera, I. Rodríguez, and N. Morelos, coords., *Teotihuacán 1980–1982: Nuevas Interpretaciones*, pp. 345–375. Mexico City: Instituto Nacional de Antropología e Historia.

Lorenzo, J. L. 1977. Agroecosistemas prehis-

tóricas. In E. Hernández Xolocotzi, ed., *Agroecosistemas de México*, pp. 1–20. Chapingo, Mex.: Colegio de Posgraduados.

Lozano-García, S., B. Ortega-Guerrero, M. Caballero-Miranda, and J. Urrutia-Fucugauchi. 1993. Late Pleistocene and Holocene paleoenvironments of Chalco Lake, central Mexico. *Quaternary Research* 40: 332–342.

Lozano-García, S., and B. Ortega-Guerrero. 1994. Palynological and magnetic susceptibility records of Lake Chalco, central Mexico. *Paleogeography, Paleoclimatology, Paleoecology* 109: 177–191.

Maderey, L. E. 1980. Intensidad de la precipitación en el Valle de México. *Boletín del Instituto de Geografía* (Universidad Nacional Autónoma de México) 10: 46–48.

McClung de Tapia, E. 1980. *Informe Preliminar sobre el Análisis de Restos Botánicos Carbonizados, Procedentes de la Excavación de Canales Prehispánicos de Riego en el Municipio de Otumba, Estado de México*. Manuscript, Instituto de Investigaciones Antropológicas, Universidad Nacional Autónoma de México.

———. 1990a. Analysis of paleoethnobotanical remains from TA-80: Preliminary results. In T. H. Charlton and D. L. Nichols, eds., *Preliminary Report on Recent Research in the Otumba City-State*, pp. 223–231. Mesoamerican Research Report 3. Iowa City: Department of Anthropology, University of Iowa.

———. 1990b. Ecología, agricultura y ganadería durante la Colonia. *Historia General de la Medicina en México*, vol. 2, pp. 60–77. Mexico City: Academia Nacional de Medicina, Facultad de Medicina, Universidad Nacional Autónoma de México.

———, coord. 1996. *Informe Técnico del Proyecto Cambios Paleoambientales y sus Efectos Sociales en Teotihuacán 1995–6, Primera Parte, Subproyecto: El Paleoambiente de la Región de Teotihuacán*. Informe Técnico 4, Laboratorio de Paleoetnobotánica y Paleoambiente. Mexico City: Instituto de Investigaciones Antropológicas, Universidad Nacional Autónoma de México.

————. 1997. Mesoamerica: Agriculture and ecology. *Encyclopedia of Mexico,* pp. 812–821. Chicago: Fitzroy-Dearborn.

McClung de Tapia, E., and D. Martínez Yrizar. 1996. *Plant Resources from Postclassic Xaltocan, Estado de México.* Informe Técnico 3, Laboratorio de Paleoetnobotánica y Paleoambiente. Mexico City: Instituto de Investigaciones Antropológicas, Universidad Nacional Autónoma de México.

McClung de Tapia, E., M. C. Serra Puche, and A. Limón de Dyer. 1986. Formative lacustrine adaptation: Botanical remains from Terremote-Tlaltenco, D.F., México. *J. Field Archaeology* 13: 99–113.

McClung de Tapia, E., E. Ibarra Morales, J. Zurita-Noguera, and M. Meza Sanchez. 1995. *Prehistoric Human Impact in the Teotihuacan Region of Central Mexico.* Paper presented during the International Union for Anthropological and Ethnological Sciences Inter-Congress on Biodemography and Human Evolution, Florence, Italy, 19–27 April.

McClung de Tapia, E., J. Zurita, E. Ibarra, J. Cervantes, and M. Meza. 1998. Cronología de procesos geomorfológicos en el Valle de Teotihuacán. In R. Brambila and R. Cabrera, coords., *Los Ritmos de Cambio en Teotihuacán: Reflexiones y Discusiones de su Cronología,* pp. 503–518. Mexico City: Instituto Nacional de Antropología e Historia.

Millon, R. 1957. Irrigation in the Valley of Teotihuacan. *American Antiquity* 23: 160–166.

————. 1973. *The Teotihuacan Map, Part 1: Text.* Austin: University of Texas Press.

Nichols, D. L. 1980. *Prehispanic Settlement and Land Use in the Northwestern Basin of Mexico, the Cuauhtitlan Region.* Ph.D. dissertation, Pennsylvania State University. Ann Arbor, Mich.: University Microfilms.

————. 1982. A Middle Formative irrigation system near Santa Clara Coatitlan in the Basin of Mexico. *American Antiquity* 47: 133–144.

————. 1987. Risk and agricultural intensification during the Formative period in the northern Basin of Mexico. *American Anthropologist* 89: 596–616.

————. 1988a. Reply to Feinman and Nicholas: There is no frost in the Basin of Mexico? *American Anthropologist* 91: 1023–1026.

————. 1988b. Infrared aerial photography and Prehispanic irrigation at Teotihuacan: The Tlajinga canals. *J. Field Archaeology* 15: 17–27.

————. 1997. *Production Intensification and Regional Specialization: Maguey Fibers and Textiles in the Aztec City-State of Otumba.* Paper presented at the Sixty-second Annual Meeting of the Society for American Archaeology, Nashville, Tenn.

Nichols, D. L., and C. D. Frederick. 1993. Irrigation canals and chinampas: Recent research in the northern Basin of Mexico. *Research in Economic Anthropology* Suppl. 7: 123–150.

Nichols, D. L., M. W. Spence, and M. D. Borland. 1991. Watering the fields of Teotihuacan: Early irrigation at the ancient city. *Ancient Mesoamerica* 2: 119–129.

Niederberger, C. 1975. *Zohapilco: Cinco Milenios de Ocupación Humana en un Sitio Lacustre de la Cuenca de México.* Mexico City: Instituto Nacional de Antropología e Historia.

————. 1987. *Paléopaysages et Archéologie Pré-Urbaine du Bassin de Mexico.* Collection Etudes Mésoaméricaines 1–2. Mexico City: Centre d'Etudes Mexicaines et Centraméricaines.

Palerm, A. 1961. Sistemas de regadío prehispánico en Teotihuacán y en el Pedregal de San Angel. *Revista Interamericana de Ciencias Sociales* segunda época 1(2).

Parsons, J. R. 1974. The development of a prehistoric complex society: A regional perspective from the Valley of Mexico. *J. Field Archaeology* 1: 81–108.

————. 1976. The role of chinampa agriculture in the food supply of Aztec Tenochtitlan. In C. E. Cleland, ed., *Cultural Change and Continuity,* pp. 233–257. New York: Academic Press.

————. 1983. *Archaeological Settlement Pattern Data from the Chalco, Xochimilco, Iztapalapa, Texcoco, and Zumpango Regions, Mexico.* Museum of Anthropology Technical Reports 14. Ann Arbor: University of Michigan.

————. 1990. Critical reflections on a decade

of full-coverage regional survey in the Valley of Mexico. In S. K. Fish and S. A. Kowalewski, eds., *The Archaeology of Regions: A Case for Full-Coverage Survey*, pp. 7–31. Washington, D.C.: Smithsonian Institution Press.

———. 1991. Political implications of Prehispanic chinampa agriculture in the Valley of Mexico, In H. Harvey, ed., *Land and Politics in the Valley of Mexico: A Two Thousand Year Perspective*, pp. 17–42. Albuquerque: University of New Mexico Press.

Parsons, J. R., E. M. Brumfiel, M. Parsons, V. Popper, and M. Taft. 1981. *Late Prehispanic Chinampa Agriculture on Lake Chalco, Xochimilco, Mexico*. Manuscript on file, Museum of Anthropology, University of Michigan, Ann Arbor.

Pastrana, A., and P. Fournier. 1997. Cuicuilco desde Cuicuilco, *Actualidades Arqueológicas* 3(13): 7–9, Mexico.

Popper, V. 1988. *Reconstructing the Prehispanic Vegetation of Lake Chalco*. Paper presented at the Twelfth Ethnobiology Conference, Riverside, California.

Reyna Robles, R. M., and L. González Quintero. 1978. Resultados del análisis botánico de formaciones troncocónicas en Loma Torremote, Cuauhtitlan, Estado de México. In *Arqueobotánica: Métodos y Aplicaciones*, pp. 33–42. Colección Científica 63. Mexico City: Departamento de Prehistoria, Instituto Nacional de Antropología e Historia.

Rojas Rabiela, T. 1988. *Las Siembras de Ayer: la Agricultura Indígena del Siglo 16*. Mexico City: Secretaría de Educación Pública, Centro de Investigación y Estudios Superiores de Antropología Social.

Rojas Rabiela, T., P. A. Strauss, and J. Lameiras. 1974. *Nuevas Noticias Sobre las Obras Hidráulicas y Coloniales en el Valle de México*. Mexico City: Secretaría de Educación Pública, Instituto Nacional de Antropología e Historia.

Ruvalcaba Mercado, J. 1985. *Agricultura India en Cempoala, Tepeapulco, y Tulancingo, Siglo 16*. Mexico City: Departamento del Distrito Federal, Unión de Ciudades Capitales Iberoamericanas.

Rzedowski, J., and G. C. de Rzedowski. 1979. *Flora Fanerogámica del Valle de México*, vol. 1. Mexico City: Editorial CECSA.

Rzedowski, J., G. Guzmán, A. Hernández, and R. Muñiz. 1964. Cartografía de la vegetación de la parte norte del Valle de México. *Anales de la Escuela Nacional de Ciencias Biológicas del Instituto Politécnico Nacional* 13: 31–57.

Sanders, W. T. 1965. *The Cultural Ecology of the Teotihuacan Valley*. University Park: Department of Sociology and Anthropology, Pennsylvania State University.

———. 1985. Tecnología agrícola, economía, y política: Una introducción. In T. Rojas Rabiela and W. T. Sanders, eds., *Historia de la Agricultura: Epoca Prehispánica–Siglo 16*, pp. 9–52. Mexico City: Instituto Nacional de Antropología e Historia.

Sanders, W. T., and B. J. Price. 1968. *Mesoamerica: The Evolution of a Civilization*. New York: Random House.

Sanders, W. T., and R. S. Santley. 1977. A prehispanic irrigation system near Santa Clara Xalostoc in the Basin of Mexico. *American Antiquity* 42: 582–588.

Sanders, W. T., J. R. Parsons, and R. S. Santley. 1979. *The Basin of Mexico: Ecological Processes in the Evolution of a Civilization*. New York: Academic Press.

Serra, M. C. 1988. *Los Recursos Lacustres en la Cuenca de México durante el Formativo*. Coordinación General de Estudios de Posgrado, Instituto de Investigaciones Antropológicas. Mexico City: Universidad Nacional Autónoma de México.

Steward, J. H. 1955. *The Theory of Culture Change*. Urbana: University of Illinois Press.

Tolstoy, P., S. K. Fish, M. W. Bosenbaum, K. B. Vaugnn, and C. E. Smith, Jr. 1977. Early sedentary communities of the Basin of Mexico. *J. Field Archaeology* 4: 91–106.

Wittfogel, K. 1957. *Oriental Despotism: A Comparative Study of Total Power*. New Haven: Yale University Press.

CHARLES S. SPENCER

6 PREHISPANIC WATER MANAGEMENT AND AGRICULTURAL INTENSIFICATION IN MEXICO AND VENEZUELA
Implications for Contemporary Ecological Planning

In this essay I examine three archaeological cases that document how Prehispanic agriculturalists intensified production by coming up with workable solutions to water supply problems. In Mexico, the inhabitants of two regions in the arid southern highlands constructed irrigation systems that marshaled scarce, seasonally available runoff from mountain slopes for use in nearby agricultural fields; in both regions, agriculture would have been vastly less productive if farmers had relied upon local rainfall alone. In western Venezuela, Prehispanic people dealt with an overabundance of water during the six-month rainy season by constructing drained-field systems that allowed for longer and much more productive growing seasons than would have otherwise been possible in this area of seasonal flooding. In the concluding section I discuss how these archaeological data could be of practical use to ecological planners and policymakers who are seeking to promote strategies of sustainable devlopment.

BACKGROUND

Although there are many impediments to a sustainable global economy, few are more daunting than the growing imbalance between human populations and the production of food. The situation is especially worrisome in much of the developing world, where agricultural improvements have slowed in recent years but fertility rates continue to be high, in contrast to a few industrialized nations that are approaching population stabilization (Flavin 1997:16–19);

as a consequence, a "widening gap between the demand and the supply of grain can be seen in nearly all of the more populous developing countries" (Brown 1996:106). If present trends continue, the annual shortfall of grain in eleven of the largest developing countries (whose total population is two-thirds of all humanity) will increase from 38 million tons in 1990 to a projected 407 million tons in 2030 (Brown 1996:106–107, table 7-2). A deficit of such magnitude is expected to have an array of serious consequences, from widespread hunger and disease to social disintegration and political violence (Renner 1996:35–51). While it is clear that reducing the rate of population growth in the developing world should be a high priority, the most effective response to food scarcity is probably not fertility reduction alone, since "fundamental reforms in many developing countries in education, health care, and the status of women, along with a massive global reordering of priorities" (Brown et al. 1991:99), are necessary for birth-control programs to be successful, and such reforms will probably take many years. A more promising, albeit more complex, approach is to find ways of increasing food production while simultaneously applying the brakes to human population growth (Brown 1996:119–135).

Because croplands provide most of the earth's food (either directly through grains, fruits, and vegetables or indirectly through livestock fed with crops), cropland expansion would seem to be a logical response to food scarcity. Yet the prospects for a major increase are not bright. The amount of cropland in the world peaked in 1980, at about 710 million hectares, and has since declined; on a per capita basis, global cropland area shrank dramatically from 0.24 ha/person in 1950 to 0.12 ha/person in 1990 (Gardner 1997:44) and is projected to decline still further, to 0.08 ha/person by 2030 (Brown 1996:71). Compounding the problem has been a reduction over the past decade in the rate of increase in crop yields. Crop yields were raised over the three and half decades following 1950 through the introduction of new, high-yielding varieties and strains, but since 1985 the increase in yields has slowed, falling well behind the rate of population growth (Gardner 1997:44). The combination of growing populations, shrinking croplands, and slowing yield improvements has led to a sharp global decline in the per capita grain harvest, from 346 kg/person in 1984 to 295 kg/person in 1995 (Brown 1996:36).

Given these trends, food production will probably have to be increased through agricultural intensification: "with little prospect of expanding the cultivated area, satisfying future food needs thus depends on raising the productivity of existing cropland" (Brown et al. 1991:85). Irrigation, which can allow more reliable cropping in areas of scarce or highly variable water supplies, is one potentially effective technique of agricultural intensification (Postel 1992). Yet, here too, population growth is pulling ahead: the amount of irrigated cropland per capita was 0.037 ha/person in 1950, hit a peak of 0.048 ha/person in 1979, but in 1993 had declined to 0.044 ha/person (Brown 1996: figure 5-2). Irrigation will be able to enhance food production over the long term only if water is consistently deployed with maximal benefit to crops but with minimal waste. "Just as satisfying future food needs depends on using land more intensively, so too it depends on using water more efficiently. With the amount of fresh water produced by the hydrological cycle essentially fixed by nature, raising efficiency is the key to producing more food where water is already scarce" (Brown et al. 1991:86–87).

Achieving greater efficiency in water usage, however, will probably be hindered by human-caused climate changes that alter the global distribution of water. A number of environmental scientists have predicted that among the major consequences of global warming will be a climatic regime in which some areas receive far less rain and others far more than is presently the case (Silver and DeFries 1990:67–75); future irrigation projects will have to be flexible enough to deal with such perturbations. Researchers recently have been urged to investigate whether small-scale irrigation strategies tailored to local conditions might have a higher probability of success over the long term than large-scale projects run by centralized bureaucracies (Mabry 1996; Stone 1997). Crumley (1994) has suggested that archaeology is particularly well suited to the task of providing diachronic data on the successes and failures of various intensification strategies in other times and places.

THE PURRÓN DAM, TEHUACÁN, MEXICO

The first case study is situated in the arid Arroyo Lencho Diego, in the Valley of Tehuacán, Mexico (figure 6.1). During the late 1960s, members of the Tehuacán Archaeological-Botanical Project investigated a massive water-control facility, called the Purrón Dam (figure 6.2; Brunet 1967; Woodbury and Neely 1972). Habitation sites associated with the structure were mapped and sampled by the author through controlled, intensive surface collecting in June–August of 1976 (Spencer 1979, 1993). These investigations indicate that the dam was in use between 700 B.C. and A.D. 250 and, over this period, underwent four major enlargements, with the final construction stage bringing the dam to undeniably monumental proportions: some 400 m long, 20 m high, and 100 m thick at the base (figures 6.3, 6.4). The reservoir associated with this fourth stage had an estimated capacity of 2,640,000 m^3 (Woodbury and Neely 1972:82–99). The human population in the Arroyo Lencho Diego also showed dramatic growth, from an estimated 45–60 persons at 700 B.C. to 975–1,190 by A.D. 250 (Spencer 1979). The water stored by the Purrón Dam was undoubtedly used to irrigate the substantial alluvial fan (measuring approximately 425 ha) that lies between the dam and the Río Salado (figure 6.5; Spencer 1979:63).

It is important to recognize that rainfall-based farming is not reliable in this locality. Although average rainfall is 400–500 mm per year, mostly during the June–September rainy season, there is considerable interyear variability, and extremely dry years are not uncommon. Moreover, a strong rainshadow effect is produced by wind flowing off the high peaks of the Sierra Madre Oriental to the east. When evapotranspiration is taken into account, there is an 80 percent probability of useful rainfall of only 132–168 mm per year (Byers 1967:54). This is well below the amounts needed for successful rainfall farming (Kirkby 1973), and all agriculture in the Lencho Diego area today is dependent upon irrigation. Unfortunately, the Río Salado (which flows the length of the Tehuacán Valley) lives up to its name and is so salty that contemporary farmers in the

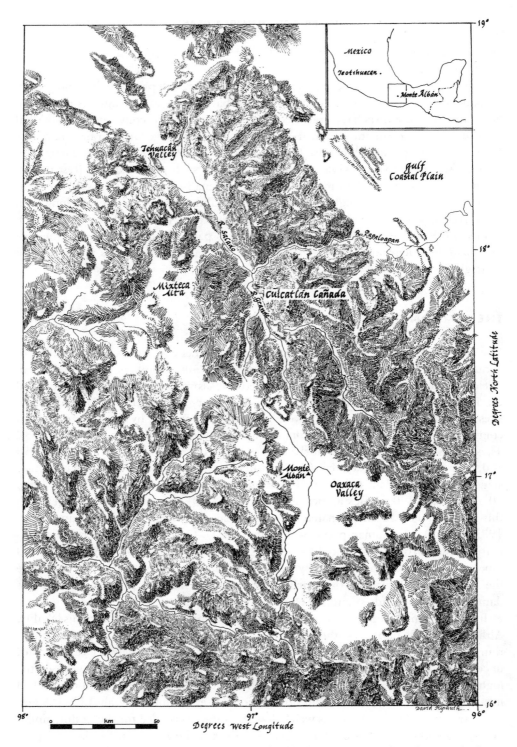

FIGURE 6.1. The southern Mexican highlands, including the Tehuacán Valley, the Cañada de Cuicatlán, and adjacent regions.

FIGURE 6.2. The Purrón Dam, looking north across the southeastern end, where the dam has been cut by erosion since its abandonment around A.D. 250.

Lencho Diego area, most of whom live in the small town of San Rafael (see figure 6.5), do not use its water for irrigation. Instead, they rely upon runoff that courses through the tributary arroyos during the rainy season (from May to October, primarily). This water originates as rainfall on the higher slopes of the Sierra Madres to the east, which have a longer and wetter rainy season than the Tehuacán Valley proper. This runoff is not salty like that of the Salado, into which it eventually drains, but harnessing it for agricultural purposes presents a significant problem.

In numerous visits to the locality between 1975 and 1998, I have seen no evidence that the local inhabitants were irrigating with runoff from the Arroyo Lencho Diego itself. Instead, the fields on the alluvial fan of the Arroyo Lencho Diego (see figure 6.5) were being irrigated with water that originated in a tributary arroyo more than 12 km to the north, near the town of Coxcatlán. The water arrives at San Rafael by way of a concrete-lined canal that was built by the irrigation commission of the Mexican federal government, the Secretaría de Recursos Hidráulicos (SRH). The present farmers of San Rafael use this water to cultivate maize and sugarcane on the alluvial fan. We are relatively certain, however, that the amount of land currently being irrigated by the SRH canal is less than that which was kept under irrigation two millennia ago with water from the Purrón Dam. The reason for this is that the SRH canal enters the Lencho Diego area at an elevation that is too low to provide water to all parts (especially the

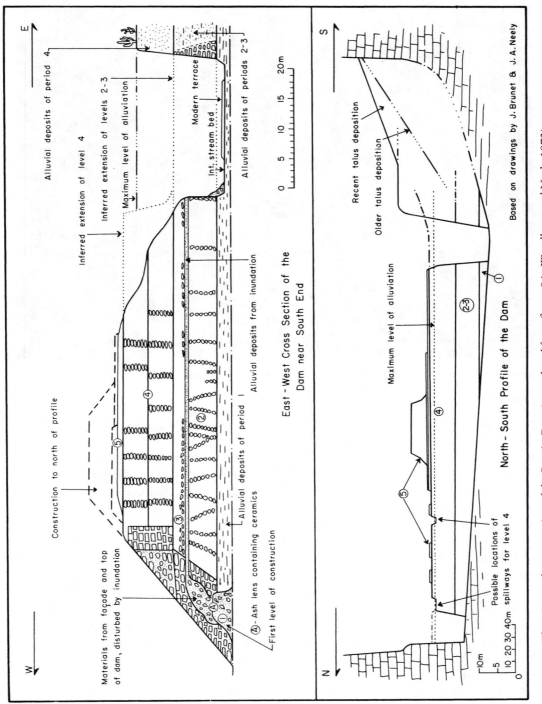

FIGURE 6.3. Elevation and cross-section of the Purrón Dam (reproduced from figure 8 in Woodbury and Neely 1972).

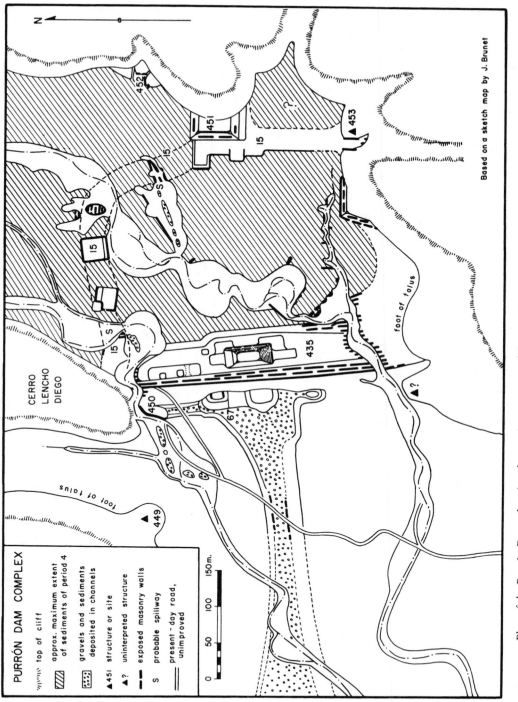

FIGURE 6.4. Plan of the Purrón Dam, showing the maximum extent of the reservoir, the presumed cofferdam (shown as site 15), and other associated features (reproduced from figure 9 in Woodbury and Neely 1972).

Based on a sketch map by J. Brunet

PURRÓN DAM COMPLEX

- top of cliff
- approx. maximum extent of sediments of period 4
- gravels and sediments deposited in channels
- ▲451 structure or site
- ▲? uninterpreted structure
- exposed masonry walls
- S probable spillway
- present-day road, unimproved

0   50   100   150 m.

CERRO LENCHO DIEGO

foot of talus

foot of talus

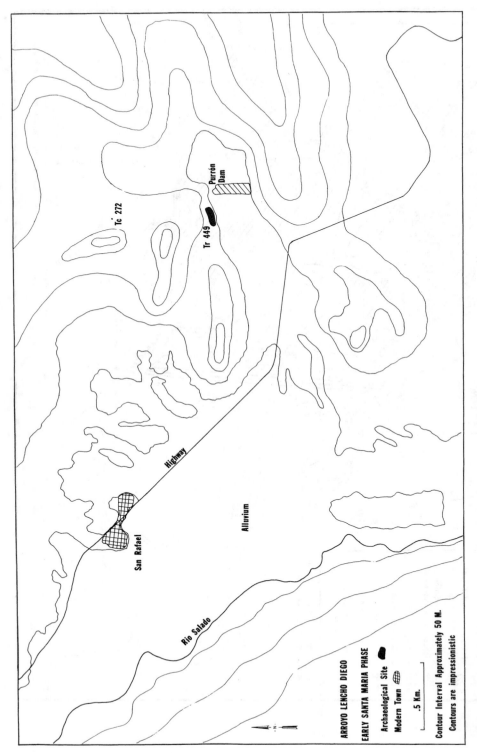

FIGURE 6.5. Early Santa María phase settlement in the Arroyo Lencho Diego.

higher elevations) of the alluvial fan, whereas the Purrón Dam reservoir is about 50 m higher in elevation than the SRH canal and was capable of supplying water through gravity flow to all parts of the alluvium (see figure 6.5).

The current climatic regime of the Tehuacán Valley has been in force since the on-set of the Holocene, according to the paleoenvironmental data from the Tehuacán Archaeological-Botanical Project (Byers 1967). It is thus not surprising that the first sedentary village in the Arroyo Lencho Diego, the Early Santa María phase site of Tr449, was accompanied by the first efforts at water management here, Level 1 of the Purrón Dam (see figures 6.3 and 6.5). The Level 1 dam, dated to about 700 B.C., was a structure 6 m wide, 2.8 m high, and 175 m long; it contained an estimated 2,940 m³ of fill and created a reservoir with an estimated capacity of 37,000 m³ (Woodbury and Neely 1972:82–99). Tc272 (Purrón Cave), which was occupied during the Archaic period, continued in use as a temporary collecting-hunting camp.

Tr449 was a 1-ha occupation consisting of 9–12 residences, with an estimated population of 45–60 persons (Spencer 1979:24, figure 2.4). Although there was no evident variation in household size, two of the houses in the central part of the site were associated with stone foundations while the remaining 7–10 were not. The surface collections suggest that the inhabitants of these two houses enjoyed somewhat higher social standing than their neighbors (Spencer 1993:50–51, table 1). Level 1 of the Purrón Dam comprised 2,940 m³ of fill. I have calculated that a workforce of just 3–8 people could have built Level 1 of the dam in a single dry season (Spencer 1979:56–57). This project could easily have been the work of the two households with stone foundations at Tr449 and perhaps one or two other adjacent households. Interestingly, the carrying capacity estimates I have calculated suggest that the Level 1 reservoir could not have irrigated sufficient land to provide all the inhabitants of Tr449 with all of their sustenance (Spencer 1979: table 2.13). The Early Santa María subsistence data from Purrón Cave indicate a varied strategy that mixed horticulture with hunting and col-lecting (MacNeish et al. 1972:95–115). What the Purrón Dam probably did was in-crease the reliability of domesticated plant production, so that yields were sufficient to sustain year-round life in the arroyo for the first time.

Level 2 of the Purrón Dam was built around 600 B.C. (Woodbury and Neely 1972: 93). This was a massive construction of 123,000 m³, bringing the dam to a length of 400 m, a width of 100 m, and a total height of 8 m (see figures 6.3, 6.4). A reservoir with an estimated capacity of 1,430,000 m³ was created (ibid.:82–99). This task ob-viously called for the coordinated efforts of a large labor force. It appears to have been a two-step process; first, a cofferdam and associated spillway (Tr15) were constructed in a single dry season to divert water away from the construction effort and thus allow the main dam to be built over more than one year (see figure 6.4). This cofferdam, which contains about 40,000 m³ of fill, would have required an estimated workforce of 41–104 people for its construction, a figure almost identical to the labor require-ments (41–106) of building Level 2 proper of the dam over three dry seasons (Spencer 1979:56–58, table 2.8).

While the construction of Level 2 demanded a much larger labor force than Level 1, it is also the case that the arroyo's human population was increasing, probably the result of internal processes of growth as well as immigration. In the Middle Santa María phase (ca. 600–450 B.C.), Tr449 was abandoned and two new habitation sites comprising 30–34 households were founded (Tr452 and Tr67), along with a single ceremonial mound (Tr450) and Purrón Cave (Tc272), which continued to serve as a temporary camp (figure 6.6). If each Middle Santa María phase household contributed 2–3 members to the construction effort, a workforce of 60–102 workers would have been available, a figure consistent with the estimated manpower requirements of building the Tr15 cofferdam followed by the Level 2 dam over three dry seasons.

I have calculated that the water impounded by Level 2 of the dam could have irrigated far more land than would have been necessary just to produce food for the arroyo's population during the Middle Santa María phase (Spencer 1979:63–74). The carrying capacity estimates of the irrigation-based regime during this phase range from 50–80 households (single cropping) to 100–161 households (double cropping); as noted, the Middle Santa María phase population is estimated at 30–34 households. To what purposes were these vast quantities of water being put? If we examine the floral data from Purrón Cave (Tc272), we can see that Level 2 of the dam was associated with a dramatic increase in domesticated plants over Level 1 (MacNeish et al. 1972:378–432; Spencer 1979:67). Among them were not only the basic staples (maize, squash, etc.) but also a series of plants that would have grown in the *tierra caliente* ("hot country") environment of the Arroyo Lencho Diego (850–950 m above sea level), but not at the higher elevations of the central Tehuacán Valley (1,500–1,700 m) or the Puebla Basin (2,000 m). These plants include cotton (*Gossypium hirsutum*), fruits such as white zapote (*Casimiroa edulis*), black zapote (*Diospyros digyna*), ciruela (*Spondias mombin*), chupandilla (*Cyrtocarpa procera*), and the coyol palm nut (*Acrocomia mexicana*, recently reclassified as *Acrocomia aculeata*), as shown in table 12 of Woodbury and Neely (1972:93). Archaeological data indicate that these tropical plants were trade items in Prehispanic Mexico. For example, deposits containing black zapote and coyol palm nut fragments were excavated at the important Middle-Late Santa María site of Quachilco, about 50 km north of the Arroyo Lencho Diego in the central Tehuacán Valley, a habitat where neither plant thrives (Smith 1979:225).

It therefore appears that Level 2 of the Purrón Dam was built large enough so that land could be irrigated not only to meet the nutritional needs of the arroyo's population but also to permit the cultivation of certain tropical plants for use as trade items. In exchange for their tropical plants, the arroyo's inhabitants received goods that were not locally available. In our surface collections the most evident of these nonlocal items was obsidian, the nearest sources of which are in the Puebla Basin, more than 100 km to the north. Obsidian was just one of a number of important trade items that moved through exchange networks that connected many population centers in ancient Mesoamerica (Pires-Ferreira 1975).

The Purrón Dam continued in use during the Late Santa María phase (450–150 B.C.), when two more sites (Tr451 and Tr131) appeared in the locality (figure 6.7), bringing the total estimated population to 96–105 households, or 480–525 people at 5 persons/household (Spencer 1979:26, 36–39). Level 3 of the dam, little more than a repair job, was built at the interface between the Late Santa María and Early Palo

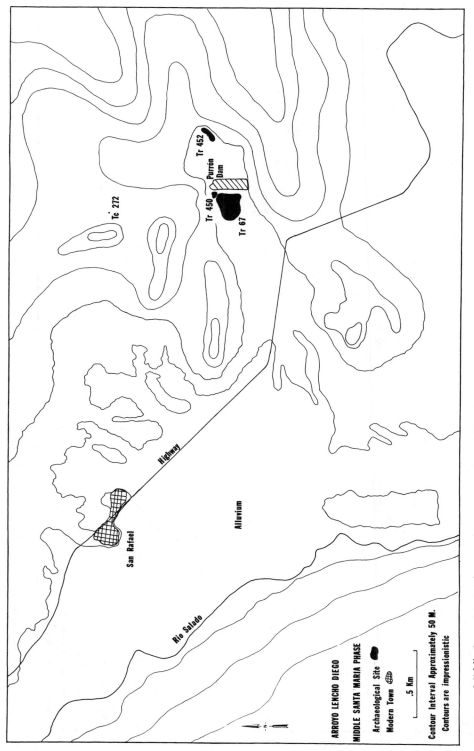

FIGURE 6.6. Middle Santa María phase settlement in the Arroyo Lencho Diego.

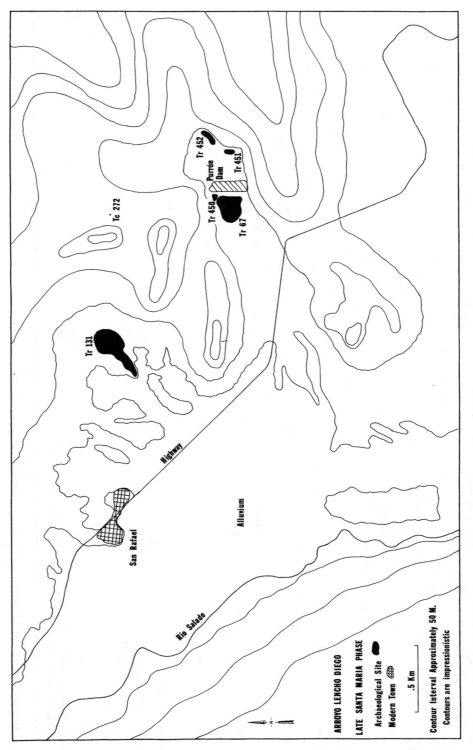

FIGURE 6.7. Late Santa María phase settlement in the Arroyo Lencho Diego.

Blanco phases (Woodbury and Neely 1972:90–91). Level 4 was then constructed during the first half of the Early Palo Blanco phase (150 B.C.–A.D. 250), which brought the dam to the overall size that we see today.

During the Early Palo Blanco phase, human occupation in the arroyo shifted to a nearby hilltop (figure 6.8). Protected by a double defensive wall, the inhabitants of Tr73/79 were able to survive the widespread instability that characterized this time period (Spencer 1979:71–73). The total estimated population grew to 195–238 households, or 975–1,190 people at 5 persons/household (ibid.: table 2.1). At the end of the Early Palo Blanco phase, the habitation sites and the Purrón Dam were abandoned, and the locality saw no more than sparse occupation until the Colonial period. Even the modern town of San Rafael has but a small fraction of the population that lived in the arroyo during the heyday of the Purrón Dam; in the 1990 census, San Rafael had a population of only 260 (INEGI 1995:1187).

Just why the arroyo was abandoned around A.D. 250 is not well understood, but we suspect it was due to larger political forces, such as rising tensions between Monte Albán and Teotihuacán, and particularly the apparently successful penetration by the latter into the Tehuacán region (Drennan and Nowack 1984). I see no reason to think that a failure of the Purrón Dam irrigation system was at fault. My estimates of local carrying capacity (Spencer 1979: table 2.13) indicate that the productive potential of the irrigation-based agricultural regime was more than adequate to support the local population throughout the 850 years when Levels 2–4 of Purrón Dam were in use. Even when local population was at its height in the Early Palo Blanco phase, the estimated carrying capacity of the irrigation-based regime ranges from 145–233 households (single cropping) to 290–467 households (double cropping), figures that are quite compatible with the independent population estimate of 195–238 households for that time period.

THE CAÑADA DE CUICATLÁN, OAXACA, MEXICO

The second case study is located about 50 km south of the Arroyo Lencho Diego, in the Cañada de Cuicatlán, Oaxaca (see figure 6.1). The Cañada's elevation of 500–700 m places it well within the tierra caliente zone, and precipitation here averages less than 300 mm per year (Hopkins 1984:5). When evapotranspiration rates are accounted for, the useful precipitation is only 130–140 mm per year, less than is required for reliable rainfall agriculture (Hunt and Hunt 1974:136–137). On the other hand, various forms of irrigation are possible with water that originates as rainfall in the high mountains that surround the Cañada and traverses the region by way of numerous small tributary streams and two major rivers, the Río Grande and the Río de las Vueltas. In contrast to the Río Salado in Tehuacán, these rivers and streams all contain water that is not too salty for irrigation.

Cultivation in the Cañada is limited to four localities, each of which is associated with an alluvial fan: at Quiotepec, Cuicatlán, San José el Chilar, and Dominguillo (figure 6.9). During our 1977–1978 field project (Redmond 1983; Spencer 1982; Spencer

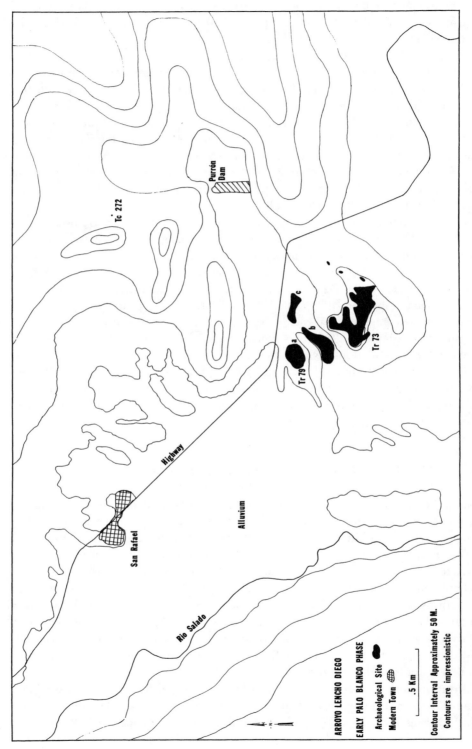

FIGURE 6.8. Early Palo Blanco phase settlement in the Arroyo Lencho Diego.

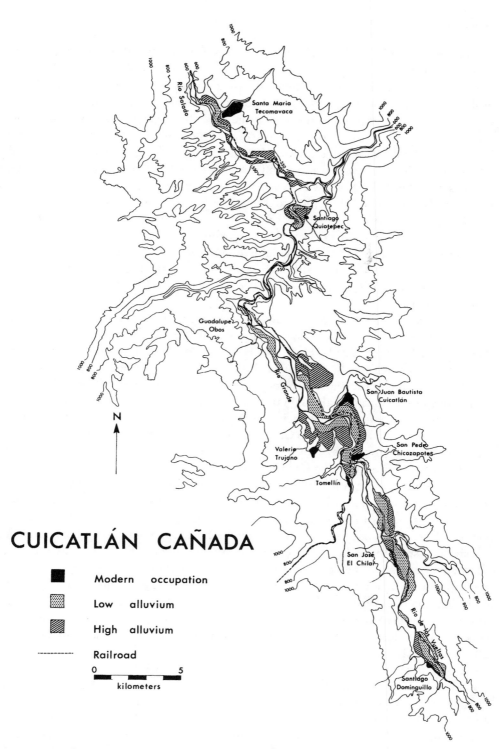

CUICATLÁN CAÑADA

- ■ Modern occupation
- ▨ Low alluvium
- ▧ High alluvium
- ---------- Railroad

0 ___ 5
kilometers

FIGURE 6.9. The Cañada de Cuicatlán, showing the four alluvial fans at Quiotepec, Cuicatlán, San José el Chilar, and Dominguillo.

and Redmond 1997), Redmond and I noted that there are two distinct kinds of alluvium in each locality. Low alluvium is prime bottomland along the riverbanks; soil fertility is high and sustained by annual flooding, usually after the fall harvest. Fields in the low alluvium often can be cultivated without elaborate canal irrigation. High water table farming is possible in some areas, while floodwater irrigation and simple diversionary dam irrigation also are used. In the latter case, simple dams built of tree limbs, brush, and river cobbles are used to deflect water from the river channel and divert it onto the adjacent low alluvium (Hopkins 1984:77; Redmond 1983: plate 8). There is a total of 997 ha of low alluvium in the four alluvial fans of the Cañada, with maize yields that range from 1,000 to 2,000 kg of husked maize per planting. Many fields in the low alluvium, moreover, can produce two maize crops per year. The high alluvium is a zone of older deposits that lie farther from the riverbank and overlook the low alluvium. The 737 ha of high alluvium in the Cañada are usually farmed by means of true canal irrigation. Although maize yields on the high alluvium are rarely greater than 500–1,000 kg of husked maize per year, double-cropping is possible in many locations.

At the time of our 1977–1978 fieldwork, there were many small-scale canal irrigation systems and one large-scale system in the Cañada; these systems also have been discussed by Hopkins (1984), Hunt and Hunt (1974), and Hunt (1994). The large-scale system, known locally as the Canal de Matamba, was being used to irrigate a significant portion of the high alluvium, and the low alluvium as well, on the east side of the river in the Cuicatlán alluvial fan (Hunt 1994). The concrete-lined main canal is 24 km long and brings water from a reservoir created by the Matamba Dam, located on the upstream portion of the Río Grande about 3 km east of San Pedro Chicozapotes. The entire Matamba canal system was built in the 1960s by the SRH. Hunt (1994) has criticized the engineering of the Canal de Matamba, noting that it is oversized, capable of bringing much more water than can actually be used by farmers in the Cuicatlán locality; the expansion of irrigation agriculture here since the building of the canal has been limited more by the amount of irrigable land at elevations reachable by the canal than by the quantity of water transported by the canal. Hunt also faulted the Canal de Matamba system for its high maintenance costs. Throughout the rest of the Cañada, the high alluvium was being irrigated in 1977–1978 with relatively small-scale canal systems that brought water to the fields from nearby tributary streams. We noted, however, that not all of the high alluvium in the Cañada was being cultivated in this fashion. Indeed, in the Dominguillo locality we found archaeological evidence of more extensive canal irrigation on the high alluvium during the Lomas phase (200 B.C.–A.D. 200) and the Trujano phase (A.D. 200–1000) than was the case at the time of our fieldwork, and this was still so when I visited the locality in 1997.

Our 1977–1978 fieldwork documented prehistoric settlement in the Cañada from about 600 B.C. to the early sixteenth century. There were twelve archaeological sites dating to the first phase of occupation, the Perdido phase (600–200 B.C.), with a total habitation area of 38 ha. All the Perdido phase sites are located on high alluvial terraces or low piedmont spurs directly overlooking stretches of low alluvium. Because our survey found no irrigation facilities associated with any of these sites, we have inferred that the inhabitants were probably not farming the high alluvium (which does require true canal irrigation) at this time, but instead were cultivating only the low alluvium, perhaps using simple techniques such as diversionary dam irrigation. Consistent with that interpretation is our analytical conclusion that the estimated populations of all the Perdido phase sites are comfortably below the calculated carrying capacities of just the low alluvium in their localities. For example, at the site where we conducted excavations, Llano Perdido (Cs25) in the Dominguillo alluvial fan, there are 142.65 ha of low alluvium, which Redmond's calculations suggest could have supported 814 people (1983: table 5). This is much higher than the population estimates for this 2.25 ha site, which range from a minimum of 26–66 persons (at 10–25 persons/ha) to a maximum of 140–195 persons, extrapolating from the density of excavated residences at 5 persons/household (ibid.; Spencer 1982:144). In short, there would have been no need to carry out canal irrigation agriculture on the high alluvium in the Perdido phase.

During the Lomas phase (200 B.C.–A.D. 200) major changes came to the Cañada de Cuicatlán. The Lomas phase is contemporaneous with the famous Building J "conquest slab" inscriptions at Monte Albán, the capital of the early Zapotec state in the Oaxaca Valley. One of these inscriptions has been interpreted as referring to the Cañada as a place conquered by Monte Albán (Marcus 1976). The data from the 1977–1978 fieldwork support the hypothesis of a Zapotec subjugation of the Cañada during the Lomas phase (Redmond 1983; Spencer 1982; Spencer and Redmond 1997).

At the onset of the Lomas phase, all of the Perdido phase sites on the Cañada's high alluvium were suddenly abandoned, and new settlements were established on nearby piedmont ridges. In the Cuicatlán, El Chilar, and Dominguillo alluvial fans, there were fourteen sites, none of which was more than 5 ha in size. These fourteen sites yield a total of 27.30 ha of occupation. A strikingly different pattern, however, is seen in the Quiotepec area, at the northern end of the Cañada where a narrow mountain pass is the only natural route north to Tehuacán. Here on the smallest of the Cañada's alluvial fans (92.40 ha of low and high alluvium), where there had previously been only a single 1.5 ha Perdido phase village, seven new settlements were established in the Lomas phase, totaling 44.34 ha of occupation. These settlements sprawled across both sides of the only natural pass into the Cañada and also occupied strategic mountain peaks. Heavily fortified, the Quiotepec sites were undoubtedly designed to control movement through the Cañada's northern frontier (Redmond 1983:91–120). The Quiotepec installation also marks the northern limit of Lomas phase pottery, some of which is very similar to contemporaneous Monte Albán Ic–II pottery from the Oaxaca Valley (ibid.: 86).

We estimate that a total of 1,940–2,050 people lived on the Quiotepec alluvial fan in the Lomas phase (388–410 house mounds at 5 persons/house mound; Redmond 1983: table 16). This is a much larger population than is seen in any other Prehispanic period in this locality (ibid.: tables 4, 35, 40), and it is even higher than at present; the human population of Quiotepec in the 1990 census was just 261 (INEGI 1995:2689). Sustaining the extraordinary Lomas phase occupation almost certainly required that food be brought in from outside the Quiotepec locality. Indeed, our calculations suggest that there were more than four times as many people residing here during the

Lomas phase than could have been supported by farming the Quiotepec alluvial fan (Redmond 1983:105–106). In the Cañada as a whole, the Lomas phase occupation is 71.64 ha, with an estimated population of 707–4,125 (ibid.: tables 16, 26). This is comfortably below the regional carrying capacity for the Lomas phase, calculated to be 9,625 by Redmond (ibid.: table 18). I should point out that Redmond utilized both the low alluvium and the high alluvium of each locality in making these calculations, because we recovered firm archaeological evidence for the use of canal irrigation on the high alluvium during the Lomas phase.

In the Dominguillo locality, there was an abandonment of the Perdido phase settlement (Llano Perdido) on the high alluvium and the establishment of a Lomas phase village on a nearby ridge (Loma de La Coyotera), a pattern repeated throughout the central and southern Cañada at this time (figure 6.10). Our excavations at Llano Perdido produced evidence that the abandonment of this site was attended by violence. For example, we recovered tremendous quantities of burned daub and adobe, suggesting that the community was burned to the ground. Furthermore, we discovered the remains of a woman and a child lying on the floor of a residence (Burial 1 in House 1); they probably perished when the village was destroyed by a Zapotec military force.

The succeeding Lomas phase settlement of Loma de La Coyotera occupied a nearby ridge and covered 3.04 ha, only slightly larger than the Perdido phase community on the high alluvium. Yet, the Lomas phase village differed from its predecessor in a number of ways. The basic unit of residence changed from the large compound of earlier times to a pattern of one or two residences on a terrace carved into the hillside. The Lomas phase pattern is much more like the "household clusters" (Winter 1976) or "household units" (Flannery 1983) observed at Formative period sites in the Valley of Oaxaca. We suspect that the Zapotec may have enacted a deliberate policy of breaking up the traditional residential compounds to disrupt interfamilial relationships; this would have allowed for more effective control over the conquered Cañada population. Like the other thirteen settlements located on hilltops in the central and southern Cañada, the public sector of Loma de La Coyotera in the Lomas phase consisted of a single plaza, in this case with two small mounds on a 3 m tall platform. This actually represented a reduction in the amount of ceremonial construction compared to the preceding Perdido phase occupation at Llano Perdido, the rich ritual life of which was attested to by abundant ceremonial paraphernalia; ceremonial artifacts also show a precipitous decline in our Lomas phase deposits. Public space at Loma de La Coyotera was dominated by the fearsome presence of the Zapotec state, in the form of a skull rack that we excavated in the middle of the Lomas phase plaza (Spencer 1982: 234–242).

At Loma de La Coyotera we recorded 35–38 residential terraces and a total of 55–60 house mounds or house foundations for the Lomas phase. The total population of Loma de La Coyotera at this time is estimated to be 275–300 (55–60 house mounds plus 35–38 residential terraces at 5 persons/residence) (Redmond 1983: table 26). This is well below the Lomas phase carrying capacity of 1,210 (ibid.: table 18) of the Dominguillo locality, which was calculated using both the low alluvium (142.65 ha) and the high alluvium (59.2 ha) in the locality.

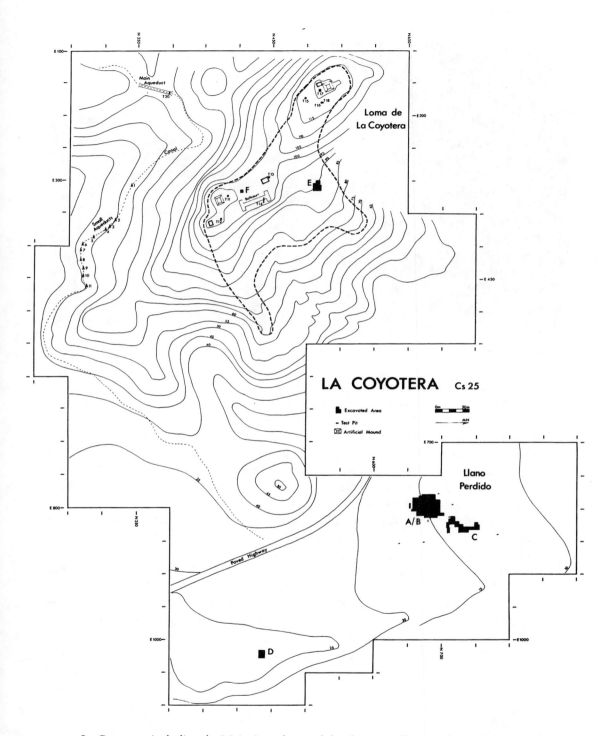

FIGURE 6.10. La Coyotera, including the Main Aqueduct and the eleven smaller aqueducts; the extent of Lomas phase occupation at the site is also indicated.

FIGURE 6.11. The Main Aqueduct at La Coyotera, looking southeast.

An elaborate irrigation facility was found during our survey of Loma de La Coyotera. We mapped the route of an ancient canal that brought water down from a tributary barranca behind the site, over a dozen gullies and depressions where twelve different aqueducts were constructed of stone, and then channeled the water onto the expanse of high alluvium where the previous Perdido phase community of Llano Perdido had been located (see figure 6.10). Stone-lined feeder canals were located in a corner of our Llano Perdido excavations, stratigraphically above the Perdido phase deposits (Spencer and Redmond 1997:528). We carried out an excavation in the Main Aqueduct (figures 6.11, 6.12), which revealed that the earliest canal here, the canal bed labeled C3 on figure 6.12, was constructed in the Lomas phase (200 B.C.–A.D. 200), when the Llano Perdido site on the high alluvium was abandoned in favor of the ridgetop. The canal beds labeled C2 and C1 both date to the Trujano phase (A.D. 200–1000).

With the introduction of canal irrigation, the Cañada's 740 ha of high alluvium could have been brought under cultivation (joining the 997 ha of low alluvium), resulting in an enormous expansion of the agricultural regime. Yet, as at La Coyotera, the local population residing at all the settlements in the central and southern Cañada still remained far below the carrying capacity of just the low alluvium in their localities (Redmond 1983:123–126). It therefore seems likely that Lomas phase production was expanded largely in response to demands by the Zapotec state for tribute. The items produced as tribute probably included basic foodstuffs, which would have been particularly important for the provisioning of the frontier facility at Quiotepec. However, there is also evidence that certain tropical fruits and nuts, the kinds of items that were previously exchanged between the Cañada and the Oaxaca Valley, were now being cultivated here on a very large scale, possibly in response to tribute demands. On a residential terrace at Loma de La Coyotera, excavated midden deposits contained extremely high densities of tropical fruit and nut remains, especially the coyol palm (*Acrocomia aculeata*), the black zapote (*Diospyros digyna*), and the ciruela (*Spondias mombin*). These items occurred in much higher densities than was the case in the earlier Perdido phase midden deposits (Spencer 1982:228). Much of the newly irrigated high alluvium was probably sown with these tierra caliente species, all of which require considerable amounts of water (Smith 1979:241). With the increase in the production of these plants at La Coyotera came a decrease in the variety of craft activities performed. Our Lomas phase deposits showed a sharp drop in the relative frequencies of craft-related artifacts (such as spindle whorls, sherd disks, and loom weights) from Perdido times (Spencer 1982:230). This pattern may reflect a policy of enforced economic specialization in the central and southern Cañada, enacted by Zapotec administrators to increase the relative labor time spent in tribute production.

The canal at La Coyotera continued to be used during the succeeding Trujano phase (A.D. 200–1000), when our data indicate that the Cañada was no longer under Zapotec control (Redmond 1983:145–154; Spencer 1982:253–255). During the Trujano phase, a ballcourt and additional mounds were built at La Coyotera, and the total population of the Dominguillo locality grew to an estimated 1,345–1,675, based on 269–335 house mounds at 5 persons/house mound (Redmond 1983: table 35). The midpoint of this range is 1,510; this is compatible with the carrying capacity estimate of 1,541 people, which Redmond calculated could have been supported by the low and

Main Aqueduct Profile   (Test 20)

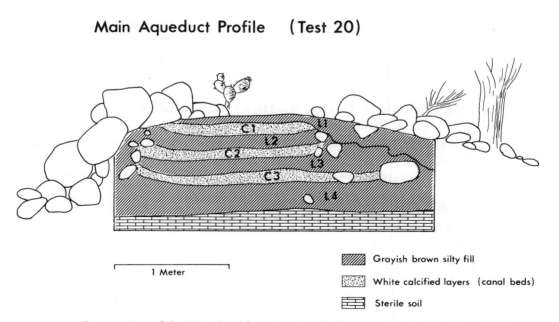

1 Meter

Grayish brown silty fill

White calcified layers (canal beds)

Sterile soil

FIGURE 6.12. Cross-section of the Main Aqueduct, showing the three canal beds (C1, C2, and C3).

high alluvial zones taken together (ibid.: table 36). The Trujano phase population in the Dominguillo locality was much higher than that seen at present; in the 1990 census, the population of Dominguillo was 477 (INEGI 1995:2689). Since the Trujano phase population in the Dominguillo locality was greater than the 1,264 people that Redmond computed could have been sustained with the low alluvium alone, it seems likely that the high alluvium in the Trujano phase was being used less for specialized fruit and nut production and more, along with the low alluvium, for basic subsistence. Our Trujano phase deposits, in fact, contain few examples of the nonstaple tropical fruits and nuts that were so abundant in the Lomas phase proveniences (Spencer and Redmond 1997:505–529, 548–568). At the beginning of the succeeding Iglesia Vieja phase (A.D. 1000–1520), both the Loma de La Coyotera site and the irrigation system were abandoned, and settlement shifted to a ridge overlooking the eastern bank of the Río de las Vueltas (ibid.: 569–598), although occupation returned to the western side of the river during the Colonial period.

It is noteworthy that as recently as 1997 there was little irrigation on the high alluvium of the Dominguillo locality (Spencer, personal observation), which from 200 B.C. to A.D. 1000 had been extensively irrigated by the Loma de La Coyotera canal system. The only portion of the high alluvium currently under irrigation lies to the north of the Llano Perdido site (amounting to less than one-fourth of the 59.2 ha of all the high alluvium in the locality). A canal of modest scale built by the SRH brings water to these fields from a takeoff point several kilometers upstream on the main Río de las Vueltas. Local informants told us that just prior to our fieldwork in 1977–1978, the SRH sought to construct another long canal that would have transported water from an even higher takeoff point on the Río de las Vueltas and channel it onto the high alluvial terrace where the Llano Perdido site is situated. The canal was never built, apparently because the inhabitants of Dominguillo refused to cooperate with the project engineers. Today, the high alluvium that was irrigated for 1,200 years by the Coyotera canal system is largely scrubland; part of it sees occasional use as a soccer field.

LA TIGRA DRAINED FIELDS, BARINAS, VENEZUELA

The third case study is located in the llanos, or humid savanna grasslands, of the western Venezuelan state of Barinas. The average annual rainfall here exceeds 2,500 mm, most of it falling between May and December (Gravina et al. 1989:22, cited by Gassón 1998:2). In the llanos there is often too much water for optimal farming; disastrous seasonal floods are frequent, especially toward the end of the rainy season. Farmers usually manage to cultivate a single crop between the desultory onset of the rainy season and the high point of seasonal flooding; double-cropping is rarely attempted.

Between 1983 and 1988, Redmond and I carried out a regional-scale archaeological project in a portion of the Río Canaguá drainage of Barinas (figure 6.13; Spencer and Redmond 1992). Although most of our study region is currently a sparsely populated area of large ranches and a few small farms, 103 Prehispanic sites dating to four different time periods were located on survey, 10 of which were eventually excavated. One

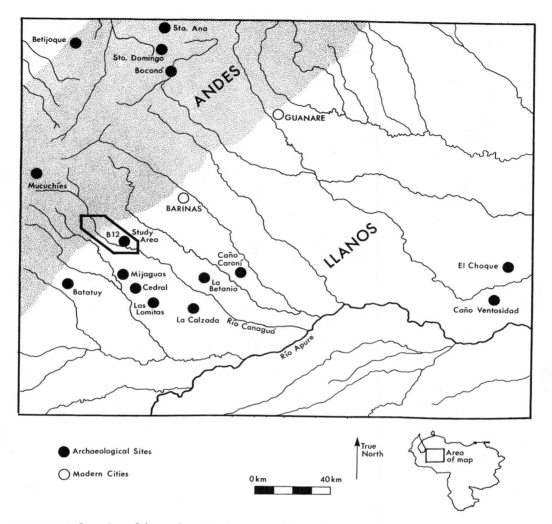

FIGURE 6.13. Location of the study region in western Venezuela.

of these was the site of La Tigra (B27), a drained-field facility covering 35 ha, which we have dated to the Late Gaván phase (A.D. 550–1000).

We located 32 Late Gaván phase occupations on the llanos during our survey, with a total occupation area of 124 ha (figure 6.14; Spencer et al. 1994). We were able to discern a regional settlement hierarchy of three levels according to site size and associated mounded architecture. The Gaván site, B12, is at the top of this hierarchy. Covering at least 33 ha, it was by far the largest site of its time and also contained an impressive assortment of earthworks, including two large mounds (12 m and 10 m tall) on opposite sides of a plaza that measures 500 m across. A number of internal elongated earthen features were also recorded, as well as 134 smaller earthen mounds, all

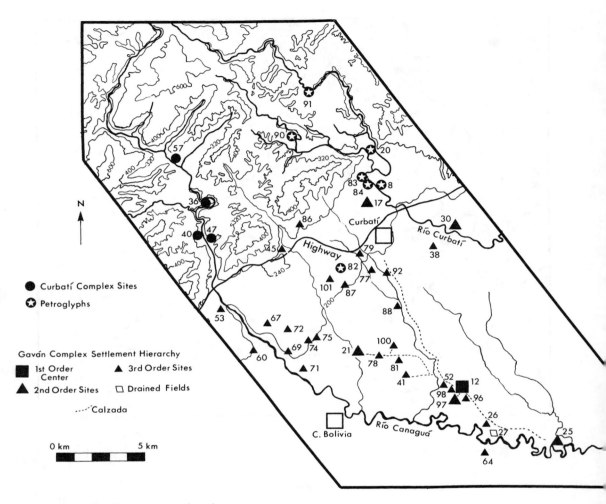

FIGURE 6.14. Late Gaván regional settlement patterns.

of which appear to have been house mounds. Variations in house-mound height, house floor area, associated artifacts, and burial patterns undoubtedly reflect considerable social differentiation by this time period (Spencer and Redmond 1992). We estimate the total population of B12 to range from a minimum of 134 households to a maximum of 200 households, or approximately 670–1,000 people at 5 persons/household.

Radiating out from the B12 site are three *calzadas* ("causeways") that link it to a number of smaller Late Gaván sites (see figure 6.14). We have interpreted this network of calzadas as a manifestation of substantial regional centralization and cohesion by this phase. Five of these sites can be interpreted as second-tier occupations in the Late Gaván regional settlement hierarchy: B97, B21, B25, B17, and B30. These sites range in size from 6–10 ha and have two to four mounds that reach 2–6 m in height. There are also 26 habitation sites in the third level of the Late Gaván regional settlement hierarchy. No mounded architecture was noted at any of these sites, which vary in size from 0.5 to 4.4 ha.

The calzada that issues from B12 toward the southeast passes alongside B96 and B26, both third-order village sites, before skirting the edge of a drained-field facility, B27 (figure 6.15). B27 is situated in the floodplain (*vega*) of the Río Canaguá, a soil type that is more fertile and far easier to till than the surrounding savanna clays. Two remnant river bends (now oxbow lagoons) and a *caño* (the Caño Colorado) were linked together by a network of canals, creating approximately 35 ha of drained fields (figure 6.16). Such a combination of natural features and artificial construction, as Zucchi (1985) has argued, was highly characteristic of Prehispanic agroengineering projects in the Venezuelan llanos. The canals themselves at La Tigra are variable in size, the widest being some 8 m across, the majority 4–6 m, and the narrowest just a meter in width. Canal depths vary from 30 cm up to 2.0 m, with most in the 0.5–1.5 m range. We mapped a total length of 2,960 m of canals with alidade, plane table, and stadia rod (figure 6.17). An additional 832 m had to be located less precisely (with compass, tape measure, and aerial photograph) because of extremely dense vegetation. In addition to the clearly artificial canals, there were modified sections in the course of

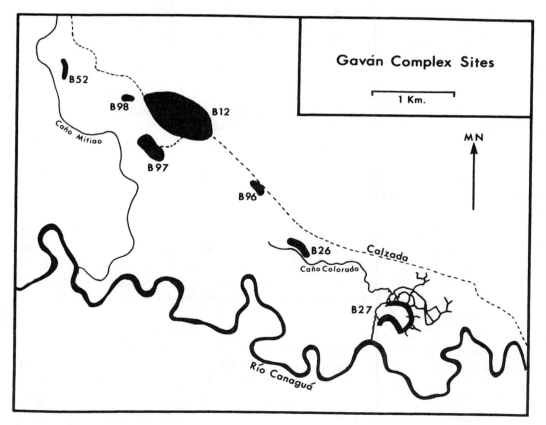

FIGURE 6.15. The Gaván Locality.

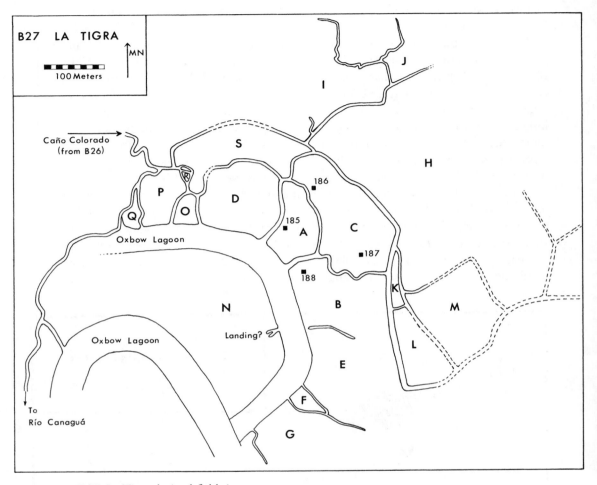

FIGURE 6.16. B27, La Tigra drained-field site.

the Caño Colorado; we estimated the total length of modified sections to be 534 m. Thus, a grand total of 4,326 m is produced by combining the lengths of canals and modified caño sections. The two oxbow lagoons appear largely unmodified, although there is one possible boat landing (see figure 6.16). The northern lagoon is 630 m long, 40–50 m wide, and 1–2 m deep, while the southern lagoon is 360 m long, 50–60 m wide, and 1–2 m deep.

It seems likely that the B26 site was the home village of those who worked the drained fields at B27 (see figure 6.15). If each of the estimated 12–18 households at B26 (Spencer et al. 1994: table 4) contributed 2 members to the effort (Kowalewski 1980:156), the available labor force would have numbered 24–36. Kirkby's data (1973: table 10) from the Valley of Oaxaca, Mexico, indicate that a traditional farmer with a digging stick can keep 2 ha under cultivation. Accordingly, we would estimate that the postulated workforce from B26 could farm 48–72 ha, more than enough to keep the 35 ha of fields at B27 under cultivation.

The present owner of the modest farm on which the drained fields are located told us that he established his farm more than 30 years ago. When he first cleared the brush, he discovered the canal network and quickly realized that it was an ancient water management facility. He cleaned out the canals somewhat and noticed that the network could be used to promote drainage during the rainy season, when plants are threatened by flooding. Further, at the beginning and end of the rainy season, when plants are threatened by dry spells, the canal system could be used for irrigation. The facility thus served to minimize crop loss through dehydration or flooding and also permitted an extension of the effective growing season. He said that when he was younger he farmed a portion of the B27 system assisted by some family members. He emphasized maize and obtained regular yields considerably higher than those reaped by other small-scale farmers in the area. He recalled that two successful harvests per year were possible because of the canal system and that the total yield was some 3,600 kg of husked maize per hectare per year. This is a relatively high yield, comparing quite favorably with, for instance, the harvests reaped on the most fertile river levee land in the southern Gulf Coast region of Mesoamerica (Coe and Diehl 1980:80).

A palynological analysis of one of the soil samples excavated at B27 revealed a predominance of maize pollen (Spencer et al. 1994: tables 2, 3), indicating that the drained-field system was largely planted in maize. Using a series of assumptions about

FIGURE 6.17. In the B27 drained fields, a field assistant holds a stadia rod in the canal between Field A and Field D.

Prehispanic cultivation practices and yields, we have estimated that the B27 facility during the Late Gaván phase could have supported 72–114 households, or some 360–570 people at 5 persons/household (ibid: table 6). Although estimates such as these have to be viewed with some caution, it is nonetheless clear that this range greatly exceeds the estimated population size of the nearby B26 village site (12–18 households; 60–90 people at 5 persons/household). We therefore propose that the drained-field system was capable of generating a considerable surplus beyond the subsistence needs of the people who farmed it. Also, because both B27 and B26 are connected by calzada to B12, we would go further to suggest that this surplus was regularly sent to the first-order center, probably as part of the operation of the regional political economy.

Late Gaván society thrived for four and a half centuries. We estimate that the total population of all the Late Gaván sites was 503–751 households, or 2,515–3,755 people at 5 persons/household (see figure 6.14). The estimated population of the six sites in the Gaván Locality is 185–275 households, or 925–1,375 people at 5 persons/ household (see figure 6.15). Then, around A.D. 1000, the Late Gaván settlement system was abandoned. We found widespread evidence of burning in the uppermost occupation levels of the B12 site, suggesting that hostilities played a key role in the collapse of Late Gaván society. The attackers possibly came from the Río Anaro drainage to the south. There, archaeologists have located another settlement system focused on Cedral, a site that was three times larger than B12 and perhaps the capital of a more powerful regional polity (see figure 6.13).

From A.D. 1000 to the present, the Gaván Locality has seen but a small fraction of the human population that lived here in Late Gaván times. We observed during our 1983–1988 fieldwork that the area was mainly in the hands of a single large-scale cattle rancher who kept about 10 permanent employees; there were also 7 small farmsteads, bringing the total contemporary population in the Gaván Locality to a maximum of about 55. Aside from the aforementioned reuse of La Tigra fields, agriculture in the area was nonintensive and spotty, and humans were greatly outnumbered by cattle.

CONCLUSION

Although our three Prehispanic systems of agricultural intensification constitute a diverse collection—from seasonal water storage to stream diversion to drained fields—they have one simple but striking feature in common: they worked. Furthermore, the data show that these systems operated successfully and continuously over the "long term" (by any reasonable definition of the phrase), sustaining the people that used them for four centuries in Barinas, nine centuries in Tehuacán, and just over a millennium in the Cañada. In each case, when the local system was abandoned it probably came about because of extralocal political realignments; insufficient productive potential on the local level does not seem to have been at fault.

It is also notable that all three Prehispanic systems utilized their local environments more effectively, and supported larger local populations, than do the systems of production that are currently in place. In Tehuacán, the canal built by the SRH enters the Arroyo Lencho Diego area at too low an elevation to reach all the potentially irrigable land; the Prehispanic Purrón Dam, by contrast, was built 50 m higher and was able to provide water by gravity feed to all parts of the alluvial zone, sustaining a local population that 800 years ago was four times larger than is presently seen. In the Cañada, the elaborate 24 km canal built by the SRH is capable of transporting more water than can actually be used in the farmland areas that it reaches, while at the same time arable lands elsewhere in the canyon are left unirrigated. Yet, between 200 B.C. and A.D. 1000, small-scale irrigation systems such as the one at La Coyotera brought water from nearby streams to lands that are not currently under irrigation. In the case of Dominguillo, the ancient irrigation system supported up to three times as many people as now inhabit the locality. In Barinas, the Gaván locality is currently occupied by some five dozen farmers and ranchers, along with the ubiquitous cattle who feast on the savanna grass. One thousand years ago, however, the locality had six villages with an estimated 925–1,375 people. A system of drained-field agriculture not only sustained the small (12–18 households) village that farmed the fields but also yielded a substantial surplus that helped support the region's first-order center, a bustling town of 134–200 households. It hardly seems rash to conclude that contemporary planners should take a look at these Prehispanic systems and consider reintroducing them to their local habitats.

At the same time, these cases contain a cautionary note for bureaucrats fond of large-scale, centralized approaches to agricultural intensification. Our three Prehispanic systems were all relatively small in scale, requiring only modest amounts of labor for their initial construction and drawing strictly upon local sources for their water supply; there is no evidence that an elaborate bureaucracy was required to build or operate any of them. The long-term success of these Prehispanic systems can be used to support the proposition that small-scale, local solutions are potentially more sustainable than large-scale, centralized strategies.

Yet, I should point out that our three Prehispanic systems were not completely local, disconnected, and isolated; in each case, the local system was linked in important ways to extralocal networks. The common ingredient was surplus production; all three systems were capable of production levels well beyond the needs of the local farmers who cultivated the irrigated or drained fields. In the Tehuacán and Cañada cases, at least part of this surplus took the form of tropical fruits and nuts. In Tehuacán, the tropical products were evidently exchanged for obsidian and other desired items not available locally. In the Cañada, tropical fruits and nuts were probably exacted as tribute by the imperialistic Zapotec state during the Lomas phase. In Barinas, it seems that surplus maize was sent from the drained fields at La Tigra to the regional center of Gaván, where it was probably stockpiled for later use. The potential for surplus production would have imparted flexibility to each system, an extra margin of productivity that could be called upon when circumstances warranted. Growing local populations, crop shortfalls, and other problems could be dealt with through modification, not drastic reconstruction, of the available system. The original decisions to build the systems so large, however, were evidently prompted not by local food scarcity per se but by other incentives and needs, such as the opportunity to engage in long-distance trade in Te-huacán, the need to meet tribute demands (at least in the Lomas phase) in the Cañada, and the desire to participate in the regional polity of the Late Gaván phase in Barinas.

Drawing upon these Prehispanic examples, I would respectfully submit the following recommendations to contemporary planners in Mexico, Venezuela, and other developing-world countries where rapidly growing populations are creating a need for sustainable strategies of intensified agricultural production:

1. Planners should avoid large-scale, centralized approaches and instead promote numerous small-scale, locally managed systems of agricultural intensification that are closely tailored to the specifics of their environmental settings.
2. These small-scale systems should nonetheless be capable of surplus production well beyond local subsistence needs.
3. Such surpluses can be used to connect the local agricultural systems and their associated populations to extralocal economic networks; the benefits that local people derive from such contacts provide important incentives for the construction of the systems of intensified agriculture in the first place, and the extra margin of productive potential serves as insurance against future subsistence needs.
4. Lasting connections between the local communities and extralocal entities will be forged most effectively through exchange ties. It hardly needs stating that imperialistic tribute exaction (the Zapotecs' long-term success with this strategy notwithstanding) is no longer acceptable; however, a modern equivalent of the Pax Zapoteca might be accomplished through international organizations, in order to minimize the likelihood of exchange-system disruption because of political instability.

Of course, such strategies of agricultural intensification should be carried out in conjunction with—not as a substitute for—effective birth-control programs; it is the combination of increasing agricultural production and slowing population growth that offers the greatest hope of reducing food scarcity in the developing world. It is also clear that the achievement of long-term sustainability will require active collaboration among many players—farmers, agronomists, entrepreneurs, demographers, bankers, politicians, and others—so that truly comprehensive development strategies can be devised and put to work. Archaeologists should have a role in this too; more research needs to be conducted on ancient methods of water management and agricultural intensification. From archaeology we can learn which strategies and techniques have worked in the past, under what conditions, and for how long; such data deserve our attention as we try to plan wisely for what lies ahead.

REFERENCES

Brown, L. R. 1996. *Tough Choices: Facing the Challenge of Food Scarcity*. New York: W. W. Norton.

Brown, L. R., C. Flavin, and S. Postel. 1991.

Saving the Planet: How to Shape an Environmentally Sustainable Global Economy. New York: W. W. Norton.

Brunet, J. 1967. Geologic studies. In D. S.

Byers, ed., *The Prehistory of the Tehuacán Valley, 1: Environment and Subsistence,* pp. 66–90. Austin: University of Texas Press.

Byers, D. S. 1967. Climate and hydrology. In D. S. Byers, ed., *The Prehistory of the Tehuacán Valley, 1: Environment and Subsistence,* pp. 48–65. Austin: University of Texas Press.

Coe, M. D., and R. A. Diehl. 1980. *The People of the River: In the Land of the Olmec,* vol. 2. Austin: University of Texas Press.

Crumley, C. L. 1994. Historical ecology: A multidimensional ecological orientation. In C. Crumley, ed., *Historical Ecology: Cultural Knowledge and Changing Landscapes,* pp. 1–16. Santa Fe: School of American Research Press.

Drennan, R. D., and J. Nowack. 1984. Exchange and sociopolitical development in the Tehuacán Valley. In K. Hirth, ed., *Trade and Exchange in Early Mesoamerica,* pp. 147–156. Albuquerque: University of New Mexico Press.

Flannery, K. V. 1983. The Tierras Largas phase and the analytical units of the early Oaxaca village. In K. V. Flannery and J. Marcus, eds., *The Cloud People: Divergent Evolution of the Zapotec and Mixtec Civilizations,* pp. 43–45. New York: Academic Press.

Flavin, C. 1997. The legacy of Rio. In L. R. Brown et al., eds., *State of the World, 1997: A Worldwatch Institute Report on Progress toward a Sustainable Society,* pp. 3–22. New York: W. W. Norton.

Gardner, G. 1997. Preserving global cropland. In L. R. Brown et al., eds., *State of the World, 1997: A Worldwatch Institute Report on Progress toward a Sustainable Society,* pp. 40–59. New York: W. W. Norton.

Gassón, R. A. 1998. *Prehispanic Intensive Agriculture, Settlement Pattern, and Political Economy in the Western Venezuelan Llanos.* Ph.D. dissertation, Department of Anthropology, University of Pittsburgh.

Gravina, G. O., C. Alvarado, Y. Oballos, J. Pereyra, and F. Vargas. 1989. *Caracterización de Suelos de la Reserva Forestal de Ticoporo, Barinas.* Merida, Venezuela: Universidad de Los Andes.

Hopkins, J. W. 1984. *Irrigation and the Cuicatec Ecosystem: A Study of Agriculture and Civilization in North Central Oaxaca.* Memoirs of the University of Michigan Museum of Anthropology 17. Ann Arbor.

Hunt, R. C. 1994. Irrigation in Cuicatlán: The question of Río Grande waters. In J. Marcus, and J. F. Zeitlin, eds., *Caciques and Their People: A Volume in Honor of Ronald Spores,* pp. 163–187. Anthropological Papers 89. Ann Arbor: University of Michigan Museum of Anthropology.

Hunt, R. C., and E. Hunt. 1974. Irrigation, conflict, and politics: A Mexican case. In T. Downing and M. Gibson, eds., *Irrigation's Impact on Society,* pp. 129–157. Tucson: University of Arizona Press.

INEGI (Instituto Nacional de Estadística, Geografía e Informática). 1995. *Conteo de Población y Vivienda 1995: Resultados Definitivos: Tabulados Básicos,* vols. 1–33. Aguascalientes, Mexico.

Kirkby, A. V. T. 1973. *The Use of Land and Water Resources in the Past and Present Valley of Oaxaca, Mexico.* Memoirs of the University of Michigan Museum of Anthropology 5. Ann Arbor.

Kowalewski, S. 1980. Population-resource balances in Period I of Oaxaca, Mexico. *American Antiquity* 45: 151–156.

MacNeish, R., M. Fowler, A. García Cook, F. Peterson, A. Nelken-Terner, and J. Neely. 1972. *The Prehistory of the Tehuacán Valley, 5: Excavations and Reconnaissance.* Austin: University of Texas Press.

Mabry, J. B., ed. 1996. *Canals and Communities: Small-Scale Irrigation Systems.* Tucson: University of Arizona Press.

Marcus, J. 1976. The iconography of militarism at Monte Albán and neighboring sites in the Valley of Oaxaca. In H. B. Nicholson, ed., *The Origins of Religious Art and Iconography in Preclassic Mesoamerica,* pp. 123–139. Los Angeles: Latin American Center, University of California.

Pires-Ferreira, J. W. 1975. *Formative Mesoamerican Exchange Networks with Special Reference to the Valley of Oaxaca.* Memoirs of the University of Michigan Museum of Anthropology 7. Ann Arbor.

Postel, S. 1992. *Last Oasis: Facing Water Scarcity.* New York: W. W. Norton.

Redmond, E. M. 1983. *A Fuego y Sangre: Early Zapotec Imperialism in the Cuicatlán Cañada, Oaxaca.* Memoirs of the University of Michigan Museum of Anthropology 16. Ann Arbor.

Renner, M. 1996. *Fighting for Survival: Environmental Decline, Social Conflict, and the New Age of Insecurity.* New York: W. W. Norton.

Silver, C. S., and R. S. DeFries. 1990. *One Earth, One Future: Our Changing Global Environment.* Washington, D.C.: National Academy of Sciences Press.

Smith, J. 1979. Carbonized botanical remains from Quachilco, Cuayucatepec, and La Coyotera: A preliminary report. In R. D. Drennan, ed., *Prehistoric Social, Political, and Economic Development in the Area of the Tehuacán Valley: Some Results of the Palo Blanco Project,* pp. 217–246. Technical Reports 11. Ann Arbor: University of Michigan Museum of Anthropology.

Spencer, C. S. 1979. Irrigation, administration, and society in formative Tehuacán. In R. D. Drennan, ed., *Prehistoric Social, Political, and Economic Development in the Area of the Tehuacán Valley: Some Results of the Palo Blanco Project,* pp. 13–109. Technical Reports 11. Ann Arbor: University of Michigan Museum of Anthropology.

———. 1982. *The Cuicatlán Cañada and Monte Albán: A Study of Primary State Formation.* New York: Academic Press.

———. 1993. Human agency, biased transmission, and the cultural evolution of chiefly authority. *J. Anthropological Archaeology* 12: 41–74.

Spencer, C. S., and E. M. Redmond. 1992. Prehispanic chiefdoms of the western Venezuelan llanos. *World Archaeology* 24: 134–157.

———. 1997. *Archaeology of the Cañada de Cuicatlán, Oaxaca.* Anthropological Papers 80. New York: American Museum of Natural History.

Spencer, C. S., E. M. Redmond, and M. Rinaldi. 1994. Drained fields at La Tigra, Venezuelan llanos: A regional perspective. *Latin American Antiquity* 5: 119–143.

Stone, G. D. 1997. Review of *Canals and Communities: Small-Scale Irrigation Systems,* ed. by J. Mabry. Tucson: University of Arizona Press. In *American Antiquity* 62: 764–766.

Winter, M. 1976. The archaeological household cluster in the Valley of Oaxaca. In K. V. Flannery, ed., *The Early Mesoamerican Village,* pp. 25–31. New York: Academic Press.

Woodbury, R. B., and J. A. Neely. 1972. Water control systems of the Tehuacán Valley. In F. Johnson, ed., *The Prehistory of the Tehuacán Valley, 4: Chronology and Irrigation,* pp. 81–153. Austin: University of Texas Press.

Zucchi, A. 1985. Recent evidence for pre-Columbian water management systems in the western llanos of Venezuela. In I. S. Farrington, ed., *Prehistoric Intensive Agriculture in the Tropics,* pp. 167–180. BAR International Series 232. Oxford: British Archaeological Reports.

NICHOLAS DUNNING
TIMOTHY BEACH

7 | STABILITY AND INSTABILITY IN PREHISPANIC MAYA LANDSCAPES

When Spaniards first arrived on the shores of the Yucatán
Peninsula in the early 1500s they encountered a civilization
already 2,500 years old. This landscape was much altered
by centuries of human activity including alternating periods of relative stability and instability. Like much of the
New World, the Maya Lowlands were far removed from
their pristine or prehuman past (Denevan 1992). Despite
severe population declines accompanying the collapse of
Classic Maya civilization between A.D. 800 and 1000,
population was again growing and expanding in many regions by A.D. 1500. Even regions largely abandoned since
A.D. 800 still showed many effects of earlier environmental
change. Nevertheless, the Spaniards regarded the region as
largely a wilderness, a view that shaped their activities and
helped to mask the true character of the civilization that
had long preceded them in this tropical region.

In this essay we review the nature of environmental
change that accompanied the course of Prehispanic Maya
civilization. The essay begins with a brief overview of changing thought regarding fundamental aspects of ancient Maya
settlement, agriculture, and environmental impact. It then
reviews the environmental variability of the Maya Lowlands and the complex and changing adaptations made by
Maya peoples in this tropical region. We also include case
studies of the Río de la Pasión region of Guatemala and
Three Rivers region of Belize where we have been studying ancient Maya environmental archaeology and cultural
ecology since 1990.

CHANGING VIEWS ON THE ANCIENT MAYA

Like that of most Europeans arriving in the New World in the sixteenth and seventeenth centuries, the Spaniards' understanding of the Yucatán was overshadowed by religious and attendant practices operated according to theologically informed politics and economics (e.g., Sanford 1969; Mugerauer 1995). This view not only justified the conversion and reorganization of native populations but also included a view of the New World environment as a mostly untamed wilderness, given by God for human use. In colonial Yucatán, conflicting perceptions of nature led to a fundamental misunderstanding concerning land use and land rights (Clendinnen 1987; McAnany 1995). In a time when population was declining rapidly under the scourge of Old World diseases, long- or forest-fallow agriculture was becoming the prevalent mode of Maya cultivation. The Maya term for fallow land was *k'ax,* a word that also indicates secondary forest. To the Spaniards, *k'ax* became simply "forest," unused wilderness needing to be put to good uses such as cultivation but also conservation for shipbuilding and construction (McAnany 1995:64–68; Simonian 1995:29). The convenient misinterpretation of *k'ax* fit the Spanish program of social and economic reorganization, which included population relocation and the appropriation of inherited or ancestral lands. In this process a simplified form of swidden cultivation was taken to be the typical Maya agriculture, a misunderstanding that significantly distorted early scientific views of Prehispanic Maya cultural ecology.

Scientific thinking on ancient Lowland Maya agriculture and its attendant environmental impacts has grown and matured in the past several decades. While our knowledge about the ancient Maya world has grown exponentially, what we still do not know is daunting. We can also view our changes in thinking, as Turner (1993) has recently discussed, in Kuhnian terms of paradigm shifts or changing orthodoxies. As noted above, early scientific thinking about the Prehispanic Maya was strongly influenced by the effects of centuries of colonial and postcolonial changes to the Maya world. Until the mid to late 1960s, a view prevailed that ancient Maya civilization represented an anomaly in world history: a complex civilization supported by rotating, long-fallow (swidden) agriculture in a relatively homogeneous, environmentally limiting tropical forest setting. This view also saw ancient Maya population densities as necessarily low and dispersed, and Maya cities as largely vacant ceremonial centers. For example, Cowgill (1962) studied indigenous long-fallow milpa systems in the central Petén and estimated that the southern Maya Lowlands could support a permanent population of ca. 40 to 80 persons per km^2. In contrast to this old orthodoxy, evidence kept mounting for much more densely settled rural and urban populations as well as for intensive agricultural land use manifest in field walls, terraces, and visible wetland fields. Thus, that old, deterministic orthodoxy, derived from the available information of the period, was overturned in the 1970s by a view that recognized the environmental heterogeneity of the Maya Lowlands and the variety of adaptive agricultural systems that were used to transform the region into a largely intensively cultivated landscape (e.g., Harrison and Turner 1978).

More recently, the new orthodoxy has also been challenged as too anthrocentric. Recent scholarship indicates that both relatively permanent elements within the lowland environment (e.g., groundwater hydrology) and factors of environmental change (e.g., environmental degradation or climate change) may have significantly constrained Maya agriculture as well as influenced the course of Maya civilization. Nowhere has the debate between the new orthodoxy and its challengers been more heated than in the discussion over the island field systems of northern Belize and adjacent areas. Here patterned wetland features have been viewed as entirely natural, entirely artificial, or some chronological combination in origin (cf. Adams et al. 1981, 1990; Harrison 1990, 1996; Jacob 1995; Pohl et al. 1990, 1996; Pope and Dahlin 1989, 1993; Pope et al. 1996; Turner and Harrison 1983). We know much more information with the addition of more narrowly focused and trained scholarship about ancient Maya wetlands and all other aspects of the agroenvironment, but we are far from achieving any consensus about wetland agriculture or one about the extent of environmental degradation.

What is becoming abundantly clear is that no single model of adaptive systems can be applied ubiquitously across the Maya Lowlands. Clearly, ancient Maya agriculture was highly varied in nature, adapting to a mosaic of environments (see Dunning 1992, 1996; Fedick 1996) and responding to shifting political and economic pressures (see Pyburn 1996). Elsewhere (Dunning et al. 1998a), we have reviewed the spatial variation in ancient Maya agriculture in greater detail. Figure 7.1 shows this variation as environmental and agricultural regions within the Maya Lowlands. Even this mosaic of environmental regions, however, is a coarse representation of the actual heterogeneity of the lowland landscape.

THE MAYA LOWLANDS ENVIRONMENT

The cultural region of the Maya Lowlands includes the Yucatán Peninsula, other low-lying contiguous areas of Central America, and a small area of higher elevation in the Maya Mountains. The chief environmental factors that create a mosaic of habitats across this region include variation in rainfall, soil, geomorphic processes, and the slope gradients and drainage caused by structural geology.

Annual rainfall totals generally grade from a low of about 500 mm in the northwest to a high near 2,500 mm in the southern extremes of the lowlands, though the higher elevation of the Maya Mountains induces more orographic rainfall as well as a rainshadow to their west. Year-to-year variation is quite high, necessitating inventive risk management planning by Maya farmers. Further, the distribution of rainfall is highly uneven throughout the year, with a pronounced and variable winter dry season generally from November through April.

Most of the Maya Lowlands are underlaid by carbonate rocks, chiefly limestone. This simple fact has had profound implications for human settlement because limestone effects peculiar water movement and soil development and fertility. These parent

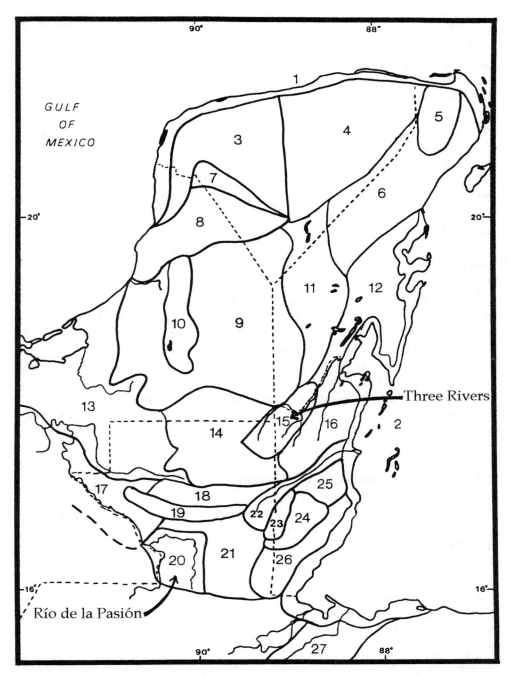

FIGURE 7.1. Adaptive regions of the Maya Lowlands. 1, North Coast. 2, Caribbean Reef and Eastern Coastal Margin. 3, Northwest Karst Plain. 4, Northeast Karst Plain. 5, Yalahau. 6, Coba-Okop. 7, Puuc–Santa Elena. 8, Puuc–Bolonchen Hills. 9, Central Hills. 10, Edzna–Silvituk Trough. 11, Quintana Roo Depression. 12, Uaymil. 13, Río Candelaria–Río San Pedro. 14, Petén Karst Plateau. 15, Three Rivers. 16, Río Hondo. 17, Lacandon Fold. 18, Petén Itza Fracture. 19, Libertad Anticline. 20, Río de la Pasión. 21, Dolores. 22, Belize River Valley. 23, Vaca Plateau. 24, Maya Mountains. 25, Hummingbird Karst. 26, Karstic Piedmont. 27, Motagua and Copán Valleys. (Adapted from Dunning et al. 1998a.)

material and climatic factors largely explain the panregional distribution of soils across the lowlands. Soils in the north tend to be shallow, well drained, clayey, and calcareous. Soils in the south tend to be deeper, poorly drained, clayey, and calcareous, but also more leached. Within the various subregions of the lowlands, geologic structure is also an important factor creating variation in the spatial distribution of soils. Faulting and karst dissolution have created a landscape with enough drainage variation to have produced a mosaic of land surfaces varying from well-drained uplands with generally shallow but fertile Rendoll soils (dark, productive, carbonate-rich soils) to areas of seasonal inundation with deeper, Vertisol soils (shrinking and swelling clay) and some perennially flooded wetlands with waterlogged Histosol soils (peat or muck). Geologic structure also largely governs the regional availability of both surface and subsurface water. Surface and spring-fed rivers drain the margins of the southern half of the peninsula, but perennial surface water is virtually absent and groundwater largely inaccessible throughout much of the hilly central portions of the Yucatán and surrounding Karst Plain. Along many coastal margins, cenotes, springs, and caves make water more readily accessible.

Natural vegetation across the Maya Lowlands follows rainfall patterns and the seasonality of water supply, though edaphic factors such as soil drainage and chemistry, cultural factors, and lithologic factors are also influential (Hartshorn 1988; Flores and Espejel Carvajal 1994; Murphy and Lugo 1995). According to Greller (this volume), much of the Yucatán falls into the category of tropical semideciduous forest, or "bosque tropical subcaducifolio" of Rzedowski (1981). Following a transect from the northwest coast southward through the Yucatán, the dry beach ridges of the northwest coast merge into a 10–20 km swath of swampy, mangrove and palm, estuarine wetlands along the coast. This zone then merges into a thorn woodland ("bosque espinosa") and savanna complex in the seasonally inundated fringes of the region (see figure 7.1). These grade to the south and east, as precipitation increases, into the semideciduous woodlands of Yucatán State and surroundings, which in turn grade into tropical semideciduous forest and tropical lowland evergreen rain forest (or "bosque tropical perennifolio") in the areas of the highest rainfall. This tropical semideciduous forest still has a long dry season in the northern Petén, and the forest breaks up into a savanna zone (probably induced by edaphic factors) around La Libertad in the north-central Petén. The tropical lowland evergreen rain forest, or "tropical moist forest," of the southern half of the Petén and Belize is not strictly a tropical rain forest, but this forest type has diverse species assemblages nonetheless (Hartshorn 1988).

ENVIRONMENTAL AND CULTURAL HISTORY

The Pleistocene-Holocene transition in the Maya Lowlands was manifest in a shift from dry savanna conditions to more mesic forest conditions, which were fully established by 5000 B.C. (Leyden et al. 1993, 1998; Rice 1993; Hodell et al. this volume). During the middle Holocene, optimum forest starts to change toward a more herbaceous and secondary assemblage, often with the presence of maize pollen, in most places between 2000 and 1000 B.C., though 1,000 to 1,600 years earlier in some places (Jones 1994; Pohl et al. 1996; Islebe et al. 1996; Dunning et al. 1998a). In the Petén Lakes District this transition corresponds to the rapid deposition of thick Maya clay, which along with the secondary ecological assemblage dominated the region until the time of the Spanish *entradas* (figure 7.2). In general, the arrival of Maya people is marked across the lowlands by this combination of primary forest clearance, maize and disturbance taxa, and accelerated soil erosion. Pollen and sediment evidence from the central Petén lakes indicates a pattern of escalating environmental disturbance for the course of Preclassic through Classic Maya civilization (ca. 500 B.C.–A.D. 900) (see figure 7.2). Several scholars have estimated Late Classic (A.D. 550–800) population densities at ca. 200 km^2 for the end of the Late Classic in the central Maya Lowlands (Culbert and Rice 1990). By this time the central Petén landscape was essentially deforested, with most available land given over to agriculture. In addition to the need for agricultural land, regional forests were undoubtedly depleted by a growing demand for construction materials and firewood (for Copán, see Abrams and Rue 1988). In denuded areas with appreciable slopes, soil erosion was increasingly stripping sloping land surfaces to bedrock. Despite years of research in the central Petén lakes area, however, we still know comparatively little about the precise nature of Classic Maya agriculture because few vestiges of field walls, terraces, or wetland fields have yet to be identified here. What we do know is that both population levels and the complex civilization they supported crashed dramatically around the Petén lakes between A.D. 800 and 900. In the aftermath of the Postclassic (A.D. 1000–1500), the greatly diminished regional population continued to affect forest assemblages through extensive swidden cultivation and silviculture, but many climax tropical forest species returned.

Several studies have now analyzed the oxygen isotopes and geochemistry of sediment cores to show a long period of aridity in the Yucatán around A.D. 200 to 1300 that intensifies during the Maya collapse about A.D. 700 to 900 (Hodell et al. 1995, this volume; Curtis et al. 1996; Whitmore et al. 1996). We do not yet know how extensive or intensive this drought was, because evidence thus far is conflicting. Several authors suggest aridity may also be a factor in the Petén (e.g., Islebe et al. 1996), and another drought may coincide with the end of the Yucatán's Postclassic florescence (A.D. 1368 to 1429), but Leyden et al. (1996) could find no evidence for it in the dry, northwest Yucatán. Hodell et al. (1995) have suggested that the drought periods in the Maya Lowlands relate to pan-Caribbean climatic cycles, but it is also likely that human modifications of the environment would have also contributed to regional drought problems, particularly the effects of prolonged deforestation on local soil water balances and transpiration rates (see Leyden et al. 1998).

While the general trajectory of cultural and environmental history outlined above for the Petén lakes region applies loosely to many other parts of the Maya Lowlands (e.g., Copán, see Wingard 1996), it is reckless to use it as a general model for the entire area (Dunning et al. 1998a). Many regions also experienced significant cultural disruptions toward the end of the Late Preclassic period (ca. A.D. 200). For example, the huge urban centers of El Mirador and Nakbe on the central Petén karst plateau were largely abandoned at this time, possibly because of earlier severe environmental degradation

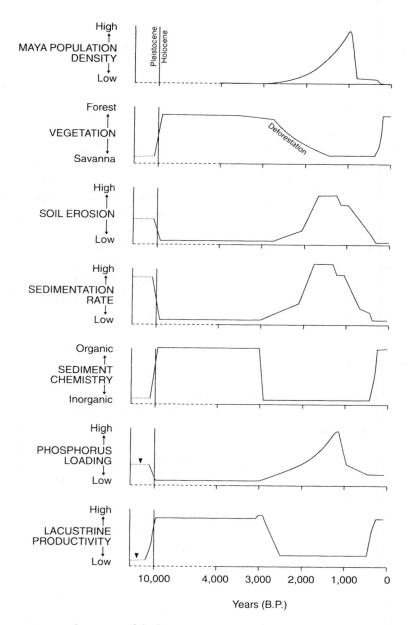

FIGURE 7.2. Summary of the long-term impacts of Maya settlement on the terrestrial and lacustrine environments of the central Petén lakes. *Source:* After Rice 1993.

(Hansen 1995; Jacob 1995). Changing water table levels may have influenced cycles of abandonment of wetland field systems in northern Belize at the end of the Late Preclassic (Rejmankova et al. 1995). Water table changes may have similarly adversely affected the Yalahau region of northern Quintana Roo at end of the Early Classic, ca. A.D. 500 (Fedick and Hovey 1995). In other regions more long-term agricultural success was achieved with the intensive cultivation of hilly uplands. Massive urban population concentration was achieved at the city of Caracol on the Vaca Plateau through the use of elaborate terrace-based agriculture (Chase and Chase 1998). Terrace agriculture also sustained dense rural populations across large areas of the Central Hills (Turner 1983). In some areas human ingenuity and labor overcame significant environmental obstacles. For example, the Maya sustained a dense urban population at the city of Chunchucmil on the semiarid northwest Karst Plain perhaps in part by importing organic soil over considerable distances to establish and maintain raised garden beds (Beach 1998a).

Nevertheless, no region of the Maya Lowlands was immune from at least periodic failure, and regional population levels often waxed and waned, sometimes catastrophically. The Río de la Pasión and Three Rivers regions both experienced tremendous shifts of human success and failure associated with changing levels of environmental and social perturbation. We examine these two regions in succession below.

RÍO DE LA PASIÓN REGION

The Río de la Pasión lowlands of the southwestern Petén District of Guatemala are a region of sharply divided terrain (figure 7.3). The region is largely underlaid by a series of late Cretaceous–early Tertiary carbonate rocks which are formed into a series of uplands (horsts) and troughs (grabens) by normal faults. The Petexbatun Escarpment and its adjacent trough that contains the Laguna and Río Petexbatun is one of these prominent horst and graben, fault-block structures. The combination of the structural lows of the grabens, backflows from the Río de la Pasión as it debouches from the highlands, the high regional water table, and approximately 2,000 mm of precipitation that arrives mostly between May and November results in prolonged inundation of much of the low-lying areas. This inundation often produces a 10 m annual fluctuation in water levels in these lowlands, thus limiting agriculture and occupance.

Between 1990 and 1996 we studied the interrelationship of ancient settlement and environment in the region as part of the Vanderbilt University Petexbatun Regional Archaeological Project (Demarest 1997; Dunning et al. 1997). As part of that research we undertook a coring program at Laguna Tamarindito to test a series of interrelated hypotheses: (1) that the Petexbatun region underwent progressively greater deforestation and environmental disturbance as Maya population grew and expanded from Preclassic through Late Classic times, as evident elsewhere in the Petén (figure 7.2; summarized in Rice 1993); (2) that initial forest clearance for agriculture generated serious soil erosion, as suggested by studies in local watersheds (Dunning and Beach 1994; Beach and Dunning 1995; Beach 1998b); and (3) that soil erosion around La-

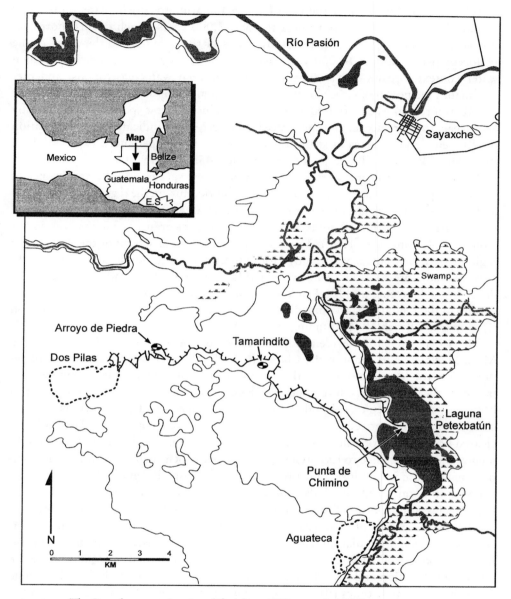

FIGURE 7.3. The Petexbatun region (modified from Killion et al. 1991).

guna Tamarindito was partially checked by the construction of terraces and dams, as suggested by archaeological operations (Dunning et al. 1998b). We extracted cores from the lake in 1991 and 1995. A core 235 cm deep taken in 1991 proved to contain an uncomplicated Holocene paleoecological record. Many similarities as well as some notable differences exist between this core and those from numerous cores of central Petén lakes taken since the 1960s.

While we initially obtained the 1991 Laguna Tamarindito core to better understand human-environment interactions over the past 3,000 years, much of this short core proved to be much older. The lowermost levels of the core date to the onset of the Holocene period almost 10,000 years ago. Samples from this time indicate an open water context: organic-rich sediments, abundant deep water gastropods, and predominant pine pollen of either local or distant origin. By 5500 B.C., the Tamarindito-Petexbatun region was colonized by tropical deciduous forest species (exemplified by Moraceae, Combretaceae, and Burseraceae; figure 7.4). As indicated by the oxidation of sediments and changes in snail populations, the lake experienced two periods of significant lake-level reductions, at approximately 4500 and 2900 B.C.

The clearance of primary tropical deciduous forest species and the appearance of *Zea* pollen provide evidence of the emergence of agriculture in the region between 2000 and 1000 B.C. These dates correspond with other areas in the Maya Lowlands and Central America, although some regions saw forest clearance a millennium earlier (Jones 1994; Pohl et al. 1996). The distinct physiographic setting of the Petexbatun region may have made it unattractive for extremely early agriculturalists, who probably favored more easily cultivable river and lake margins. We found sedimentary evidence in the lake core and local sinkholes that the onset of forest clearance and agriculture in the region was accompanied by significant acceleration of local soil erosion (Dunning and Beach 1994; Dunning et al. 1997).

The 1991 core shows evidence of two major episodes of forest clearance, in the Maya Preclassic and again in the Classic (probably Late Classic) periods. Severe depression of the Moraceae curve at these points in the core suggests significant deforestation of the high climax forest; nevertheless, nonclimax arboreal species persist in both periods. Archaeological evidence indicates that the first of these periods, the Preclassic, was characterized by a relatively small, scattered population that probably cleared broad areas for cultivation. This period is also marked by an increase in sedimentation and a change to less organic, more mineral-rich sediments, indicating a significant increase in soil loss. Our investigation of the small Aguada Catolina on the Petexbatun Escarpment also indicates severe disturbance of the soil cover associated with Preclassic agriculture (Beach and Dunning 1995; Dunning and Beach 1994).

The second period of deforestation, the Late Classic, accompanied enormous increases in regional population density and the onset of intensive cultivation. This bimodal deforestation curve is different from that of the Petén lakes region, where the rate of deforestation and disturbance seems to have increased steadily from Preclassic through Late Classic times (Rice 1993), a pattern we expected to see repeated in the Petexbatun. But the apparent Early Classic regrowth of high forest in the region parallels archaeological data that indicate a possible decline in regional population during this period. Data from the Tamarindito cores also indicate that high tropical forest recolonized the region rapidly following its general population abandonment between A.D. 800 and 1000. This rapid Postclassic forest recovery, compared with the central Petén lakes region, probably results from at least two related factors: (1) forest cover was not as decimated here during the Late Classic, and (2) there was less Postclassic human disturbance in the Río de la Pasión region. Forests in this wetter region may also have greater resilience.

Lagunita Tamarindito, Guatemala
1991 Core, Summary Pollen Diagram

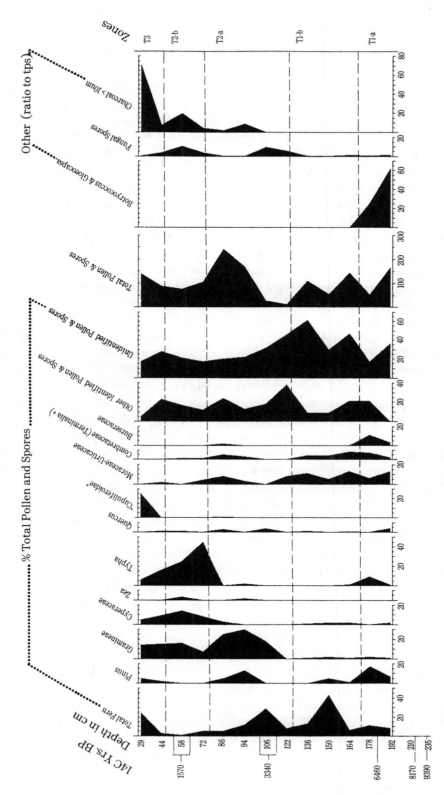

FIGURE 7.4. Pollen in a sediment core taken from Laguna Tamarindito in 1991. *Source: After Dunning et al. 1998b.*

We found only equivocal evidence for significant drying of local climate in both the Preclassic and Classic periods, as has been detected in cores from elsewhere in the Maya Lowlands (Leyden et al. 1998; Curtis et al. 1996; Whitmore et al. 1996; Hodell et al. 1995). Increased charcoal quantities and a shift in snail species could indicate climatic drying, but other explanations like human caused fire and hydrologic change are equally possible. Continuing analysis on the 1995 core taken from Laguna Tamarindito may help isolate evidence for or against climatic desiccation.

In some parts of the Maya Lowlands, such as the Copán valley, the spread of agriculture onto hillsides led to significant soil loss (Rue 1987; Wingard 1996). In the Petexbatun region we noted that many hillsides were at least partially protected by a variety of soil conservation terraces and dams, including many on the hills near Laguna Tamarindito. These Late Classic soil conservation methods appear to have been quite effective, because despite intensification of agriculture and higher human populations during this time, the Laguna Tamarindito core shows only a minor increase in sedimentation, and the sediments were relatively organic compared with the dense mineral clays from the central Petén lakes during this time.

Pollen data indicate considerable regional forest cover throughout the Late Classic, although this assemblage may indicate more of a managed forest landscape, because some of the tree species associated with primary forest virtually disappear. Pollen from the 1995 core also indicate the presence of significant numbers of fruit-producing and other economic trees during the Maya occupation of the region (Rue et al. n.d.). The regional fauna data confirm that considerable amounts of forest cover remained in the region during the Late Classic period (Emery 1997). Both the archaeofaunal data and stable isotope and trace element analyses of human skeletons for dietary information indicate that animal protein remained abundant during the Late Classic, although the quantities consumed varied by social status (Wright 1994; Wright and White 1996). The growing discrepancy in quality of diet is a phenomenon even more pronounced elsewhere in the Maya world during the Late Classic (Lentz 1991).

Generally, our ecological data suggest the dominance of household-level agricultural management during the Late Classic in the Petexbatun region, although some evidence exists for communal projects (e.g., the reservoir dam and check dam systems at Tamarindito). It is noteworthy that individual households in some instances appear to have had substantial disparities in the size and quality of their landholdings. There appear to have been efforts late in the Classic period toward using walls to close off larger tracts of land in some areas, such as the vicinity of Aguateca (Killion and Dunning 1992; Dunning et al. 1997).

Competition for prime upland areas must have been intense during the Late Classic period. Proximity to the important centers of the Petexbatun Escarpment and defensibility clearly influenced the desirability of land. During the Late Classic the escarpment land around and between the major centers was partially divided by a field wall system. Some of these areas were naturally defensible, and others had defensive systems nearby. On the other hand, adjacent low-lying areas (both well drained and poorly drained) were less densely populated and were not divided by walls, nor were they easily defensible. This juxtaposition of contrasting landscapes suggests that the pervasive climate of intersite warfare during the Late Classic may have given added value to the more defensible real estate atop the escarpment (Demarest 1997). In sum, the interplay of environmental factors, such as the existence of agriculturally valuable land along the Petexbatun Escarpment, and social factors, such as prolonged intersite hostilities, probably spurred the development of the distinctive, complex, and divided regional landscape. Persistent warfare in the region may have contributed to abandonment of large portions of the region after A.D. 800 and a reconcentration of population around the large urban center of Seibal to the east (Demarest 1997).

The latest chapter of this region's cultural-ecological history has been its deforestation since the early 1980s as landless Maya have moved back onto these ancient lands and turned even national parks into milpas (Beach 1998b). One of the effects of this deforestation has been soil erosion, and hence we studied current soils and soil erosion to understand the contemporary chapter of Maya history. We compared the soils of primary forest slopes with those of milpas that had been slashed and burned from primary forests since the 1980s. The milpa slopes had soils that were 8 to 18 cm thinner, truncated horizons, and physical evidence of erosion. Where ancient Maya terraces are present in the slashed and burned landscape, however, they are still holding 2.7 to 3.6 times more soil than on the surrounding hillslopes. The presence of these relict terraces in stark contrast with modern eroded slopes testifies to potential erodibility and the potential sustainability of this landscape under intensive land uses.

THREE RIVERS REGION

The Three Rivers region includes the eastern margins of the large Petén karst plateau, a hydrologically elevated limestone area characterized by rugged free-draining uplands and seasonally inundated, clay-filled depressions (bajos) (Dunning et al. 1998b). It also includes the generally low-lying valleys of the Río Bravo and Booth's River, areas of low, limestone ridges and large, perennial wetlands (figure 7.5). Since 1992 we have been studying the complex interplay of cultural and environmental processes in this region, with much of our work focusing on the lands around and between the major sites of La Milpa and Dos Hombres. The site center of La Milpa (figure 7.6) is situated atop a topographically prominent ridge on the upland plateau far from a perennial stream; the Dos Hombres site center occupies a low but locally prominent rise amid the Río Bravo lowlands.

Soils on the limestone uplands are fertile, but these shallow clay Mollisols are vulnerable to erosion in areas with steep slopes (Dunning et al. 1996). Bajo and lowland soils are deep clay Vertisols, Mollisols, and organic mucks (Histosols). These soils are also often fertile but subject to significant drainage limitations or shrink-swell (argilloturbation) problems. Regional native vegetation ranges from what Browkaw and Mallory (1993) call perennial swamp forest and grasslands in the lowlands to tropical wet-dry deciduous forest across the uplands. Greller's smaller scale taxonomy (this volume) would include all of this in the tropical semideciduous forest category. The seasonally dry upland forests reflect the powerful influence of this droughty regional climate. The severity of the dry season poses a significant obstacle to human occupation of the karstic uplands, where perennial water sources are few and far between.

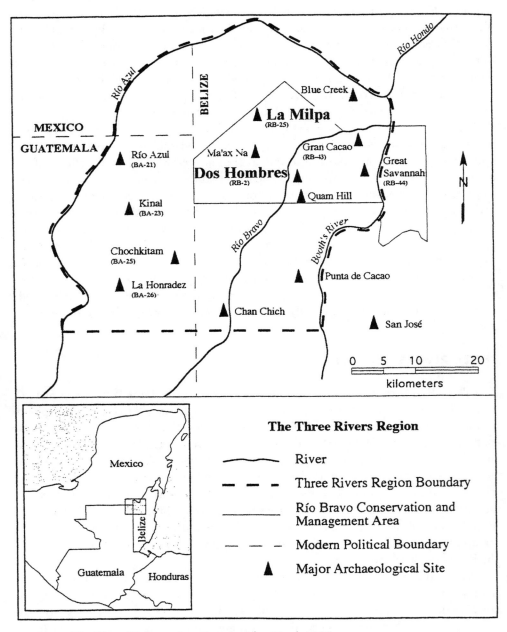

FIGURE 7.5. The Three Rivers region. *Source:* After Houk 1996.

The earliest cultural remains thus far uncovered in the Three Rivers region are ephemeral, nonstructural artifact deposits dating to the Middle Preclassic period (900– 300 B.C.). Based on comparative data from other parts of the Maya Lowlands, we presume this period was characterized by small groups of farmers, who may already have begun significantly altering the local environment by clearing large areas of forest using swidden cultivation (cf. Dunning et al. 1998a; Jones 1994; Piperno 1994; Pohl et al. 1996).

Urbanization and associated landscape modifications came to the region during the Late Preclassic–Protoclassic (300 B.C.–A.D. 250) (Adams 1995). At La Milpa and Dos Hombres significant investment in monumental architecture took place during this time. However, the nucleated settlements at both La Milpa and Dos Hombres appear to have remained quite small during the Late Preclassic, apparently clustering around the small concentration of monumental architecture (Houk 1996; Tourtellot et al. 1994, 1995). A generally light, dispersed rural population is also indicated for this time (Robichaux 1995; Tourtellot et al. 1995).

To date our best pollen record for the Three Rivers region is from a short sediment core taken from a shallow lake near Dos Hombres. This core dates from approximately 500 B.C. to A.D. 1000 and shows numerous disturbance indicators, abundant maize pollen, and patchy tree cover in all strata (Dunning et al. 1999). Our understanding of local environmental changes is also based on geoarchaeological investigations (Dunning et al. 1996, 1999). For example, numerous soil pits and trenches in a small bajo at the mouth of Drainage 1 at La Milpa point to significant environmental disturbance during the Late Preclassic and later periods (figure 7.7). A profile from Operation D09 is illustrative (figure 7.8). Unit 6 in this profile is a buried topsoil horizon dating to about 1500 B.C. Buried topsoils of roughly similar dates have been found in many depositional contexts in the region, which implies a regional paleosol that dates to the Preclassic. Most astonishing, trenches excavated in the Far West Bajo at the mouth of

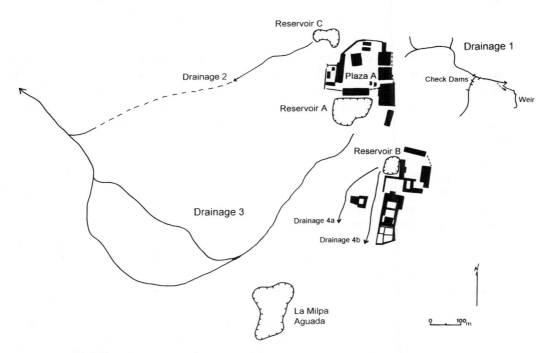

FIGURE 7.6. La Milpa site center and associated drainages. *Source:* After Scarborough et al. 1992.

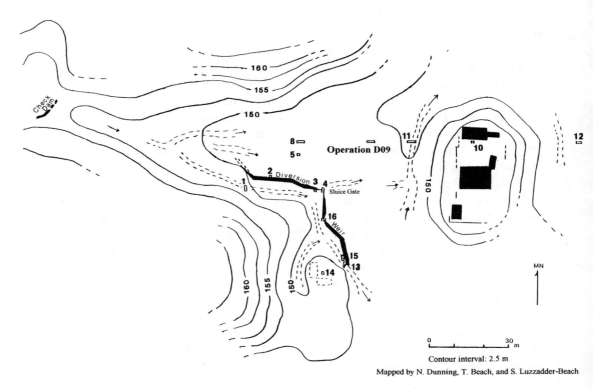

FIGURE 7.7. The mouth of Drainage 1 at La Milpa, Belize, with the location of excavation units. Dashed lines indicate visible surface channels.

Drainage 3 revealed a buried peaty horizon dating to Preclassic times and containing pollen predominantly from wetland and aquatic plants indicative of a perennial wetland and shallow lake that was filled in and covered over by in-washing sediment (Dunning et al. 1999). Unit 5 represents clayey sediments that buried the topsoil, probably as the result of deforestation and erosion beginning in the Middle and Late Preclassic. Unit 4 is an incipient topsoil horizon that probably began to form when the site was largely abandoned between A.D. 500 and 700. Unit 3 is a colluvial layer deposited as the result of later Late Classic and Terminal Classic erosion. The severe distortion and fragmentation of Unit 4 and lower strata are the result of significant argilloturbation. The most likely explanation for this shrink and swell activity is the onset of severe drying and wetting extremes, probably reflecting deforestation or possibly regional climatic drying during the Late Classic–Terminal Classic (Rice 1993; Hodell et al. 1995). We will return to these latter points below. For the moment, it is important to note that the regional Preclassic population (relatively small compared to the Classic), probably practicing extensive forms of forest clearance and agriculture, generated tremendous environmental disturbance. Although this type of agriculture is usually viewed as having only mild or sustainable environmental impacts (see Beach 1998b), we think much of the region's sloping upland terrain suffered severe erosion. This conclusion is in line with better documented findings elsewhere in the Maya Lowlands that suggest that the Maya of the ensuing Early Classic period may have inherited a severely eroded landscape from their ancestors (Jacob 1995; Dunning 1995; Dunning et al. 1998a; Dunning and Beach n.d.; Rice 1993; Holley et al. n.d.).

The Early Classic period (A.D. 250–550) in the Three Rivers region remains poorly understood. Dynastic stelae also began appearing at La Milpa at this time, as they did at other major centers across the Maya Lowlands. Where legible, the texts on such stelae tie the founding of Classic royal dynasties to the first years of the Early Classic, marking a notable shift to political states based on institutionalized royal succession and the increased stability of the political system (Grube 1995). The Early Classic growth and transformations at La Milpa may well also relate to changing patterns of basic resource control. A necessary adaptation for urbanization in the seasonally dry Maya Lowlands was toward the use of reservoirs and other forms of water control as part of a long-term modification of local watersheds (Scarborough 1993). Taking advantage of heavy rainfall for seven months of the year, the Maya could impound enough water in urban areas to survive the dry season. In the Preclassic, such reservoir systems typically occupied natural topographic depressions. However, as Classic Maya civilization took shape, reservoir construction shifted to the upper reaches of local watersheds, that is, the rugged karst ridges on which urban centers were being built.

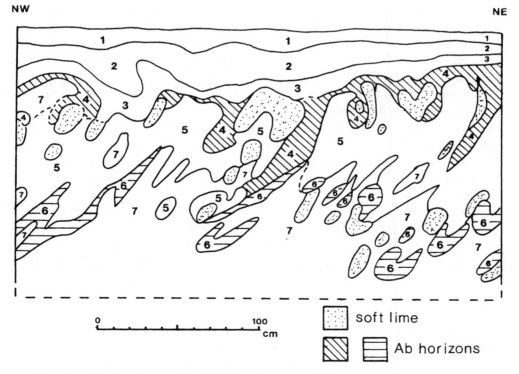

FIGURE 7.8. Profile of northern wall of La Milpa Operation D09.

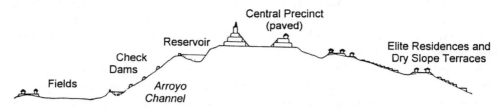

FIGURE 7.9. Model of a modified convex watershed in an urban center during the Late Classic period in the southern Maya Lowlands. *Source:* After Dunning 1995, modified from Scarborough 1993.

At La Milpa the upper portions of local watersheds draining the site core were dammed to create reservoirs (see figure 7.6). The release of water from these reservoirs may have been controlled through the combined use of sluice gates and check dams (Scarborough et al. 1992, 1995). As Drennan (1988) has noted, the intensification of land use is often most pronounced in settings like the Maya Lowlands because the naturally dispersed population live in close proximity to their fields and gardens. The prevalence of seasonally arid conditions in the region, however, would have acted to combat the tendency toward population dispersal by making a secure source of water more desirable, a trend that would have engendered further intensification of land use. In other words, the need to secure water would have significantly increased the investment in local, fixed space or place. Thus, the latter years of the Late Preclassic and beginning of the Early Classic saw a focusing of Maya settlements across the Petén karst plateau on local watersheds. The built environment of temples, palaces, and courtyards at the heart of these watersheds came to be the critical source of life-sustaining water (figure 7.9).

Between 700 and 900 A.D., both La Milpa and Dos Hombres were transformed by massive monumental building programs in their site centers, bringing them into closer conformity with the form of major centers across the Petén plateau during the Late Classic (Houk 1996). Evidence strongly suggests that the resurrection of La Milpa and Dos Hombres was at least as much accomplished by a large migration of people from the neighboring Petén region as by any internal population growth (Tourtellot et al. 1995).

As mentioned above, these migrants may have inherited an environment still significantly degraded by Preclassic–Early Classic deforestation and agriculture. Our investigations in La Milpa Drainages 1 and 3, for example, suggest many upland areas were seriously eroded in earlier times and probably had only minimally redeveloped soil cover by the mid eighth century. We should note that this denudation had little effect on agricultural production, except for accelerated runoff, because these watersheds drain the site center.

Settlement mapping and excavations at La Milpa have revealed areas of extremely high settlement densities and a remarkable array of linear stone features (Tourtellot et al. 1994, 1995; Hammond et al. 1996). In some segments of the landscape the terraces are the best built cross-channel, footslope, and box terraces we have observed in the Maya Lowlands. They have been constructed to use the natural soil and water conservation attributes of stream channels and alluvial fans. In contrast, many others of these features are crude in form and constructed of rough chich stones that are usually low-grade chert. They all appear to date to the last century of the Late Classic and first century of the Terminal Classic (A.D. 700–900). While some of these stone alignments form apparent residential boundary walls and others comprise simple terraces, many defy easy explanation. These structures do not indicate the tidy, rationalized, economic landscape documented for terraced lands of the Río Bec (Turner 1983). Walls that do form terraces are usually of the footslope type, positioned to capture soil accumulating at the base of a long slope (Dunning and Beach 1994). In short, the landscape suggests that by some time in the Terminal Classic, segments of the farming population of La Milpa were trying to salvage remnants of soil and soil moisture from largely denuded upland surfaces. One explanation for this landscape is that it began as a rationally managed or engineered landscape (Scarborough 1993) but devolved into a landscape of desperation (Tourtellot et al. 1995). Where deep clay soil had accumulated in upland depressions, Maya farmers faced other problems: devastating rainy-season runoff and flooding and dry-season desiccation and clay contraction, likely exacerbated by deforestation (Rice 1993). Based on the evidence for degradation, the long-term abandonment of the city and region may attest to the severity of land degradation and depopulation that occurred in the Terminal Classic. It should be noted, however, that we are beginning to find more and more evidence for well ordered and engineered conservation features that have survived almost completely intact until today.

DISCUSSION

Our model for the relationship of the ancient Maya to the environment is still evolving. Periods of relative landscape stability were clearly punctuated by episodes of instability, often in the form of human-induced degradation, followed by gradual recovery. We have evidence so far that shows remarkable knowledge about soil and water conservation. The Late Classic Maya around La Milpa and the Petexbatun built intricate terraces and dams in several landscape situations that expertly use soil moisture, divert runoffs, dam water in reservoirs, minimize erosion, and maintain soil platforms. In the Late Classic at Tamarindito, pollen evidence also indicates comparatively low soil erosion and a human-modified forest that provided economic goods and protected the landscape from degradation during the era of the highest population. Faunal evidence also indicates no diminution of animal protein during this period. Based on these sources of evidence, the ancient Maya provide an anti-Malthusian example of indigenous sustainable agriculture in a tropical environment. These conclusions, however, contrast with findings for increased soil erosion and sedimentation at both the central Petén lakes and at Copán. At these sites, population increased and spread up onto steeper slopes without the benefit of extensive conservation techniques.

We have evidence for this same pattern occurring around La Milpa and the Petexbatun, but, based on sedimentation rates and regional buried soils, it seems to have been in the Preclassic when populations were first exploiting steeper slopes for cultivation. After their initial failures with these steeper lands, the Late Classic Maya may have learned and spread their growing population deliberately onto steeper lands with terraces and cover crops.

On the other hand, the actual influences of environmental change on the Maya are still unknown. The droughts of the northern and central Yucatán based on seemingly incontrovertible geochemical and oxygen isotope data (Hodell et al. this volume) may have influenced the Late Classic period collapse, but the collapse occurred in the Petén to the south, where we still await straightforward evidence for drought. Indeed, evidence for a regional, Classic period drought is mixed: missing from a spotty core in the northwest Yucatán only about 120 km away from the drought at Chichancanob (Whitmore et al. 1996), missing from Lake Petén Itza in the Petén (Curtis et al. 1998), but perhaps evident as dry periods in Mexico and Costa Rica (Hodell et al. this volume). Clearly, we need pollen and oxygen isotope data from across the Maya Lowlands and Mesoamerica to find the chronological and spatial margins of these large periods of desiccation.

The latest chapter in the Petén and other parts of the Maya Lowlands is being written by a wave of immigrants who have entered this landscape during the last two decades. The Petén population has increased by more than an order of magnitude over the last three decades, changing from 25,207 in the mid 1960s to an estimated 500,000 or more by the mid 1990s (CONAP 1996). Like the pioneers of the Preclassic, they are bringing deforestation and accelerated soil erosion, sometimes eroding soils to bedrock (Beach 1998b). They show us how easily land denudation can occur on this landscape, as it did in the Preclassic, and how much effort must be exerted to maintain the measure of sustainability that occurred in the Late Classic.

REFERENCES

Abrams, E. M., and D. J. Rue. 1988. The causes and consequences of deforestation among the prehistoric Maya. *Human Ecology* 16: 377–395.

Adams, R. E. W. 1995. A regional perspective on the Lowland Maya of the northeast Petén and northwestern Belize. Paper presented at the 60th Annual Meeting of the Society for American Archaeology, Minneapolis.

Adams, R. E. W., W. E. Brown Jr., and T. P. Culbert. 1981. Radar mapping, archaeology, and ancient Maya land use. *Science* 213: 1457–1463.

Adams, R. E. W., T. P. Culbert, W. E. Brown, Jr., P. D. Harrison, and L. J. Levi. 1990.
Rebuttal to Pope and Dahlin. *J. Field Archaeology* 17: 241–244.

Beach, T. 1998a. Soil constraints on northwest Yucatán, Mexico: Pedo-archaeology and Maya subsistence at Chunchucmil. *Geoarchaeology* 13 (8): 759–791.

———. 1998b. Soil catenas, tropical deforestation, and ancient and contemporary soil erosion in the Petén, Guatemala. *Physical Geography* 19: 378–405.

Beach, T., and N. Dunning. 1995. Ancient Maya terracing and modern conservation in the Petén rain forest of Guatemala. *J. Soil and Water Conservation* 50: 138–145.

———. 1997. An ancient Maya reservoir and

dam at Tamarindito, El Petén, Guatemala. *Latin American Antiquity* 8: 20–29.

Browkaw, N. V. L., and E. P. Mallory. 1993. *Vegetation of the Rio Bravo Conservation and Management Area, Belize.* Monomet, Mass.: Manomet Bird Observatory.

Chase, A. F., and D. Z. Chase. 1998. Scale and intensity in Classic period Maya agriculture: Terracing and settlement of the "garden city" of Caracol, Belize. *Culture and Agriculture* 20: 60–77.

Clendinnen, I. 1987. *Ambivalent Conquests: Maya and Spaniard in Yucatán, 1517–1570.* Cambridge: Cambridge University Press.

CONAP (Consejo Nacional de Areas Protegidas). 1996. *El Estado de la Biossfera Maya en 1996.* Guatemala City: CONAP, USAID, and the Peregrine Fund.

Cowgill, U. M. 1962. Soil fertility and the ancient Maya. *Transactions of the Connecticut Academy of Arts and Sciences* 42: 1–56.

Culbert, T. P., and D. S. Rice, eds. 1990. *Precolumbian Population History in the Maya Lowlands.* Albuquerque: University of New Mexico Press.

Curtis, J. H., D. Hodell, and M. Brenner. 1996. Climate variability on the Yucatán Peninsula (Mexico) during the past 3,500 years, and its implications for Maya cultural evolution. *Quaternary Research* 46: 37–47.

Curtis, J. H., M. Brenner, D. A. Hodell, R. A. Balser, G. A. Islebe, and H. Hooghiemstra. 1998. A multi-proxy study of Holocene environmental change in the Maya Lowlands of Petén, Guatemala. *J. Paleolimnology* 19: 139–159.

Demarest, A. A. 1997. The Vanderbilt University Petexbatun regional archaeological project, 1989–1994. *Ancient Mesoamerica* 8: 169–185.

Denevan, W. M. 1992. The pristine myth: The landscape of the Americas in 1492. *Annals of the Association of American Geographers* 82: 369–385.

Drennan, R. D. 1988. Household location and compact versus dispersed settlement in Prehispanic Mesoamerica. In R. Wilk and W. Ashmore, eds., *Household and Commu-nity in the Mesoamerican Past*, pp. 273–293. Albuquerque: University of New Mexico Press.

Dunning, N. P. 1992. Umwelt, Siedlungweise, Ernaehrung, und Lebensunderhalt im Maya-Tiefland waehrend der Klassik (250–900 n. Chr.). In E. Eggebrecht, A. Eggebrecht, and N. Grube, eds., *Die Welt der Maya*, pp. 92–106. Mainz am Rhein: Verlag Phillip von Zumbrow.

———. 1995. Coming together at the temple mountain: Environment, subsistence, and the emergence of Classic Maya segmentary states. In N. Grube, ed., *The Emergence of Classic Maya Civilization*, pp. 61–70. Acta Mesoamericana 8. Möckmühl, Germany: Verlag von Flemming.

———. 1996. An examination of regional variability in the Prehispanic Maya agricultural landscape. In S. L. Fedick, ed., *The Managed Mosaic: Ancient Maya Agriculture and Resource Management*, pp. 53–68. Salt Lake City: University of Utah Press.

Dunning, N. P., and T. Beach. 1994. Soil erosion, slope management, and ancient terracing in the Maya Lowlands. *Latin American Antiquity* 5: 51–69.

———. N.d. *An Ancient Maya Landscape: Environmental Archaeology and Ancient Settlement of the Petexbatun Region, Guatemala.* Nashville, Tenn.: Vanderbilt University Press. In preparation.

Dunning, N. P., S. Luzzadder-Beach, T. Beach, P. Hughbanks, and F. Valdez Jr. 1996. Ancient Maya soil use and abuse: Pedoarchaeological investigations in the Programme for Belize. Paper presented at the 61st Annual Meeting of the Society for American Archaeology, New Orleans.

Dunning, N. P., T. Beach, and D. Rue. 1997. The paleoecology and ancient settlement of the Petexbatun region, Guatemala. *Ancient Mesoamerica* 8: 185–197.

Dunning, N. P., T. Beach, P. Farrell, and S. Luzzadder-Beach. 1998a. Prehispanic agrosystems and adaptive regions in the Maya Lowlands. *Culture and Agriculture* 20: 87–106.

Dunning, N. P., D. Rue, T. Beach, A. Covich, and A. Traverse. 1998b. Human-environment interactions in a tropical watershed: The paleoecology of Laguna Tamarindito, Petén, Guatemala. *J. Field Archaeology* 25: 139–151.

Dunning, N. P., V. L. Scarborough, F. Valdez Jr., S. Luzzadder-Beach, T. Beach, and J. G. Jones. 1999. Temple mountains, sacred lakes, and fertile fields: Ancient Maya landscapes in northwestern Belize. *Antiquity* 73: 650–660.

Emery, K. F. 1997. The Maya collapse: A zooarchaeological perspective. Ph.D. dissertation, Cornell University, Ithaca, N.Y.

Fedick, S. L. 1996. Conclusion: Landscape approaches to the study of ancient Maya agriculture and resource use. In S. L. Fedick, ed., *The Managed Mosaic: Ancient Maya Agriculture and Resource Management*, pp. 335–347. Salt Lake City: University of Utah Press.

Fedick, S. L., and K. Hovey. 1995. Ancient Maya settlement and use of wetlands at Naranjal and the surrounding Yalahau region. In S. L. Fedick and K. Taube, eds., *The View from Yalahau: 1993 Arachaeological Investigations Northern Quintana Roo, Mexico,* pp. 89–100. Latin American Studies Program Field Report Series 2. Riverside: University of California.

Flores, J. S., and I. Espejel Carvajal. 1994. *Tipos de Vegetación de la Peninsula de Yucatán.* Etnoflora Yucatánese 3. Mérida: Universidad Autónoma de Yucatán.

Grube, N. 1995. Transformations of Maya society at the end of the Preclassic: Processes of change between predynastic and dynastic periods. In N. Grube, ed., *The Emergence of Classic Maya Civilization,* pp. 1–7. Acta Mesoamericana 8. Möckmühl, Germany: Verlag von Flemming.

Hammond, N., G. Tourtellot, S. Donaghey, and A. Clarke. 1996. Survey and excavation at La Milpa, 1996. *Mexicon* 18: 86–91.

Hansen, R. D. 1995. Nuevas perspectivas de los origines y colapso de la civilizacion Maya. Paper presented at the Encuentro de los Investigadores de la Cultura Maya 5, Universidad Autónoma de Campeche, Mexico.

Harrison, P. D. 1990. The revolution in ancient Maya subsistence. In F. S. Clancy and P. D. Harrison, eds., *Vision and Revision in Maya Studies,* pp. 99–113. Albuquerque: University of New Mexico Press.

———. 1996. Settlement and land use in the Pulltrouser Swamp archaeological zone, northern Belize. In S. L. Fedick, ed., *The Managed Mosaic: Ancient Maya Agriculture and Resource Use,* pp. 177–191. Salt Lake City: University of Utah Press.

Harrison, P. D., and B. L. Turner II, eds. 1978. *Pre-Hispanic Maya Agriculture.* Albuquerque: University of New Mexico Press.

Hartshorn, G. S. 1988. Tropical and subtropical vegetation of Meso-America. In M. G. Barbour and W. D. Billings, eds., *North American Terrestrial Vegetation,* pp. 365–390. New York: Cambridge University Press.

Hodell, D. A., J. H. Curtis, and M. Brenner. 1995. Possible role of climate in the collapse of Maya civilization. *Nature* 375: 391–394.

Holley, G. R., R. Dalan, W. I. Woods, and H. Watters. N.d. Implications of a buried Preclassic site in western Belize. In S. R. Ahler, ed., *Mounds, Modoc, and Mesoamerica: Paper in Honor of Melvin L. Fowler.* Springfield: Illinois State Museum Scientific Paper Series.

Houk, B. A. 1996. The archaeology of site planning: An example from the Maya site of Dos Hombres, Belize. Ph.D. dissertation, University of Texas, Austin.

Islebe, G. A., H. Hooghiemstra, M. Brenner, J. H. Curtis, and D. A. Hodell. 1996. A Holocene vegetation history from lowland Guatemala. *Holocene* 6: 265–271.

Jacob, J. S. 1995. Archaeological pedology in the Maya Lowlands. In *Pedological Perspectives in Archaeological Research,* pp. 51–79. Special Publication 44. Madison, Wisc.: Soil Science Society of America.

Jones, J. G. 1994. Pollen evidence for early settlement and agriculture in northern Belize. *Palynology* 18: 205–211.

Killion, T. W., and N. P. Dunning. 1992. Land use, land holding, and war among the Late Classic Maya: A study of prehistoric wall systems on the Petexbatun Escarpment, Petén, Guatemala. Paper presented at the 58th Annual Meeting of the Society for American Archaeology, Pittsburgh, Pa.

Lentz, D. L. 1991. Maya diets of the rich and poor: Paleoethnobotanical evidence from Copán. *Latin American Antiquity* 2: 269–287.

Leyden, B. W., M. Brenner, D. A. Hodell, and J. H. Curtis. 1993. Late Pleistocene climate in the Central American lowlands. In P. K. Swart, K. C. Lohmann, J. McKenzie, and S. Savin, eds., *Climate Change in Continental Isotopic Records,* pp. 165–178. Geophysical Monograph 78. Washington, D.C.: American Geophysical Union.

Leyden, B. W., M. Brenner, T. Whitmore, J. H. Curtis, D. R. Piperno, and B. H. Dahlin. 1996. A record of long- and short-term climatic variation from northwest Yucatán: San José Chulchacá. In S. L. Fedick, ed., *The Managed Mosaic: Ancient Maya Agriculture and Resource Management,* pp. 30–51. Salt Lake City: University of Utah Press.

Leyden, B. W., M. Brenner, and B. Dahlin. 1998. Cultural and climatic history of a Lowland Maya city: Cobá, Quintana Roo, Mexico. *Quaternary Research* 49: 111–122.

McAnany, P. A. 1995. *Living with the Ancestors: Kinship and Kingship in Ancient Maya Society.* Austin: University of Texas Press.

Mugerauer, R. 1995. *Interpreting Environments: Tradition, Deconstruction, and Hermeneutics.* Austin: University of Texas Press.

Murphy, P. G., and Lugo, A. E. 1995. Dry forests of Central America and the Caribbean. In S. H. Bullock, H. Mooney, and E. Medina, eds., *Seasonally Dry Tropical Forests.* New York: Cambridge University Press.

Piperno, D. R. 1994. On the emergence of agriculture in the New World. *Current Anthropology* 35: 637–639.

Pohl, M. D., P. Bloom, and K. Pope. 1990. Interpretation of wetland farming in northern Belize: Excavations at San Antonio Río Hondo. In M. D. Pohl, ed., *Ancient Maya Wetland Agriculture: Excavations on Albion Island, Northern Belize,* pp. 187–254. San Francisco, Calif.: Westview Press.

Pohl, M. D., K. O. Pope, J. G. Jones, J. S. Jacob, D. R. Piperno, S. D. deFrance, D. L.

Lentz, J. A. Gifford, M. E. Danforth, and J. K. Josserand. 1996. Early agriculture in the Maya Lowlands. *Latin American Antiquity* 7: 355–372.

Pope, K. O., and B. H. Dahlin. 1989. Ancient Maya wetland agriculture: New insights from ecology and remote sensing research. *J. Field Archaeology* 16: 87–106.

———. 1993. Radar detection and ecology of ancient Maya canal systems: Reply to Adams et al. *J. Field Archaeology* 20: 379–383.

Pope, K. O., M. D. Pohl, and J. S. Jacob. 1996. Formation of ancient Maya wetland fields: Natural and anthropogenic processes. In S. L. Fedick, ed., *The Managed Mosaic: Ancient Maya Agriculture and Resource Use,* pp. 165–176. Salt Lake City: University of Utah Press.

Pyburn, K. A. 1996. The political economy of ancient Maya land use: The road to ruin. In S. L. Fedick, ed., *The Managed Mosaic: Ancient Maya Agriculture and Resource Use,* pp. 236–250. Salt Lake City: University of Utah Press.

Rejmankova, E., K. O. Pope, M. D. Pohl, and J. M. Rey-Benayas. 1995. Freshwater wetland plant competition in northern Belize: Implications for paleoecological studies of Maya wetland agriculture. *Biotropica* 27: 28–36.

Rice, D. S. 1993. Eighth-century physical geography, environment, and natural resources in the Maya Lowlands. In J. A. Sabloff and J. A. Henderson, eds., *Lowland Maya Civilization in the Eighth Century* A.D., pp. 11–63. Washington, D.C.: Dumbarton Oaks.

Robichaux, H. R. 1995. Ancient Maya community patterns in northwestern Belize: Peripheral zone survey at La Milpa and Dos Hombres. Ph.D. dissertation, Austin: University of Texas.

Rue, D. J. 1987. Early agriculture and Postclassic occupation in western Honduras. *Nature* 326(6110): 285–86.

Rue, D. J., J. G. Jones, and N. Dunning. N.d. Pollen analysis of the Laguna Tamarindito 1995 B core. Appendix in N. P. Dunning and T. Beach, *An Ancient Maya Landscape: Environmental Archaeology and Ancient Settlement of the Petexbatun Re-*

gion, Guatemala. Nashville, Tenn.: Vanderbilt University Press. In preparation.

Rzedowski, J. 1981. Vegetacion de Mexico. Mexico, D.F.: Editorial Limusa.

Sanford, C. L. 1969. The Quest for Paradise. Urbana: University of Illinois Press.

Scarborough, V. L. 1993. Water management for the southern Maya Lowlands: An accretive model for the engineered landscape. In V. L. Scarborough and B. Isaac, eds., Economic Aspects of Water Management in the Prehispanic New World, pp. 17–68. Research in Economic Anthropology Supplement 7. Greenwich, Conn.: JAI Press.

Scarborough, V. L., M. E. Becher, J. L. Baker, G. Harris, and J. D. Henz. 1992. Water management studies at La Milpa, Belize. Department of Anthropology, University of Cincinnati. Mimeograph.

Scarborough, V. L., M. E. Becher, J. L. Baker, G. Harris, and F. Valdez Jr. 1995. Water and land at the ancient Maya community of La Milpa. Latin American Antiquity 6: 98–119.

Simonian, L. 1995. Defending the Land of the Jaguar: A History of Conservation in Mexico. Austin: University of Texas Press.

Tourtellot, G. III, J. J. Rose, N. Grube, S. Donaghey, and N. Hammond. 1994. More light on La Milpa: Maya settlement archaeology in northwestern Belize. Mexicon 16: 119–124.

Tourtellot, G. III, N. Hammond, and J. Rose. 1995. Reversals of fortune: Settlement processes at La Milpa, Belize. Paper presented at the First International Symposium of Maya Archaeology, San Ignacio, Belize.

Turner II, B. L. 1983. Once Beneath the Forest: Prehistoric Terracing in the Rio Bec Region of the Maya Lowlands. Boulder, Colo.: Westview Press.

———. 1993. Rethinking the "new orthodoxy": Interpreting ancient Maya agriculture and environment. In K. W. Mathewson, ed., Culture, Form, and Place: Essays in Cultural and Historical Geography, pp. 57–88. Geoscience and Man 32.

Turner II, B. L, and P. D. Harrison. 1983. Pulltrouser Swamp and Maya raised fields: A summation. In B. L. Turner II and P. D. Harrison, eds., Pulltrouser Swamp: Ancient Maya Habitat, Agriculture, and Settlement in Northern Belize, pp. 246–269. Austin: University of Texas Press.

Whitmore, T. J., M. Brenner, J. H. Curtis, B. H. Dahlin, and B. W. Leyden. 1996. Holocene climatic and human influences on lakes of the Yucatán Peninsula: An interdisciplinary, palaeolimnological approach. Holocene 6: 273–287.

Wingard, J. D. 1996. Interactions between demographic processes and soil resources in the Copán Valley, Honduras. In S. L. Fedick, ed., The Managed Mosaic: Ancient Maya Agriculture and Resource Use, pp. 207–235. Salt Lake City: University of Utah Press.

Wright, L. E. 1994. The sacrifice of the earth? Diet, health, and inequality in the Pasion Maya Lowlands. Ph.D. dissertation, University of Chicago.

Wright, L. E., and C. D. White. 1996. Human biology in the Classic Maya collapse: Evidence from paleopathology and paleodiet.

CHARLES M. PETERS

8 PRECOLUMBIAN SILVICULTURE AND INDIGENOUS MANAGEMENT OF NEOTROPICAL FORESTS

According to most accounts, the practice of silviculture was first introduced to the tropics in the late 1800s by colonial foresters working in Asia. German foresters drew up management plans for Burmese teak as early as 1860 (U Kyaw Zan 1953), and the first tropical forestry training center was founded in 1878 in Dehra Dun, India (Lamprecht 1989). British foresters working in Peninsula Malaysia around the turn of the century developed silvicultural prescriptions to enhance the productivity of *Palaquium gutta,* an important latex-producing tree in local dipterocarp forests (Wyatt-Smith 1963), and the first textbook of tropical silviculture (Troup's three-volume *Silviculture of Indian Trees*) was published in 1922. In 1926, the regeneration improvement felling (RIF) system was formulated in Malaysia. In addition to being the first in a long line of tropical shelterwood systems, the RIF is usually considered to be the first documented and codified example of a silvicultural system for the tropics (Taylor 1962).

What is overlooked in this historical treatment of tropical silviculture is the fact that the indigenous populations of Central and South America have been using, manipulating, and managing tropical forests for several thousand years. Long before Columbus arrived in the New World, indigenous foresters were already skilled in "the art of producing and tending a forest," i.e., they were, as defined by Smith et al. (1997), practicing silviculture. In spite of the enduring myth that Neotropical forests were a vast, pristine wilderness in 1492, there is considerable evidence to suggest that the structure and composition of many forests have been deliberately molded by anthropogenic forces (Roosevelt 1989; Balée 1989; Denevan 1992). In the same way that German foresters used selective thinning, weeding, and enrichment planting to maximize the density and growth of teak trees in Burma, Precolumbian woodsmen used their own silvicultural techniques to favor the growth of desirable forest species.

The basic problem in trying to reconstruct the forest management practices of ancient cultures is the lack of direct evidence. There are no bark-paper codices describing Mayan silviculture, or pollen records documenting the subtle manipulation of forest composition, or fossilized plant remains showing the variety and abundance of plant resources obtained from managed forests. What we are left with is simply the observation that these same cultures are skilled forest managers today and that many of the forests they inhabit exhibit conspicuous imprints of past silvicultural treatment. These two pieces of the puzzle, however, can provide convincing glimpses of what silviculture might have been like in the New World prior of Columbus.

The purpose of this article is threefold. The first and perhaps most important objective is to challenge the common assumption that the tropical forests of the New World were virgin, untouched plant communities during Precolumbian times. The second objective is to provide an overview of the variety of indigenous forest management in Central and South America, to speculate on the types of silvicultural interventions that might have been used a thousand years ago, and to present examples of tropical trees and forests that seem to have been produced through the conscious activities of Precolumbian silviculturists. The final objective is to highlight the potential of indigenous silviculture in the search for more sustainable ways of using tropical forests. By necessity, these management systems are parsimonious, low input, and very effective. They are the culmination of hundreds of years of trial and error, and there is a lot to be learned from them.

INDIGENOUS FOREST MANAGEMENT IN THE NEOTROPICS

In contrast to more conventional forms of forest exploitation in the tropics, indigenous systems of silviculture can be very hard to detect. There are no marked stumps, no bulldozer roads, no skid trails, and no straight lines of neatly planted seedlings. If nontimber resources are the product of interest, there may, in fact, be no visible evidence that forest management is occurring on the site. To the untrained eye, the managed and the pristine can easily merge into one.

Given the relative invisibility of these practices, it is not surprising that the subtle manipulation of forest vegetation by indigenous communities went unnoticed for so many years. The early studies that did focus on forest use were usually more concerned with the slashing and burning of small agricultural plots than with what happened after the plots were "abandoned." Much of this work took a rather dim view of forest farming (e.g., FAO 1957; Webb 1960; Watters 1971). Once researchers began to take a closer look at the swidden plots created by traditional communities, however, they found a surprisingly complex and diverse mixture of annual and perennial crops. Far from being the result of "an inefficient and destructive form of land-use" (FAO 1957), local agroforestry systems were found to produce a multitude of useful plant resources, to protect and enrich the soil, to provide important wildlife habitat, and to accelerate the recovery of forest vegetation on the site (Carneiro 1961; Harris 1971; Hecht 1982). With further investigation, it was discovered that local communities actively manage fallow regrowth following shifting cultivation, enriching it with useful species and consciously directing the course of forest succession on the site (Alcorn 1984a; Denevan et al. 1984; Posey 1982). Finally, several studies have shown that indigenous populations also use silvicultural techniques to control the composition and structure of intact forests (Gordon 1982; Alcorn 1983; Gómez-Pompa et al. 1987).

What is clear from all of these studies is that indigenous populations in the Neotropics have evolved a diverse array of techniques for managing trees and forests. For the purposes of this essay, we can group all of these techniques into three main silvicultural systems based on the successional status of the vegetation being managed and the nature and intensity of the interventions employed: (1) homegarden systems, (2) managed fallow systems, and (3) managed forest systems. Homegardens and managed forests are essentially polycyclic systems (sensu Dawkins 1958) in which only a small fraction of the growing stock is removed in each harvest. Canopy cover is continuous, and the forest contains a wide range of different age classes. Managed fallow systems, on the other hand, can be either monocyclic or polycyclic, depending on whether the fallow is ultimately cleared and cycled back into agriculture or maintained in forest and gradually transformed into a managed forest. The actual intensity of management usually increases from managed forest to fallow to homegarden. Each of these systems is discussed in detail below.

HOMEGARDENS

Homegardens, also known as dooryard gardens or kitchen gardens, are diverse, multistoried mixtures of trees, shrubs, vines, and herbaceous plants maintained as an annex to the house (figure 8.1). In addition to edible fruits and other food crops, homegardens are a repository of medicinal plants, spices, ornamentals, and other utilitarian species frequently used by the household. Because of their proximity and easy access, homegardens are highly managed. Family members spend spare time here tending their plants, and children learn to identify and care for useful species by observing the actions of their parents.

The structure and composition of a Neotropical homegarden is highly variable and reflects the cultural background, needs, motivation, and horticultural proficiency of the household that has created it. There is no characteristic or standard homegarden. Ribereños (floodplain farmers) in the Peruvian Amazon manage 168 plant species in their homegardens, yet none of these species is present in all of the gardens and more than half (90 species) are found in only one or two gardens (Padoch and de Jong 1991). Similar patterns of heterogeneity were noted by Alcorn (1984b), who recorded 182 plant species in the gardens of Huastec Maya in Mexico. The Costa Rican homegar-

FIGURE 8.1. Maya homegarden in Quintana Roo, Mexico. Large, buttressed trees are *Brosimum alicastrum*.

dens studied by Price (1983), on the other hand, contained an average of only 16 species of plants.

The origin of the plant material managed in homegardens is also variable, and local households blend together dynamic mixtures of wild plants, semidomesticates, and domesticated stock. Novel species and recently introduced domesticates appear to be especially prized. Studies of Maya homegardens in Yucatán reveal that 26 percent of the species are nonnatives introduced to the region after the sixteenth century (Barrera 1980). The remaining species are either native to the local flora (61 percent) or Neotropical elements not found locally in the wild (13 percent). Similarly, one-third of the species found in Choco gardens in Panama are introduced domesticates (Covich and Nickerson 1966), and the Ka'apor of Brazil include 36 introduced species in their homegardens (Balée 1994). Twenty-five of the nonnative domesticates managed by the Ka'apor are of Old World origin (e.g., *Citrus* spp., *Mangifera indica, Coffea arabica,* and *Colocasia esculenta*). Apparently, the first trials of exotic germ plasm are usually conducted in homegardens where the owner can keep a close eye on things.

The creation and maintenance of homegardens involve several silvicultural operations. Perhaps the most fundamental of these is periodic weeding or brushing, which keeps the garden open, reduces the competition from secondary vegetation, and provides easy access to certain plants. The selection of which plants or genotypes to keep and which to remove is made spontaneously during the weeding process. Although some form of weeding is usually employed in all homegarden systems, the importance of this management operation varies from place to place. The Ka'apor, for example, refer to their homegardens as *kar*, which means an area that has been intensively weeded (Balée 1994), while some Huastec Maya families create gardens with large areas of secondary vegetation that are weeded and cleaned only at irregular intervals (Alcorn 1984a). These thickets, which appear wild to the untrained eye, contain a large concentration of medicinal species, food items, and ritual and utilitarian plants. Padoch and de Jong (1991) found neatly weeded gardens adjacent to gardens that were totally overgrown with weeds in the same Peruvian village.

Planting is also an important activity in the management of homegardens. The intensity of this operation ranges from casual dispersal of seeds into the garden and subsequent protection of the seedling to careful sowing and tending of certain species in nurseries for later transplanting. The Choco produce seedlings for their homegardens in old cans and dishes placed on stumps so that they won't be trampled by livestock or children (Covich and Nickerson 1966). The seedlings are transplanted into the garden when they are of sufficient size to survive the disturbance. Maya foresters in Yucatán construct elevated, compost-filled seedbeds known as *ka'anché* for germinating and growing planting stock (Vargas 1983). The Ka'apor, who employ a less labor intensive strategy at times, report that several of the tree species in their homegardens (e.g., *Theobroma grandiflorum* and *T. speciosum*) are planted by first "swallowing the seed" (Balée 1994). Simple planting and weeding, at whatever intensity, are the core operations that shape the floristic composition of a homegarden. Desirable species are introduced by planting cuttings, seeds, or seedlings or by selectively sparing the volunteers that recruit themselves into the garden. Undesirable stems are removed through periodic weeding.

Most homegardens are fertilized by a continual input of household refuse, organic material from periodic weedings, and ashes from kitchen fires. In communities with livestock, manure may also be added. Whether by conscious purpose or unintended consequence, the addition of this organic waste and compost enriches the soil and enhances the long-term productivity of the garden. Given the low nutrient status of many tropical soils and the low light levels, the high herbivory, and the intense competition for resources in the understory of a tropical forest, homegardens provide a growth environment that is far superior to that confronted by wild species.

A final aspect of homegardens relevant to the present discussion concerns what happens to them after the owners move. Most homegardens are quickly swamped by fast-growing, secondary vegetation once the house is abandoned. The herbaceous cultivars and the nonnative domesticates are the first plants to succumb to this onslaught. The larger trees, with their crowns above the successional fracas and their roots below it, exhibit the highest survivorship and continue to grow and reproduce. In cases where the owners of the garden return periodically to the site to harvest, these trees may be lightly weeded and the competing brush cut back to provide better access. With time, the surviving relicts of the homegarden become engulfed by the developing forest. These trees must be able to establish seedlings under forest conditions for there to be a second or third generation of homegarden progeny. It is at this point that many of the introduced and domesticated tree species are lost. They may continue to flower and produce fruit, but their seedlings are ineffective in securing a permanent place in the forest. The native tree species, especially those obtained from local forests, are a different story. These species were able to maintain themselves in the forest before the homegarden was created, and they continue to do so in the absence of human intervention. The salient difference is that now the trees are growing in an exceptionally fertile microsite at atypically high densities, and now the original seed dispersers and pollinators have returned because there are fewer people in the forest. The net result is that distribution and abundance of many of the founder populations created in homegardens may actually increase over time. This general pattern has been observed for *Spondias mombin* (Balée 1994), *Brosimum alicastrum* (Peters 1989), *Couma macrocarpa* (Gordon 1982), *Astrocaryum vulgare* (Wessels-Boer 1965), and a variety of other useful forest species in the Neotropics (Balée 1989; Gómez-Pompa and Kaus 1990).

MANAGED FALLOWS

Fallows are tracts of forest that are being left to recover after several years of cultivation. In contrast to homegardens, which temporarily arrest succession, managed fallow systems are designed to facilitate and enrich the successional process. The spontaneous growth of secondary vegetation is viewed as a welcome consequence of farming, not as a weed problem (Alcorn 1981). Through subtle substitutions in the species assemblage of the developing vegetation and gradual manipulations of forest structure, managed fallow systems can produce lasting, if almost imperceptible changes, in the forest.

Monocyclic fallows are the most common variant of this system in the Neotropics. As it is generally practiced, small plots of forest are felled and burned and the clearing is planted with agricultural crops such as corn or manioc. Other useful species, both domesticates and semidomesticates, are also introduced at this time. After one or two years of crop production, the site fills with young secondary growth that has been enriched with fruit trees, construction materials, and medicinal plants. In short fallow systems, such as employed by the Huastec Maya, the managed successional sere is allowed to develop for 4 to 8 years before it is cleared to start another agricultural cycle (Alcorn 1984). The Totonac of Veracruz, Mexico, fallow their fields for a slightly longer period (10 to 12 years) to extend the productive life of the vanilla vines planted on the site (Kelly and Palerm 1952). Many indigenous communities in South America manage their fallows using long-rotation systems. For example, the Bora of northeastern Peru (Denevan et al. 1984), the Runa of Ecuador (Irvine 1989), and the Kayapó of central Brazil (Posey 1984) all maintain enriched forest regrowth for 20 years or more before clearing it to replant. The ribereños of Tamshiyacu, Peru, mold their fallows into high-density stands of umarí (*Poraqueiba sericea*) and Brazil nut (*Bertholletia excelsa*), both of which are valuable market fruits, and they may leave these trees standing for 25 to 50 years (Padoch et al. 1985).

Polycyclic fallow systems start out the same way as monocyclic systems, but the fallow is allowed to continue growing until mature forest is produced. Fruits and fibers and medicinal plants are periodically harvested as the forest develops, but there is never a final harvest cut or felling to clear the plot as with monocyclic systems. Over time, polycyclic managed fallows become managed forest orchards. For example, many of the te'lom and pet kotoob managed forests in Mexico (described below) were once polycyclic managed fallows.

The creation of a managed fallow starts during the early phases of the agricultural cycle when the forest is cleared. Some useful tree species are usually spared during the clearing operation to insure their presence in the fallow. The Bora, for example, refrain from cutting valuable timber species and certain palms (Denevan et al. 1984); the Yucatec Maya will spare a tree because of its fruits, fibers, medicinal properties, or nitrogen-fixing ability (Gómez-Pompa and Kaus 1990). Other subtle management techniques such as directional felling, coppicing, and slash piling are also used during the clearing process. The felled trees and slash are allowed to dry for several weeks and then burned to prepare the site for planting.

The primary agricultural crop is planted a week or so after burning. A variety of domesticated and semidomesticated tree species may also be introduced at this time. Edible fruit trees such as uvilla (*Pourouma cecropiaefolia*), peach palm (*Bactris gasipaes*), pacae (*Inga* spp.), and cashew (*Anacardium occidentale*) are common components of young swiddens in Amazonia (figure 8.2). These fast-growing, heliophilic trees are usually aggregated in the plot, either in the center to facilitate access or along the

FIGURE 8.2. Young managed fallow near village of Tamshiyacu in Peruvian Amazonia. A few cassava (*Manihot esculenta*) plants share the site with pineapples, bananas, and peach palms (*Bactris gasipaes*). Large leaves in the foreground are young umarí (*Poraqueiba sericea*) plants. (Photo by C. Padoch.)

perimeter so that they don't interfere with the primary crops. The young plot is periodically weeded to slow the growth of secondary vegetation, and some farmers fertilize their plants by adding ashes from household cooking fires or by mounding and burning the stubble left after harvesting. Although the frequency of weeding and fertilization appears to decrease once the annual cropping phase has finished, additional species continue to be added to the plot by planting, transplanting, or selectively favoring the growth of certain secondary species. Species introduced after the cropping phase are usually long-lived, shade-tolerant nondomesticates that are able to survive the highly competitive conditions in the understory of the developing fallow. Common examples include hogplum (*Spondias mombin*), Brazil nut (*Bertholletia excelsa*), and wild cacao (*Theobroma speciosum*).

After five or six years, the fallow contains a mixture of successional vegetation and harvestable tree species of either planted or spontaneous origin. These managed plots can be very diverse. Of the 645 plants collected in young fallows created by the Bora, 207 plants from 188 species were identified as useful (Unruh and Alcorn 1988). Other farmers, like the ribereños in Peru who plant Brazil nuts (Padoch et al. 1985) or the Kalapalo of central Brazil who manage for *Caryocar brasilensis* (Basso 1974), opt to create high-density stands of only one or two species in their fallows.

Silvicultural operations shift from intensive planting and tending to periodic harvesting and occasional slash weeding as the fallows increase in age. Local farmers return to their fallows during fruiting season to harvest, and they may slash the weeds around the base of the tree to aid in the collection of fruits. Animals are also attracted to these sites because of the fruit trees and hunting is another incentive to visit and casually manage older fallows. Depending on local traditions and priorities, one of three things can happen to an old fallow. In many cases, it will be felled and burned to start the agricultural cycle again (i.e., monocyclic systems). Alternatively, a decision will be made to continue managing the woody vegetation on the site, and it will be gradually converted to a managed forest or orchard (i.e., polycyclic systems). Finally, the old fallow will simply be abandoned when the village picks up and moves to another locale. It should be noted that villagers may return to these old sites for many years to hunt, to camp, and to harvest fruits.

Managed fallows produce long-term changes in the floristic composition of the forest. After several cycles of clearing, burning, and farming, the mix of species in the fallow regrowth becomes progressively simplified. Successional vegetation continues to swamp the site after clearing and must be weeded, but the overall species richness of the vegetation gradually declines. Several factors are responsible for this subtle biotic impoverishment. Repeated burning destroys many of the buried seeds in the soil (Ewel et al. 1981), and selective weeding reduces the density of many nonuseful taxa. The dispersal of large-seeded forest species is diminished because of lack of seed trees, lack of dispersal agents, or lack of appropriate regeneration niches. Over time, the net effect is that the overall species diversity of the fallow declines while the relative contribution of harvestable, useful trees increases. These floristic changes have only a minimal impact on the recovery and function of the tropical forest ecosystem (Uhl and Jordan

1984; Uhl et al. 1990), and they are exactly the result that a successful silvicultural system should produce, i.e., increase the density of desirable species by decreasing the density of undesirable ones. As was noted previously for homegardens, managed fallow systems produce small plots of forest enriched with useful species. Many of these species are native forest trees, and some of them will be able to maintain their local dominance on the old fallow site indefinitely.

MANAGED FORESTS

Managed forest systems are the most overlooked and least studied form of indigenous plant management. Unlike homegardens or managed fallows, which are highly visible and spatially defined, managed forests get lost in what is usually considered natural or primary forest. In most cases, the only evidence that some form of management is taking place is the distribution and abundance of useful trees in the forest.

Managed forests can be produced from old fallows, young fallows, homegardens, or intact forest. In each case, silvicultural treatment removes the unwanted stems through weeding and selective felling and introduces new stems through enrichment planting, coppicing, and protection of desirable volunteer species. Fertilization, pruning, and mulching are employed in some systems to enhance the productivity of important market species. Ecologically, managed forest systems represent the endpoint of the successional process on a site. Most of the favored species are shade-tolerant canopy trees adapted for growth and regeneration under a closed canopy or in small canopy gaps. Selective felling and the occasional windthrow provide the canopy openings required to maintain these species. Pioneer species that colonize the larger canopy gaps will be tolerated or removed. Rarely, however, will managed forests be cleared or the canopy drastically opened to initiate secondary succession and do something else with the site.

The Huastec Maya of Mexico create complex managed forests known as te'lom, or "the place of trees" (Alcorn 1984a,b). These forests, which are usually developed from old fallows, are most commonly found on slopes and ridges, where they control erosion and protect the watershed, or along streams, where they provide a shady riparian environment. Alcorn (1983) estimates that about 25 percent of community land is maintained in managed forest. Silvicultural practices in these forests are limited to casual weeding, enrichment planting, occasional selective felling of unproductive fruit trees, and protection of desirable wild or volunteer species. Desirable species that do not recruit themselves into the te'lom, or do so in the wrong place, will be transplanted from the forest or a homegarden. Undesirable plants will be removed by weeding or felling. After several years of irregular weeding and planting, the floristic composition of the forest becomes enriched with useful species. Different parts of the forest are managed for different groups of resources with the result that a te'lom is composed of several distinct stands. A managed forest, for example, may contain a small stand of trees useful for construction, a parcel of avocado trees, a stand of copal trees (*Protium copal*), a commercial coffee grove, and a patch of firewood trees. The species composition, size, and location of each stand varies from site to site and reflects the management objectives of the individual farmer.

Floristic inventories have shown that te'lom may contain more than 300 plant species (Alcorn 1983). In addition to edible fruits, construction materials, medicinal plants, and firewood, plants used for cordage, fish poison, tool handles, dyes, soap, incense, and tanning are also produced in Huateca managed forests. There are 81 food plants in these forests, including important Old World domesticates such as mangos, oranges, and coffee. Coffee and the native ornamental palm *Chamaedorea elegans* are common commercial products.

A similar type of managed forest called pet kot was constructed by the Yucatec Maya. The system takes its name from the low wall of stones (*pet,* circular; *kot,* wall of loose stones) that characteristically surrounds the forest plot (Barrera-Vázquez 1980). These enriched patches of tall forest stand out in stark contrast to the surrounding low deciduous forest, which is the dominant vegetation in the area. Pet kot are created from either old fallows or natural forest through the continual enrichment of the site with desirable species. The introduced species may come from local forests, from more distant and humid forests of the Maya realm, or from homegardens. Common tree genera found in pet kot include *Brosimum, Spondias, Pithecellobium, Malmea, Bursera,* and *Sabal,* which also occur as dominant elements in homegardens of the region (C. E. Smith and Cameron 1977). The creation of pet kot near areas of shifting cultivation would have provided local farmers with a shady, resource-rich place to stay while they were tending their fields. Most of the weeding, thinning, and planting operations required to maintain the pet kot undoubtedly took place during these visits.

Although the local Maya no longer make pet kot, other types of enriched forest orchards are created on the slopes and bottom of sinkholes (cenotes) and in scattered patches of deep soil (rejoyas) in karst areas (Gómez-Pompa 1987). These managed forests are of particular interest because of their small size, careful placement in especially moist and fertile microsites, and high density of native and introduced fruit species (e.g., *Citrus, Persea, Musa, Manilkara,* and *Annona*). The blending of water, fertile soils, fruit trees, and forests in a droughty, karst environment reflects a considerable degree of silvicultural acumen.

Ribereño farmers in the Amazon estuary of Brazil actively manage floodplain forests to favor the density and productivity of several economically important forest resources (e.g., *Euterpe oleracea, Mauritia flexuosa,* and *Theobroma cacao;* figure 8.3). Management involves a number of operations designed to enrich the forest and enhance fruit production (Anderson et al. 1985). Enrichment occurs through selective weeding and thinning of competitors as well as through planting and protection of desirable volunteers. Cacao, mango, genipap (*Genipa americana*), and cupuaçu (*Theobroma grandiflorum*) are frequently planted, while *Spondias mombin, Hevea brasiliensis,* and *Virola surinamensis,* a valuable timber species, are desirable and carefully protected volunteers. Woody vines, the spiny murumuru palm (*Astrocaryum murumuru*), and firewood trees are usually the first victims of selective thinning. Large trees are

FIGURE 8.3. Managed floodplain forest dominated by açaí palms (*Euterpe oleracea*) on Ilha das Onças in the Amazon estuary.

frequently girdled rather than felled to minimize damage to the surrounding trees. Fertilizing and thinning techniques are used to enhance the productivity of commercial species. For example, the decaying leaves and inflorescences of the *Euterpe* palm are used as mulch and piled at the base of reproductive trees. Local residents also selectively thin the multiple stems of the *Euterpe* palm when harvesting palm hearts, and this practice, when judiciously applied, enhances the overall production of fruits on the remaining stems (Anderson 1990).

There are numerous varieties of managed forest systems in the Amazon estuary, yet each one appears to be successful in enhancing the forest resources of interest. Compared to unmanaged floodplain forests, managed forests have a more open canopy, fewer total stems, a lower species richness, and a greater density of useful stems (Anderson et al. 1995). The exact species mix favored by the farmer seems to depend on the age of the forest, the flooding regime, and the distance to market. In each case, however, the structure and composition of the forest is notably altered

Managed forest systems are subtle, but they can produce lasting changes. Certain trees and species are selectively removed, and the longer that management continues, the less frequently these unwanted volunteers will crop up in the forest. Over time, as seed trees and soil seed banks are depleted, these directional floristic changes can become persistent. As long as the communities that initially shaped the forest are present, the te'lom, pet kot, or managed floodplain forest will be maintained and continually refined, because of its cultural, ecological, and economic value. If the site is abandoned, many of the abundant canopy species in the managed forest will probably maintain themselves. This persistence is the result of long-term silvicultural treatment. Through selective weeding and thinning, the regeneration of desirable canopy species is favored over less useful taxa. After several decades of this treatment, the canopy species develop all-age populations with seedlings, saplings, juveniles, poles, and adults. In the absence of human intervention, these canopy species will continue growing on the site for several generations even if the level of regeneration is drastically reduced. If the site is abandoned and then later cleared for agriculture by another cultural group, a successional process will be initiated, and a whole different suite of species may occupy the site. In all probability, however, the second community to farm the site possesses its own form of traditional silviculture and will, like its predecessors, spare many of the valuable canopy trees when the forest is cleared. Although the old forest temporarily will be lost, many of the genotypes produced by the previous management system will play a major role in shaping the new forest that develops on the site.

IMPRINTS OF PRECOLUMBIAN SILVICULTURE

In spite of the number of indigenous management systems that are practiced today in the Neotropics, it would be erroneous to assume that these same systems were applied in the same way and at the same intensity during Precolumbian times. Forest management systems evolve based on the needs, available resources, and cultural traditions of a particular group of people at a given point in time. Clearly, life in the New World prior to European contact was quite a bit different from today's. For one thing, the native population in the Americas was several orders of magnitude higher then (Borah 1970; Meggers 1971; Parsons 1975). Denevan (1992) estimates that there were more than 50 million native people during Precolumbian times, with 17 million in Mexico, 6 million in Central America, and almost 9 million in lowland Amazonia. Within 100 years following the arrival of the Europeans, 90 percent of that population was lost to disease, enslaved, or killed (Cook and Borah 1971). A wealth of silvicultural experience was also lost at this time.

The social organization of indigenous communities was much different during Precolumbian times. Many of the populations were concentrated in large settlements containing thousands of people with complex economies based on intensive use of resources and extensive trade networks (Turner 1976; Roosevelt 1989). Chiefdoms extending over tens of thousands of hectares were established, and elaborate earthworks were constructed for permanent cultivation, water control, defense, habitation, and burial (Roosevelt 1987). In contrast, the indigenous populations of today live in small, independent communities and largely practice a subsistence level of agriculture and forest resource exploitation. Given their population density, sociopolitical organization, and intensity of resource use, Precolumbian indigenous communities would have had a significantly larger impact on the forest than their present-day descendants. They probably applied many of the same silvicultural systems, e.g., homegardens, managed fallows, and managed forests, but they would have done so over larger areas of forest, for longer periods of time, at a much higher intensity. It is really not surprising that hundreds of years of indigenous management have left a permanent imprint on the forests of the Neotropics.

ENRICHMENT OF USEFUL SPECIES

The most noticeable residual effects of Precolumbian forest management are the high-density aggregations of useful species found in many Neotropical forests. These floristic anomalies have been found in the vicinity of human settlements in numerous areas of "primary" forest. The nomadic Guajá in eastern Amazonia always seem to locate their temporary villages near dense stands of the babassu palm (*Attalea speciosa*) that are surrounded by species-rich forest. The edible seeds and fleshy pulp of the palm play an important role in the diet of the Guajá. Excavations in some of these villages yielded potsherds and ceramic manioc griddles of the neighboring Ka'apor whose ancestors used to live on the site (Balée 1987). Other old Ka'apor sites also contain dense enclaves of babassu. The Ka'apor are skilled forest managers (Balée 1994), and the scattered populations of babassu within the forest are thought to be relicts from their old managed fallows. Babassu seedlings can easily dominate burned forest clearings because of their cryptogeal germination, i.e., the apical meristem initially buries itself in the ground, rather than growing up, so that is protected from fire, and adult palms were probably spared during the clearing process. Over the past several hundred years, the Ka'apor have created permanent, two- to three-hectare stands of babassu palms in many of the forests of northern Maranhão and eastern Pará. Although their relations with the Guajá have historically been rather hostile, the vestiges of their silvicultural practice have been a lasting gift to this tribe.

Several other useful Amazonian palm species occur in dense aggregations in areas of previous human occupation. *Astrocaryum vulgare,* for example, forms high-density stands in old secondary forests in eastern Amazonia (Balée and Gély 1989), and the species is also common on archaeological sites in coastal Pará. Wessels-Boer (1965) reports that *A. vulgare* never occurs in undisturbed forests in Suriname. The palm is used locally as a source of fiber. Enriched pockets of *Elaeis oleifera*, a native oil palm, and *Acrocomia aculeata,* a source of edible fruits, also have been reported from old settlement sites along the lower Amazon.

Clumps of fruit trees are another important indicator of indigenous management in Amazonian forests. *Grias peruviana* is a common edible fruit in the seasonally flooded forest of Peruvian Amazonia. The species is planted in homegardens and is occasionally managed in fallows by present-day ribereños (Peters et al. 1989). The species aggressively maintains itself on these sites after abandonment and has been observed to form populations of 400–500 adult trees/ha along the lower Ucayali River in Peru (Peters unpublished data). Groves of *Platonia insignis* trees, an important native fruit for the Ka'apor, have been reported from old settlement sites in the Ka'apor reserve in Maranhão (Balée 1989), and Basso (1974) found populations of *Caryocar brasiliensis* "extending for several miles" around old Kalapalo villages in central Brazil. The fruits of this species contain an oily pulp and a seed that is used for food. Taken together, dense clusters of fruit trees, monospecific palm stands, and other types of anthropogenic forests are estimated to comprise over 11 percent of Brazilian Amazonia (Balée 1989).

Gordon (1982) makes note of several anomalous distributions of tree species in the Bocas del Toro region of Panama. In addition to isolated groves of trees at abandoned habitation sites, which he terms "archaeological disclimaxes," plantings along the edges of trails that penetrate remote areas of forest have produced notable patterns of species enrichment. The fruit trees *Couma macropcarpa* and *Manilkara bidentata* and *Pseudolmedia spuria,* a source of edible fruits and bark cloth, occur in especially high densities in the forest. *Brosimum alicastrum* and *Dipteryx panamensis* are important trees in local arboriculture, and mixed groves of these species are frequently encountered on old Térreba settlements.

The Olmec and Maya civilizations occupied tropical forest habitats in southeastern Mexico and northern Central America for a combined period of over 3,000 years and reached population densities of 400–500 people/km^2 in some areas (Turner 1976). The forests of this region have been exploited and managed so intensively and for so long that it is very difficult to find plant communities that have not been shaped by Precolumbian foresters. Two of the most widespread forest trees in the region, *Manilkara zapota* and *Brosimum alicastrum,* were important food and construction materials for the ancient Maya. The fact that dense aggregations of both species have been reported growing on ruin sites is particularly suggestive of an anthropogenic origin for these populations (figure 8.4; Thompson 1930; Bartlett 1935; Lundell 1937). Other important forest resources in Maya subsistence, such as *Protium copal* (incense), *Ceiba pentandra* (sacred, fiber), *Dialium guianense* (edible fruit), *Haematoxylon campechianum* (dye), and *Swietenia macrophylla* (construction), are also dominant elements of the local flora. The recent discovery of relict cacao groves in karst sinkholes in northern Yucatán (Gómez-Pompa et al. 1990) provides further evidence of the intensity and sophistication with which the Maya managed their forests. Agronomically, it should not be possible to grow *Theobroma cacao* in a region with a six-month dry season (Purseglove 1968).

SELECTION OF FAVORABLE GENOTYPES

The management of plant populations involves the manipulation of the number and distribution of individuals. Some individuals are selectively favored or planted, and their numbers increase. Other individuals are removed through weeding, felling, or thinning, and their populations decrease. This change in the relative abundance of different species is the criterion by which we judge the success of existing management systems. This is also how we usually try to document the lasting impact of indigenous silviculture. Populations of forest trees, however, are notably intractable entities. High-density populations created by conscientious management may not necessarily stay high-density populations. Conversely, it is possible that low-density, managed popula-

FIGURE 8.4. *Brosimum alicastrum* trees growing on the ruins of Cobá in Quintana Roo, Mexico.

tions may convert themselves into high-density aggregations after abandonment in response to other ecological factors completely unrelated to management. All of these factors confound the study of Precolumbian silviculture. Fortunately, there are additional aspects of forest management that engender more permanent changes than the density of a particular species in the forest. When farmers are selecting individuals to favor, they are also selecting certain genotypes. After several cycles of cutting and planting, a considerable degree of artificial selection can occur in managed forests. From this perspective, the imprint of management can come into sharp focus, because the genotypes produced by forest farmers are markedly different from those produced by natural selection.

A useful example of this approach is provided by the case of *Brosimum alicastrum* in Mexico. This multipurpose species (Peters and Pardo-Tejeda 1982) is cultivated in Maya homegardens and forms high-density forest populations in many parts of Mexico (Rzedowski 1963; Gómez-Pompa 1973). The occurrence of dense aggregations of *B. alicastrum* on or near Maya ruin complexes was initially interpreted as evidence of Maya silviculture, i.e., the stands were relicts of ancient Maya orchards (Puleston 1968, 1982; Folan et al. 1979). Later work, however, suggested that the spatial relationship between *B. alicastrum* and Maya ruins was primarily the result of normal ecological processes (Miksicek et al. 1981; Lambert and Arnason 1982). As explained by Peters (1983), *B. alicastrum* forms high-density aggregations around Maya ruins because large quantities of the fruit are eaten and dispersed by the *Artibeus* bats that roost there. The bats fly to the forest, collect the fruit, bring it back to their roost, eat the fruit, and then discard the undamaged seed. The continual input of bat-dispersed seed has maintained the clumps of *B. alicastrum* around ruins for hundreds of years.

This observation, of course, does not negate the possibility that the species was used and managed by the Classic Maya. In fact, if we examine the behavior, rather than the density or spatial location, of these populations, we are presented with strong evidence of deliberate genetic improvement. This is especially notable in the phenology, productivity, and breeding systems of the *B. alicastrum* trees that cluster around the ruins at Tikal in Guatemala. For example, phenological data from various parts of its distribution indicate that fruit production by *B. alicastrum* is usually annual, with peak seedfall occurring at the onset of the rainy season (Pennington and Sarukhán 1968; Croat 1978; Peters and Pardo-Tejeda 1982). In contrast, several authors have reported that the trees at Tikal bear fruit twice a year (Gonzales 1939; Puleston 1968), while others have suggested that fruiting is continuous throughout the year with three periods of peak abundance (Coelho et al. 1976; Schlichte 1978).

Quantitative data on fruit production by *B. alicastrum* are very scarce. Using estimates from Puleston (1968) and Coelho et al. (1976), an average annual production of approximately 110 kg/tree can be calculated for the populations at Tikal (see Peters 1983). Data from a detailed ecological study of the species in Veracruz, Mexico (Peters 1989), under environmental conditions almost identical to that of Tikal (i.e., similar elevation, mean temperature and precipitation, and substrate) reveal that the maximum annual production of fruits by a large tree in this region is only 65 kg, or approximately half of that derived for the Tikal trees.

The original description of *B. alicastrum* (Swartz 1797) reports that male and female flowers are found on separate trees, i.e., the species is dioecious, and more recent taxonomic treatments have confirmed this (Berg 1972). Reports from the Tikal region (Standley and Steyermark 1946; Puleston 1968), however, have consistently described the tree as monoecious, with male and female flowers on the same tree. The salient difference between these two breeding systems is that every individual in a monoecious population can potentially produce fruit, while fruit production is limited to female trees in dioecious populations.

What we seem to be presented with in Tikal, therefore, is a dense population of *B.*

alicastrum in which fruit is produced in extremely large quantities at frequent intervals by every mature individual. The curious thing is that similar forests of the tree in other regions do not exhibit this behavior. One possible explanation is that the regeneration of the species is severely limited for some reason at Tikal, and such an abundant reproductive output is necessary to maintain the population. A second interpretation, however, is that such a high level of fruit production is in excess of what is actually required, and the atypical reproductive behavior of the *B. alicastrum* trees at Tikal are manifestations of relict genotypes deliberately selected for by the continued use and management of the species. What appears to be suggested is the conscious mixing of diverse genotypes with the objective of producing an abundant, year-round supply of seeds.

THE POTENTIAL OF INDIGENOUS FOREST MANAGEMENT

Indigenous people throughout the Neotropics have developed sophisticated systems for managing forest resources. These systems produce timber and nontimber resources; they preserve valuable sources of domesticated, semidomesticated, and wild germ plasm, and they do so without irreparably destroying the forest matrix within which they are created. Every new homegarden, managed fallow, and managed forest is one more replication in a long-term experiment of silvicultural treatment that spans hundreds of years, hundred of sites, and hundreds of species. The basic understanding of the regeneration, growth, treatment, yield, and use of tropical trees that has accumulated in this vast and on-going experiment eclipses anything that European foresters have learned during their hundred years in tropical forests. The management protocols of indigenous silviculture are based on oral traditions rather than textbooks, access is by trail rather than roads, and weeds are removed selectively by hand rather than en masse by herbicide. These systems are a living, evolving embodiment of Occam's razor: never opt for something complicated when something simple will produce the same result.

Current rates of deforestation in the Neotropics have made people very aware of the difference between good forest management and bad forest management. In most people's mind, clearing thousands of hectares of Amazonian forest to put in a cattle pasture would profile as bad forest management. Selectively high-grading a stand of *Virola surinamensis* for timber with little regard for the regeneration of the species would also qualify as bad forest management. Managing hundreds of plant species in old fallows or stocking small tracts of forest with 300 species of fruits, medicinal plants, and timber trees would seem, from most perspectives, to be good forest management. If we really want to manage the forests of the Neotropics on a sustainable, long-term basis, it would seem that the study of indigenous management systems merits a much higher priority than it is currently afforded. Clearly, the easiest and quickest way to learn how to manage a tropical forest is to ask the people who have been doing it for the longest time.

ACKNOWLEDGMENTS

I would like to thank Arturo Gómez-Pompa, Christine Padoch, and Janis Alcorn for several enlightening discussions about indigenous systems of forest management. I thank David Lentz for his patience and persistence in motivating me to write about the links between Neotropical forests and Precolumbian silviculture. Many of the ideas in this paper were developed with support from the U.S. Man and the Biosphere Program and the John D. and Catherine T. MacArthur Foundation, and the assistance of both institutions is gratefully acknowledged. This essay is dedicated to all of the Precolumbian silviculturists who were so subtle and so successful in shaping their forests that today we enclose them in parks and call them pristine.

REFERENCES

Alcorn, J. B. 1981. Huastec noncrop resource management: Implications for prehistoric rain forest management. *Human Ecology* 9: 395–417.

———. 1983. El te'lom huasteco: Presente, pasado, y futuro de un sistema de silvicultura indígena. *Biótica* 8: 315–331.

———. 1984a. *Huastec Mayan Ethnobotany.* Austin: University of Texas Press.

———. 1984b. Developmental policy, forests, and peasant farms: Reflections on Huastec managed forests' contribution to commercial production and resource conservation. *Economic Botany* 38: 389–406.

Anderson, A. B. 1990. Extraction and forest management by rural inhabitants in the Amazon estuary. In A. B. Anderson, ed., *Alternatives to Deforestation,* pp. 65–85. New York: Columbia University Press.

Anderson, A. B., A. Gély, J. Strudwick, G. L. Sobel, and M. G. C. Pinto. 1985. Um sistema agroflorestal na várzea do estuário amazônico (Ilha das Onças, Município de Barcarena, Estado do Pará). *Acta Amazônica* 15: 195–224.

Anderson, A. B., P. Magee, A. Gély, and M. A. G. Jardim. 1995. Forest management patterns in the floodplain of the Amazon estuary. *Conservation Biology* 9: 47–61.

Balée, W. 1987. Cultural forests of the Amazon. *Garden* 11: 12–14.

———. 1989. The culture of Amazonian forests. In D. A. Posey and W. Balée, eds., *Resource Management in Amazonia: Indige-nous and Folk Strategies,* pp. 1–21. New York Botanical Garden: Advances in Economic Botany 7.

———. 1994. *Footprints of the Forest.* New York: Columbia University Press.

Balée, W., and A. Gély. 1989. Managed forest succession in Amazonia: The Ka'apor case. In D. A. Posey and W. Balée, eds., *Resource Management in Amazonia: Indigenous and Folk Strategies,* pp. 129–158. Advances in Economic Botany 7. New York: New York Botanical Garden.

Barrera, A. 1980. Sobre la unidad de habitacíon tradicional campesina y el manejo de los recursos bióticos de el área maya yucatanense, 1: Arboles y arbustos de las huertas familiares. *Biótica* 5: 115–128.

Barrera-Vázquez, A. 1980. *Diccionario Maya Cordemex.* Mérida, Mexico: Ediciones Cordemex.

Bartlett, H. H. 1935. A method of procedure for field work in tropical American phytogeography based upon botanical reconnaissance in parts of British Honduras and the Petén forests in Guatemala. In *Botany of the Maya Area,* pp. 1–26. Publ. 461. Washington, D.C.: Carnegie Institution.

Basso, E. B. 1974. *The Kalapalo Indians of Central Brazil.* New York: Holt, Rinehart & Winston.

Berg, C. C. 1972. Brosimae. *Flora Neotropica Monograph* 7: 161–208.

Borah, W. 1970. The historical demography of Latin America: Sources, techniques, controversies, yields. In P. Duprez, ed., *Population and Economics*, pp. 78–103. Winnipeg: University of Manitoba Press.

Carneiro, R. L. . 1961. Slash and burn cultivation among the Kuikuru and its implications for cultural development in the Amazon Basin. *Antropologica Supplement* 2: 47–67.

Coelho, A. M., L. S. Coelho, C. A. Bramblett, S. S. Bramblett, and L. B. Quick. 1976. Ecology, population characteristics and sympatric association in primates: A sociobioenergetic analysis of howler and spider monkeys in Tikal, Guatemala. *Yearbook of Physical Anthropology* 20: 96–135.

Cook, S., and W. Borah. 1971. *Essays in Population History: Mexico and the Caribbean.* Berkeley: University of California Press.

Covich, A. P., and N. H. Nickerson. 1966. Studies of cultivated plants in Choco dwelling clearings, Darien, Panama. *Economic Botany* 20: 285–301.

Croat, T. B. 1978. *Flora of Barro Colorado Island.* Stanford, Calif.: Stanford University Press.

Dawkins, H. C. 1958. *The Management of Natural Tropical High-forest with Special Reference to Uganda.* Paper 34, Imperial Forestry Institute, University of Oxford.

Denevan, W. M. 1992. The pristine myth: The landscape of the Americas in 1492. *Annals of the Association of American Geographers* 82: 369–385.

Denevan, W. M., J. M. Treacy, J. B. Alcorn, C. Padoch, J. Denslow, and S. Flores Paitan. 1984. Indigenous agroforestry in the Peruvian Amazon: Bora Indian management of swidden fallows. *Interciencia* 9: 346–357.

Ewel, J., C. Berish, B. Brown, N. Price, and J. Raich. 1981. Slash and burn impacts on a Costa Rican wet forest site. *Ecology* 62: 816–829.

FAO. 1957. Shifting cultivation. *Unasylva* 11: 9–11.

Folan, W. J., L. A. Fletcher, and E. R. Kintz. 1979. Fruit, fiber, bark, and resin: Social organization of a Mayan urban center. *Science* 204: 697–701.

Gómez-Pompa, A. 1973. Ecology of the vegetation of Veracruz. In A. Graham, ed., *Vegetation and Vegetational History of Northern Latin America,* pp. 73–48. Amsterdam: Elsevier.

——. 1987. On Maya silviculture. *Mexican Studies* 3: 1–17.

Gómez-Pompa, A., E. Flores, and V. Sosa. 1987. The "pet kot": A man-made tropical forest of the Maya. *Interciencia* 12: 10–15.

Gómez-Pompa, A., and A. Kaus. 1990. Traditional management of tropical forests in Mexico. In A. B. Anderson, ed., *Alternatives to Deforestation,* pp. 45–63. New York: Columbia University Press.

Gómez-Pompa, A., J. Salvadore Flores, and M. A. Fernandez. 1990. The sacred cacao groves of the Maya. *Latin American Antiquity* 1: 247–257.

Gonzales, A. 1939. El ramon o capomo. *Boletín Platanero Agrícola* 2: 221–222.

Gordon, B. L. 1982. *A Panama Forest and Shore: Natural History and Amerindian Culture in Bocas del Toro Province, Panama.* Pacific Grove, Calif.: Boxwood Press.

Harris, D. R. 1971. The ecology of swidden cultivation in the upper Orinoco rain forest, Venezuela. *Geographical Review* 61: 475–495.

Hecht, S. B. 1982. Agroforestry in the Amazon Basin: Practice, theory, and limits of a promising land-use. In S. B. Hecht. ed., *Amazonia: Agriculture and Land Use Research,* pp. 331–371. Cali, Colombia: Centro Internacional de Agricultura Tropical.

Irvine, D. 1989. Succession management and resource distribution in an Amazonian rain forest. In D. A. Posey and W. Balée, eds., *Resource Management in Amazonia: Indigenous and Folk Strategies,* pp. 223–237. Advances in Economic Botany 7. New York: New York Botanical Garden.

Kelly, I., and A. Palerm. 1952. *The Tajín Totonac.* Smithsonian Institution, Institute of Social Anthropology Publ. 113. Washington, D.C.: GPO.

Lambert, J. D. H., and T. Arnason. 1982. Ramon and Maya ruins: An ecological, not an economic, relation. *Science* 216: 298–299.

Lamprecht, H. 1989. *Silviculture in the Tropics.* Eschborn: Deutsche Gesellschaft für Technische Zusammenarbeit.

Lundell, C. L. 1937. *The Vegetation of Petén.* Washington, D.C.: Carnegie Institution.

Meggers, B. J. 1971. *Amazonia: Man and Na-*

ture in a Counterfeit Paradise. Arlington Heights, Ill.: AHM Publishing.

Miksicek, C. H., K. J. Elsesser, I. A. Wuebber, K. O. Bruhns, and N. Hammond. 1981. Rethinking ramon: A comment on Reina and Hill's lowland Maya subsistence. American Antiquity 46: 916–918.

Padoch, C., and W. de Jong. 1991. The house gardens of Santa Rosa: Diversity and variability in an Amazonian agricultural system. Economic Botany 45: 166–175.

Padoch, C., J. Chota Inuma, W. de Jong, and J. Unruh. 1985. Amazonian agroforestry: A market-oriented system in Peru. Agroforestry Systems 3: 47–58.

Parsons, J. J. 1975. The changing nature of New World tropical forests since European colonization. In The Use of Ecological Guidelines for Development in the American Humid Tropics, pp. 28–38. Morges, Switzerland: International Union for Conservation and Natural Resources Publications.

Pennington, T. D., and J. Sarukhán. 1968. Manual para la Identificacíon de Campo de los Principales Arboles Tropicales de México. Mexico, D.F.: Instituto Nacional de Investigaciones Forestales.

Peters, C. M. 1983. Observations on Maya subsistence and the ecology of a tropical tree. American Antiquity 48: 610–615.

———. 1989. Reproduction, Growth, and the Population Dynamics of Brosimum alicastrum Sw. in a Moist Tropical Forest of Veracruz, Mexico. Ph.D. dissertation, Yale University, New Haven, Conn.

Peters, C. M., and E. Pardo-Tejeda. 1982. Brosimum alicastrum (Moraceae): Uses and potential in Mexico. Economic Botany 36: 166–175.

Peters, C. M., M. J. Balick, F. Kahn, and A. B. Anderson. 1989. Oligarchic forests of economic plants in Amazonia: Utilization and conservation of an important tropical resource. Conservation Biology 3: 341–349.

Posey, D. A. 1982. The keepers of the forest. Garden 6: 18–24.

———. 1984. A preliminary report on diversified management of tropical forest by the Kayapó Indians of the Brazilian Amazon. Advances in Economic Botany 1: 112–126.

Price, N. 1983. The Tropical Mixed Garden: An Agroforestry Component of the Small Farm. Turrialba, Costa Rica: Centro Agronomico Tropical de Investigacion y Ensenanza.

Puleston, D. E. 1968. Brosimum alicastrum as a Subsistence Alternative for the Classic Maya of the Central Southern Lowlands. Master's thesis, University of Pennsylvania, Philadelphia.

———. 1982. The role of ramon in Maya subsistence. In K. Flannery, ed., Maya Subsistence: Studies in Memory of Dennis E. Puleston, pp. 349–366. New York: Academic Press.

Purseglove, J. W. 1968. Tropical Crops: Dicotyledons. London: Longman Group.

Roosevelt, A. 1987. Chiefdoms in the Amazon and Orinoco. In R. D. Drennan and C. A. Uribe, eds., Chiefdoms in the Americas, pp. 153–185. Lanham, Md.: University Presses of America.

Roosevelt, A. 1989. Resource management in Amazonia before the conquest: Beyond ethnographic projection. In D. A. Posey and W. Balée, eds., Resource Management in Amazonia: Indigenous and Folk Strategies, pp. 30–62. Advances in Economic Botany 7. New York: New York Botanical Garden.

Rzedowski, J. 1963. The northern limit of tropical rain forests in continental North America. Vegetatio 11: 173–198.

Schlichte, H. J. 1978. A preliminary report on the habitat utilization of a group of howler monkeys (Alouatta villosa-pigra) in the National Park of Tikal, Guatemala. In G. G. Montgomery, ed., The Ecology of Arboreal Folivores, pp. 551–560. Washington, D.C.: Smithsonian Institution Press.

Smith, C. E., and M. L. Cameron. 1977. Ethnobotany in the Puuc, Yucatán. Economic Botany 31: 93–110.

Smith, D. M., B. C. Larson, M. J. Kelty, and P. M. S. Ashton. 1997. The Practice of Silviculture. 9th ed. New York: John Wiley & Sons.

Standley, P. C., and J. A. Steyermark. 1946. Flora of Guatemala, pp. 13–16. Botanical Series 24. Chicago: Field Museum of Natural History.

Swartz, O. 1797. Flora Indiae occidentalis aucta atque illustrata sive descriptiones *plantarium in prodromo recensitarum.* Erlangae.

Taylor, C. J. 1962. *Tropical Forestry.* Oxford: Oxford University Press.

Thompson, J. E. S. 1930. *Ethnology of the Maya of Southern and Central British Honduras.* Anthropology Series 17(2), Publ. 274. Chicago: Field Museum of Natural History.

Troup, R. S. 1922. *Silviculture of Indian Trees,* vols. 1–3. Oxford: Oxford University Press.

Turner II, B. L. 1976. Population density in the Classic Maya Lowlands: New evidence for old approaches. *Geographical Review* 66: 73–82.

Uhl, C., and C. F. Jordan. 1984. Succession and nutrient dynamics following forest cutting and burning in Amazonia. *Ecology* 65: 1476–1490.

Uhl, C., D. Nepstad, R. Buschbacher, K. Clark, B. Kauffman, and S. Subler. 1990. Studies of ecosystem response to natural and anthropogenic disturbances provide guidelines for designing sustainable land-use systems in Amazonia. In A. B. Anderson, ed., *Alternatives to Deforestation,* pp. 24–42. New York: Columbia University Press.

U Kyaw Zan. 1953. Note on the effect of 70 years of treatment under the selection system and fire protection in the Kangyi Reserve. *Burmese Forester* 3: 11–18.

Unruh, J., and J. B. Alcorn. 1988. Relative dominance of the useful component in young managed fallows at Brillo Nuevo. In W. B. Denevan and C. Padoch, eds., *Swidden-Fallow Agroforestry in the Peruvian Amazon,* pp. 47–52. Advances in Economic Botany 5. New York: New York Botanical Garden.

Vargas, C. 1983. El ka'anché: Una práctica hortícola Maya. *Biótica* 8: 151–173.

Watters, R. F. 1971. *Shifting Cultivation in Latin America.* FAO Forestry Development Paper 17. Rome: Food and Agriculture Organization of the United Nations.

Webb, W. L. 1960. Restrictive and incentive control of shifting cultivation. In *Proceedings of the Fifth World Forestry Congress,* Seattle, pp. 2021–2025.

Wessels-Boer, J. G. 1965. *Palmae: Flora of Suriname,* vol. 5. Leiden: E. J. Brill.

Wyatt-Smith, J. 1963. *Manual of Malayan Silviculture for Inland Forests.* Malayan Forest Record 23. Kuala Lumpur.

GAYLE J. FRITZ

9 | NATIVE FARMING SYSTEMS AND ECOSYSTEMS IN THE MISSISSIPPI RIVER VALLEY

The Mississippi River is about 2,300 miles long, and the history of native agriculture in its valley goes back more than 4,000 years. Given this considerable range of both space and time, the development of many different farming systems could be expected. When Europeans first entered the valley, native farmers from Lake Itasca to Lake Pontchartrain grew maize, beans, squashes, and sunflowers as their primary crops, but agricultural strategies varied significantly due to both cultural and environmental factors. The archaeological record shows that differences between the northern, central, and southern reaches of the valley were even more marked in the more distant past, and that developmental pathways were not parallel. My purpose is to examine through time those diverse agricultural systems and the broader ecosystems into which they fit.

THE UPPER, CENTRAL, AND LOWER MISSISSIPPI VALLEY SUBREGIONS: ENVIRONMENTAL AND HISTORICAL BACKGROUND

The Mississippi River valley is divisible into archaeological subregions, with the upper, central, and lower valleys having their own traditions of research (Kidder 1992; Kidder and Fritz 1993; Morse and Morse 1983; Phillips et al. 1951; S. Williams and Brain 1983). Although this region supports a number of vegetation types, distinct environmental boundaries from north to south do not exist, and cultural frontiers shifted through time and oscillated in strength. For

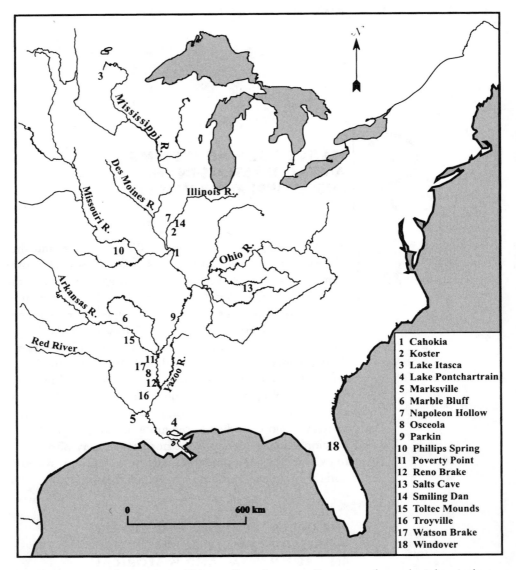

FIGURE 9.1. Sites in the Mississippi River valley relevant to discussion of precolonial agriculture.

convenience, I divide the Upper Valley from the Central Valley at about the mouth of the Des Moines River, separating the states of Iowa and Missouri (figure 9.1). The equally arbitrary line between the Central and Lower valleys can be drawn at the mouth of the Arkansas River.

The three cultural regions have different boundaries from Greller's (this volume) floristic provinces and vegetation zones. In terms of floristic provinces, the major division is at the mouth of the Ohio River, with the Appalachian Province to the north and the Atlantic and Gulf coastal plains to the south. The northern half of the Mississippi Valley is bordered closely by the North American Prairies Province to the west. Major vegetation zones are divided at the mouth of the Missouri River, with a deciduous oak forest–woodland savanna complex following the Valley northward into central Minnesota and Wisconsin, and the deciduous forest and oak-pine zone extending to the south into central Louisiana. The grassland vegetation zone lies close to the Mississippi Valley's west as far south as southwestern Missouri, where the deciduous oak forest–woodland savanna complex curves toward the southwest. At the extreme southern end of the Mississippi Valley, a zone of southern mixed hardwoods lies just to the north of a narrow coastal strip of evergreen oak–palm–magnolia (Greller, this volume).

People living in and near the Mississippi River at different latitudes obviously faced different climatic, edaphic, and biotic exigencies, but environmental factors did not dictate cultural responses in predictable ways. The long frost-free period in the south (averaging 229 days per year in northeast Louisiana) did not motivate inhabitants of the Lower Valley to embark upon an early shift from foraging to farming. The shorter growing season (ca. 160 days) in Iowa, Wisconsin, and Minnesota did not impose a barrier to the Precolumbian farmers, who adopted the hardy, short-season ancestor to Northern Flint maize, built ridges and hills of soil to drain frost away from the roots, and developed appropriate storage techniques.

Fertile soils were available along the entire length of the Mississippi River valley, but bottomland gardens and fields were at risk of flooding, and wider sections of the valley have the additional disadvantage to hoe farmers of having large tracts of heavy backswamp clays. Many southeastern American Indian farmers at the time of European contact and during late precolonial times preferred higher, well-drained terraces along major and minor streams, as well as natural levees in the large bottomland expanses today called the American Bottom area (the hemispherical lowland in Illinois across the river from St. Louis), the Mississippi Delta (the inland delta of northwestern Mississippi drained by the Yazoo and Sunflower rivers), and the Tensas-Mississippi River lowlands of northeastern Louisiana. At the north end of the Mississippi Valley, however, tillable glacial soils exist in interfluvial zones, and tracts of nonprairie soils can be found in upland areas to the east and west of the Mississippi River valley in Illinois and Missouri.

The Mississippi Valley is also rich in animal resources and in wild plant foods, many of which were used in varying combinations together with agricultural produce throughout the entire span of Native American occupation, continuing in places today, and along the full reach of the valley. Improvements in fishing technology may have been instrumental in the spread of the first cultivated plant, the native eastern gourd (*Cucurbita pepo* ssp. *ovifera*), which might have been grown for net floats before people favored larger fruits and sweet flesh. The abundance of resources (especially fish, birds, mammals, nuts, and other plant foods) concentrated in recently stabilized riverine settings in parts of the Central Valley during the middle Holocene (ca. 6000–3000 B.C.) may have attracted increasingly sedentary groups of people to areas where domestication of native seed-bearing plants occurred (Smith 1992). However, the abundance of plant and animal foods in the Lower Valley, especially its southern end, and their availability throughout most of the year seem to have allowed foragers in Louisiana and Mississippi to practice increasingly intensified use and management of wild resources, enabling them to participate in mound-building and long-distance trade between 3000 B.C. and A.D. 200 with no signs of agriculture apart from probable cultivation of gourds. In the Upper Valley, the eastward expansion of bison herds beginning at approximately A.D. 1400 appears to have resulted in a narrower range of crops being grown, along with a shift by most members of the community to seasonal bison hunting away from fields and winter villages.

CULTURAL HISTORICAL TERMINOLOGY

Table 9.1 includes the names of significant cultural and temporal units used throughout the essay. *Oneota* is the name given by archaeologists to the late prehistoric farmers of the Upper Valley who grew maize and other crops beginning at about A.D. 1000. Their contemporaries in the Central Valley are called *Mississippian*, a label sometimes modified by the adjective *Middle*. Oneota and Mississippian times are preceded by the Woodland period, which spans approximately two millennia, between 1000 B.C. and A.D. 1000. The Woodland period is divided into three subperiods: Early, Middle, and Late. Middle Woodland (ca. 100 B.C. to A.D. 300 or 400) was the time of Hopewellian mortuary ceremonialism and exchange. The Woodland period is preceded by the Archaic, which is when the transition to agriculture began.

The most recent archaeological culture in the Lower Valley, beginning at A.D. 1200, goes by the designation *Plaquemine*. The maize-growing Plaquemine societies evolved out of the earlier Coles Creek culture, which began near the end of the Late Woodland period and overlapped with both Emergent Mississippian and Early Mississippian in the Central Valley. Troyville (ca. A.D. 400–700), Marksville (ca. A.D. 1–400), and Tchefuncte (ca. 700–1 B.C.) are the main Woodland period cultures in the Lower Valley. Poverty Point is the terminal Late Archaic period mound-building manifestation of Louisiana and surrounding states.

DEVELOPMENTAL TRAJECTORIES

THE CENTRAL MISSISSIPPI VALLEY

As things stand now, the earliest gardening and farming developments occurred in the Central Valley, and the earliest crops were indigenous gourds and temperate, seed-bearing plants. Prime habitation sites in recently stabilized river valleys had become occupied by increasingly sedentary bands of people who fished, hunted, and harvested native plant foods during the Middle Archaic period (J. A. Brown and Vierra 1983). Increased sedentarism and intensified use of riverine resources were accompanied by increases in population density at the junctions of the most desirable ecosystems. The Koster site, for example, sits on well-drained colluvium at the edge of the Illinois River

TABLE 9.1. CULTURAL-HISTORICAL TERMS APPLIED TO VARIOUS TIME PERIODS IN THE MISSISSIPPI RIVER VALLEY

Date	Upper Mississippi Valley	Central Mississippi Valley	Lower Mississippi Valley
1600			
1400	Oneota	Misissippian	Plaquemine
1200			
1000	————	————	Coles Creek
800	Late Woodland	Emergent Mississippian	
600		Late Woodland	
400			Troyville
200	Middle Woodland	Middle Woodland (Hopewell)	Marksville
A.D.			
B.C.			
200			Tchefuncte
400			
600	Early Woodland	Early Woodland	
800			
1000	————	————	Poverty Point
1200			- - - - - - - - -
1400			
1600	Late Archaic	Late Archaic	Late Archaic
1800			
2000			
2200			

valley. The river's wide terraces, backwater sloughs, and main channel are located to the west of the site, and a wooded oak-hickory zone follows the slope and valley margin immediately east of the site (N. B. Asch et al. 1972). This location afforded access to fish, waterfowl, deer, smaller mammals, turkeys, hickory nuts, pecans, walnuts, acorns, tubers, and seeds and greens from a wide variety of forbs and grasses (Zawacki and Hausfater 1969).

Both the Koster site and a nearby base camp, the Napoleon Hollow site, have yielded fragments of *Cucurbita pepo* gourd rind that have been directly dated by the accelerator mass spectrometer (AMS) radiocarbon technique to ca. 5000 B.C. (Conard et al. 1984). These currently stand as candidates for the earliest cultivated plants in the Mississippi River valley, although a hot debate rages over whether they were being gathered from the wild or intentionally propagated (Asch 1994; Smith et al. 1992; Cowan and Smith 1993). I believe we cannot rule out the possibility that the range of a native (see Decker-Walters 1990, 1993), wild-weedy pepo gourd (*C. pepo* ssp. *ovifera* var. *texana* sensu lato) extended as far north as central Illinois during the middle Holocene. It seems more likely, however, that Archaic foragers had extended the range of gourds native to the Gulf Coast region (see Newsom et al. 1993 for documentation of late Pleistocene pepo gourds in Florida) and were planting them in Illinois and elsewhere for use as net floats, bobbers, cups, rattles, containers, and food. The recent discovery of archaeological *C. pepo* rind fragments in central Maine (Peterson and Asch Sidell 1996) and northern Pennsylvania (Hart and Asch Sidell 1997) dating between 3000 and 4000 B.C. makes it much easier to argue for the spread and cultivation of gourds by Archaic fisher-gatherer-hunters. People were not necessarily selecting for larger fruits or thicker rind, however, and cultivation would have required only low-level management, not farming behavior. None of these regions exhibited so-called advances in the direction of food production for thousands of years, and the inhabitants of central Maine, in fact, remained hunter-gatherers until European contact.

The earliest archaeological specimens of eastern *C. pepo* to exhibit signs of morphological change come from the Phillips Spring site in western Missouri and date to 2000–2300 B.C. *Cucurbita pepo* seeds from Phillips Spring are slightly larger than those of any wild populations that have been studied (King 1985; B. D. Smith 1997). In addition, the nonnative bottle gourd (*Lagenaria siceraria*) had been passed into the midcontinent by that time, possibly by fishing-oriented groups on the Gulf and Atlantic coasts who could collect drift gourds. The 7,000-year-old bottle gourd from the Windover site in eastern Florida is rather small, 18 cm × 10 cm (Doran et al. 1990:355). Newsom et al. (1993:91) discuss "how early bottle gourd (*Lagenaria siceraria*) might have first appeared in Florida . . . and the eastern United States by water-dispersal along the Gulf Coast from tropical regions," so that the Windover specimen might have been gathered from the beach or from a feral population by the Middle Archaic foragers who buried it with one of their dead. Although archaeologists before 1960 generally believed that the Mississippi River served as an avenue for trade and possibly even colonization from Mesoamerica, that scenario has been abandoned by most researchers knowledgeable about eastern North America (Griffin 1966, 1980; B. D. Smith 1984).

Seeds from several taxa of native plants begin to occur in greater numbers at some sites in Illinois and Tennessee during the Late Archaic period, especially between 3000 and 1000 B.C., with definite signs of domestication dating to approximately 2300 B.C. (D. L. Asch and Asch 1985; Chapman and Watson 1993; Crites 1993). Sunflower (*Helianthus annuus* var. *macrocarpus*) and its close relative sumpweed (*Iva annua* var. *macrocarpa*) were being subjected to selection pressures that resulted in increased seed size, clearly visible in the archaeological record (D. L. Asch and Asch 1978; Yarnell

1978, 1981). Sumpweed is at home in the river valleys of the Midwest, where the earliest domesticates are found, and it spreads as an adventive species into disturbed upland areas. The annual sunflower, however, appears to have originated in the Southwest and spread into eastern North America as a weed in association with the ancestors or neighbors of the people who domesticated it (Heiser 1976). Although sumpweed was originally adapted to floodplain settings, and sunflower was native to dry western grasslands, they both ended up as crops with large, oily seeds in midwestern gardens by 2300 B.C.

These two composites (members of the Asteraceae family) were joined by a domesticated version of chenopod or goosefoot (*Chenopodium berlandieri* ssp. *jonesianum*) no later than 1500 B.C. Wild-growing chenopod thrives in both floodplain and disturbed upland settings across much of eastern North America, and the ancestors of Archaic people who domesticated it had probably been harvesting the nutritious greens and seeds for thousands of years. An increase in archaeological abundance during the third millennium B.C. leads D. L. Asch (1994) to infer that it was being cultivated, but the seeds had not yet changed phenotypically. At 1500 B.C., however, selection for drastically thinner seed coats had produced a crop with truncate-sided seeds and smooth, very thin testae (B. D. Smith 1985; Smith and Cowan 1987). The early domesticates probably also produced somewhat denser and larger infructescences, their seeds may well have ripened more simultaneously, and seeds might not have dispersed by shattering as freely as wild and weedy relatives.

Within the densely populated major river valleys, early gardening practices can be seen as a strategy for increasing storable produce by members of sedentary or nearly sedentary societies in restricted territories who knew that supplies of staple wild foods such as hickory nuts would fluctuate from year to year. By planting seeds in plots near hamlets and by favoring plants with larger seeds, thinner coats, and denser or non-shattering seed heads, the risk of running out of food during the winter or spring could be minimized. In years of plenty, surpluses could be traded to neighboring groups, either to insure reciprocity when the need arose or in exchange for a type of nonfood material item or nonmaterial enhancement of status.

By the end of the Late Archaic period, 3,000 years ago, people who grew sunflowers, sumpweed, chenopod, and native squashes were distributed across lowland and upland zones in the Midwest and Midsouth (Chapman and Watson 1993; Gremillion 1993; Fritz 1994c; Watson 1985). Unfortunately, we have only the vaguest notions of how the early native seed gardening systems really worked. An exceptional window into the past comes from the Marble Bluff site in northern Arkansas, where, in 1934, excavators from the University of Arkansas Museum found a boxed-in crevice containing the charred remains of three bags of thin-testa chenopod seeds, a bag of cultigen squash seeds mixed with cultigen sunflower seeds, a one-gallon cache of mixed seed types that may have been stored in a basket, and various possible gardening tools including antlers and perforated mussel shells. Relying on three overlapping radiocarbon age ranges on seeds from different bags, I view this cache as a single storage event of 3,000-year-old seed stock and digging tools (Fritz 1997). It shows a diverse cropping system that includes both fully domesticated and weedy types of plants. In the Buffalo

River drainage, where Marble Bluff is situated, garden plots could have been cleared by girdling and chopping trees on stream terraces or by burning vegetation on slopes and plateau surfaces.

Only two members of the eastern North American seed crop complex—sunflower and squash—were still widely grown at Contact. Because the others are all extinct as cultivated varieties, we do not have ethnographic sources to suggest how they were processed or prepared. Before cooking pots were common (and this was the case during the Late Archaic and Early Woodland periods across much of the Central Valley), I would guess that seeds were parched shortly after harvesting and that the starchy seeds were frequently ground into flour, maybe mixed with acorn meal. Whole seeds could have been added to stews or cooked as gruels by stone boiling.

The best direct evidence for early consumption comes from Salts Cave in Kentucky and dates to around 500 B.C. Sunflower, sumpweed, squash, chenopod, and maygrass seeds (*Phalaris caroliniana*) dominated the diets of the gypsum miners who left their feces in the depths of the Salts-Mammoth cave system (Gardner 1987; Yarnell 1969, 1974). Some evidence for seed parching has been observed. Sunflower and sumpweed fruits were eaten shell and all, and any attempt to spit out the brittle pericarp must have been cursory. I like to think of the cavers' rations as prehistoric trail mix, but like modern gorp, it was not necessarily typical of the way the same seeds would be served in a meal at home.

Increased commitment to farming and production of larger quantities of native grains are correlated (at least in some regions) with the cultural phenomenon known as the Hopewellian Exchange Network, which dates to the Middle Woodland period (ca. 100 B.C. to A.D. 300 or 400). Between central Illinois and southern Tennessee and branching up tributaries to the west and east, archaeologists have found Middle Woodland sites with pits and middens containing high frequencies of native cultigens. Not all of the sites were necessarily tied into or affected by Hopewellian exchange or ideology, but at those that were, people probably placed a high value on storable grains. Community leaders wishing to impress trade partners and other visitors would need surplus food for feasts. Trading parties would benefit from portable rations. Families living in the small hamlets scattered along rivers and streams seem to have been more highly motivated than before to produce and store the carbohydrate and protein-rich mixture of native crops for their own use and for communal functions.

Tobacco (*Nicotiana* sp.) and maize (*Zea mays* ssp. *mays*) seem to have entered the Eastern Woodlands by separate routes at this time, approximately 2,000 years ago. Until recently, archaeologists assumed that all tobacco grown in the Eastern Woodlands was *Nicotiana rustica,* a plant that biologists consider native to Andean South America, although it is more weedy than wild there (Goodspeed 1954). The earliest archaeological tobacco from well-dated contexts comes from Illinois, consisting of four seeds from Smiling Dan and one from Meridian Hills, all from Middle Woodland pits dating between A.D. 100 and 200 (D. L. Asch 1994:45). Pipes, however, appear 1,500 years earlier, and although other substances may have been smoked before tobacco was introduced, the record for tobacco might well get pushed back (ibid.). Careful inspection, using scanning electron microscopy, of archaeological seeds from the Mississippi and Missouri River valleys has created doubt as to the taxonomic affiliation of early precolonial tobacco in this region. Specimens from several sites fail to exhibit micromorphological characters expected of *N. rustica,* conforming more closely to either *N. quadrivalvis* or *N. multivalvis,* formerly members of the *N. bigelovii* complex native to California and Oregon (Haberman 1984). Unfortunately, the intraspecific variability makes it extremely difficult to classify small assemblages, especially if they are old and imperfectly preserved, but the picture has become more complicated, so the origin of the earliest tobacco in North America now seems open to debate. The question, as posed by D. L. Asch (1994:47), has become: "Did smoking and tobacco use in North America originate in the West, with both the use of pipes and a western species of *Nicotiana* spreading eastward about 3,500 years ago? Later, South American *N. rustica* could have displaced the species grown originally in the east. This is one of the scenarios for the history of tobacco use in eastern North America which remains to be tested." Whichever came first, it now seems feasible that at least two species of tobacco were cultivated in the Mississippi River valley in Precolumbian times.

Nobody disputes the Mexican origins of maize, which is represented by 2,000-year-old AMS-dated fragments from Illinois (Riley et al. 1994) and slightly younger fragments from sites in Ohio and Tennessee (Fritz 1993). Claims that maize pollen from Alabama and Florida is significantly earlier (1500 B.C., for example, for a single grain from Shelby Pond in Alabama; Fearn and Liu 1995) are on shaky chronological ground because individual pollen grains cannot be dated directly, and the possibility of contamination is very real (see the recent discussions by Rowley-Conwy 1995 and Crawford et al. 1997). The evidence currently favors a scenario in which maize was introduced into the Mississippi River valley from the Southwest rather than from eastern Mexico or South America, although alternatives have been proposed (e.g., Riley et al. 1990). Hopewellian trade routes extended to the west well into the Rocky Mountains (Griffin 1965) but not to Mesoamerica or the Caribbean region. Furthermore, molecular biological evidence shows that eastern maize is more closely related to Southwestern varieties (which are related to races such as the Chapalotes of northwestern Mexico) than to any maize in eastern Mexico or South America (Doebley 1990:20; Doebley et al. 1986).

Middle Woodland farmers apparently did not perceive maize to be a superior food, since they ate very little of it (Bender et al. 1981; Lynott et al. 1986) and continued to produce their native seed crops instead. This situation persisted into the Late Woodland period (A.D. 300–800), with sites across the Central Mississippi Valley yielding thousands or even millions of carbonized seeds from large pits dug for storage. Maygrass (*Phalaris caroliniana*) is found by this time at sites well outside its modern range, with no climatic explanation, sometimes in clear storage context or in masses of winnowed grains and in association with domesticated chenopod, other native crops, and eventually with maize. The archaeological caryopses show no size increase or other phenotypic signs of domestication, but all the evidence points to its status as an intentionally propagated (i.e., cultivated) crop in both the Upper and Central Valley (D. L. Asch and Asch 1985; Cowan 1978; Fritz 1986). Erect knotweed (*Polygonum erectum*) also occurs in storage context, in masses of charred achenes, and in high frequencies together with known domesticates (Fritz 1986, 1994c; Lopinot et al. 1991). Furthermore, the proportion of knotweed achenes with smooth, thin pericarps increases to 80–100 percent by A.D. 1200, whereas this thin-coated morph is poorly represented in the wild, and a slight increase in achene size through time has been documented (N. B. Asch and Asch 1985; Fritz 1987). Little barley (*Hordeum pusillum*) may not have had its range extended or seed size increased (see Hunter 1992, however, for evidence of morphological change as well as range extension), but an increase in abundance in features with other crops indicates that it was harvested and processed along with them and, therefore, was probably a part of the garden complex. Hopewell-related trade and mortuary ceremonialism ceased at the end of the Middle Woodland period, but human population sizes were growing (Buikstra et al. 1986), ceramic technology was improving (Braun 1987), and the stage was being set for the rise of Mississippian chiefdoms.

Cultivation of maize was intensified at about A.D. 800 (Johannessen 1993), and the incorporation of maize as a staple in the diets of many Central Valley dwellers was complete by A.D. 1000. Native crops were intensified along with maize, indicating that societies in the process of becoming more complex were trying to increase their harvests using an even more diverse multicrop system, rather than switching to maize (Lopinot 1991, 1994). Surprisingly, garden beans (*Phaseolus vulgaris)* were not incorporated into Central Valley agriculture at this time. A new species of squash, *Cucurbita argyrosperma* ssp. *argyrosperma,* was adopted by A.D. 1000. These cushaw squashes were almost certainly acquired from sources in the Southwest, since the variety present in the East after European contact is similar to land races grown by Indian people in Arizona and New Mexico but different from those found in central, eastern, and southern Mexico (Fritz 1994b).

Maize's rise to prominence in many parts of the Southeast is correlated with the formation of Mississippian chiefdoms, and several researchers implicate the high-ranking leaders themselves with making demands for surplus grains and eventually, for granaries filled primarily with maize (Rose et al. 1991; Scarry 1993). Farming strategies were probably affected by increased political integration, and some of the surplus field produce probably became tribute. Field allocation policies may have been affected as well.

A less diverse, maize-dominated farming system became entrenched in the Central Valley by A.D. 1400. Beans, which were also probably carried across the plains from the Southwest, were a major crop by that time, but all the native seed crops except sunflower had declined. The members of De Soto's expedition who entered the southern end of the Central Valley in A.D. 1540 described Mississippian chiefdoms flourishing on their agricultural bases (Clayton et al. 1993). Archaeobotanical assemblages from late prehistoric and protohistoric components at sites such as Parkin in northeast Arkansas confirm that the trinity of maize, beans, and squashes had ascended (M. L. Williams 1994).

THE UPPER MISSISSIPPI VALLEY

No distinct cultural or environmental boundaries exist between the southern end of the Upper Mississippi Valley and the northern end of the Central Valley; therefore, many similarities can be found in the agricultural developments through time and patterns across space. In the northern end of the Upper Valley, however, the shorter growing season imposes obvious constraints on agriculture. Indigenous farmers in Wisconsin and neighboring states managed to produce large harvests of maize, beans, and squash during late prehistory by planting cultivars capable of maturing in 120 or fewer frost-free days and by constructing ridged fields in which risk of damage from frost or soil moisture was reduced. Remnants of precolonial ridged fields can be found in both floodplains and in tillable upland areas, although "in some glaciated areas . . . , this distinction between uplands and floodplains may be relatively insignificant" (Gallagher 1992:96).

Archaeologists recently have learned a great deal about agricultural developments in Iowa, Wisconsin, and northern Illinois, and much of this information is summarized by Arzigian (1987, 1993), Gallagher (1992), Gallagher and Arzigian (1994), and King (1993). Arzigian (1993:115–123) reports fragments of *Cucurbita pepo* rind along with domesticate-sized sumpweed and sunflower seeds from Middle Woodland components of six sites in the Prairie du Chien region along the Mississippi River in southwestern Wisconsin. Increased quantities of chenopod and knotweed seeds also came from some of these sites, but they were not distinguishable from seeds of wild populations. Late Woodland and Early Mississippian (Oneota) times saw increased production of native seed crops (maygrass, sunflower, sumpweed, and probably chenopod, knotweed, and little barley) in southern Wisconsin, northern Iowa, and northwestern Illinois.

Small amounts of maize have come from a few components predating A.D. 800, but rapid intensification transformed the scene at approximately A.D. 900–1000. After that date, a characteristically diverse mix of native seed crops, maize, squash, and wild plant resources can be found along the rivers and streams of Wisconsin, Iowa, and northern Illinois. These northern farmers may have helped breed the short-season Eastern Eight row variety of maize that became the historic Northern Flint race. If not, they acquired it from farther east (cf. Crawford et al. 1997) and grew substantial amounts of it at the same time that varieties with higher row numbers (and presumably longer growing seasons) dominated the Central Valley. The transition to heavier maize production in southwestern Wisconsin was not accompanied by major changes in procurement of animal resources, according to Arzigian (1993:136):

> With the addition of maize, the only change in the utilization of faunal resources was the intensified use of mussels at some riverine sites in the Mississippi trench. It may be significant that, although maize had poor quality protein, mussels are good sources of complementary amino acids. . . . Later, in Oneota diets, beans served this same role as complements to maize; however, no beans were found at any of the sites considered here.

Archaeologists debate the dietary significance of maize and other cultigens for Oneota peoples, with some scholars characterizing Upper Mississippi Valley subsistence as too heavily reliant on wild plants and animals to be called maize-dependent (J. A. Brown 1982; Gibbon 1982; but see Gallagher and Arzigian 1994). We need more stable carbon isotope data and more quantified information from flotation samples to speak with confidence about relative dietary contributions, but recent scholars seem to have shifted in the direction of viewing late prehistoric agriculture here as relatively intensive (King 1993). Gallagher and Arzigian (1994) combine multiple lines of evidence, including the extensive ridged field systems and maize hills, to argue convincingly that maize agriculture was productive and reliable in spite of the northerly geographic position. They offer the concept of intensification with diversification (as opposed to intensification with specialization) to describe the varied and flexible economy that was "geared toward increasing security and minimizing risk" in a noncentralized, unstratified society (ibid.:184).

The height of diversity in Oneota fields and gardens occurred between A.D. 900 and 1300, when native seed crops and the probably native pepo squash were well represented along with maize and, eventually, common beans. A shift in overall Oneota subsistence patterning is evident at A.D. 1400–1600. At this time, bison hunting became a major seasonal focus, larger winter villages housed people who went off in groups to hunt bison during the warmer months, and crop diversity decreased. The importance of native seeds declined. Morphological characteristics of crops such as chenopod, sumpweed, and sunflower reverted to a more wild or weedy state (William Green, personal communication, 27 April 1995).

THE LOWER MISSISSIPPI VALLEY

Journeying downstream to the Lower Mississippi Valley, we find a much different sequence of agricultural development. The biggest surprise coming out of Lower Valley research in recent years has been the lateness of the transition from foraging to food production. Archaeologists recently recognized that mound-building began very early—5,000 to 6,000 years ago—in Louisiana (Saunders 1994; Saunders et al. 1997). These mounds reflect a more highly structured level of ritual organization and possibly a lower level of residential mobility than previously envisioned for the nonagricultural people who built them. A few *Chenopodium* seeds were recovered from Watson Brake, one of the earliest mound sites in Louisiana, but they have not been shown to exhibit the classic signs of domestication outlined by D. L. Asch and Asch (1985), Fritz (1984, 1986, 1994c), Fritz and Smith (1988), Gremillion (1993), B. D. Smith (1985, 1995), and Wilson (1981). Because wild-weedy chenopod plants bear a low percentage (average 3 percent) of seeds with thin and relatively smooth seed coats, the presence of only a few thin-testa seeds in a small archaeological assemblage or in an assemblage dominated by thick-coated seeds cannot be taken as proof of cultivation. Similarly, the mere presence of a few members of the genus *Chenopodium* does not indicate early stages in a transition to agriculture, because wild chenopods and many other native ruderal plants with nutritious leaves and seeds were harvested for millennia across many continents, and yet evidence for chenopod domestication exists for only three regions: *C. quinoa* in the Andes, *C. berlandieri* ssp. *nuttalliae* in Mesoamerica, and *C. berlandieri* ssp. *jonesianum* in the midwestern part of what is now the United States (Sauer 1993; B. D. Smith 1985).

The small amount of subsistence information currently available from mounds in Louisiana predating 1500 B.C., as well as from their counterparts in Florida, indicates that the abundant natural resources of coastal zones and forested river bottomlands made it possible for fisher-gatherer-hunters to engage in earth-moving activities requiring more intergroup coordination and involving more ritual elaboration than we believed possible even a few years ago (Gibson 1994; Russo 1994; Saunders et al. 1997). Early mound builders in North America took advantage of the availability of acorns, pecans, and hickory nuts; fruits such as persimmons, grapes, and blackberries; greens; edible roots; and tubers in the deciduous forests, southern mixed hardwoods, and evergreen oak-palm-magnolia vegetation zones (see Greller, this volume). The rivers, bayous, estuaries, and coastal waters also offered fish, turtles, and waterfowl, while turkeys, deer, bear, and dozens of species of small to medium sized mammals could be found in the forests. Management activities to enhance productivity of various nut- and fruit-producing trees may well have begun, but the clearings created for Late Archaic settlements and ceremonial centers in Louisiana and Florida did not seem to trigger the same movement to cultivation of native seed plants as in the Central Mississippi Valley.

By 1200 B.C. at the latest, Lower Valley residents were probably living in repeatedly occupied base camps and participating in periodic community rituals and trade fairs at major mound centers (Gibson 1994; Jackson 1989). Poverty Point–style earthworks and proximity to Mexico led archaeologists of past decades to speculate that an early maize-dominant agricultural base had been laid in Louisiana (Willey and Phillips 1958). The numerous Archaic mounds now confirmed as predating Poverty Point, however, obliterate the need for outside influences and provide solid evidence for in situ developments. No contemporaneous (or earlier) earthworks on the scale of Watson Brake are known to exist in Mesoamerica, and no exchange of durable goods between the two regions can be documented.

Flotation recovery at the Poverty Point center and associated sites (1500–800 B.C.) has failed to produce evidence for early farming in the Lower Mississippi Valley. The available database is limited, but evidence currently indicates that hickory nuts, pecans, acorns, and fruits served as the major plant foods. Small seeds, including *Chenopodium*, were evidently gathered in relatively low quantities, but no signs of domestication have been noticed (Jackson 1989; Ramenofsky 1986; Shea 1978; Ward 1998). A system of intensified fishing, hunting, and probably nut management through periodic understory burning of acorn and pecan groves enabled people in the rich alluvial bottomlands to develop and maintain the elaborate-looking Poverty Point culture.

The next florescence of mound-building activities in the Lower Valley is called the Marksville culture, and it dates to the Middle Woodland period. Marksville mound sites are contemporaneous with Hopewellian centers in Illinois, Ohio, Tennessee, and elsewhere, and traders were actively paddling the waters of the Mississippi River and its tributaries. In the Midwest, as stated above, more and more native seed-bearing crops (chenopod, sunflower, sumpweed, and others) were produced during the Middle Woodland period, but evidence for similar premaize gardening or farming activities has not been found in the Lower Mississippi Valley. A paltry amount of flotation has been conducted at Marksville sites, and future investigations may yield unanticipated results, but for now it appears that residents of the Lower Valley persisted in their patterns of harvesting wild and managed resources, with cultivation of gourds and maybe other specialty plants constituting a small gardening component, but not on the scale of their trading partners to the north.

Rituals involving Hopewellian exchange and burial in the Lower Valley were discontinued after A.D. 300, but the succeeding Late Woodland period witnessed the rise of another mound-building society in Louisiana, called Troyville (Belmont 1982). Limited flotation at the Reno Brake site in Tensas Parish showed no shift to agriculture in Troyville times (Fritz and Kidder 1993), but the contemporaneous Taylor Mounds site in southeastern Arkansas yielded domesticated chenopod along with frequencies of erect knotweed, indicating that cultivation of native seed crops had picked up by this time at the northern end of the Lower Valley (M. L. Brown 1996).

At approximately A.D. 700 in the lower Arkansas River valley, near the modern city of Little Rock, the culture known as Plum Bayou had developed a distinctive pattern of mound-based ceremonialism along with a subsistence pattern in which native crops probably played a major role approaching their economic significance in the nearby Central Mississippi Valley (Rolingson 1990). Maize is also present in flotation samples from the Toltec Mounds site, the premier Plum Bayou center, in low frequencies from most parts of the site that have been studied, but in distinctly higher frequencies in the Mound S midden, a concentrated deposit of deer bones and other remains interpreted as a likely feasting event (C. J. Smith 1994). This supports earlier suggestions that at least some people began growing more maize in conjunction with social and ritual developments. Although the presence of maize is very important and the significance of native seed crops notable, Plum Bayou subsistence continued to include substantial quantities of acorns, hickory nuts, fruits, and many other wild resources.

When T. R. Kidder and I embarked upon a subsistence-focused research project at the Osceola site in Tensas Parish, Louisiana, we expected to find a similar pattern of subsistence among these Coles Creek culture mound builders who were closely related to their Plum Bayou neighbors. By A.D. 700 or 800, Coles Creek people in northeastern Louisiana and northwestern Mississippi built platform mounds around plazas and appeared to be on a trajectory toward Mississippian lifeways. However, our excavations and concerted efforts at flotation recovery at Osceola and three additional Coles Creek sites revealed a different pattern (Kidder 1993; Kidder and Fritz 1993). These sites are all situated on natural levees in the wide bottomland zone lying between the Tensas and Mississippi River channels, where mixed bottomland forests (Greller, this volume) once prevailed. Acorn shell dominates the Coles Creek archaeobotanical assemblages to an extent that is, in my experience, unparalleled in the Mississippi Valley. Pecans, persimmons, grapes, and other fruits are very well represented. Thin *Cucurbita* gourd rind is not uncommon. Small native seeds are more frequent in early Coles Creek contexts (ca. A.D. 750–950) than later, but the sumpweed seeds are all wild in size, and the maygrass might have been uncultivated. Most chenopod seeds recovered so far from Tensas Parish have thick seed coats, but a minority appears to represent the eastern domesticate *C. berlandieri* ssp. *jonesianum*. This type of pattern might reflect casual, low-level cultivation of a few crops, but most plant foods were evidently procured by management of oak trees and fruit trees and perhaps by encouragement of wild seed-bearing plants. Fragments of starchy material that might be some edible tuber (e.g., *Smilax* spp., *Ipomoea pandurata*, *Nelumbo lutea*, or *Sagittaria* spp.) are also relatively common and might have been gathered from the wild, encouraged or managed in some way, or even cultivated.

The sophisticated system of fishing, hunting, and the harvesting of managed wild plant resources supplemented by cultivation of gourds and, at least occasionally, by native seed crops persisted until the end of the first millennium A.D. in northeast Louisiana. Maize becomes visible in the archaeobotanical record at approximately A.D. 900, although a few fragments of maize from questionable contexts might push the timing back if they can be validated by direct dating (Kidder 1992). As in the Central Valley, the acquisition of maize does not seem to have been a case of love at first sight. Complex fisher-hunter-gatherers in Louisiana had almost certainly been in contact with their neighbors to the west, north, and east who grew increasing quantities of maize after A.D. 750, yet they themselves grew little if any before A.D. 900 and only limited amounts between A.D. 1000 and 1200.

Why did they finally, at about A.D. 1200, begin producing more than before? We have insufficient data from settlement pattern studies and bioarchaeology to rule out stress resulting from population pressure, but I currently favor sociopolitical explanations over demographic and ecological ones. The northern end of the Coles Creek region, in the Yazoo Delta of Mississippi, underwent accelerating Mississippianization after A.D. 1000, as a result of intensified contacts with evolving and expanding societies who were headed by hereditary chiefs. These Mississippian people made shell-tempered pottery, cleared the mixed bottomland forest trees from larger patches of well-drained natural levee soil, and planted fields dominated by maize. The entire Coles Creek region may have been pulled into the agricultural movement because of demands for maize at ceremonies where alliances were negotiated or because of a perceived need to conform to the cultural standards of more powerful rivals (Fritz 1998).

Whether because of stress, status, or other factors, the Lower Mississippi Valley underwent the final transition to maize agriculture between A.D. 1200 and 1400. Use of acorns and native fruits, especially persimmon, did not disappear after maize was intensified (Kidder et al. 1993). I reconstruct the Plaquemine landscape as one where maize fields alternated with groves of nut and fruit trees along the natural levees of the alluvial bottomlands (Fritz 1994a).

AGRICULTURAL ECOSYSTEMS AND OTHER ANTHROPOGENIC PATCHES

At European contact, maize-based agriculture extended all the way from the upper to the lower reaches of the Mississippi River. This convergent pattern masks the variable pathways followed by groups in the different regions as their relatively similar late precolonial farming systems developed (table 9.2). The timing of initial stages of the transition from foraging to farming differed greatly from region to region. Domestication of native seed-bearing plants was well under way 4,000 years ago in the Midwest, but societies in the Lower Valley grew few crops until near the end of the first millennium A.D.

Strategies for successful food production had been crucial for social and economic success for centuries before maize was introduced, and for quite a few more centuries before it was a major crop. The sunflower and the ovifera-group squashes and gourds were important cultigens up through colonial times and remain major economic players today. Admittedly, we have difficulty assessing the importance of native crops in terms of percentages of total dietary calories or nutrients. Caloric significance might have varied by age or gender or according to specialized task group as well as varying across time and space. Regional differences in percentages of calories contributed by maize in late precontact times cannot be charted at this time with precision either, although there is hope that ongoing stable isotope research will improve the situation.

Gardens and fields constitute the most obvious anthropogenic ecosystem other than human towns and villages on the precolonial landscape in the Eastern Woodlands. The native, temperate crops originally came from very different plant communities. Wild gourds depend on floodwaters to disperse their seeds to sand and gravel bars. The wild sunflower is adapted to relatively xeric grasslands (mixed grass prairie, short grass prairie, and desert grassland), as is little barley. Wild sumpweed and maygrass are at home along the edges of floodplains that transect numerous vegetation zones. Chenopod and erect knotweed grow naturally in disturbed floodplain settings and also in openings on forested terraces. Sunflower, chenopod, sumpweed, and little barley are ruderal plants that can colonize disturbed ground within various vegetation zones. As crops, however, the same species were no longer restricted to their previous niches. Several were significantly altered genetically. Early food producers may have recognized microhabitats within the setting of the garden or field that would be best for a given species, but in general, plants that formerly grew in separate natural communities (with some overlap) became crops growing together in an environment created by people. Even maize, a plant that originally came from thousands of miles away and traveled with people through territory totally unlike the Mississippi River valley, eventually flourished in fields that we assume were once intercropped with chenopod, sumpweed, sunflower, maygrass, and the temperate ovifera-type squash. Although species richness in agricultural settings declined through the Mississippian period when some native crops were abandoned, when beans were finally accepted, and when maize was intensified, even the sixteenth-century fields were described and depicted by Europeans as mixtures of maize, beans, squashes, and sunflowers. Furthermore, we believe that a suite of tolerated, encouraged, transplanted, and semicultivated plants inhabited the fields, their edges, and clearings made for other purposes (Yarnell 1982). When fields were in fallow or abandoned, successional stages of vegetation included additional useful resources.

Agricultural ecosystems were not the only anthropogenic environments of the Mississippi River valley. Europeans described some wooded areas as orchards (Hammett

TABLE 9.2. DIFFERENT TRAJECTORIES TO FARMING FOLLOWED BY NATIVE PEOPLES IN THE MISSISSIPPI VALLEY

Date	Upper Mississippi Valley	Central Mississippi Valley	Lower Mississippi Valley
1700			
1500	Corn/beans/squash specialization	Corn/beans/squash specialization	Corn/beans/squash specialization
1300			Intensification of corn
1100			
900	Intensification of corn	Intensification of corn	*Introduction of corn*
700			
500	*Introduction of corn?*		Fisher/ gatherer/ hunters (with gourds)
300			
100	Intensification of native seed crops		
A.D.		*Introduction of corn*	
B.C.			
100			
300		Intensification of native seed crops	
500			
700			
900	?		
1100			
1300			Use of gourds
1500		Early domestication of native seed-bearing plants	
1700			
1900			
2100			
2300		Cultivation of gourds	

1992), and these managed groves of nut and fruit trees had probably been prominent features of the landscape since Archaic times. The managed groves were located within several kilometers of villages and towns. Productivity of hickory nuts, butternuts, walnuts, pecans, and acorns could be enhanced by optimal spacing and removal of poor producers. Competition with squirrels and other wildlife could be reduced by removing brush and litter through understory burning.

Farther away from the villages, prescribed burning would also have been practiced in forests and prairies used repeatedly for hunting. Although researchers (usually not anthropologists) have speculated that burning for communal hunts and other purposes was frequent and widespread enough to create vast expanses of parklike openings in the Eastern Woodlands (Day 1953; Guffey 1977), I follow Hammett (1992:35) in inferring that "extremely large, expansive open areas were the result of occasional natural wild fires," and that burning in hunting camps was restricted to patches less than three miles in width. More frequent and more widespread burning may well have been practiced on the prairies to the east and west of the Mississippi River's main channel, especially in the Upper Valley area and northern end of the Central Valley.

The impact made by Indian peoples of the past on their environments is a topic of great interest to researchers in many fields. In the American Bottom, there are indications that heavy demands on available supplies of wood may have resulted in excessive cutting of trees and consequent problems, such as a rising water table that limited the choice of habitation sites and arable soils (Lopinot and Woods 1993; Milner 1990). Climate, however, was also a factor (Lopinot 1994). The demise of large civic-ceremonial centers, such as Cahokia, cannot be attributed solely, if at all, to over-exploitation or mismanagement of resources.

Use of fire to alter vegetation was probably the most effective method of managing all but the smallest localities away from actual settlements. Appreciation for the benefits and successful application of controlled burning must be balanced with realistic assessments of population density and with awareness of how settlements and activity areas were distributed across the landscape. Estimates of precontact population density are error-prone and notoriously variable. Archaeologists' calculations of the number of people who lived at Cahokia, the largest site north of Mexico, at its zenith range from fewer than 10,000 to more than 40,000 (Milner 1990; Pauketat and Lopinot 1997). The site had been virtually abandoned for hundreds of years before the first Europeans came to Illinois, and no other Mississippian center had a population of more than a few thousand. Most mound centers were occupied either by a few hundred people or by small groups of caretakerlike specialists who were supported by a dispersed populace living in hamlets and farmsteads scattered along nearby stream terraces.

Using estimates contributed by a team of anthropologists knowledgeable about particular geographical regions, Douglas Ubelaker has calculated the size of North American populations at the time of initial European contact. He figures the minimum population of the American Southeast at 1492 to have been 155,800 and the maximum to have been 286,000, with a population density between 17 and 31 people per 100 km^2 (Ubelaker 1988:291). His most likely estimate for the Southeast is 204,400, with a population density of 22 per 100 km². Figures for the Northeast are 205,000–503,200 (best estimate 357,700 people), at densities of 11–26 people per 100 km² (best estimate 19). At these densities, farmers and hunters could have had significant impacts on localized territories, but the affected areas would have been separated by even larger tracts of land not cleared or burned by human agents. With periodic shifting, vegetation in once highly affected areas would grow back. Parts of the Central Mississippi Valley between A.D. 1450 and 1540—before European diseases—for example, are known as the "Vacant Quarter" (S. Williams 1990) due to the paucity of late Mississippian centers and associated settlements where earlier mound and village sites had been numerous. Chiefdom-level polities such as those in the Central and Lower Mississippi Valley after A.D. 900 or so were notoriously unstable (Anderson 1994). More powerful chiefdoms rose and fell, with centers of population concentration and buffer zones between them shifting through time. Therefore, no known locality in the Mississippi River valley was subjected to continuous and intensive human impact for more than a few centuries, with Cahokia and the surrounding American Bottom being, as usual, at the top of the scale.

A FINAL CONSIDERATION OF AGRICULTURAL SYSTEMS

Agricultural technology in the Mississippi River valley clearly evolved through time and varied across space. Systems were geared to localized soils, vegetation, and climatic variables. Choice of digging tools depended not only on soils and on which resources were locally available but also on whether the raw material or finished tools could be acquired through trade. Large hoes made of Mill Creek chert from southern Illinois, for example, were imported into the American Bottom while Cahokia was an active trading center during Mississippian times, but not later (Kelly 1991). Other types of stone used for hoes, such as the siliceous siltstone from the Arkansas River valley, were not traded as widely beyond their source. Hardwood digging sticks were evidently manufactured and used in most of the greater region during many time periods. Bison scapula hoes were most common in late Oneota societies.

It is likely that great variability also applies to discussions of the sizes and layouts of fields and the strategies for allocation of farmland. Even for Mississippian farmers within the Central Valley, field distribution would have varied according to overall settlement patterning. Some groups were dispersed up and down river valleys in farmsteads and small hamlets, while others congregated in fortified towns. In a town of several hundred inhabitants, a farmer might have maintained a dooryard garden but walked some distance to her plot in a communal field where most of the maize was grown. Members of an extended family with its own fields in a string of farmsteads within a dispersed community would not have realized the same advantages of an infield-outfield system.

A few general commonalities exist throughout the region. Women did virtually all of the farming in the Mississippi Valley at Contact, and most of us suspect this was always the case. If so, women deserve credit for domesticating North American native crops (Watson and Kennedy 1991), and for breeding the Northern Flint maize that when crossed with Mexican Dents after Contact produced Corn Belt dent varieties that played a large role in making the United States a prosperous nation.

Native farming systems were very productive, but wild food resources remained important everywhere in the Mississippi Valley. The gathering and processing of nuts, acorns, fruits, greens, tubers, and seeds of wild (or managed) herbaceous plants allowed subsistence strategies to be flexible, diverse, and stable. All American Indian societies in the Eastern Woodlands retained a gathering—and of course a very crucial hunting—component. In our search today to preserve biological diversity and to understand and promote sustainable systems of food production, the successful native farming strategies from Lake Itasca to Lake Pontchartrain can serve as valuable sources of information.

REFERENCES

Anderson, D. G. 1994. *The Savannah River Chiefdoms: Political Change in the Late Prehistoric Southeast.* Tuscaloosa: University of Alabama Press.

Arzigian, C. 1987. The emergence of horticultural economies in southwestern Wisconsin. In W. F. Keegan, ed., *Emergent Horticultural Economies of the Eastern Woodlands,* pp. 217–242. Center for Archaeological Investigations Occasional Paper 7. Carbondale: Southern Illinois University.

———. 1993. *Analysis of Prehistoric Subsistence Strategies: A Case Study from Southwestern Wisconsin.* Ph.D. dissertation, Department of Anthropology, University of Wisconsin, Madison. Ann Arbor, Mich.: University Microfilms.

Asch, D. L. 1994. Aboriginal specialty-plant cultivation in eastern North America: Illinois prehistory and a post-contact perspective. In W. Green, ed., *Agricultural Origins and Development in the Midcontinent,* pp. 25–86. Report 19, Office of the State Archaeologist. Iowa City: University of Iowa.

Asch, D. L., and N. B. Asch. 1978. The economic potential of *Iva annua* and its prehistoric importance in the lower Illinois Valley. In R. I. Ford, ed., *The Nature and Status of Ethnobotany,* pp. 300–341. Anthropological Papers 67. Ann Arbor: University of Michigan Museum of Anthropology.

———. 1985. Prehistoric plant cultivation in west-central Illinois. In R. I. Ford, ed., *Prehistoric Food Production in North America,* pp. 149–203. Anthropological Papers 75. Ann Arbor: University of Michigan Museum of Anthropology.

Asch, N. B., and D. L. Asch. 1985. Archeobotany. In M. D. Conner, ed., *The Hill Creek Homestead and the Late Mississippian Settlement in the Lower Illinois Valley,* pp. 115–170. Research Series 1. Kampsville, Ill.: Center for American Archeology, Kampsville Archeological Center.

Asch, N. B., R. I. Ford, and D. L. Asch. 1972. *Paleoethnobotany of the Koster Site: The Archaic Horizons.* Reports of Investigations 24. Springfield: Illinois State Museum.

Belmont, J. S. 1982. The Troyville concept and the Gold Mine site. *Louisiana Archaeology* 9: 65–97.

Bender, M. M., D. A. Baerreis, and A. L. Steventon. 1981. Further light on carbon isotopes and Hopewell agriculture. *American Antiquity* 46: 346–353.

Braun, D. P. 1987. Coevolution of sedentism, pottery technology, and horticulture in the central Midwest, 200 B.C.–A.D. 600. In W. F. Keegan, ed., *Emergent Horticultural Economies of the Eastern Woodlands,* pp. 153–181. Center for Archaeological Investigations Occasional Papers 7. Carbondale: Southern Illinois University.

Brown, J. A. 1982. What kind of economy did the Oneota have? In G. E. Gibbon, ed., *Oneota Studies,* pp. 107–112. Publications in Anthropology 1. Minneapolis: University of Minnesota.

Brown, J. A., and R. K. Vierra. 1983. What happened in the Middle Archaic? An introduction to an ecological approach to Koster Site archaeology. In J. L. Phillips and J. A. Brown, eds., *Archaic Hunters and Gatherers in the American Midwest,* pp. 165–195. New York: Academic Press.

Brown, M. L. 1996. Plant remains from the Taylor Mounds (3DR2), southeastern Arkansas. Second-year (M.A.) paper, Department of Anthropology, Washington University, St. Louis, Mo.

Buikstra, J. E., L. W. Konigsberg, and J. Bullington. 1986. Fertility and the development of agriculture in the prehistoric Midwest. *American Antiquity* 51: 528–546.

Chapman, J., and P. J. Watson. 1993. The Archaic Period and the flotation revolution. In C. M. Scarry, ed., *Foraging and Farming in the Eastern Woodlands,* pp. 27–38. Gainesville: University Press of Florida.

Clayton, L. A., V. J. Knight, and E. C. Moore, eds. 1993. *The De Soto Chronicles.* Tuscaloosa: University of Alabama Press.

Conard, N., D. L. Asch, N. B. Asch, D. Elmore, H. E. Gove, M. Rubin, J. A. Brown, M. D. Wiant, K. B. Farnsworth, and T. G. Cook. 1984. Accelerator radiocarbon dating of evidence for prehistoric horticulture in Illinois. *Nature* 308: 443–446.

Cowan, C. W. 1978. The prehistoric distribution and use of maygrass in eastern North America: Cultural and phytogeographical implications. In R. I. Ford, ed., *The Nature and Status of Ethnobotany,* pp. 263–288. Anthropological Papers 67. Ann Arbor: University of Michigan Museum of Anthropology.

Cowan, C. W., and B. D. Smith. 1993. New perspectives on a wild gourd in eastern North America. *J. Ethnobiology* 13: 55–72.

Crawford, G. W., D. G. Smith, and V. E. Bowyer. 1997. Dating the entry of corn (*Zea mays*) into the lower Great Lakes region. *American Antiquity* 62: 112–119.

Crites, G. D. 1993. Domesticated sunflower in fifth millennium B.P. temporal context: New evidence from middle Tennessee. *American Antiquity* 58: 146–148.

Day, G. 1953. The Indian as an ecological factor in the northeastern forest. *Ecology* 34: 329–346.

Decker-Walters, D. S. 1990. Evidence for multiple domestications of *Cucurbita pepo.* In D. M. Bates, R. W. Robinson, and C. Jeffrey, eds., *Biology and Utilization of the Cucurbitaceae,* pp. 96–101. Ithaca, N.Y.: Cornell University Press.

———. 1993. New methods for studying the origins of New World domesticates: The squash example. In C. M. Scarry, ed., *Foraging and Farming in the Eastern Woodlands,* pp. 91–97. Gainesville: University Press of Florida.

Doebley, John. 1990. Molecular evidence and the evolution of maize. *Economic Botany* 44(3 suppl.): 6–27.

Doebley, John, M. M. Goodman, and C. W. Stuber. 1986. Exceptional genetic divergence of Northern Flint corn. *American J. Botany* 73: 64–69.

Doran, G. H., D. N. Dickel, and L. A. Newsom. 1990. A 7,290-year-old bottle gourd from the Windover site, Florida. *American Antiquity* 55: 354–360.

Fearn, M. L., and K. Liu. 1995. Maize pollen of 3500 B.P. from southern Alabama. *American Antiquity* 60: 109–117.

Fritz, G. J. 1984. Identification of cultigen amaranth and chenopod from rockshelter sites in northwest Arkansas. *American Antiquity* 49: 558–572.

———. 1986. *Prehistoric Ozark Agriculture: The University of Arkansas Rockshelter Collections.* Ph.D. dissertation, University of North Carolina, Chapel Hill. Ann Arbor, Mich.: University Microfilms.

———. 1987. The trajectory of knotweed domestication in prehistoric eastern North America. Paper presented at the 10th Annual Meeting of the Society of Ethnobiology, Gainesville, Fla.

———. 1993. Early and Middle Woodland period paleoethnobotany. In C. M. Scarry, ed., *Foraging and Farming in the Eastern*

Woodlands, pp. 39–56. Gainesville: University Press of Florida.

———. 1994a. Coles Creek and Plaquemine landscapes. Paper presented to the 51st Southeastern Archaeological Conference, Lexington, Ky.

———. 1994b. Precolumbian *Cucurbita argyrosperma* ssp. *argyrosperma* (Cucurbitaceae) in the Eastern Woodlands of North America. *Economic Botany* 48: 280–292.

———. 1994c. In color and in time: Prehistoric Ozark agriculture. In W. Green, ed., *Agricultural Origins and Development in the Midcontinent,* pp. 105–126. Report 19, Office of the State Archaeologist. Iowa City: University of Iowa.

———. 1997. A three-thousand-year-old cache of crop seeds from Marble Bluff, Arkansas. In K. J. Gremillion, ed., *People, Plants, and Landscapes,* pp. 42–62. Tuscaloosa: University of Alabama Press.

———. 1998. The development of native agricultural economies in the Lower Mississippi Valley. In V. P. Steponaitis, ed., *The Natchez District in the Old, Old South,* pp. 23–47. Southern Research Report 11. Chapel Hill: Center for the Study of the American South, Academic Affairs Library, University of North Carolina.

Fritz, G. J., and T. R. Kidder. 1993. Recent investigations into prehistoric agriculture in the Lower Mississippi Valley. *Southeastern Archaeology* 12: 1–14.

Fritz, G. J., and B. D. Smith. 1988. Old collections and new technology: Documenting the domestication of *Chenopodium* in eastern North America. *Midcontinental J. Archaeology* 13: 3–28.

Gallagher, J. P. 1992. Prehistoric field systems in the upper Midwest. In W. I. Woods., ed., *Late Prehistoric Agriculture,* pp. 95–135. Studies in Illinois Archaeology 8. Springfield: Illinois Historic Preservation Agency.

Gallagher, J. P., and C. M. Arzigian. 1994. A new perspective on late prehistoric agricultural intensification in the Upper Mississippi River valley. In W. Green, ed., *Agricultural Origins and Development in the Midcontinent,* pp.171–188. Report 19, Office of the State Archaeologist. Iowa City: University of Iowa.

Gardner, P. S. 1987. New evidence concerning the chronology and paleoethnobotany of Salts Cave, Kentucky. *American Antiquity* 52: 358–367.

Gibbon, G. E. 1982. Oneota origins revisited. In G. Gibbon, ed., *Oneota Studies,* pp. 85–90. Publications in Anthropology 1. Minneapolis: University of Minnesota Press.

Gibson, J. L. 1994. Before their time? Early mounds in the Lower Mississippi Valley. *Southeastern Archaeology* 13: 162–186.

Goodspeed, T. H. 1954. *The Genus Nicotiana.* Waltham, Mass.: Chronica Botanica.

Gremillion, K. J. 1993. Evidence of plant domestication from Kentucky caves and rockshelters. In W. Green, ed., *Agricultural Origins and Developments in the Midcontinent,* pp. 87–104. Report 19, Office of the State Archaeologist. Iowa City: University of Iowa.

Griffin, J. B. 1965. Hopewell and the dark black glass. *Michigan Archaeology* 11: 115–155.

———. 1966. Mesoamerica and the eastern United States. In R. Wauchope, ed., *Handbook of Middle American Indians,* vol. 4, pp. 111–131. Austin: University of Texas Press.

———. 1980. The Mesoamerican–southeastern U.S. connection. *Early Man* 2(3): 12–18.

Guffey, S. Z. 1977. A review and analysis of the effects of pre-Columbian man on the eastern North American forests. *Tennessee Anthropologist* 2(2): 121–137.

Haberman, T. W. 1984. Evidence for aboriginal tobaccos in eastern North America. *American Antiquity* 49: 269–287.

Hammett, J. E. 1992. Ethnohistory of aboriginal landscapes in the southeastern United States. *Southern Indian Studies* 41: 1–50.

Hart, J. P., and N. Asch Sidell. 1997. Additional evidence for early cucurbit use in the northern Eastern Woodlands of the Allegheny Front. *American Antiquity* 62: 523–537.

Heiser, C. B., Jr. 1976. *The Sunflower.* Norman: University of Oklahoma Press.

Hunter, A. 1992. *Utilization of Hordeum pusillum (Little Barley) in the Midwest United States: Applying Rindos' Co-Evolutionary*

Model of Domestication. Ph.D. dissertation, University of Missouri, Columbia. Ann Arbor, Mich.: University Microfilms.

Jackson, H. E. 1989. Poverty Point adaptive systems in the Lower Mississippi Valley: Subsistence remains from the J. W. Copes site. *North American Archaeologist* 10: 173–204.

Johannessen, S. 1993. Farmers of the Late Woodland. In C. M. Scarry, ed., *Foraging and Farming in the Eastern Woodlands*, pp. 57–77. Gainesville: University Press of Florida.

Kelly, J. E. 1991. The evidence for prehistoric exchange and its implications for the development of Cahokia. In J. B. Stoltman, ed., *New Perspectives on Cahokia: Views from the Periphery*, pp. 65–92. Madison, Wisc.: Prehistory Press.

Kidder, T. R. 1992. Timing and consequences of the introduction of maize agriculture in the Lower Mississippi Valley. *North American Archaeologist* 13: 15–41.

———. 1993. *1992 Archaeological Test Excavations in Tensas Parish, Louisiana*. Archaeological Report 2. New Orleans: Center for Archaeology, Tulane University.

Kidder, T. R., and G. J. Fritz. 1993. Subsistence and social change in the Lower Mississippi Valley: The Reno Brake and Osceola sites, Louisiana. *J. Field Archaeology* 20: 281–297.

Kidder, T. R., G. J. Fritz, and C. J. Smith. 1993. Emerson. In T. R. Kidder, ed., *1992 Archaeological Test Excavations in Tensas Parish, Louisiana*, pp. 110–137. Archaeological Report 2. New Orleans: Center for Archaeology, Tulane University.

King, F. B. 1985. Early cultivated cucurbits in eastern North America. In R. I. Ford, ed., *Prehistoric Food Production in North America*, pp. 73–98. Anthropological Papers 75. Ann Arbor: University of Michigan Museum of Anthropology.

———. 1993. Climate, culture, and Oneota subsistence in central Illinois. In C. M. Scarry, ed., *Foraging and Farming in the Eastern Woodlands*, pp. 232–254. Gainesville: University Press of Florida.

Lopinot, N. H. 1991. Archaeobotany. In N. H. Lopinot, L. S. Kelly, and G. R. Mil-

ner, *The Archaeology of the Cahokia ICT-II: Biological Remains*, pp. 1–268. Illinois Cultural Resources Study 13. Springfield: Illinois Historic Preservation Agency.

———. 1994. A new crop of data on the Cahokian polity. In W. Green, ed., *Agricultural Origins and Development in the Midcontinent*, pp. 127–154. Report 19, Office of the State Archaeologist. Iowa City: University of Iowa.

Lopinot, N. H., G. J. Fritz, and J. E. Kelly. 1991. The archaeological context and significance of *Polygonum erectum* achene masses from the American Bottom region. Paper presented to the 14th Annual Meeting of the Society of Ethnobiology, St. Louis, Mo.

Lopinot, N. H., and W. I. Woods. 1993. Wood overexploitation and the collapse of Cahokia. In C. M. Scarry, ed., *Foraging and Farming in the Eastern Woodlands*, pp. 206–231. Gainesville: University Press of Florida.

Lynott, M. J., T. W. Boutton, J. E. Price, and D. E. Nelson. 1986. Stable carbon isotopic evidence for maize agriculture in southeast Missouri and northeast Arkansas. *American Antiquity* 51: 51–65.

Milner, G. R. 1990. The late prehistoric Cahokia cultural system of the Mississippi River valley: Foundations, florescence, and fragmentation. *J. World Prehistory* 4: 1–43.

Morse, D. F., and P. A. Morse. 1983. *Archaeology of the Central Mississippi Valley*. New York: Academic Press.

Newsom, L. A., S. D. Webb, and J. S. Dunbar. 1993. History and geographic distribution of *Cucurbita pepo* gourds in Florida. *J. Ethnobiology* 13: 55–74.

Pauketat, T. R., and N. H. Lopinot. 1997. Cahokian population dynamics. In T. R. Pauketat, ed., *Cahokia: Domination and Ideology in the Mississippian World*, pp. 103–123. Lincoln: University of Nebraska Press.

Peterson, J. B., and N. Asch Sidell. 1996. Mid-Holocene evidence of *Cucurbita* sp. from central Maine. *American Antiquity* 61: 685–698.

Phillips, P., J. A. Ford, and J. B. Griffin. 1951. Archaeological survey in the lower Mississippi alluvial valley, 1940–1947. Peabody Museum of Archaeology and Ethnology Papers 25. Cambridge, Mass.: Harvard University.

Ramenofsky, A. F. 1986. The persistence of Late Archaic subsistence-settlement in Louisiana. In S. W. Neusius, ed., *Foraging, Collecting, and Harvesting: Archaic Period Subsistence and Settlement in the Eastern Woodlands*, pp. 289–312. Center for Archaeological Investigations Occasional Paper 6. Carbondale: Southern Illinois University.

Riley, T. J., R. Edging, and J. Rossen. 1990. Cultigens in prehistoric eastern America: Changing paradigms. *Current Anthropology* 31: 525–541.

Riley, T. J., G. R. Walz, C. J. Bareis, A. C. Fortier, and K. E. Parker. 1994. Accelerator mass spectrometer (AMS) dates confirm early *Zea mays* in the Mississippi River valley. *American Antiquity* 59: 490–498.

Rolingson, M. A. 1990. The Toltec Mounds site: A ceremonial center in the Arkansas River lowland. In B. D. Smith, ed., *The Mississippian Emergence*, pp. 27–50. Washington D.C.: Smithsonian Institution Press.

Rose, J. C., M. K. Marks, and L. L. Tieszen. 1991. Bioarchaeology and subsistence in the central and lower portions of the Mississippi Valley. In M. L. Powell, P. S. Bridges, and A. M. Wagner Mires, eds., *What Mean These Bones? Studies in Southeastern Bioarchaeology*, pp. 7–21. Tuscaloosa: University of Alabama Press.

Rowley-Conwy, P. 1995. Making first farmers younger: The West European evidence. *Current Anthropology* 36: 346–353.

Russo, M. 1994. A brief introduction to the study of Archaic mounds in the Southeast. *Southeastern Archaeology* 13: 89–93.

Sauer, J. D. 1993. *Historical Geography of Crop Plants: A Select Roster*. Boca Raton, Fla.: CRC Press.

Saunders, R. 1994. The case for Archaic Period mounds in southeastern Louisiana. *Southeastern Archaeology* 13: 118–134.

Saunders, J. W., R. D. Mandel, R. T. Saucier, E. T. Allen, C. T. Hallmark, J. K. Johnson, E. H. Jackson, C. M. Allen, G. L. Stringer,

D. S. Frink, J. K. Feathers, S. Williams, K. J. Gremillion, M. F. Vidrine, and R. Jones. 1997. A mound complex in Louisiana at 5400–5000 years before the present. *Science* 277: 1796–1799.

Scarry, C. M. 1993. Variability in Mississippian crop production strategies. In C. M. Scarry, ed., *Foraging and Farming in the Eastern Woodlands*, pp. 78–90. Gainesville: University Press of Florida.

Shea, A. B. 1978. Botanical remains. In P. M. Thomas Jr. and L. J. Campbell, *The Peripheries of Poverty Point*, pp. 245–260. New World Research Reports of Investigations 12. Pollack, La.

Smith, B. D. 1984. Mississippian expansion: Tracing the historical development of an explanatory model. *Southeastern Archaeology* 3: 13–32.

———. 1985. The role of *Chenopodium* as a domesticate in pre-maize garden systems of the eastern United States. *Southeastern Archaeology* 4: 51–72.

———. 1992. *Rivers of Change: Essays on Early Agriculture in Eastern North America*. Washington, D.C.: Smithsonian Institution Press.

———. 1995. *The Emergence of Agriculture*. New York: Scientific American Library.

———. 1997. The initial domestication of *Cucurbita pepo* in the Americas 10,000 years ago. *Science* 276: 932–934.

Smith, B. D., and C. W. Cowan. 1987. Domesticated *Chenopodium* in prehistoric eastern North America: New accelerator dates from eastern Kentucky. *American Antiquity* 52: 355–357.

Smith, B. D., C. W. Cowan, and M. P. Hoffman. 1992. Is it an indigene or a foreigner? In Bruce D. Smith, *Rivers of Change: Essays on Early Agriculture in Eastern North America*, pp. 67–100. Washington, D.C.: Smithsonian Institution Press.

Smith, C. J. 1994. Analysis of plant remains from Mound S at the Toltec Mounds site. *Arkansas Archeologist* 35: 51–76.

Ubelaker, D. H. 1988. North American Indian population size, A.D. 1500–1985. *American J. Physical Anthropology* 77: 289–294.

Ward, H. 1998. The paleoethnobotanical record of the Poverty Point culture: Implications of past and current research. *Southeastern Archaeology* 17: 166–174.

Watson, P. J. 1985. The impact of early horticulture in the upland drainages of the Midwest and Midsouth. In R. I. Ford, ed., *Prehistoric Food Production in North America*, pp.73–98. Anthropological Papers 75. Ann Arbor: University of Michigan Museum of Anthropology.

Watson, P. J., and M. C. Kennedy. 1991. The development of horticulture in the Eastern woodlands of North America: Women's role. In J. M. Gero and M. W. Conkey, eds., *Engendering Archaeology: Women and Prehistory*. Oxford: Basil Blackwell.

Willey, G. R., and P. Phillips. 1958. *Method and Theory in American Archaeology*. Chicago: University of Chicago Press.

Williams, M. L. 1994. Plant remains from Locus 3 at the Parkin site. *Arkansas Archeologist* 35: 77–85.

Williams, S. 1990. The vacant quarter and other late events in the Lower Valley. In D. H. Dye and C. A. Cox, eds., *Towns and Temples along the Mississippi*, pp. 170–180. Tuscaloosa: University of Alabama Press.

Williams, S., and J. P. Brain. 1983. *Excavations at the Lake George Site, Yazoo County, Mississippi, 1958–1960*. Peabody Museum of Archaeology and Ethnology Papers 7. Cambridge, Mass.: Harvard University.

Wilson, H. D. 1981. Domesticated *Chenopodium* of the Ozark Bluff dwellers. *Economic Botany* 35: 233–239.

Yarnell, R. A. 1969. Contents of human paleofeces. In P. J. Watson, ed., *The Prehistory of Salts Cave, Kentucky*, pp. 41–54. Reports of Investigations 16. Springfield: Illinois State Museum.

———. 1974. Plant food and cultivation of the Salts Cavers. In P. J. Watson, ed., *Archeology of the Mammoth Cave Area*, pp. 113–122. Orlando, Fla.: Academic Press.

———. 1978. Domestication of sunflower and sumpweed in eastern North America. In R. I. Ford, ed., *The Nature and Status of Ethnobotany*, pp. 289–299. Anthropological Papers 67. Ann Arbor: University of Michigan Museum of Anthropology.

———. 1981. Inferred dating of Ozark Bluff dweller occupations based on achene size of sunflower and sumpweed. *J. Ethnobiology* 1: 55–60.

———. 1982. Problems of interpretation of archaeological plant remains of the Eastern Woodlands. *Southeastern Archaeology* 1: 1–7.

Zawacki, A. A., and G. Hausfater. 1969. *Early Vegetation of the Lower Illinois Valley*. Reports of Investigations 17. Springfield: Illinois State Museum.

SUZANNE K. FISH

10 | HOHOKAM IMPACTS ON SONORAN DESERT ENVIRONMENT

Current archaeological interest in the environmental legacy of past societies reflects an intellectual heritage in the social sciences that came to the fore with the influential publication *Man's Role in Changing the Face of the Earth* (Thomas 1956). In recent years, cultural practices aimed at selectively manipulating plant species and modifying vegetational structure have been highlighted in studies of foraging societies and gained recognition as key processes in the transition from hunting and gathering to farming economies (e.g., Hillman and Harris 1989; Keeley 1995; Smith 1992). Archaeological studies of ancient agriculture confirm it as foremost among human endeavors that transform the natural world from its inception in prehistoric times.

Direct impacts of cultivation include the removal and replacement of natural vegetation, manipulation of topography and soil, and, in arid lands, the diversion and concentration of supplemental water. Other intentional and unintentional activities of farmers create anthropogenic landscapes profoundly shaped by secondary effects. Additional far-reaching consequences result from the capacity of agriculture to support relatively large populations in the same location over time.

The following discussion examines the environmental role of the Hohokam who inhabited the southern deserts of the southwestern United States. Participants in this archaeologically defined culture were not the first agriculturalists in the low-elevation basins of southern Arizona. Preceramic cultivators and early pottery-making populations preceded the Hohokam by more than 1,000 years. Newly emerging information reveals that these initial farmers already occupied villages for prolonged intervals and possessed a range of technologies that included the capability of constructing moderate-sized canals (e.g., Huckell 1996; Mabry 1998). From the third to the fifteenth centuries A.D., however, the Hohokam far surpassed their predecessors by constructing more massive irrigation networks than any built by the high cultures of Mesoamerica and by extending agricultural landscapes well beyond settlements and well-watered fields.

Hohokam productivity supported a developmental trajectory that attained uppermost levels of regional complexity within the maximally hot, dry confines of the northeastern Sonoran Desert. More than those of other southwestern cultures, Hohokam occupations must have tested the delicate balance between fragile arid ecosystems and the impingements of sustained agriculture. Even within the desert setting, the manner and magnitude of impact by Hohokam societies cannot be characterized in a singular way. Natural environmental processes are more regular and subject to uniformitarian approaches than are the overlay of human interventions that affect them. Human impacts must be evaluated with reference to population, settlement, and economy. An understanding of Hohokam impacts rests on the accurate reconstruction of relevant cultural practices, their scale, and their consequences within the environmental variation of the Sonoran Desert homeland.

SONORAN DESERT ENVIRONMENT OF THE HOHOKAM

The Sonoran Desert encompasses almost 310,000 km² of desert scrub and enclaves of upland vegetation types within Arizona, Sonora, southeastern California, and Baja California (McGinnies 1981:41). This corresponds to the Northwestern Coastal Plain (Sonoran) Province in the Mexican Xerophytic Region described by Greller in this volume. Within the Hohokam range in southern Arizona, two vegetational subdivisions of the Sonoran Desert biome predominate: the Arizona Upland on the northeast edge and the western-lying Lower Colorado Valley (Turner and Brown 1994). At the extreme east of the Hohokam occupations in the San Pedro drainage, the Sonoran Desert interdigitates with elements of the Chihuahuan Desert to form a generic desert of creosote bush (*Larrea*), mesquite (*Prosopis*), acacia (*Acacia*), and yucca (*Yucca*), while lacking the distinctively Sonoran columnar cactus, the saguaro (*Cereus giganteus*, formerly *Carnegiea gigantea*). At slightly higher elevations (above 1,200 m), semidesert grassland, interior chaparral, and madrean evergreen woodlands occur, all but the latter within 100 km of any settlement within Hohokam country. Within a 200 km reach of any Hohokam village, and usually much closer, Rocky Mountain subalpine conifer forest above 1,900 m is a vegetation formation that the Hohokam visited to obtain coniferous timbers (P. R. Fish et al. 1992; Dean et al. 1996; Wilcox and Shenk 1977).

Within the Sonoran Desert thornscrub in Arizona that was home to most Hohokam, four vegetation communities are located in the Lower Colorado Subdivision and three within the Arizona Upland Subdivision (Turner and Brown 1994). The paloverde (*Cercidium*)-cacti mixed scrub series is the Arizona Upland subunit best known on the upper basin slopes in the Tucson area, the classic Sonoran Desert landscape. Structurally complex, with a diversity of life-forms, from towering saguaros to a variety of tree legumes, shrubs, herbaceous annuals, and root perennials, 90 percent of this series appears on basin slopes and sloping plains. It is the best-watered, least desertlike series in the Arizona desert scrub formation. The jojoba (*Simmondsia chinensis*)–mixed shrub series is often adjacent at the Sonoran Desert's upper limits, particularly in the northeastern reaches. At the northern boundaries, crucifixion thorn (*Canotia holacantha*) codominates with creosote at moderate elevations in the Verde, Gila, and Hassayampa valleys, mixing with chaparral at slightly higher elevations. Near Safford, on the eastern edge of the Sonoran Desert in Arizona, this crucifixion thorn–creosote series also mixes with Chihuahuan elements to form generic desert scrub.

At lower elevations, particularly in the west, the creosote-bursage (*Ambrosia*) series forms the dominant community in the Lower Colorado Subdivision. Structurally simple and poorer in perennial species diversity compared to the Arizona Upland, this community can nevertheless be rich in ephemeral annuals that flower and seed after rains. On lowlands subject to flooding and salinization, saltbush (*Atriplex*) dominates with other compact shrubs. The saltbush series extends up the Gila River as far as the well-known Hohokam site of Snaketown, south of Phoenix.

Close to the Lower Colorado River itself, at the western edge of Hohokam occupations, a creosote–big galleta grass (*Hilaria rigida*) series occurs on sandy soils. The most complex series in the Lower Colorado Subdivision is a mixed scrub assembly occurring on the granitic and volcanic mountains that edge the basins. Although little-leaf paloverde (*Cercidium microphyllum*) and saguaros may be rare or absent, other tree legumes, desert lavender (*Hyptis*), beargrass (*Nolina*), agave, and jojoba may be present.

A bimodal distribution of rainfall in the Hohokam area contrasts with winter-dominant rainfall to the west and summer-dominant rainfall to the east. Although total precipitation is nonetheless low, this seasonal balance of rainfall in the Sonoran Desert has been associated both with the greater structural diversity of the Sonoran Desert vegetation and with the arborescent character of many of its perennials, compared to the predominance of shrubs in the neighboring Chihuahuan, Great Basin, and Mohave deserts (Turner and Brown 1994:182). As a result of the vegetational advantages of rainfall in two yearly seasons rather than one, gatherable resources are particularly abundant. Distinctively large Sonoran Desert life-forms include such important economic plants as the leguminous mesquite, paloverde, and ironwood (*Olneya*) and relatively large cacti such as cholla (*Opuntia*), prickly pear (*Opuntia*), hedgehog (*Echinocereus*), and saguaro. Less seasonally balanced rainfall and lower annual amounts correspond with diminished desert arboreals. Higher precipitation in a broad arc to the north, east, and south of the greater Phoenix Basin is correlated with increased density and diversity in these key food species. Wildlife densities and diversities also increase along this gradient of decreasing aridity (Vander Wall and MacMahan 1984).

Wetland and riparian areas make up far less than 0.1 percent of the Sonoran Desert lands today, but they were more extensive in the past (Dobyns 1981; Minckley and Clark 1984; Hendrickson and Minckley 1984; Rea 1983, 1997). Oasis marshes and riparian forests (bosques) contributed disproportionately to habitat complexity and overall diversity of available plant species. The resource richness of these zones made them invaluable for human hunting and gathering.

In the sense that they circumscribe the distribution of unequivocally and distinctively Hohokam styles, two factors can be considered key environmental correlates of Hohokam culture. These are location within the northeast boundaries of the Sonoran Desert and, with few exceptions, an elevational range below 1,065 m. Residence and subsistence activity was concentrated on the floors and slopes of basins and frequently extended onto the adjacent lower slopes of moderate-sized desert mountains. Even where larger mountain masses occur, as in the Santa Catalinas near Tucson, utilization of higher-elevation slopes and valleys was minor when compared with upland habitation and farming by other southwestern cultures (S. K. Fish and Nabhan 1991:30).

AN OVERVIEW OF HOHOKAM CULTURE AND AGRICULTURE

The prehistoric Hohokam of southern Arizona achieved one of the premier modifications of New World desert environments by constructing the most massive irrigation system north of Peru (Doolittle 1990:79). Hohokam inhabitants of the Phoenix Basin extended over 500 kilometers of main trunk lines from the perennial Salt and Gila rivers. In other narrower basins, Hohokam farmers irrigated from intermittent rivers and used a variety of additional techniques tapping ephemeral watercourses and surface runoff. These combined methods greatly expanded the productive boundaries and cultural imprint of Hohokam occupations (figure 10.1).

The Hohokam endured as an archaeologically recognizable entity for more than a millennium, from approximately A.D. 200 until 1450 or later. Members of this cultural tradition inhabited 120,000 km² of generally linear basins separated by moderate-sized mountain ranges (P. R. Fish 1989). Densities of regional population reflected the greater productive capacities of irrigated land along the perennially flowing rivers. Nevertheless, stylistic and organizational trends progressed in tandem throughout the Hohokam domain.

The Hohokam were masterful agriculturalists in the Sonoran Desert. Studies of subsistence remains throughout Hohokam territory reveal a consistently strong component of crop domesticates, indigenous species that were tended or cultivated, and a broad array of wild resources (Bohrer 1991; S. K. Fish and Nabhan 1991; Gasser and Kwiatkowski 1991). Remains of Hohokam settlements, water management devices, and fields cover substantially greater areas than were inhabited by succeeding historic farmers of Indian and Hispanic heritage, who employed many of the same techniques. The ecological parameters of prehistoric subsistence were irreversibly altered in the early contact era; a similarly direct extractive and productive orientation, without domesticated animals as intermediaries, persists nowhere today.

During Preclassic times before A.D. 1150, pithouses were the common domestic structures and earthen-banked ballcourts the predominant form of Hohokam public architecture. Thereafter, in the Classic period, both free-standing and contiguous adobe rooms were built and often enclosed by a surrounding wall to form residential

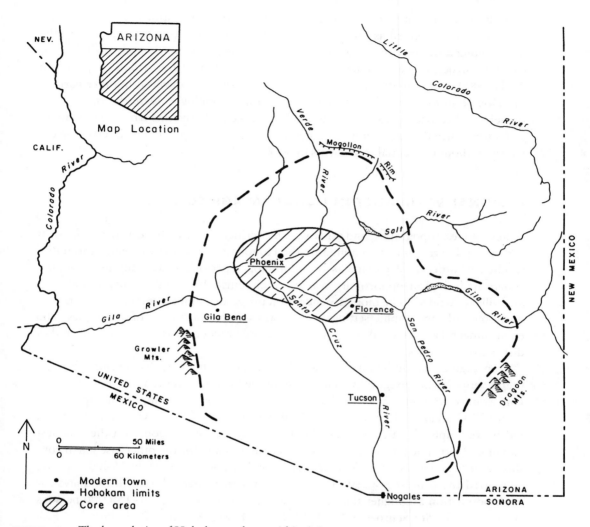

FIGURE 10.1. The boundaries of Hohokam culture within Arizona.

compounds. Platform mounds, which supported adobe buildings on their summits and were also enclosed by thick adobe walls, became the foremost public edifices in increasingly formal layouts at larger Classic sites (Gregory 1987; P. R. Fish and Fish 1991).

For most of the Hohokam cultural sequence, the primary units of territory and concentrated land use were clusters of villages and interspersed land that archaeologists term "communities." Hohokam communities consist of a central site and an outlying set of related and usually smaller settlements. Ballcourts mark the centers of the earliest communities. By A.D. 1000, these territorial entities integrating multiple sites can be widely defined in Hohokam settlement patterns (Wilcox and Sternberg 1983). During the ensuing Classic period, centers were distinguished by the presence of platform mounds for communitywide observances. Community organization in the Phoenix Basin has been interpreted as providing a framework for allocating water among settlements along a shared canal system and for mobilizing and coordinating the labor for canal construction and maintenance (Crown 1987; Doyel 1976, 1980; Gregory and Nials 1985). Community organization also was pervasive, however, in other parts of the Hohokam domain where mixed agricultural technologies were the rule. In the densely packed occupations corresponding to large-scale riverine irrigation, communities were virtually continuous; in other areas they were separated by intervening expanses without residential settlements (S. K. Fish 1996).

REGIONAL VARIABILITY IN HOHOKAM AGRICULTURE

Water sources, topography, and other variables critical to agriculture were not uniform across the desert country inhabited by the Hohokam. This environmental variability resulted in differing deployments of a common repertoire of farming technologies. Where hydrological opportunity was more circumscribed, individuals, households, and villages tended to be more dependent on mixed techniques. Methods of water acquisition and topographic settings were distinctive for each class of techniques. The environmental and cultural consequences of these practices must also be considered as distinctive outcomes.

Converging just southwest of modern Phoenix, the Salt and Gila rivers flow through the Phoenix Basin topography that is among the lowest, hottest, and driest in Hohokam country (see figure 10.1). Annual precipitation is 180 mm, and temperatures above 38°C occur on 90 or more days (Sellers et al. 1985). These two perennial rivers, fed by vast upland watersheds outside the desert, in conjunction with the broadest expanse of irrigable basin floor, more than compensated for the otherwise harsh conditions of the Phoenix Basin. This area of 2,000 km², often termed the Hohokam core, supported the densest populations and largest settlements. Almost all estimates for peak population are higher than 25,000 and range up to 100,000–150,000 persons (Haury 1976:356; Schroeder 1960:20; Doyel 1991:265–266; P. R. Fish and Fish 1991:155–157).

Although always less than 385 mm, annual precipitation increases with elevation in areas of Hohokam occupation to the north, east, and south of the Phoenix Basin, where maximum temperatures also are slightly moderated. In these regional sectors, intermittently flowing rivers originate primarily in watersheds within the desert rather than in external highlands. Valley and floodplain morphologies also restrict the width of irrigable land in those locations where riverine canals can be filled. Inhabitants of the vast remainder of the Hohokam tradition outside the Phoenix Basin practiced a smaller scale of irrigation and depended heavily on techniques for capturing the floodwaters of ephemeral drainages and overland runoff.

Whether based on riverine irrigation or runoff technologies, the sustainability of Hohokam agriculture appears not to have been significantly inhibited by soil exhaustion or salinity. Occupations at large sites in the Phoenix Basin often span hundreds of years, while irrigation networks were present in the earliest ceramic phases and developed toward a maximum extent in late prehistoric times (Masse 1981; Nicholas and Feinman 1989; Howard 1993). Two analyses (Ackerly et al. 1987; Nials et al. 1989) of overall system capacity in the Phoenix Basin concur in a total for irrigated acreage ranging between 12,100 and 24,200 ha. Multiple excavated examples of canal interruption, abandonment, and rebuilding have increased appreciation of the dynamic and labor-intensive nature of these systems.

Zones of active deposition from ephemeral drainages likewise witnessed long-term continuity in settlement, although occupations at individual small sites within these zones were often of lesser duration. Floodwater farmers of alluvial fans on lower basin slopes, broader arroyo bottoms, and upper basin pediments watered their crops by diverting storm flows onto fields at the sides of shallow drainages. As active channels shifted across the surfaces of alluvial fans and new fans became hydrologically active, Hohokam settlements likewise shifted within the topographically defined zones of depositional enrichment. A third broad category of Hohokam agricultural techniques obtained water from overland runoff rather than from channelized flow in drainages. A variety of simple constructions were designed to intercept and concentrate storm runoff from sheet flows and shallow rills on broad surfaces. Archaeological remains associated with these techniques are widely scattered across the slopes of Hohokam basins, but the scale and implications of this form of cultivation are only now beginning to be recognized. The most extensive arrays of cobble features supplied by surface runoff are termed "rockpile fields," but diversionary walls, contour terraces, hillside terraces, grids, and other constructions were also employed to direct and retain surface flow.

OCCUPATIONAL AND AGRICULTURAL SUSTAINABILITY

The nature of agricultural water used by the Hohokam was central to their achievement of long-term population and settlement continuity. In restricted, high-elevation locales of the northern Southwest where rainfall alone was sufficient for farming or where clear mountain streams furnished supplemental water, eventual soil exhaustion may have induced progressive clearing of short-term fields, prolonged fallowing, and the sequential abandonment of arable locales (cf. Kohler and Matthews 1988). The Hohokam were not similarly subject to such processes. Although water sources are highly localized in the low southern deserts, those supplying Hohokam cultivation renewed soil fertility with each wetting.

Canals carried suspended sediment from the undammed flow of the Salt and Gila rivers to the perimeters of irrigated cultivation, up to 10 km inland from the river channel and for linear distances up to 30 km. Some of this rich soil and detritus was delivered directly to fields through irrigation, and some appears to have been spread beyond the banks of canals in the course of periodic channel cleanings. Ribbons of distinctive soil types of waterborne origin still mark the paths of both Hohokam canals tapping perennial rivers and those filled from intermittently flowing watercourses (Dart 1986).

On alluvial fans of lower basin slopes, broader arroyo bottoms, and upper basin pediments, farmers simultaneously watered and fertilized their crops by diverting storm flows in shallow drainages. Organic debris concentrated in these waters by rapid runoff following seasonal rains offset a major deficiency of desert soils (Bryan 1929; Nabhan 1979, 1986). Farmers could even improve the coarse texture of soil in new fields by diverting flows to deposit fine-grained sediments prior to field use (S. K. Fish et al. 1992).

Highly localized sources of domestic and agricultural water in Hohokam basins must have created a strong impetus for extractive practices that conserved the long-term productive potential of the environment. Limited situations for water diversion or canal headings, and investment in the construction of canals and runoff features, added to locational constraints. Movement to new locations was not a simple alternative if the local environment became depleted or degraded, and mobility options further decreased as populations grew.

Several natural and cultural patterns contributed to the Hohokam ability to maintain stable occupations and an adequate reserve of natural resources in their fragile environment. Pit ovens, or hornos, were fuel-efficient cooking facilities of Hohokam villages. Food was placed in these pits along with coals and heated stones, then covered and left to cook slowly over long periods. Shared horno usage by several households was a typical pattern that would have further minimized fuel consumption (Sires 1987: 180). Ceramics were fired at relatively low temperatures. Pithouse, wattle and daub, and adobe architectural styles utilized a minimum of structural wood. Materials from alternative species of desert plants such as the ribs of saguaro cactus and ocotillo (*Fouquieria*) were frequently substituted.

Compared to prehistoric groups of the northern Southwest, the Hohokam had a much lower need of fuel for heating homes, with long months of hot weather and average annual temperatures between 10 and 18°C (Sellers, Hill, and Sanderson 1985). Destruction of leguminous trees, the most common desert species, was likely minimized in deference to the dietary importance of their abundant and nutritious beans. Among historic Piman groups, the geographic successors and probable descendants of the Hohokam, desirable trees such as mesquite were often left standing in fields and grew densely in hedgerows benefiting from agricultural water (Rea 1981; Castetter and Bell 1942). Charred seeds and wood charcoal from lengthy prehistoric occupations attest to the consistent availability of such trees, if not always in the same proportions. Wide-ranging sources of additional fuels included riparian trees, driftwood, and woody desert shrubs.

ANTHROPOGENIC CHANGE

The Hohokam created distinctively anthropogenic settings about their settlements with culturally altered distributions of plants and animals. Agricultural activity produced a variety of vegetational effects (Bohrer 1970; Gasser 1982; S. K. Fish 1984, 1985; Miksicek 1984, 1988). Farming on floodplains in the environs of drainages probably created the least divergence from surrounding plant communities of naturally disturbed riparian habitats. Even in these situations, manipulation through such documented aboriginal practices as selective removal of unwanted species, reseeding and tending of utilized plants, or introduction of nonlocal taxa could have altered distributions toward advantageous ends.

Canals transporting water laterally as well as downstream for miles beyond drainage sources had a greater potential for creating biotic conditions different from the locally prevailing ones. Riverine canals traversed flatter portions of lower basins among xeric saltbush or creosote communities that contrasted with irrigated fields in the absence of both supplemental water and surface disturbance. Preparation of the soil to increase infiltration, concentration of surface runoff, and the diversion of ephemeral drainages into basin slope or hillside terrace agricultural complexes also generated contrastive growth conditions, although augmenting moisture less than canal irrigation. Even without management, weedy plants of these latter two agricultural contexts would have differed in densities from adjacent uncultivated land.

In large part, insights into Hohokam creation and management of anthropogenic vegetation derives from ethnographic studies of plant use by Piman-speaking groups who intensively harvest wild plants as well as crops in field, ditch, hedgerow, abandoned field, and dooryard garden microhabitats (Rea 1981, 1983; Crosswhite 1981; Nabhan et al. 1983). Such practices serve as analogues for the Hohokam capacity for transporting, transplanting, watering, or otherwise managing wild plants for their desirable products. The presence of particular weedy or semicultivated species in the archaeobotanical record is not definitive evidence for origin in anthropogenic plant communities, but patterns of quantitative contrast and contextual correspondence have been increasingly identified in recent studies. These provide the basis for reconstructing environmental alterations that included both intentional manipulation and unintentional enhancement of species other than cultigens.

Supplemental water in fields must have increased biomass production over surrounding vegetation. Pollen samples from prehistoric agricultural contexts differ from natural vegetation in the distribution of weedy taxa. They reveal a rich weedy flora alongside crops responding to agriculturally enhanced conditions (S. K. Fish 1984, 1985). The Hohokam likely followed ethnographic southwestern practices of permitting desirable weeds to remain in fields and sometimes even scattering seeds to insure a sufficient supply (Whiting 1939; Crosswhite 1981; Bye 1979). Among weedy plants of Hohokam fields were species such as chenopods, amaranths, and spiderling (*Boerhavia*) that furnish edible seeds and greens. In seasons when water was inadequate to mature cultigens successfully, these secondary resources of fields may have constituted a lesser but welcome harvest.

Modified environments created by the Hohokam also featured transplanted desert species receiving more directed attention. The agave or century plant is now recognized as a mainstay of Hohokam cultivation. Less conclusive evidence suggests additional candidates for this cultivated category that include cholla, prickly pear, and little barley (Bohrer 1991; S. K. Fish and Nabhan 1991). Such plantings in poorer fields, field borders, and fallow ground may have supplemented the production of natural stands near long-term residence. A continuum of active intervention involving a variety of plants is probable. Proposed species include mesquite, hedgehog cactus, wolfberry (*Lycium* spp.), hog potato (*Hoffmanseggia densiflora*), amaranth (*Amaranthus*), chenopods (*Chenopodium berlandieri, C. murale, Monolepis nuttalliana, Atriplex wrightii*), tobacco (*Nicotiana trigonophylla*), and spiderling (S. K. Fish and Nabhan 1991).

A major proportion of animals hunted by the Hohokam also conforms to the culturally modified environs of their settlements (e.g., Szuter and Bayham 1989; Szuter and Gillespie 1994). Bones of large animals such as deer are infrequent compared to quantities for small species. Rabbits and rodents, characteristic of agricultural habitats, were key components of Hohokam cuisine. Animals attracted to the vegetation and water in fields could be conveniently trapped or hunted during agricultural tasks. At sites occupied for long intervals, trends in consumption reflect increasing percentages of species that prefer environmental situations corresponding to anthropogenic vegetation. In multiple instances, an earlier predominance of cottontail was superseded by greater reliance on jackrabbit.

Numerous combinations of happenstance and design are embodied by the array of noncultigens now thought to have been productively enhanced or concentrated in anthropogenic vegetation by the Hohokam. Among this group are plants such as mesquite that may have been differentially spared in field clearing, become dense in hedgerows, thrived on canal seepage, or been tended and selectively harvested in adjacent natural settings. Species such as cholla may have been transplanted to dooryard gardens or fields as a crop, to out-of-the-way spots among habitations and fields, or employed as residential fencing. Chenopods and spiderling are representative of weedy herbaceous plants that may have received focused attention or none at all in fields and other culturally disturbed habitats (S. K. Fish and Nabhan 1991).

NEW INSIGHTS INTO LAND-EXTENSIVE CULTIVATION

The scale of runoff-dependent techniques is one of the surprising outcomes of ongoing Hohokam research. Although not replicating the extent of Hohokam systems, historic canals of the Pima Indians furnish a means for visualizing former irrigation. Likewise, the vestigial floodwater farming of recent times by Piman cultivators and traditional farmers of northern Mexico serves as a guide to prehistoric methods. The ethnographic record is mute, however, with regard to more land-extensive farming practices of the Hohokam. Archaeological studies provide the only insights into the areally diffuse Hohokam techniques that disappeared in the postcontact era.

Cobble features are the key to comprehending an agriculturally engineered landscape in southern Arizona that extended well beyond the confines of habitations, irrigation, floodwater fields, and other cultivated land that depended on channelized sources of water. Subsumed under the general term of rockpile fields or complexes, these features consist of rounded heaps and linear arrangements of unshaped rock. In most parts of the Hohokam domain, distributions of such features have been shown to fill appreciable segments of what otherwise would be considered voids in settlement pattern. In place of empty or natural areas with solely extractive potential, rockpile distributions demonstrate a significant modification of land surfaces and managed productivity.

Archaeological studies and replicative experiments in recent years have revealed much about the agricultural functions of rockpiles (S. K. Fish et al. 1985, 1992). Rockpiles enhance the growth environment for crops planted in them. Excavated cross-sections reveal that cobbles often cap mounds of soil beneath. The uneven, porous texture of piled rocks permits greater penetration of rainfall than does the surrounding hard-packed and impermeable ground surface and also increases the interception of rapid, transitory surface runoff. The rocks then act as a mulch, slowing evaporation of soil moisture by blocking capillary action and preserving higher moisture levels beneath. Gauged experiments show higher levels of moisture in rockpiles than in surrounding soil for days to weeks following rainfalls of varying size. Suspended nutrients in overland runoff and accumulations of windblown soil appear to have been sufficient to maintain fertility (S. K. Fish et al. 1992).

Concentrations of annuals and perennials and the presence of moisture-dependent lichens and mosses demonstrate the continuing response of natural vegetation to the microhabitat of prehistoric rockpiles. This modern plant response has been quantified by comparing root biomass in soil directly beneath rockpiles and in adjacent controls. Root weight in rockpile soils is on average 80 percent higher than in nearby soils (S. K. Fish et al. 1985).

Fields consist of contiguous sets of these agricultural features (figures 10.2, 10.3). The rockpiles rarely exceed 1.5 m in diameter and 75 cm in height. Contour terraces and checkdams of one to several cobble courses are often interspersed in small fields and are always present in large complexes. Field sizes range from clusters of as few as ten rockpiles to arrays of rockpiles and linear features covering many hectares.

Roasting pits are consistently present in rockpile fields, ranging from 3 m to more than 30 m in diameter. The typically large pits of extensive fields have the shape of broad, shallow basins and a complicated stratigraphy chronicling the intrusions and accretions of seasonal reuse over many years. Pits are filled with ash, charcoal, fire-cracked rock, and occasional artifacts. Flotation of fill has identified agave, a desert succulent, as the crop in many fields. Cooking up to 48 hours in sealed pits converted carbohydrates stored in the base of the agave into a sugary, nutritious food. Stone tools on field surfaces include knifelike implements resembling tools used ethnographically to sever agave leaves during harvest of the plants. Other specialized tools for removing leaf pulp to extract fiber, the second product of the agave, are recovered in associated habitation sites (S. K. Fish et al. 1992).

The cultivation of drought-resistant agave in rockpile fields permitted expanded cropping on large expanses of otherwise marginal slopes in Hohokam basins. The moisture-enhanced microhabitats of rockpiles, terraces, and checkdams benefited these succulents that were cultivated at somewhat hotter and lower elevations than natural populations. Modern experimental plantings also indicate that rockpiles significantly

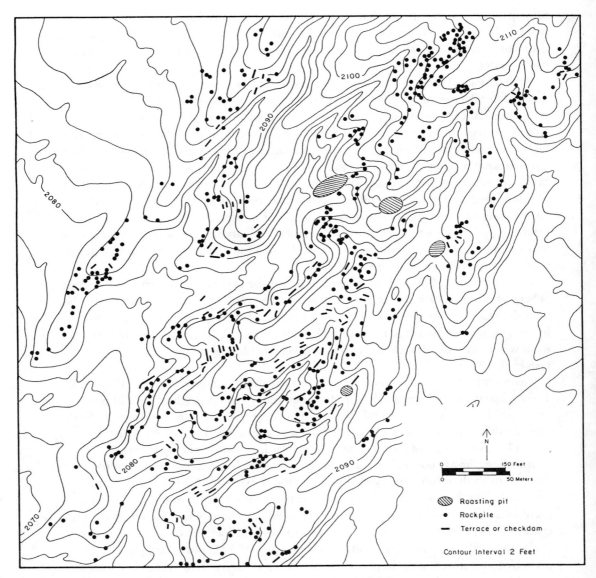

FIGURE 10.2. Agricultural features in a segment of a Tucson rockpile field.

FIGURE 10.3. Agricultural features in a segment of a Tucson rockpile field.

inhibit rodent predation, which is preferentially focused on the bases and roots and is accomplished by digging beneath the plant (ibid.).

THE REGIONAL SCALE OF ROCKPILE TECHNIQUES

Recognition of the form and function of dispersed agricultural features such as rock-piles is not a sufficient basis for assessing environmental or economic impact without a means for evaluating areal extent (S. K. Fish et al. 1990). Estimates are difficult at a regional scale because archaeologists have only sporadically noted these relatively un-obtrusive remains. The failure to record such agricultural traces systematically is un-doubtedly related to the lack of an ethnographic framework for understanding their significance.

Entries in the Arizona State Museum site files (the largest and most complete re-pository for archaeological records in Arizona) furnish data for examining scale. A criterion of at least ten cobble features minimizes the inclusion of nonagricultural ac-tivities that might create one or a few heaps of rock. In spite of sporadic recording, 556 locations fitting the minimal criterion have been reported since the early 1950s. The 556 sites represent about 2 percent of the 25,000 entries in the Arizona State Museum files (S. K. Fish and Fish in press).

The regional distribution of fields is a significant pattern that emerges despite the incomplete nature of available records. In general, the distribution of rockpile fields parallels the outline of Hohokam and culturally related occupations of the Classic pe-riod (A.D. 1150–1450). A gap coinciding with urban Phoenix reflects modern land disturbance as well as the competing prehistoric alternative of large-scale irrigation. Stone agricultural features of various sorts were constructed throughout the remainder of Arizona and the Southwest, but the Hohokam domain coincides with a generally well-delimited and continuous distribution of fields in which rounded heaps or rock-piles are a predominant form.

In light of pervasive inconsistencies in the manner of quantifying and bounding rockpile occurrences, areal magnitude must be approached cautiously. The total area reported for all 556 locations is 8,350 ha or about 78 km^2. Average size of fields is not a meaningful figure when individual complexes range from tens of square meters to over 100 ha, and the extremes are unquestionably biased by highly divergent recording procedures. Rockpiles are concentrated in an area between Phoenix and Tucson. In the area surrounding Phoenix, they tend to become more common toward the outer edges of the massively irrigated core. They occur widely in conjunction with Classic period communities along the Santa Cruz, San Pedro, and Gila rivers (S. K. Fish and Fish in press).

Seventy-eight square kilometers of rockpile fields is an appreciable sum, but it is an admittedly minute total compared to the thousands of square kilometers encompassed by the overall distribution of the features in southern and central Arizona. A striking disparity among data sources is a clue to the weakness of current records for estimating the true areal magnitude of these agricultural techniques. Only five archaeological projects were responsible for recording nearly 60 percent of the total area in rockpile fields documented over the past four decades (ibid.). Each of these five investigations was an archaeological survey with the broad, continuous coverage necessary to delineate sets of complexes exceeding several hundred hectares, and each entailed an explicit commitment to document agricultural remains systematically. The disproportionate results from this handful of projects strongly suggest that similar arrays have been overlooked or seriously underreported by many other investigations.

EXPANDING CONCEPTS OF AGRICULTURAL LAND USE

Ecological parameters of land-extensive practices can be examined in order to assess environmental implications. Multiple lines of evidence identify agave as the primary crop in a majority of intensively investigated fields. Pollen of corn, cotton, and cucurbits from some fields indicates other, and possibly intercropped, cultigens in a minority of situations that often afford access to sources of supplemental water in addition to surface runoff. The typical location of rockpile fields on dry basin slopes and the location of almost every large complex in such situations, however, are commensurate with drought-resistant agave as the main crop. In the manner of the small leguminous trees and large cacti that distinguish Sonoran Desert vegetation, but unlike annual crops, agaves can make cumulative use of the region's seasonally bimodal rainfall, divided between summer and winter months.

Environmental modifications in rockpile fields take several forms. Pollen assemblages show a proliferation of weedy species in response to soil disturbance and the manipulation of surface runoff. Surficial characteristics of the land were altered by the gathering and piling of rocks. Porous rockpiles and linear features along contours and across small channels intercepted transitory runoff that otherwise would have been rapidly lost from desert hydrological systems. Nevertheless, it is unlikely that runoff captured by rockpile constructions significantly reduced amounts available for other methods of farming.

The ultimate impact of rockpile farming was its contribution to the support of relatively large Hohokam populations in persistent settlements and a correspondingly heightened pressure on the plant and animal resources of their arid surroundings. It enabled cropping on vast stretches of marginal land without other sources of supplemental water. At the same time, localized crop plants in dispersed rock features would have presented no serious conflict with the continued presence of useful wild species within and around rockpile fields.

The cultivation of agave augmented overall Hohokam subsistence by: (1) providing an additional, relatively low-maintenance crop; (2) avoiding seasonal bottlenecks in agricultural labor through an alternative schedule for tending, harvesting, and processing; (3) permitting field storage of harvestable, semimature plants that could temporarily counteract catastrophic shortages of annual crops; and (4) insuring a low-level, productive stability on large expanses of land because agave crops were not lost in seasons of poor precipitation. These land-extensive practices can be seen as an aridland version of agricultural intensification, particularly when fields at a distance from habitations necessitated increased travel.

Dispersed agricultural features serve as a reminder that much of what otherwise might be considered the natural setting of prehistoric settlements was in fact a socially structured landscape that is only partially demarcated by durable remains. Occasional fieldhouses and upright stones resembling ethnographic field boundary markers occur within rockpile complexes. These suggest systems of individual and collective tenure.

Another clue to the socially structured nature of these expansive segments of agricultural landscape comes from regularities in the ratios of communal roasting pits to cultivated area in large fields (S. K. Fish and Fish in press). Ratios were calculated for sets of fields in three widely separated Hohokam communities in the Tucson region. Average amount of cultivated land per huge roasting facility in each case fell into a restricted range between 13 and 17 ha. This modal tendency strongly suggests commonly held concepts about the organization of land and tenure in rockpile fields and the proper size of groups roasting their harvests in shared pits. Communal harvesting and roasting also enabled efficient expenditure of scarce desert fuels that were already taxed by long-term residence.

Rockpile complexes are difficult to date on the basis of highly diffuse surface scatters of mostly undecorated ceramics. Only a minority of well-studied fields can be assigned an age with reasonable certainty. Hohokam rockpile technology appears to have been employed as early as A.D. 600 (S. K. Fish et al. 1992). Relatively small rockpile fields occur widely in conjunction with habitation sites clearly dating between A.D. 750 and 1150. However, very large rockpile fields that substantially expanded the agricultural landscape at a distance from villages appear to be a hallmark of the Classic period after A.D. 1150 and to coincide with dense, aggregated populations.

Where systematically recorded, rockpile features disclose a previously unsuspected scale of land-extensive agriculture as a routine component of settlement systems. Recognition significantly expands the scale of culturally altered Hohokam landscapes. In view of the prominent role of earthen and brush constructions in the ethnographic agriculture of the Sonoran Desert, rockpile technology may well represent the most durable and archaeologically visible remains among a broader but related set of runoff techniques.

ZONAL LAND USE IN A HOHOKAM COMMUNITY OF THE TUCSON BASIN

The overall impact of Hohokam agriculture can best be judged in the framework of the territorial organization of multisite communities. It was within these entities that individuals and groups of farmers determined the layout of land use and the proportional emphasis on different techniques. A comprehensively investigated study area in the northern Tucson Basin near Marana, Arizona, exemplifies the typical duration of Hohokam settlement, varied components of the subsistence base, community development, and responses to increasing population (S. K. Fish et al. 1992). The Marana community, reaching maximum size in the early Classic period (ca. A.D. 1150 to 1300), integrated topographic zones presenting reciprocal annual threats to agriculture in the Southwest: floodplains of the primary drainages flooded destructively with too much rain, and upper basin slopes yielded poorly with too little rain (cf. Lightfoot and Plog 1984; Abruzzi 1989). The greatest degree of productive specialization and exchange within this Marana community, coinciding with highest population levels of the prehistoric sequence, would have served to diffuse localized effects of low and unpredictable precipitation.

Annual rainfall is between 225 and 300 mm in the Tucson Basin. Mountains rim the basin and divide it from adjacent drainage regimes. Elevational diversity is repeated on either side of the floodplain of the Santa Cruz River, but the eastern valley slope occupies much of the northern basin interior as it rises toward the Tortolita Mountains (figures 10.4, 10.5). A foreshortened slope occurs to the west below the smaller Tucson Mountains. Full-coverage survey of 390 km² encompassing the Marana community has revealed settlement patterns dating from the Late Archaic adoption of agriculture through all subsequent ceramic periods (see figure 10.4).

One of two zonal sources for permanent water is found in springs and canyon streams along the flanks of the Tortolita Mountains. Although the Santa Cruz is intermittent in the northern Tucson Basin, a second community location of persistent surface flow occurs near the end of the Tucson Mountains where intrusive bedrock maintains high water tables. Throughout the Hohokam occupation, both areas were continuously preferred for settlement. In the Preclassic period prior to A.D. 1150, separate and independent communities existed in each of these two hydrologically optimal locales. The riverine and mountain flank communities covered 56 and 70 km², respectively.

Developments in the Classic period illustrate the flexibility of community organization in integrating larger and denser populations. At this time, the two earlier communities coalesced into a single larger one incorporating 146 km². A platform mound was constructed in a new central site at the juncture of the two Preclassic axes of settlement. Additional sites appeared in the previously intervening area, and substantial population growth is apparent. Within the enlarged Classic community boundaries, six topographic zones can be defined on the basis of residence patterns, productive activities, and environmental variables. Environmental opportunities and limitations were different in each of these zones, as were methods and consequences of Hohokam land use (see figures 10.4 and 10.5).

ZONE 1. In Zone 1 at the lower edge of the basin slope above the river floodplain, the gradients of tributary streams decrease and the deposition of suspended soil forms alluvial fans. Shallow channels were easily tapped by the ditches and short canals of floodwater farmers. Ethnographic data suggest that better floodwater fields yielded as bountifully as irrigated ones (Doelle 1980:67–75), although at greater risk from spotty rainfall. Village locations over time correspond with depositionally active fans, but no

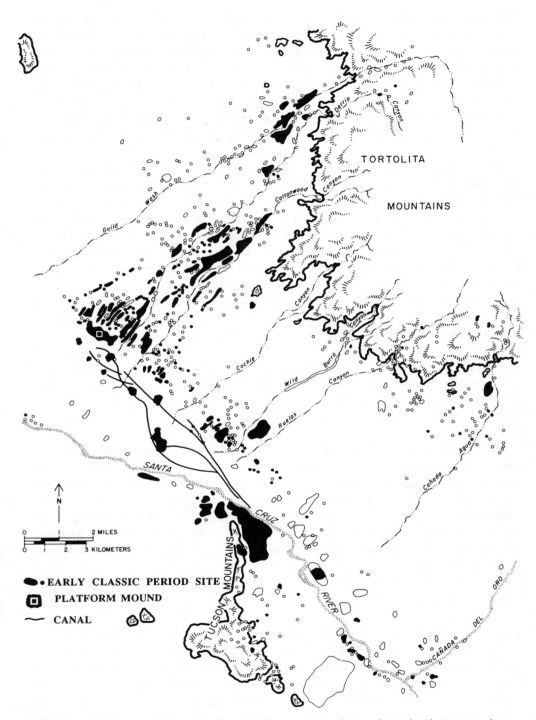

FIGURE 10.4. Hohokam settlements in the Marana community during the early Classic period (A.D. 1150–1300).

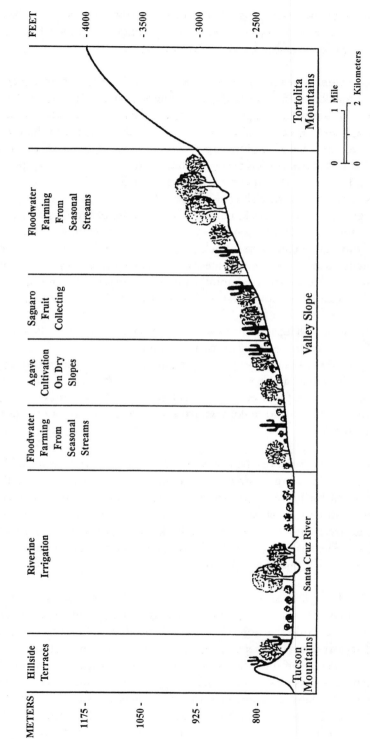

FIGURE 10.5. Idealized cross-section of the northern Tucson Basin showing zonal land use in the Marana community.

permanent water occurs in Zone 1. Preclassic settlement was within daily travel distances for domestic water from the river. At the beginning of the Classic period, a new canal permitted sites farther from the Santa Cruz.

ZONE 2. Permanent water is even more distant from Zone 2, and large drainages are too entrenched for diversion onto fields. This zone was utilized primarily for hunting and gathering prior to the Classic period. Increasing Zone 1 populations at that time transformed large segments of these dry slopes into fields for agave, a dual source of food and fiber. Rockpile field locations, up to 50 ha in extent, covered a total of 485 ha in Zone 2. Crop identity is confirmed by charred remains in roasting pits for fieldside processing. In large fields, pits up to 30 m in diameter cooked the communal harvests of groups of farmers (S. K. Fish et al. 1985, 1992).

Although a few smaller rockpile fields for agave are of Preclassic age, Classic populations expanded this cultivation onto broad tracts of previously uncultivated and agriculturally marginal land. Zone 2 fields are the most striking evidence for response to the heightened subsistence requirements of Classic community inhabitants. Fields of the largest size were constructed uphill from a densely settled segment of Zone 1 where residents had relatively poor access to land for irrigated or floodwater farming.

ZONE 3. Domestic and agricultural water is also absent in Zone 3. Residential sites are lacking. The common site type consists of huge scatters of broken pottery and few other artifacts. Like Piman groups of later times, community members appear to have returned to the same camps near dense stands of saguaro cactus year after year, harvesting the fruits for a few weeks in the early summer. Vessels for camp water supplies and for boiling the fruit to make saguaro syrup were broken and the debris accumulated over time.

ZONE 4. Shallow bedrock on the mountain flanks insures high water tables in Zone 4 drainages. The three largest streams originating in the mountains had floodplains of sufficient width for floodwater farming on bottomland and are correlated with clustered sites of all periods on surrounding ridges. Smaller drainages and surface runoff were also utilized, with low terraces and checkdams at scattered locations. Sets of rockpiles and contour terraces for agave are present in Zone 4 but do not compare in size or number with fields in Zone 2 below.

ZONE 5. The floodplain and terraces of the Santa Cruz River constitute Zone 5. The densest populations were concentrated along the stretches of high water table. Irrigation would have made this zone the foremost producer of annual crops such as corn with high moisture requirements. Due to alluvial deposition on the Santa Cruz floodplain, canals have been identified only through excavation, and the precise extent of irrigation networks is unknown. Where canals could be filled from high water tables along the river, relatively minor investment is apparent in other methods of farming. From a point where canals leave the active floodplain, their paths can be traced for more than 10 km north to the community center. Such canals would have extended fields with crops and dense stands of weedy annuals into the habitat of shrubby creosote bush and bursage associations.

ZONE 6. The compressed basin slope west of the Santa Cruz forms Zone 6. Small alluvial fans supported only minor floodwater farming. In the adjacent Tucson Mountains, rock-walled terraces that supported dwellings and small gardens were constructed on hillsides. Volcanic soils in this low range are high in clay and moisture retentive. Above the elevation affected by cold air drainage and winter freezes on the basin floor, small crops grown in the spring season on terraces could precede major summer harvests. Charred agave and tools related to its cultivation from excavated structures suggest a planting pattern still followed in Mexico, with these agaves lining the walls and annual crops on the remainder of the planting surfaces (S. K. Fish et al. 1984; Downum et al. 1994).

COMPARATIVE SCALE OF LAND-EXTENSIVE CULTIVATION

In addition to the Marana community, two more of these territorial entities north of Tucson have been mapped in detail. Rockpile distributions in one of these, the McClellan community, can be compared with distributions in the Marana community to illustrate relative areal and economic scales. As previously noted, the Marana community in the northern Tucson Basin encompasses 146 km² (table 10.1). Rockpile fields greater than 10 ha cover 5.3 km² or about 3 percent of its territory. The McClellan community farther north is only slightly smaller, at 136 km². Rockpile complexes were constructed on 14.8 km², or a higher 9 percent of its area. Such proportional measures provide a basis for comparing the areal intensity of this kind of land use between culturally meaningful territorial units.

Another measure compares the intensity of rockpile field cultivation in relation to community population levels. The figures for Marana community rockpile fields pertain to the early part of the Classic period and those for the McClellan community to a subsequent Classic interval of increasing aggregation. The cumulative area of 16,047,500 m² for all McClellan habitation sites is much greater than the 5,743,500 m² total for Marana habitation sites (see table 10.1). If the cumulative area of habitation sites is even a rough proxy for relative size of population in the two communities,

TABLE 10.1. COMPARISON OF AREA MEASURES FOR TWO CLASSIC PERIOD HOHOKAM COMMUNITIES IN THE NORTHERN TUCSON BASIN

	Marana	McClellan
Community size	146 km²	136 km²
Habitation site total	5,743,500 m²	16,047,500 m²
Habitation site area / km²	39,334 m²	117,946 m²
Rockpile field area	5,268,503 m²	9,190,737 m²
Field area / km²	36,086 m²	67,579 m²
Field area/Habitation area	0.9 m² / 1.0 m²	0.6 m² / 1.0 m²

population densities are much higher in the later McClellan community. Although rockpile fields cover a greater proportion of the McClellan land, the ratio of habitation area to field area shows a somewhat greater per capita investment in this kind of cultivation by Marana residents.

FIRE AND AGRICULTURAL LANDSCAPES

A decisive role for wildfire in the evolution and maintenance of desert and plains grasslands and open forests in the American Southwest has become an important component of management philosophy and ecological thinking. Some scholars have long proposed that the former broad, brush-free grasslands and mature, open forests of the New World originated with repeated fires set by Native Americans as much as or more than from naturally occurring conflagrations (e.g., Sauer 1944, 1956; Stewart 1956). A few current environmental historians working in the western United States have expanded this thesis to implicate Native American burning in the overall development of North American ecosystem structure (e.g., Pyne 1982; Kay 1994). More moderate views acknowledge high lightning ignition rates in the Southwest in order to put the role of aboriginal burning into perspective as a residual mechanism in total fire ecology. Such a perspective is especially relevant for the Hohokam domain; the Coronado National Forest of southern Arizona leads southwestern forests in average acreage burned annually by lightning fires, despite a long history of fuel reduction through grazing (Bahre 1991:124). Because the historic era has witnessed depletion of ground cover and fine fuels through overgrazing and active programs of fire containment (e.g., Bahre 1985, 1995; McPherson 1995; Pyne 1984; Van Devender 1995), it is clear that prehistoric human fire ecology should be viewed in the framework of a more vigorous natural fire regime than at present (S. K. Fish 1997).

Ethnographic farming practices in the southern Southwest include targeted uses of fire in farming. Charred botanical remains in prehistoric fields, canals, and reservoirs provide evidence of fire used as a management tool in Hohokam agriculture (e.g., Miksicek 1984), although additional formation processes such as the unintentional incorporation of widespread domestic refuse may also account for burned residues recovered from these contexts. Vorsila Bohrer (1994) proposed a regionally transformed desert vegetation in response to the sweeping magnitude of Hohokam burning, based on her reinterpretation of botanical data from one locale that the original analysts interpreted as indicating anthropogenic plant communities in the site environs. It is useful to examine the probable scope and impacts of Hohokam burning in light of available information on ethnographic and prehistoric agroecology and use of wild resources.

In a cross-cultural survey of foraging societies, Lawrence Keeley (1995) found a correlation between use of fire for land management, dietary reliance on seeds and nuts, intensive manipulation of wild plants, and decreasing mobility. Indeed, this combination of attributes fits those historic groups who lived at the fringes of southwestern agriculturalists, such as the Shoshone and Paiute of the Great Basin (Steward 1938; Stewart 1939) and various groups of southeastern California (e.g., Lewis 1973; Bean and Saubel 1972), among whom the most systematic and complex ethnographic burning practices have been noted. It is likely that Late Archaic foragers with similar economies, closely pre-dating the transition to farming, were the premier landscape burners of Southwest prehistory.

Differential applications of fire should be expected over time as a function of changing land use; the zonal layout of Hohokam community territory offers a primary point of reference for modeling spatial variability during a single interval. Community boundaries encompassed the agricultural sustaining area for combined member settlements. Within the radius of well-watered and intensively cultivated land about these settlements, widespread or uncontrolled fire would have posed an unwelcome threat. Nearby village structures with highly flammable roofs contained household possessions, and more important, internal or adjacent facilities held food stores. During the growing season, crops in fields also would have been at risk.

Throughout the year, landscape-scale burning might have damaged other culturally patterned vegetational elements of intensively farmed locales. As in Pima agriculture (Rea 1981, 1983), Hohokam field edges and hedgerows probably consisted of herbaceous and woody plants that were selectively spared in clearing so that they might furnish firewood and other resources, while harboring a concentrated supply of small game. As is the case today, species such as mesquite would have grown along the outer borders of unlined canals, supported by extra moisture from continuous seepage. A mesic flora associated with canal banks and channel edges would have extended semi-aquatic and riparian resources into the zone of cultivation. Like traditional farmers in northern Sonora today (Sheridan and Nabhan 1978) and the Pima in historic times (Curtin 1984:108), the Hohokam likely planted living fencerows of cottonwood and willow to stabilize field edges fronting on floodplains and to deflect nutrient-rich flood flows safely over fields.

Uses of fire for localized and specific objectives within intensively farmed zones are clearly documented for aboriginal and traditional farmers of the southern Southwest. Piled brush was sometimes burned in the initial clearing of fields (Castetter and Bell 1942:125) and possibly to remove weedy growth in subsequent plantings. Canal channels were similarly cleaned at times. Burning to clear water delivery systems may have been extended to the dense growth of cienegas (localized marshes) and drainage bottoms in locations where canals and ditches headed; an additional goal would have been to enhance the productivity of desired species of damp habitats such as cattail and bulrush (Rea et al. 1983; Nabhan et al. 1983). In all of these cases, unrestrained fires would have been destructive to residential and agricultural precincts.

Although often at greater distances from villages than irrigated or floodwater land, zones of runoff farming in rockpile fields nevertheless would have entailed the need for contained burning. Agave has been documented as a principal cultigen, and cacti have been suggested as candidates for additional tended crops (S. K. Fish and Nabhan 1991). These crops are perennials, leaving no agricultural off-season during which the burning of natural vegetation or weeds could be freely implemented. Furthermore, succulents tend to be among the least fire-tolerant of desert plants (McLaughlin and Bowers 1982; McPherson 1995:137–140; Cave and Patten 1984).

The disturbed environs of prehistoric cultivation provide secondary resources in weedy plants (S. K. Fish and Nabhan 1991). The lower needs of farmers (compared with foragers) for weedy harvests may have been largely satisfied from field margins and fallow land. The heavy demands of Hohokam agriculture can be viewed as a counterbalance to equivalent investments beyond community boundaries. It is doubtful, for instance, that irrigation and floodwater cultivators would have had time or incentive to routinely burn at a regional scale, despite potential enhancement of grasses or other edible wild species. The most intensive Hohokam manipulations of wild plants would be anticipated near the margins of cultivation rather than at greater distances. Moreover, wild resources respond variably to fire. Grasses and other seed-bearing annuals may increase, but cacti, a prominent component of Hohokam subsistence, are typically damaged.

Vegetational effects at the broadest scale are often attributed to hunting (e.g., Stewart 1956, Kay 1994). Again, fire was widely employed for this purpose in the southern Southwest and northern Mexico (e.g., Dobyns 1981:42). Practices designed to control both fire and game are illustrated by the practice of setting fire in a circle when there was no wind so that it burned toward the center (ibid.:43). Hunters waited at the perimeter for game escaping outward. Despite such means as these for maintaining control, the Pima carried out fire drives in areas more distant from villages and fields than other forms of communal hunting (Rea 1979:114–115).

Throughout the Southwest (e.g., Minnis 1985; Kohler 1992; Betancourt and Van Devender 1981) and among the Hohokam (e.g., Miksicek 1984, 1995), there is archaeological evidence for occupational depletion of woody resources. Particularly during the centuries-long occupations of large Hohokam sites, residents of dry, shrubby desert basins must have had difficulty procuring domestic fuel. Cleared land and culturally modified vegetation in the vicinity of settlements would have lengthened the radius of travel to plentiful firewood supplies. Woody growth in riparian zones, hedgerows, canal environs, and the low-elevation woodlands of adjacent mountains must have been highly valued and carefully conserved. Repetitive harvesting of deadwood probably depressed natural fire frequencies in areas of heavy wood gathering. It is also probable that burning of fuel species would have been strongly avoided to protect hard-pressed resources. High natural incidences of lightning guaranteed the Hohokam commonplace access to taxa encouraged by fire succession. Additional targeted burning rather than broad-scale applications would have been conducive to a mosaic structure more diverse and abundant than natural vegetation, while preserving scarce wood and other vulnerable resources.

HOHOKAM IMPACTS IN LONG-TERM PERSPECTIVE

The Hohokam are noted for the most enduring settlements of the prehistoric Southwest. Renewal of fields through waterborne nutrients was fundamental to their achievement of a sustainable agriculture based on a combination of technologies from the most impressive irrigation systems of aboriginal North America to simple diversions of flooding streams. Domesticated crops were supplemented by a series of tended and weedy indigenous species of enhanced productivity in agricultural landscapes. The effects of intensive land use were concentrated in the vicinity of settlements and water sources. Surrounding areas of land-extensive cropping utilized water that otherwise would have escaped from desert ecosystems in rapid downstream runoff. Uninhabited land within and beyond community boundaries offered reserves of wild resources.

The massive canals of the Phoenix Basin, supporting what may well have been the densest populations of the prehistoric Southwest, made lasting imprints on the land. Their paths could still be largely traced across basin floors prior to accelerating land development in the early part of this century. Modern soil maps still delineate bands and swaths of alluvial sediments along the courses of ancient irrigation networks. Canal segments resurrected by Pima successors of the Hohokam in the nineteenth century produced surpluses sold to the United States Army, gold-rush travelers, and others crossing the Arizona deserts to California. Thereafter, entrepreneurs from the east revamped additional prehistoric canal lines for the beginnings of large-scale irrigated agriculture surrounding the present-day cities of Phoenix and Florence.

Hohokam struggles with the powerful surges of Sonoran Desert rivers are documented in parallel sets of adjacent canal lines. These attest to the ongoing need to repair, rebuild, and realign. Centuries later, less experienced desert irrigators unwittingly rerouted desert rivers using unyielding concrete dams, occasionally with disastrous results.

Hohokam floodwater cultivators lived from the bounty of storm-driven floods but were simultaneously subject to the unpredictable consequences of uncontrolled flows. As shallow drainages flooded and shifted across alluvial fans, they buried the houses of farmers in the same rich sediments that nourished their fields. Reoccupations at successively higher intervals in the accumulating stratigraphy of desert alluvial fans are a reminder that floodwater farmers were the beneficiaries of natural processes beyond their control.

Untold thousands of rockpiles still dotting southern Arizona slopes mark the outlying expanses of agricultural landscapes virtually unchanged since the era of Hohokam planting. Rockpiles and contour terraces rest undisturbed on modern surfaces, demonstrating a remarkable geomorphic stability from the time of their construction until the present. Thin layers of water-laid sand cover the intact ashy fill of roasting pits that the Hohokam preferentially excavated into the softer soils of tertiary drainages rather than the hard-packed ground to either side. Only the dense halos of modern annuals and perennials in and around prehistoric cobble features attest to their persisting status as enhanced microhabitats for desert plants.

Neolithic societies of the world's arid regions add a broader perspective for the environmental impacts of Hohokam agriculturalists. Unlike their counterparts in the Old World, the Hohokam were direct gatherers and consumers of desert vegetation, without domesticated animals as highly efficient, but ultimately destructive, harvesters of dispersed desert biomass. As a result, fragile Sonoran Desert ground covers were not comprehensively destroyed, nor was damage to perennial woody vegetation as far-reaching.

Hohokam continuity cannot be fully explained, however, by a versatile suite of productive and extractive techniques in desert settings. The integration of the residents of multiple settlements into community organization played a critical role in their ability to meet both short- and long-term subsistence challenges. Population in individual communities must have represented a successful balance between adequate and sustainable production under arid conditions and the numbers necessary to maintain canal networks or to spread agricultural risks over a sufficient number of environmental zones. The constraints of the Sonoran Desert environment did not preclude social and economic innovation by the Hohokam, despite their continued dependence on the same repertoire of farming methods. Changing configurations of population, settlement, and land use continued to appear as dynamic and evolving societies implemented new combinations and intensities of existing techniques.

REFERENCES

Abruzzi, W. S. 1989. Ecology, resource distribution, and Mormon settlement in northeastern Arizona. *American Anthropologist* 91: 642–655.

Ackerly, N., J. B. Howard, and R. H. McGuire. 1987. *La Ciudad Canals: A Study of Hohokam Irrigation Systems at the Community Level*. Anthropological Field Studies 17. Tempe: Department of Anthropology, Arizona State University.

Bahre, C. J. 1985. Wildfire in southeastern Arizona between 1859 and 1890. *Desert Plants* 7: 190–194.

———. 1991. *Legacy of Change: Historic Human Impact on Vegetation in the Arizona Borderlands*. Tucson: University of Arizona Press.

———. 1995. Human impacts on the grasslands of southeastern Arizona. In M. P. McClaran and T. R. Van Devender, eds., *The Desert Grassland*, pp. 196–229. Tucson: University of Arizona Press.

Bean, L. J., and K. S. Saubel. 1972. *Temalpakh: Cahuilla Indian Knowledge and Usage of Plants*. Banning, Calif.: Malki Museum Press.

Betancourt, J. L., and T. R. Van Devender. 1981. Holocene vegetation in Chaco Canyon, New Mexico. *Science* 214: 656–658.

Bohrer, V. L. 1970. Paleoecology of Snaketown. *Kiva* 36: 11–19.

———. 1991. Recently recognized cultivated and encouraged plants among the Hohokam. *Kiva* 56: 227–236.

———. 1994. New life from ashes 2: A tale of burnt brush. *Desert Plants* 10: 122–126.

Bryan, K. 1929. Floodwater farming. *Geographical Review* 19: 444–456.

Bye, R. 1979. Incipient domestication of mustards in northwestern Mexico. *Kiva* 44: 237–256.

Castetter, E. F., and W. H. Bell. 1942. *Pima and Papago Indian Agriculture*. Albuquerque: University of New Mexico Press.

Cave, G. H., and D. T. Patten. 1984. Short-term vegetation responses to fire in the Upper Sonoran Desert. *J. Range Management* 37: 491–496.

Crosswhite, F. 1981. Desert plants, habitat, and agriculture in relation to the major pattern of cultural differentiation in the O'odham people of southern Arizona. *Desert Plants* 3: 47–76.

Crown, P. 1987. Classic period Hohokam settlement and land use in the Casa Grande ruin area, Arizona. *J. Field Archaeology* 14 (2): 147–162.

Curtin, L. S. M. 1984. *By the Prophet of the Earth: Ethnobotany of the Pima*. Tucson: University of Arizona Press.

Dart, A. 1986. Sediment accumulation along Hohokam canals. *Kiva* 51: 63–84.

Dean, J. S., M. C. Slaughter, and D. O. Bowden. 1996. Desert dendrochronology: Tree-ring dating prehistoric sites in the Tucson Basin. *Kiva* 62: 7–26.

Dobyns, H. E. 1981. *From Fire to Flood: Historic Destruction of Sonoran Desert Riverine Oases.* Anthropology Papers 20. Socorro, N.Mex.: Ballena Press.

Doelle, W. 1980. *Past Adaptive Patterns in Western Papagueria: An Archaeological Study of Nonriverine Resource Use.* Ph.D. dissertation, Department of Anthropology, University of Arizona, Tucson.

Doolittle, W. E. 1990. *Canal Irrigation in Prehistoric Mexico: The Sequence of Technological Change.* Austin: University of Texas Press.

Downum, C., P. Fish, and S. Fish. 1994. Refining the role of the Cerros de Trincheras in southern Arizona settlement. *Kiva* 59: 271–296.

Doyel, D. E. 1976. Classic period Hohokam in the Gila Basin. *Kiva* 43: 27–38.

———. 1980. Hohokam social organization and the Sedentary to Classic period transition. In D. Doyel and F. Plog, eds., *Current Issues in Hohokam Prehistory,* pp. 23–40. Anthropological Research Paper 23. Tempe: Arizona State University.

———. 1991. Hohokam cultural evolution in the Phoenix Basin. In G. J. Gumerman, ed., *Exploring the Hohokam: Prehistoric Desert Peoples of the Southwest,* pp. 133–161. Albuquerque: University of New Mexico Press.

Fish, P. R. 1989. The Hohokam: 1,000 years of prehistory in the Sonoran Desert. In L. S. Cordell and G. J. Gumerman, eds., *Dynamics of Southwestern Prehistory,* pp. 19–63. Washington, D.C.: Smithsonian Institution Press.

Fish, P. R., and S. K. Fish. 1991. Hohokam political and social organization. In G. J. Gumerman, ed., *Exploring the Hohokam: Prehistoric Desert Peoples of the Southwest,* pp. 84–101. Albuquerque: University of New Mexico Press.

Fish, P. R., and S. K. Fish, C. Brennan, D. Gann, and J. Bayman. 1992. Marana: Configuration of a Hohokam platform mound site. In R. Lange and S. Germick, eds., *Proceedings of the Second Salado Conference, Glob, Arizona, 1992,* pp. 62–68. Phoenix: Arizona Archaeological Society.

Fish, S. K. 1984. The modified environment of the Salt-Gila Aqueduct Project sites: A palynological perspective. In L. Teague and P. Crown, eds., *Hohokam Archaeology along the Salt-Gila Aqueduct, vol. 7: Environment and Subsistence,* pp. 39–51. Archaeological Series 150. Tucson: Arizona State Museum, University of Arizona.

———. 1985. Prehistoric disturbance floras of the lower Sonoran Desert and their implications. In B. F. Jacobs, P. L. Fall, and O. K. Davis, eds., *Late Quaternary Vegetation and Climate in the American Southwest,* pp. 77–88. Contribution Series 16. Houston, Tex.: American Association of Stratigraphic Palynologists.

———. 1996. Dynamics of scale in the southern deserts. In P. Fish and J. J. Reid, eds., *Interpreting Southwestern Diversity: Underlying Principles and Overarching Patterns,* pp. 107–114. Anthropological Research Papers 49. Tempe: Department of Anthropology, Arizona State University.

———. 1997. Modeling human impacts to Borderlands environment from a fire ecology perspective. In H. Hambre, ed., *Effects of Fire on the Madrean Province Ecosystems,* pp. 125–134. Fort Collins, Colo.: USDA Rocky Mountain Forest and Range Experiment Station.

Fish, S. K., and P. R. Fish. In press. Unsuspected magnitudes: The scale of Hohokam agriculture. In C. L. Redman, S. James, P. Fish, and J. D. Rogers, eds., *Human Impact on the Environment: An Archaeological Perspective.* Washington, D.C.: Smithsonian Institution Press.

Fish, S. K., and Gary P. Nabhan. 1991. Desert as context: The Hohokam environment. In G. P. Gumerman, ed., *Exploring the Hohokam: Prehistoric Desert Peoples of the American Southwest,* pp. 35–54. Albuquerque: University of New Mexico Press.

Fish, S. K., P. R. Fish, and C. Downum. 1984. Hohokam terraces and agricultural production in the Tucson Basin. In S. Fish and P. Fish, eds., *Prehistoric Agricultural Strategies in the Southwest,* pp. 55–71. Anthropological Research Paper 33. Tempe:

Department of Anthropology, Arizona State University.

Fish, S. K., P. R. Fish, C. H. Miksicek, and J. H. Madsen. 1985. Prehistoric agave cultivation in southern Arizona. *Desert Plants* 7(2): 107–112.

Fish, S. K., P. R. Fish, and J. H. Madsen. 1990. Analyzing regional agriculture: A Hohokam example. In S. Fish and S. Kowalewski, eds., *The Archaeology of Regions: The Case for Full-Coverage Survey,* pp. 189–218. Washington, D.C.: Smithsonian Institution Press.

———. 1992. *The Marana Community in the Hohokam World.* Anthropological Papers 56. Tucson: University of Arizona Press.

Gasser, R. E. 1982. Hohokam use of desert plant foods. *Desert Plants* 3: 216–234.

Gasser, R. E., and S. M. Kwiatkowski. 1991. Food for thought: Recognizing patterns in Hohokam subsistence. In G. P. Gumerman, ed., *Exploring the Hohokam: Prehistoric Desert Peoples of the American Southwest,* pp. 417–460. Albuquerque: University of New Mexico Press.

Gregory, D. A. 1987. The morphology of platform mounds and the structure of Classic Hohokam sites. In D. E. Doyel, ed., *The Hohokam Village: Site Structure and Organization,* pp. 183–210. Glenwood Springs, Colo.: Southwestern and Rocky Mountain Division of the American Association for the Advancement of Science.

Gregory, D. A., and F. Nials. 1985. Observations concerning the distribution of Classic period Hohokam platform mounds. In A. E. Dittert Jr. and D. E. Dove, *Proceedings of the 1983 Hohokam Symposium,* pp. 373–388. Occasional Paper 2. Phoenix: Arizona Archaeological Society.

Haury, E. W. 1976. *The Hohokam: Desert Farmers and Craftsmen.* Tucson: University of Arizona Press.

Hendrickson, D. A., and W. L. Minckley. 1984. Cienegas: Vanishing climax communities of the American Southwest. *Desert Plants* 6: 131–175.

Hillman, G. C., and D. R. Harris. 1989. *Foraging and Farming: The Evolution of Plant Exploitation.* London: Unwin Hyman.

Howard, J. B. 1993. A paleohydraulic approach to examining agricultural intensification in Hohokam irrigation systems. In V. L. Scarborough and B. Isaac, eds., *Economic Aspects of Water Management in the Prehispanic New World,* pp. 263–324. Research in Economic Anthropology Supplement 7. Greenwich, Conn.: JAI Press.

Huckell, B. B. 1996. *Of Marshes and Maize: Preceramic Agricultural Settlements in the Cienega Valley.* Anthropological Papers 59. Tucson: University of Arizona Press.

Kay, C. E. 1994. Aboriginal overkill: The role of Native Americans in structuring western ecosystems. *Human Nature* 5: 359–398.

Keeley, L. H. 1995. Protoagricultural practices among hunter-gatherers: A cross-cultural survey. In T. Price and A. Gebauer, eds., *Last Hunters, First Farmers,* pp. 243–272. Santa Fe, N.Mex.: School for American Research Press.

Kohler, T. A. 1992. Prehistoric human impact on the environment in the upland North American Southwest. *Population and Environment* 13: 255–268.

Kohler, T. A., and M. H. Matthews. 1988. Long-term Anasazi land use and forest reduction: A case study from southwest Colorado. *American Antiquity* 53: 537–564.

Lewis, H. T. 1973. *Patterns of Indian Burning in California: Ecology and Ethnohistory.* Anthropological Papers 1. Ramona, Calif.: Ballena Press.

Lightfoot, K. G., and F. Plog. 1984. Intensification along the north side of the Mogollon Rim. In S. Fish and P. Fish, eds., *Prehistoric Agricultural Strategies in the Southwest,* pp. 79–95. Anthropological Research Papers 33. Tempe: Department of Anthropology, Arizona State University.

Mabry, J. B., ed. 1998. *Archaeological Investigations of Early Village Sites in the Middle Santa Cruz Valley.* Anthropological Papers 19. Tucson, Ariz.: Center for Desert Archaeology.

Masse, W. B. 1981. Prehistoric irrigation systems in the Salt River valley, Arizona. *Science* 214: 408–415.

McGinnies, W. 1981. *Discovering the Desert: Legacy of the Carnegie Desert Botanical Laboratory.* Tucson: University of Arizona Press.

McLaughlin, S. P., and J. E. Bowers. 1982. Effects of wildfire on a Sonoran Desert plant community. *Ecology* 63: 246–248.

McPherson, G. R. 1995. The role of fire in the desert grasslands. In M. McClaran and T. Van Devender, eds., *The Desert Grassland,* pp. 130–151. Tucson: University of Arizona Press.

Miksicek, C. 1984. Historic desertification, prehistoric vegetation change, and Hohokam subsistence in the Salt-Gila Basin. In L. Teague and P. Crown, eds., *Hohokam Archaeology along the Salt-Gila Aqueduct, vol. 7: Environment and Subsistence,* pp. 53–80. Archaeological Series 150. Tucson: Arizona State Museum, University of Arizona.

———. 1988. Rethinking Hohokam paleoethnobotanical assemblages: A progress report for the Tucson Basin. In W. Doelle and P. Fish, eds., *Recent Research on Tucson Basin Prehistory,* pp. 47–56. Anthropological Papers 10. Tucson, Ariz.: Institute for American Research.

———. 1995. Temporal trends in the eastern Tonto Basin: An archaeobotanical perspective. In M. Elson and J. Clark, *The Roosevelt Community Development Study, vol. 3: Paleobotanical and Osteological Analyses,* pp. 43–84. Anthropological Papers 14. Tucson, Ariz.: Center for Desert Archaeology.

Minckley, W. L., and T. O. Clark. 1984. Formation and destruction of a Gila River mesquite bosque community. *Desert Plants* 6: 23–30.

Minnis, P. E. 1985. *Social Adaptation to Food Stress: A Prehistoric Southwestern Example.* Chicago: University of Chicago Press.

Nabhan, G. P. 1979. The ecology of floodwater farming in southwestern North America. *Agro-ecosystems* 5: 245–255.

———. 1986. Ak-chin "arroyo mouth" and the environmental setting of the Papago Indian fields in the Sonoran Desert. *Applied Geography* 6: 61–75.

Nabhan, G. P., A. M. Rea, K. L. Reichhardt, E. Melink, and C. F. Hutchinson. 1983. Papago influences on habitat and biotic diversity: Quitovac Oasis ethnoecology. *J. Ethnobiology* 2: 124–143.

Nials, F., D. Gregory, and D. Graybill. 1989. Salt River streamflow and Hohokam irrigation systems. In D. Gregory, W. Deaver, S. Fish, R. Gardiner, R. Layhe, F. Nials, and L. Teague, eds., *The 1982–1984 Excavations at Las Colinas: The Site and Its Features,* pp. 275–306. Archaeological Series 162. Tucson: Arizona State Museum, University of Arizona.

Nicholas, L., and G. M. Feinman. 1989. A regional perspective on Hohokam irrigation in the lower Salt River Valley, Arizona. In S. Upham, K. G. Lightfoot, and R. Jewett, eds., *The Socio-political Structure of Prehistoric Southwestern Societies,* pp. 199–236. Boulder, Colo.: Westview Press.

Pyne, S. J. 1982. *Fire in America.* Princeton: Princeton University Press.

———. 1984. *Introduction to Wildfire.* New York: John Wiley & Sons.

Rea, A. M. 1979. Hunting lexemic categories of the Pima Indians. *Kiva* 44: 113–119.

———. 1981. The ecology of Pima fields. *Environment Southwest* 484: 8–15.

———. 1983. *Once a River: Bird Life and Habitat Changes on the Middle Gila.* Tucson: University of Arizona Press.

———. 1997. *At the Desert's Green Edge: An Ethnobotany of the Gila River Pima.* Tucson: University of Arizona Press.

Rea, A. M., G. P. Nabhan, and K. L. Reichhardt. 1983. Sonoran Desert oases: Plants, birds, and native people. *Environment Southwest* 503: 5–9.

Sauer, C. O. 1944. A geographic sketch of early man. *Geographical Review* 34: 529–573.

———. 1956. The agency of man on the earth. In W. L. Thomas, ed., *Man's Role in Changing the Face of the Earth,* pp. 49–69. Chicago: University of Chicago Press.

Schroeder, A. 1960. *The Hohokam, Sinagua, and the Hakataya.* Archives of Archaeology 5. Salt Lake City, Utah: Society for American Archaeology.

Sellers, W. D., R. H. Hill, and M. Sanderson-Rae. 1985. *Arizona Climate: The First Hundred Years.* Tucson: Institute of Atmospheric Physics, University of Arizona.

Sheridan, T., and G. P. Nabhan. 1978. Living with a river: Traditional farmers of the Rio San Miguel. *J. Arizona History* 19: 1–17.

Sires, E. W. 1987. Hohokam architectural

variability and site structure during the Sedentary-Classic transition. In D. Doyel, ed., *The Hohokam Village: Site Structure and Organization*, pp. 171–182. Glenwood Springs, Colo.: Southwest and Rocky Mountain Division of the American Association for the Advancement of Science.

Smith, B. D. 1992. *Rivers of Change: Essays on Early Agriculture in Eastern North America*. Washington, D.C.: Smithsonian Institution Press.

Steward, J. H. 1938. *Basin-Plateau Aboriginal Sociopolitical Groups*. Bureau of American Ethnology Bulletin 120. Washington, D.C.: GPO.

Stewart, O. C. 1939. The Northern Paiute Bands. *University of California Anthropological Records* 2: 127–149.

———. 1956. Fire as the first great force employed by man. In W. L. Thomas, ed., *Man's Role in Changing the Face of the Earth*, pp. 115–133. Chicago: University of Chicago Press.

Szuter, C. R., and F. E. Bayham. 1989. Sedentism and animal procurement among desert horticulturalists in the North American Southwest. In S. Kent, ed., *Farmers as Hunters: The Implications of Sedentism*, pp. 80–95. Cambridge: Cambridge University Press.

Szuter, C. R., and W. B. Gillespie. 1994. Interpreting use of animal resources at prehistoric American Southwest communities. In R. Leonard and W. W. Wills, eds., *The Ancient Southwest Community*, pp. 67–76. Albuquerque: University of New Mexico Press.

Thomas, W. L. 1956. *Man's Role in Changing the Face of the Earth*. Chicago: University of Chicago Press.

Turner, R. H., and D. E. Brown. 1994. Tropical-subtropical desertlands. In D. E. Brown, ed., *Biotic Communities: Southwestern United States and Northwestern Mexico*, pp. 181–221. Salt Lake City: University of Utah Press.

Van Devender, T. R. 1995. Desert grassland history: Changing climates, evolution, biogeography, and community dynamics. In M. McClaran and T. Van Devender, eds., *The Desert Grassland*, pp. 68–99. Tucson: University of Arizona Press.

Vander Wall, S. B., and J. A. MacMahan. 1984. Avian distribution patterns along a Sonoran Desert bajada. *J. Arid Environments* 7: 59–74.

Whiting, A. E. 1939. *The Ethnobotany of the Hopi*. Bulletin 15. Flagstaff: Museum of Northern Arizona.

Wilcox, D., and L. Shenk. 1977. *The Architecture of the Casa Grande and Its Interpretation*. Archaeological Series 115. Tucson: Arizona State Museum, University of Arizona.

Wilcox, D., and C. Sternberg. 1983. *Hohokam Ballcourts and Their Interpretation*. Archaeological Series 160. Tucson: Arizona State Museum, University of Arizona.

JAMES L. LUTEYN
STEVEN P. CHURCHILL

11

VEGETATION OF THE TROPICAL ANDES
An Overview

The mythical El Dorado of the New World tropics, as first conceived, was correctly thought to exist in the highlands; however, it was not gold, but green. The Andes harbor one of the richest terrestrial biota on earth. This region does, or at least did, contain the greatest plant diversity found in the tropics and possibly, for the size of the area, the world.

The Andes of South America stretch over 70 degrees of latitude, some 8,000 km, from Venezuela to the southern tip of the continent, representing the longest mountain range in the world spanning more vegetation types than anywhere else. Within this range the Andes must be further subdivided so as to take into consideration the climatic differences between the tropical and nontropical parts. Broadly speaking, these areas are (1) the Northern Andes, which range from about 11°N to about 4–5°S latitude, are wholly within the tropics, and are generally quite wet on both sides of the cordilleras, (2) the Central Andes, lying between about 5°S and 27°S latitude, which contrast in their greater width and asymmetry of precipitation (here the west coast and highlands of southern Peru and Bolivia are predominantly dry due to the influence of the cold Humboldt current, while the northern Peruvian highlands and eastern slopes of the Andes are humid), and (3) the Southern Andes, which range from about 28°S to 55°S latitude, falling within the temperate zone, with temperatures decreasing toward the south, and which will not be discussed further. For the purposes of this paper the Northern and Central Andes are referred to as the tropical Andes. This region extends approximately 35 degrees of latitude from the northern reaches of the Sierra Nevada de Santa Marta (Colombia) and the Sierra Nevada de Mérida (Venezuela) to northernmost Argentina.

Topographically, the Northern Andes arise in Venezuela in the form of two mountain ranges from the Caribbean Sea running southward and parallel to each other on either side of Lake Maracaibo. They merge and become part of the Colombian Cordillera Oriental just inside the border with Colombia. Farther south the Cordillera Oriental merges with the parallel ranges of the Cordillera Central and Cordillera Occidental at the Nudo de Pasto in southernmost Colombia. Two major river valleys are separated by Colombia's three cordilleras: the Río Cauca lies between the Cordillera Occidental and Central, and the Río Magdalena between the Cordillera Central and Oriental. At the border with Ecuador, two parallel mountain ranges are reformed and run south through Ecuador with an inter-Andean valley (or a series of basins) mostly above 2,000 m elevation between them. These two ranges merge again in southern Ecuador near Loja. Geologically, the Huancabamba Depression separates the Northern and Central Andes as a whole. This depression is a major dispersal barrier for a limited number of both montane animals and plants (Duellman 1979). Three ranges form again in northern Peru only to converge again at the Nudo de Pasco. From there three mountain chains again arise, which come together at the Nudo de Vilcanota. Farther south the Andes diverge into two major ranges that run parallel to and enclose the altiplano of Peru, Bolivia, and northern Chile and Argentina. From there southward to the tip of the South American continent, the Andes form a single, narrow range.

This great mountain system, with its many parallel and transverse cordilleras, lateral valleys, equatorial position, tropical rain climate, and mostly favorable growing conditions, creates numerous ecological conditions that give rise to varying vegetational types (figure 11.1), which in turn provide the basis for the great richness and diversity of the flora. There is no single source of information on the vegetation of the Andes. There are, however, a number of general references that provide important summaries or overviews on major topics. The first substantial general treatment of the Andes was *Geoecology of the Mountainous Regions of the Tropical Americas*, edited by Troll (1968a). (More recent summaries may be seen in Little 1981, Molina and Little 1981, Brush 1982, Seibert 1983, and Davis et al. 1997.) Neotropical montane forests have been examined in general terms by Davis et al. (1997), in terms of cloud forest ecology throughout the tropics by Hamilton et al. (1995), and plant diversity by Churchill et al. (1995a). Four recent books treat the Neotropical high-elevation vegetation in terms of plant function and growth forms (Rundel et al. 1994), general aspects of the páramo, especially human influence (Balslev and Luteyn 1992), biogeography (Vuilleumier and Monasterio 1986), and diversity and geographical distribution (Luteyn 1999). General information on floristics and vegetation of individual countries includes discussions for Venezuela (Pittier and Williams 1945; Tamayo 1958, 1979; Hueck 1960; Ewel et al. 1976; Huber and Frame 1989; Bono 1996), Colombia (Cuatrecasas 1957, 1958; Espinal and Montenegro 1963; Instituto Geográfico Agustín Codazzi 1977; Forero 1989), Ecuador (Sodiro 1874; Wolf 1892; Diels 1937; Acosta-Solís 1966, 1977; Harling 1979; Cañadas Cruz 1983; Jørgensen et al. 1995; Jørgensen and León 1999), Peru

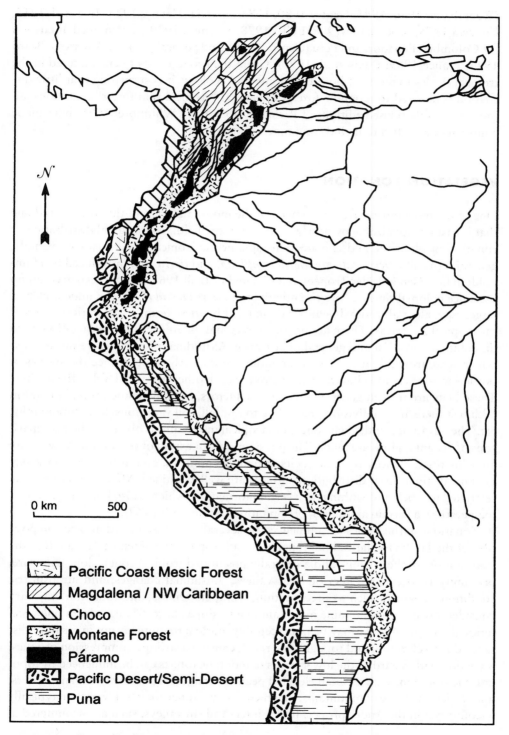

FIGURE 11.1. Generalized vegetation of the Andes and Pacific coastal region.

(Weberbauer 1936, 1945; Gentry 1980, 1993), Bolivia (Herzog 1923; Tosi et al. 1975; Unzueta 1975; Brockmann 1978; ERTS 1978; Solomon 1989), a combined discussion for Colombia, Ecuador, and Peru (Gentry 1989), and generally for the Americas (Beard 1944; Prance 1989; Davis et al. 1997). Given the broad diversity encountered in the tropical Andes in terms of both taxonomy and ecology, it is surprising that little discussion is devoted to montane forests in general "tropical rain forest" textbooks, although recently revised editions of Richards (1996) and Whitmore (1998) have made a more modest effort to include montane forests.

VEGETATION ZONATION

One of the most impressive phenomena in the mountain areas of the tropics, and one that is easily recognizable by nearly everyone, is the beltlike altitudinal distribution or zonation of plant life and climate. Throughout the tropical Americas, for example, general terms like *tierra caliente, tierra templada,* and *tierra fria,* introduced by Humboldt in his "Gemälde der Tropenländer" (cited in Troll 1968b), have been in common use since colonial times. More recently the term *tierra helada* has been added. By and large, equivalent altitudinal belts are designated by these terms in the various parts of the tropical Americas. There are of course regional differences in the vegetation, with different composition and general appearance. Altitudinal transitions are not marked but instead are gradual or at least oftentimes not clearly limited since the species of bordering zones readily intermingle or overlap each other from sea level to the treeline. These belts are framed generally by climatic factors, notably temperature, but within each belt there are subdivisions according to rainfall, soil variations, local topography, etc. The advantage of the belt concept in tropical mountains is obvious when comparisons of distant regions are made. The physiognomy or aspect of vegetation shows many amazing resemblances in similar belts of remote countries, even if the taxonomy may be quite distinct. Even distribution patterns are well explained. Many plants are quite restricted to a belt and within it to a certain formation, which is also limited by, besides temperature, a certain amount of precipitation (Hastenrath 1968).

Zonations in the Andes are influenced by several variables, including size and position of the landmass, latitudinal range (i.e., subtropical to subtemperate at either the northern or southern limits and a centrally positioned tropical zone), slope exposure, proximity to oceanic waters or river valleys, altitudinal changes, and atmospheric conditions. Features such as precipitation, wind, cloudiness, and radiation may vary considerably even on the same mountain due to slope effect. Changes from one vegetational zone to another may be imperceptibly gradual to quite abrupt in terms of distance due to elevation and topographic relief, causing a striking compression of zones. Variation with regard to the lower limits of montane forests can be influenced by topographic isolation, such as small elevated peaks or ranges, proximity to oceanic waters, and the leeward side of mountains, in which case the range may be lower than typically encountered in the Andes proper. In dry inter-Andean valleys, such as encountered in

Colombia and Ecuador, montane forests tend to be higher and zonations compressed. Elevation has been employed, in part or full, in defining Neotropical montane vegetation (Cuatrecasas 1958; Holdridge 1967; Holdridge et al. 1971). The biotic and abiotic variables cited above have contributed to a lack of agreement on a vegetational classification for montane forests among biologists. Unfortunately, these and others of the various systems have been mostly too local in application and have often blurred reality with a plethora of terms (see Stadtmüller 1987). The vegetation types of the tropical Andes and the terminology used to categorize them vary according to the author. They are so complex that this essay gives only generalities and not details. For more in-depth study of terminology as it relates to Andean vegetation, one should consult Huber and Riina 1997, or for details of life-zone classifications, one should consult Holdridge 1967 and Holdridge et al. 1971.

In this essay, we characterize the overall Andean vegetation in a simple way that uses physiognomic tendencies rather than actual discrete zones. Furthermore, the use of elevation is employed in a relative manner to describe the altitude at which a particular vegetation zone is generally found, thus a range of elevations is often given. These overall tendencies have been applied throughout mountainous regions of the tropics by others (Grubb 1974; Leigh 1975) and are herein summarized in a very general way in table 11.1. In the Northern Andes, the vegetation may be discussed under the general terms *montane forest, high-elevation vegetation, inter-Andean vegetation,* and *coastal forests.* The montane forest zone includes the Pacific, Caribbean, and Amazonian Andean slopes in Venezuela, Colombia, Ecuador, and eastern Peru and Bolivia. In southern Ecuador and Peru, however, the Pacific Andean slopes are much drier than the Amazonian slopes and do not support such a rich forest. The transition from the desert coastal regions to the higher montane vegetation is often marked by grass steppes or by shrub-herb zones that are green only in the rainy season. The lowland rain forests (or tierra caliente) and the Caribbean coast forests are covered by Daly and Mitchell (this volume). This essay emphasizes the tropical montane regions, i.e., the Northern and Central Andes and the desert coastal region.

MONTANE FOREST

In general, montane forests may be found at their lowest range from about 600 to 1,500 m and extend to a range from 3,000 to 3,500 (−4,300) m. It may be said that among the montane forests of tropical America, the Andean forests are more diverse than either those of Mexico or Central America (Gentry 1995a) and that lowland diversity with regard to woody tree species is in general significantly higher than in montane forests. What is truly characteristic about montane forest, however, is the significant number of taxa and increase in biomass contributed by epiphytes, e.g., bryophytes (liverworts and mosses), ferns and allies (particularly Lycopodiaceae), and flowering plants (primarily bromeliads and orchids). Strictly speaking, epiphytes are plants that germinate on surfaces of other plants and pass their entire life cycles without becoming

TABLE 11.1. COMPARISON OF THE SALIENT FEATURES OF THE ANDEAN VEGETATION BELTS

Formation Type	Tierra Caliente (lowland rain forest)	Tierra Templada (lower montane rain forest) Subandean Rain Forest	Tierra Fría (upper montane rain forest) Cloud Forest; Ceja	Tierra Helada (páramo)
Altitudinal range	0–1,000 m	1,000–2,500 m	2,500–3,500 m	(3,200–) 3,800–4,700 m
Average yearly temp.	23–30°C	23–16°C	15–6°C	10–1°C
Rainfall (mm)	1,800–10,000	1,000–5,000	1,000	700–2,500
Height of forest	25–45 (80) m	15–33 (45) m	1.5–18 (25) m	1.5–9 (15) m
Strata	three	two	one	one
Soils	alluvial	volcanic	humic	humic
Dominant leaf-size class of trees and shrubs	mesophyll (4,500–18,225 mm²)	notophyll (2,000–4,500 mm²) or mesophyll (2,025–18,225 mm²)	microphyll (225–2,025 mm²)	nanophyll (25–225 mm²)
Buttresses and stilt roots on trees	usually frequent and large	infrequent or small or both	usually none	none
Cauliflorous tree species	frequent	rare	none	none
Compound leaves on trees	abundant	occasional	few	none
Drip tips on leaves	abundant	frequent or occasional	few or none	none
Climbers	thick-stemmed woody species frequent	thick-stemmed woody species usually none; other species often frequent	usually very few	very few
Vascular epiphytes	frequent	abundant	frequent	occasional
Nonvascular epiphytes	occasional	occasional to abundant	very abundant	occasional
Important woody families	Leguminosae, Moraceae, Meliaceae, Lauraceae, Myristicaceae, Lecythidaceae, Arecaceae	Melastomataceae, Cunoniaceae, Betulaceae, Ericaceae	Rosaceae, Orchidaceae, Rubiaceae, Melastomataceae, Ericaceae, Cunoniaceae	Asteraceae, Poaceae, Orchidaceae, Melastomataceae, Ericaceae

connected to the ground. At least 84 vascular plant families totaling more than 23,000 species contain epiphytes (Madison 1977; Kress 1986) and virtually all are animal pollinated.

Several recent papers that give general discussions of montane habitats may be found for Colombia (Rangel and Garzón 1997a–d), Ecuador (Valencia and Jørgensen 1992; Mena et al. 1997; Neill and Palacios 1997), Ecuador and Peru (Young and Reynel 1997), Peru (Dillon et al. 1995; Young and León 1997), and Bolivia (Kessler 1995). Regional overviews may be found in Smith and Johnston 1945, Churchill et al. 1995a and papers therein, Gentry 1997, and Webster 1995.

LOWER MONTANE FOREST

This is the transitional zone between lowland and highland rain forest, the tierra templada of table 11.1. The elevational range for lower montane forest is found generally in the Andes between 900–1,000 m and 2,000–2,500 m. Physiognomically, lower montane forest trees, especially at the lower elevational range, are of virtually the same stature as those in the adjoining lowlands. Climatically, lower montane forest is often similar to the lowlands in terms of precipitation (2,000–4,000 mm per year), but due to increasing elevations it is cooler, and frequently rainfall patterns may be much higher because of orographic rains as one encounters more cloudiness and fog, such as in the area around Puya (eastern Ecuador), where rainfall measures up to 5,000 mm per year. Soils are usually richer in nutrients due to the outwash from the mountain streams and rivers.

Trees of the lower montane forest exhibit a strong lowland resemblance but have fewer species with prop roots and lianas. Typical wet lowland families such as Chrysobalanaceae (*Licania*), Lecythidaceae (*Eschweilera*), Myristicaceae, and Vochysiaceae, although occurring less frequently, extend into the initial lower montane forest. Some plants common to the lowlands, such as *Monstera* (Araceae), many Musaceae, Zingiberaceae, and Marantaceae, the palms *Astrocaryum, Bactris, Iriartea,* and *Phytelephas,* the Cyclanthaceae genera *Cyclanthus* and *Carludovica palmata,* and *Triplaris* (Polygonaceae), have their upper limits in the lower montane forest. Among the diverse families of woody plants, Lauraceae (*Nectandra* and *Ocotea*) is one of the most species-rich, followed by Melastomataceae (*Merania, Miconia, Tibouchina*) and Rubiaceae (*Cinchona, Ladenbergia, Psychotria*) and, not far behind, the Moraceae (*Cecropia, Ficus*) (Gentry 1995a). Typical tree species include *Dacroydes cupularis* (Burseraceae), *Metteniusa tessmanniana* (Icacinaceae), and *Catoblastus praemorsus* and *Dictyocaryum lamarckianum* (Arecaceae). Additional families and genera contributing to species richness include Arecaceae, Hypericaceae (including Clusiaceae and Guttiferae—*Clusia, Tovomita, Vismia*), Euphorbiaceae (*Alchornea*), Flacourtiaceae (*Banara*), Lauraceae (*Persea*), Leguminosae (*Calliandra*), Meliaceae (*Cedrela, Guarea*), Sapindaceae (*Cupania, Paullinia*), and Sapotaceae. The understory often consists of bamboos (*Bambusa*), shrubs in the Melastomataceae (*Miconia, Monochaetum, Merania, Tibouchina*), Piperaceae, and Rubiaceae, and epiphytic orchids, aroids, and ferns. At 2,000–2,500 m characteristic genera include *Brunellia, Cedrela* and *Guarea,*

Chrysochlamys, Clethra, Dendropanax and *Oreopanax, Drimys, Escallonia, Freziera* and *Laplacea, Hedyosmum, Hieronyma, Ilex, Matisia, Maytenus, Miconia* and *Topobea, Myrica, Panopsis* and *Roupala, Rhamnus, Sauraria, Symplocos, Viburnum,* and *Weinmannia* (Cuatrecasas 1958). Tree ferns and hemiepiphytes are also more prevalent growth forms. *Podocarpus* used to be widespread, but little of this forest type exists today.

The lower montane forest regions are the zones of high agricultural activity. In wet areas one encounters the cultivation of bananas, citrus, tea, rice, and sugarcane (all exotic crop plants). In the higher portions, one encounters the cultivation of corn, coffee, and coca (in Peru and Bolivia). It was once known as the cinchona zone, for here too grow naturally many of the original species of the genus *Cinchona* (Rubiaceae), which produce the bark from which quinine was made.

UPPER MONTANE FOREST

In general, upper montane forests (tierra fria of table 11.1) commence midway up any particular mountain range, reaching upward to the shrubby timberline. Elevation ranges are variable, but generally they occur from about 2,000 (–2,500) to 3,500 (–4,300) m. Climatically, the upper montane forests are cooler, cloudier, and foggier, with precipitation more often coming in the form of mist rather than rain due to the humidity gathered in clouds that are formed by the daily upslope winds. This is the primary reason for the luxuriant growth in epiphytes, the dominant life-form here, consisting of bryophytes, lichens, filmy ferns (which often cover tree trunks and branches or lie thickly underfoot), and the vascular epiphytes, especially orchids and bromeliads. Also in this area the tall tree ferns are abundant. With increasing altitude, trees are successively smaller, and fewer numbers of species are dominant. At the same time, the forest canopy becomes shorter and flatter, the trees more slender, usually with gnarled or contorted limbs, lower branching, and very compact dense crowns. Dead trunks are frequent, and decomposition in general is markedly reduced, resulting in an increased accumulation of leaf litter on the forest floor. Understory trees and shrubs are small, with poorly formed crowns on slender, crooked trunks, and leaves become smaller, harder or coriaceous, and often darker green. There are no visible strata, and the forest has a more open aspect. The decrease in species richness of trees is replaced by a greater tendency for few species to be dominants.

Cloud forest is first encountered on exposed summits and narrow ridge crests that are open to the prevailing winds, but upward it covers the entire landscape. Cloud forest is not connected with a definite altitude but rather with the cloud level itself, which is dependent upon the humidity at the foot of the mountain, i.e., the greater the humidity there, the lower the cloud level (Walter 1973). Although cloud forests are best developed in the upper montane forest, they can be found anywhere between 900 and 3,000 m or even higher, and the variety of temperature conditions which can prevail accounts for their floristic differences. The cloud forest has the lowest diurnal fluctuations of temperature of any forest type, because the abundant moisture in the form of clouds and fog acts to moderate any extreme changes. Above the cloud forest, rainfall decreases, as does temperature, and daily temperature fluctuations become more pronounced as minima gradually approach the freezing point.

In general, the upper montane forest differs from the lower montane forest in a noticeable lack of the "majority of palm genera, the Musaceae (*Heliconia*), Zingiberaceae, Marantaceae, and the genera *Monstera* and *Hevea*. On the other hand, it differs from the montaña by the greater abundance of tree ferns, Ericaceae, Lobeliaceae [Campanulaceae], and Asteraceae, and yields such genera as *Berberis, Hesperomeles, Ribes, Monnina, Fuchsia, Polylepis, Vallea, Gunnera, Viola, Geranium, Bomarea, Calceolaria,* and *Thalictrum,* which are absent or only of sparing occurrence in the montaña" (Weberbauer 1936). Weberbauer also noted that in middle and southern Peru (and Bolivia also) the western limits of many plants are the upper montane forest (*ceja de montaña*), because the wet east Andean winds do not cross onto the western sides, which are under the influence of cooler, drier air from the cold Humboldt current. Therefore, in general, "*Sphagnum*, several fern genera, *Podocarpus*, nearly all *Chusquea* species, the palms, Eriocaulaceae, Monimiaceae, Aquifoliaceae (*Ilex*), Theaceae, Marcgraviaceae, *Gaiadendron,* and *Cinchona;* most Araceae, Orchidaceae, Lauraceae, Melastomataceae, Araliaceae, Ericaceae, and Gesneriaceae; and many other types that determine the nature of the eastern flora" (Weberbauer 1936) of Peru and Bolivia are not present on the western slopes.

Important vascular plant families and genera in the upper montane forest include Actinidiaceae, Alstroemeriaceae (*Bomarea*), Aquifoliaceae (*Ilex*), Asteraceae (*Gynoxys, Liabum, Mikania, Senecio* sensu lato), Begoniaceae (*Begonia*), Betulaceae (*Alnus*), Bromeliaceae, Brunelliaceae (*Brunellia*), Buddlejaceae (*Buddleja*), Caprifoliaceae (*Viburnum*), Clethraceae (*Clethra*), Campanulaceae (*Centropogon, Siphocampylus*), Chloranthaceae (*Hedyosmum*), Clusiaceae (*Clusia*), Cunoniaceae (*Weinmannia*), Cyperaceae, Elaeocarpaceae (*Vallea*), Ericaceae (*Bejaria, Cavendishia, Macleania, Thibaudia*), Escalloniaceae, Fagaceae (e.g., *Quercus* in Colombia), Flacourtiaceae (*Abatia*), Hamamelidaceae, Juglandaceae, Lauraceae (*Ocotea*), Magnoliaceae, Melastomataceae (*Miconia*), Myricaceae (*Myrica*), Myrsinaceae (*Myrsine*), Orchidaceae, Piperaceae (*Peperomia, Piper*), Poaceae, Podocarpaceae, Rosaceae (*Hesperomeles, Prunus, Polylepis*), Scrophulariaceae, Solanaceae (*Solanum*), Styracaceae, Symplocaceae (*Symplocos*), Theaceae, and Winteraceae (*Drimys*). The upper levels also often show a belt of dense bamboos 3–4 m tall in the understory. A characteristic species (although not very common) of this higher level is the wax palm (*Ceroxylon*), with clean white waxy trunks to 40 m tall.

As was mentioned earlier, the exact altitudinal limits of the different vegetational belts are difficult to define where they become contiguous. This is especially noticeable where the upper limit of the upper montane forest reaches the high-elevation páramo vegetation (discussed below) between 3,000 m and 3,500 m. The species are often the same, and when they form a more or less closed forest association of about 10 m height, they are considered part of the Andean forest. However, when they are fragmented and form small islands of short vegetation due to edaphic or man-made reasons, we often refer to them as part of the subpáramo. Woody thickets or forest patches often extend upward in ravines, becoming smaller with increasing elevation. The upper limit depends also on exposure of the slopes and rainfall.

HIGH-ELEVATION TREELESS VEGETATION

Vegetation found above the treeline was conveniently referred to as tierra helada by early naturalists, but today it is increasingly known as tropical alpine. Although the term *alpine* is European in origin and may be objectionable to some (e.g., Luteyn 1999), its usage in this sense refers only to the belt between the tree- and snowline. There may not be a single term that is more applicable to this vegetation type, which is found not only in the Neotropics but also in Africa and Malaysia.

The higher elevations of the Northern Andes become increasingly drier and more seasonal as one moves south from the equator, and the accompanying change in vegetation is gradual but distinct. The wet páramos of the Northern Andes extend south to northern Peru (there known as *jalcas*) and receive rain and fog throughout the year. This contrasts with the drier punas of the Central Andes that extend south to northwestern Argentina, which receive much less seasonal precipitation. The vegetation of these high altitudes as a whole consists of treeless grasslands with a great variety of cushion, rosette, and caespitose herbs and shrubs. The most significant abiotic difference between tropical high-elevation and temperate alpine zones is that in tropical high elevations temperature fluctuation is diurnal, as much as 20°C during the dry season, unlike the seasonality of the temperate alpine regions. The Neotropical páramo flora is the richest high-mountain flora of the world (Smith and Cleef 1988). Both the páramo and puna were, prior to human influence, in all likelihood more restricted than at present. The wood cutting for fuel and burning to increase land for either grazing or cultivation have led to an almost certain increase in tropical alpine vegetation in South America (Luteyn 1999). Several recent papers present general discussions of páramo and/or puna habitats for Colombia (Rangel and Garzón 1997c; Cleef 1981, 1997a,b), Ecuador (Lojtnant and Molau 1982; Mena et al. 1997), and Peru (Young et al. 1997) and general regional statements by Gentry (1997), Luteyn (1992, 1999), and Luteyn et al. (1992). Discussions of human influence in páramo and puna may be found in Balslev and Luteyn 1992, Little 1981, Luteyn 1999 (and especially the references therein), Schjellerup 1989, and Seibert 1983.

PÁRAMO

Páramo vegetation is found primarily in Venezuela, Colombia, and Ecuador, with outliers in Costa Rica and Panama and northern Peru. The surface area occupied by páramo vegetation is estimated to be 2 percent or less in each of the aforementioned countries. Páramos are cold and often bleak, being wet nearly every month of the year, with precipitation varying from 500 mm to 2,000 (–3,000) mm per year. Numerous large and important rivers have their origins in the páramos, and their drainage is external toward the Caribbean and Amazon Basin. Páramo is the only high-elevation vegetation in the Andes that has been analyzed in detail with regard to plant diversity (Luteyn 1999). There are an estimated 4,697 species distributed among 865 genera and 254 families of major plant groups recorded for the páramo: lichens (465 spp.), bryophytes (835 spp.: hepatics 291, mosses 544), ferns and allies (352 spp.), conifers (2 spp.), and flowering plants (3,045 spp.: dicots 2,411, monocots 634). The dominant flowering plant families include the sunflower family (Asteraceae) with 101 genera and 858 species, Poaceae (41 gen., 227 spp.), and Orchidaceae (25 gen., 152 spp.). The sunflower genus *Espeletia* is often used as the symbol of the páramo. The three subdivisions employed in describing the páramo are subpáramo, true páramo, and superpáramo.

SUBPÁRAMO

The subpáramo is typically a transition zone between the upper montane forest and true páramo. Elevational range is from approximately 3,000 to 3,500 m. This is largely a mosaic vegetation, with the páramo grasslands intermixed with isolated thickets of shrubs and dwarf trees. Representative shrubby plants include *Aragoa, Berberis, Bejaria, Hypericum, Macleania* and *Vaccinium, Brachyotum* and *Miconia, Gynoxys* and *Diplostephium, Hesperomeles,* and the bambusoid grass *Chusquea.* Small trees in the subpáramo thickets include *Buddleja, Escallonia, Gynoxys, Hesperomeles, Myrsine, Oreopanax* and *Schefflera,* and *Weinmannia.* The subpáramo has been greatly influenced by human activities, probably due to extensive cutting and burning at the upper end of the treeline. The plant community may well be an anthropogenic formation that represents a degraded upper montane forest, a process that undoubtedly began in Precolumbian times (Laegaard 1992; Luteyn 1999).

TRUE PÁRAMO

The true páramo, which extends from ca. 3,500 to 4,100 m, is dominated primarily by grasses, in particular *Calamagrostis* and *Festuca;* shrubs such as *Baccharis, Diplostephium, Loricaria, Pentacalia, Hypericum, Gaultheria, Pernettya, Vaccinium,* and *Valeriana;* and rosette and cushion plants including *Acaena* and *Plantago.* One of the principal indicators of páramo is the woolly rosette genus *Espeletia,* a member of the sunflower family Asteraceae, and members of the tribe Espeletiinae that comprise about 150 species (Cuatrecasas 1986). Common herbaceous genera include *Acaena, Bartsia, Bomarea, Cerastium, Eryngium, Gentianella* and *Halenia, Jamesonia, Lachemilla, Lupinus, Lycopodium* sensu lato, *Paepalanthus, Senecio,* and *Sisyrinchium.*

Human activities have had a great impact on the páramo, especially in the tall-grass communities where species of *Calamagrostis* predominate. These can be transformed to short-grass communities where *Festuca, Agrostis,* and *Paspalum* predominate in the face of repeated burning and overgrazing (Luteyn 1999).

SUPERPÁRAMO

The superpáramo is a transition zone between the upper limits of true páramo and the snowline, mostly at elevations from 4,100 to 4,800 m. Generally, it has the lowest air temperatures, least rainfall, fewest soil nutrients, most frequent night frosts, and highest solar radiation of all the types of páramo. Here there is often a marked zonation between vegetated and exposed, bare ground. Even though the superpáramo landscape may appear devoid of vegetation, it actually supports a broad array of mosses, lichens, herbaceous vascular plants (such as *Arenaria* spp. and *Cerastium floccosum, Astralagus geminiflorus, Azorella pedunculata, Disterigma empetrifolium, Draba* spp. and *Eudema nubigena, Geranium multipartitum, Hypochaeris sessiliflora* and *Senecio* spp., *Luzula racemosa, Nototriche* spp., *Plantago sericea, Valeriana alpifolia, Viola pygmaea*), and numerous other grasses and even one genus of gymnosperms, *Ephedra*. Of the three types of páramo, superpáramo is the least diverse and least affected by humans, largely due to the harshness of the environment and lack of exploitable resources.

PUNA

The cold, dry puna vegetation of the Central Andes ranges from northern Peru south to northern Argentina, traversing from about 8°S to 27°S latitude. This region extends initially diagonally from the northwest to southeast and then turns southward near Lake Titicaca. Elevation of these highlands ranges from 3,300–3,700 to 5,000 m. Precipitation decreases from north to south and from east to west, becoming seasonal, varying from 150 mm to 1,500 mm per year, thus supporting a dry grassland and subxerophilous vegetation. The vegetation of the puna in the humid north or east approaches in aspect the drier páramo of central and southern Ecuador. At about 15°S, from southern Peru to northern Argentina, the high plateau above 3,600 m is called the altiplano. In the northern part of the altiplano, around the Peru-Bolivia border, water drainage is internal; elsewhere it has a more typical Amazon-Pacific drainage pattern.

A general review of the plant ecology of puna may be found in Troll 1968b, Cabrera 1968, and Young et al. 1997; Cabrera (1957) gives a detailed study of puna diversity in Argentina. The dominant life-form of the puna, as in the páramo, is the grass. The dominant flowering plant families include the Asteraceae, Poaceae, Leguminosae, Solanaceae, and Verbenaceae. One species of the pineapple family (Bromeliaceae), *Puya raimondii,* is often used as a symbol of the puna. Three subdivisions are generally acknowledged in the puna: humid, dry or tola, and desert (but see also Huber and Riina 1997).

HUMID PUNA

The humid puna begins in north-central Peru adjacent to the páramos (jalcas) and extends in a southeasterly direction throughout and across the Peruvian cordilleras to along the eastern altiplano of Bolivia to near Tarija (just north of the Argentine border). It receives between 400 mm and 1,500 mm of annual precipitation. It possesses year-round lakes and rivers, which drain outward. It supports a well-developed plant cover and is often dominated by an uninterrupted carpet of grasses (e.g., *Calamagrostis, Cortaderia, Festuca,* and *Stipa*) that grow in tussocks surrounded by numerous other grasses, herbs, sedges, lichens, mosses, and ferns. In the wettest areas, the grasses yield to clumps of sedges and rushes such as *Carex, Oreobolus, Scirpus,* and *Juncus* and to mats or cushions of *Distichia muscoides.* Other common genera of the humid puna include prostrate, rosette, and cushion plants such as *Azorella, Daucus, Baccharis, Werneria, Draba, Echinopsis, Gentiana, Geranium, Hypsela, Isoetes, Lilaeopsis, Lupinus, Nototriche, Ourisia, Oxychloe, Plantago,* and *Pycnophyllum* (Weberbauer 1945). There are also scattered cushions of cacti such as *Opuntia lagopus.* Many areas of the humid puna are farmed.

Curiously, trees of *Polylepis* spp. and shrubs such as *Buddleja, Gynoxys, Miconia, Myrsine,* and *Ribes* will move into the puna grasslands if they are left undisturbed. It appears that occupation of this type of puna took place some 10,000 years ago, and since then human activities have destroyed most of the *Polylepis* forests in the altiplano and reduced them to remnants in rocky fringes deemed unsuitable for grazing. It is unknown how often natural fires might have occurred in the region in presettlement times, although there is little indication that fires played an important role in the early Holocene ecosystem. What does seem likely is that *Polylepis* forests were once much more extensive than they are today, and the use of fires to improve agricultural potential and pasturage played a major role in altering this vast ecosystem, as in the northern páramo (Balslev and Luteyn 1992; Kessler 1995; Luteyn 1999).

DRY OR TOLA PUNA

Commencing in south-central Peru, along the western cordillera near Arequipa, the dry puna is found throughout most of southern Peru, especially along the western altiplano of Bolivia and farther south as a narrow tongue into northwestern Argentina. Rainfall ranges from 100 mm to 400 mm per year, and some of the drainage is internal. There are still lakes and rivers, but they are becoming more saline. These areas are usually dominated by extensive shrublands and thickets of the resinous tola shrubs, such as *Parastrephia* spp. and *Lepidophyllum quadrangulare* (Asteraceae), and stiff bunchgrasses of *Festuca orthophylla;* however, ground cover is usually less than 50 percent. Other common genera include *Acantholippia, Adesmia, Baccharis, Ephedra, Fabiana, Junellia, Nardophyllum, Senecio,* and *Tetraglochin.* There is some cattle raising but little agriculture in the dry puna.

DESERT PUNA

The desert puna is located mostly in the southern part of the Central Andes along the western cordillera of Bolivia and adjacent to the Atacama Desert region of northern Chile to about 27°S latitude. It receives less than 100 mm of annual precipitation. There are no rivers or lakes, no agriculture or cattle raising, and few people. There are often large flat areas of ground coated with a saline crust known as *salares.* Plants are adapted to frost, drought, and salt. The vegetation is sparse (usually less than 15 percent) and composed of xerophytic plants, including cacti, thorny shrubs, cushion plants, and bunchgrasses such as *Festuca orthophylla.* Prominent genera include *Aciachne, Adesmia, Anthobryum, Baccharis, Cereus, Chersodoma, Cortaderia, Eustephiopsis, Hemimunroa, Lampayo, Lophopappus, Margyricarpus, Oreocereus, Para-*

strephia, Pennisetum, Satureja, and *Urmenetea.* Because this formation is so dry, no agriculture is practiced here and it is of little interest to humans.

INTER-ANDEAN VEGETATIONAL TYPES

The areas between the principal mountain ranges of the Andes are called the sierras. They often consist of deep valleys and high basins or plateaus interrupted by cross-ridges (nudos). They range in elevation from about 1,000 to 3,500 m and are found from western Venezuela to south-central Ecuador; in Peru and Bolivia the higher plateaus are usually puna vegetation discussed earlier. Up to now we have been dealing with the ideal wet mountain system, but the Andes are dissected by numerous transverse and parallel mountain ranges, many with deep intermountain valleys cut by large rivers. Therefore, environments may change rapidly over short distances. Add to this the presence of wet to dry air masses and lee or rainshadow effects that help to cause a great diversity of montane vegetation. Often these valleys and plateaus act as major barriers to dispersal and species establishment (Vuilleumier 1969; Berry 1982; Molau 1988). Humans have inhabited these valleys and plateaus from Precolumbian times, and the natural vegetation has been severely altered or in many cases destroyed. As a result of many centuries of cultivation, disturbance by grazing animals, cutting of wood, and the natural lack of rainfall, the flora is restricted and depauperate. Many of these areas are still to this day the real centers of intensive agriculture in the Andes.

VALLEYS AND PLATEAUS

At the bottoms of many valleys are desert or savannalike formations consisting of cacti, thorny acacias, mimosas, and generally xerophytic vegetation. Many of the shrubs, trees, and herbs are green only during the short rainy season of two to three months. In some years the lowest parts of the valleys receive almost no rain, so that no new foliage develops. As one ascends montane valleys and moisture increases, evergreen woody plants take a more prominent role and the vegetation becomes more closed, with shrubs and trees alternating with periodic grass steppes. Along the valley or canyon walls, there are many columnar cacti and Bromeliaceae (especially *Tillandsia*) and some orchids. Climbing shrubs with tendrils or twining stems (*Cardiospermum, Mandevilla, Prestonia, Serjania*) are common. Inter-Andean desert regions are sometimes found in high basins that lie in the rain shelter of the cordilleras. In many of these areas the rainfall is less than 500 mm per year. Oftentimes with irrigation, bananas, sugarcane, and temperate fruits (like grapes, peaches, and plums) are cultivated. In general, the dry western slopes of the Peruvian Andes and the dry inter-Andean valleys are dominated by shrubby and herbaceous Asteraceae. Higher into the valleys and with further increase in moisture, especially in the form of nightly fogs, the number of larger shrubs and trees becomes greater, although most areas are under the heavy influence of cultivation. In the more northern regions of Ecuador, Colombia, and Venezuela, these higher elevation landscapes are characterized by evergreen pastures and fields of cereals, potatoes, carrots, and onions, which can be sown and harvested at any time of the year. Vegetation remnants indicate that large parts of the earlier vegetation consisted of a forest similar to that of the tierra fria. The similarity between the vegetation of the xerophytic, inter-Andean valleys and the eastern llanos and Caribbean lowlands of northern Colombia and Venezuela is seen in the large number of genera they have in common.

Throughout many of the inter-Andean valleys, depending on the moisture and elevation, one can often see the same progression of taxa: *Agave* and *Furcraea* (Agavaceae), *Schinus molle, Spondias,* and *Toxicodendron striatum* (Anacardiaceae), *Barnadesia, Bidens, Calea, Chromolaena, Mikania, Tagetes, Trixis, Vernonia, Wedelia, Wulffia* (Asteraceae), *Cynanchum* and *Gonolobus* (Asclepiadaceae), *Tabebuia* and *Tecoma stans* (Bignoniaceae), *Bombax* and *Bombacopsis* (Bombacaceae), *Cordia* and *Heliotropium* (Boraginaceae), *Aechmea, Puya,* and *Tillandsia* (Bromeliaceae), *Acanthocereus, Cereus, Melocactus, Opuntia, Pereskia* (Cactaceae), *Capparis* (Capparidaceae), *Sambucus* (Caprifoliace), *Carica* (Caricaceae) in Peru, *Evolvulus* and *Ipomoea* (Convolvulaceae), *Acalypha, Croton, Euphorbia, Hura crepitans, Jatropha, Mabea* (Euphorbiaceae), *Wigandia caracasana* (Hydrophyllaceae), *Acacia, Caesalpinia tinctoria, Calliandra, Cassia, Cercidium praecox, Desmodium, Erythrina, Mimosa, Parkinsonia aculeata, Pithecellobium, Prosopis* (Leguminosae), *Mentzelia* (Loasaceae), *Banisteriopsis, Byrsonima,* and *Malpighia* (Malpighiaceae), *Sida* (Malvaceae), *Clidemia* and *Miconia* (Melastomataceae), *Calyptranthes, Eugenia, Myrcia, Psidium* (Myrtaceae), *Aristida, Chloris, Digitaria, Panicum, Paspalum, Setaria,* and *Sporobolus* (Poaceae), *Cantua quercifolia* (Polemoniaceae), *Borreria, Faramea, Psychotria,* and *Randia* (Rubiaceae), *Dodonea viscosa, Paullinia,* and *Sapindus* (Sapindaceae), *Brugmansia* (*Datura* spp.) and *Solanum* (Solanaceae), *Byttneria* and *Waltheria* (Sterculiaceae), *Trema micrantha* (Ulmaceae), *Citharexylum, Duranta, Lantana, Lippia,* and *Petrea* (Verbenaceae).

The high plateaus were cleared of their native forests long ago and have been replaced by cultivated fields. Remnants of the original vegetation are found in ravines, on steep slopes, and other less accessible places and indicate that large parts of the earlier vegetation consisted of a low forest similar to the present upper montane forest. In Ecuador, Harling (1979) feels that the early forests were probably interspersed with a more open, savannalike land. The only forests that exist today are ones of cultivated *Pinus radiata* and *Eucalyptus* (mostly *E. globulus*), and sometimes *Cupressus*, with lesser plantations of *Casuarina equisetifolia* and *Grevillea robusta*.

Characteristic weedy plants of the plateau areas, best found along fencerows and as thickets between pastures, consist of *Agave* (Agavaceae), *Bomarea* (Alstroemeriaceae), *Baccharis, Ageratina* (*Eupatorium* sensu stricto), *Dahlia, Gynoxys,* and *Barnadesia* (Asteraceae), *Puya* (Bromeliaceae), *Siphocampylus giganteus* (Campanulaceae), *Columellia oblonga* (Columelliaceae, in Ecuador), *Coriaria thymifolia* (Coriariaceae), *Croton, Euphorbia laurina,* and *Ricinus communis* (Euphorbiacaea), *Salvia* (Lamiaceae), *Cassia* sensu lato and *Dalea caerulea* (Leguminosae), *Fuchsia* spp. (Onagraceae), *Boc-*

conia frutescens (Papaveraceae), *Calceolaria* (Scrophulariaceae), *Solanum* and *Cestrum* (Solanaceae), the fern *Pteridium aquilinum* sensu lato, and other adventives such as *Aloe vera* and *Prunus capuli*.

SOUTH ECUADOREAN SHRUB VEGETATION

Harling (1979) introduced this term for a special type of vegetation found in Cañar, Azuay, and Loja provinces (south-central Ecuador), but it also occurs in northern Peru, between 2,000 m and 3,000 m. It has the low, dry aspect of scrub vegetation that has been heavily disturbed for a long time. The most important families in this formation are Asteraceae, Bromeliaceae, Ericaceae, Melastomataceae, and Proteaceae. Dominant genera and species include *Oreopanax* (Araliaceae), *Baccharis genostelloides, Barnadesia dombeyana,* and *Diplostephium lavandulifolium* (Asteraceae), *Hypericum laricifolium* (Clusiaceae), *Coriaria* (Coriariaceae), *Weinmannia* (Cunoniaceae), *Vallea stipularis* (Elaeocarpaceae), *Bejaria, Gaultheria, Macleania pubiflora,* and *Cavendishia bracteata* (Ericaceae), *Salvia rugosa* (Lamiaceae), *Brachyotum, Miconia,* and *Tibouchina* (Melastomataceae), *Cantua quercifolia* (Polemoniaceae), *Lomatia hirsuta* and *Oreocallis* (*Embothrium*) (Proteaceae), *Streptosolen jamesonii* (Solanaceae), and *Valeriana hirtella* (Valerianaceae); the aggressive bracken fern (*Pteridium aquilinum*) as well as *Blechnum loxense* are frequently encountered. A more detailed discussion and species list may be found in Espinosa 1997.

PACIFIC COAST FORESTS AND DESERT

The Pacific coast of South America is an example of extreme environments under the effect of prevailing oceanic currents. The warm Panama current moving southward gives rise to wet northern forest in Colombia and northwest Ecuador, whereas the cold Humboldt current flowing northward from Antarctica influences a desert environment southward into Chile. These two ocean currents, in turn, meet off the coast of central Ecuador and flow westward toward the Galápagos Islands. Consequently, there is an actual change in the amount of measured rainfall from thousands of millimeters in northwest Colombia to zero in Chile. From the Panama-Colombia border to northwestern Ecuador, the Pacific coast is characterized by a very wet Chocó rain forest. Southward along the coast from this wet Chocó forest, and beginning in north-central Ecuador, there are increasingly arid formations. In central Ecuador, sandwiched between the savannalike coastal vegetation and the moist western slopes of the Andes, lie a succession of mesic and dry forests, whereas in southwestern Ecuador and adjacent Peru a generally dry savanna vegetation occurs, which includes some cacti and coastal desert formations that move higher and higher into the western-facing valleys of the Andes to about 3,500 m. Near the Ecuador-Peru border at 5°S latitude and southward for about 3,500 km to La Serena, Chile, near 30°S, the Pacific coast desert forms an almost continuous strip.

Needless to say, there are numerous vegetation types along the Pacific coast of South

America (tiny Ecuador alone has twelve life zones west of the Andes, based on Cañadas and Estrada 1978, and several more can undoubtedly be added for Peru, based on Weberbauer 1936 and 1945). Therefore, this section deals in only a very generalized way with some of the major dryland formations.

More detailed accounts may be found for Colombia (Forero and Gentry 1989; Gentry 1995b), Ecuador (Acosta-Solís 1968, 1970, 1977; Cañadas and Estrada 1978; Dodson and Gentry 1978, 1991; Harling 1979; Valverde et al. 1979; Dodson et al. 1985; Gentry 1986; Kessler 1992; Parker and Carr 1992; Espinosa 1997; Neill 1997), Ecuador and Peru (Svenson 1946; Young and Reynel 1997), Peru (Weberbauer 1936, 1945; CDC-Peru 1997; Dillon 1997), Peru and northern Chile (Rauh 1985; Rundel et al. 1991; Dillon and Hoffmann 1997), and this region in general (Gentry 1995b, 1997).

THE CHOCÓ

The Chocó region, from the Panama-Colombia border to northwestern Ecuador, is known both for its high rainfall, ranking second in the world, and for its high diversity and endemism of plants. The Chocó vegetation type is mostly lowland tropical rain forest and is summarized by Daly and Mitchell (in this volume).

SAVANNA WOODLANDS AND DRY FOREST

Savanna, often including thorny *Acacia-Mimosa* shrub woodlands, and dry forest vegetation are very extensive along the Pacific coast in southwestern Ecuador and adjacent northern Peru. In general, the vegetation is shorter closer to the coast, with thornscrub 2–3 m tall, whereas inland the dry forest reaches 20 m height. Formerly, dry forest was probably more widespread. Since climatic conditions are essentially similar for savanna woodlands and dry forest, different topographic and edaphic factors must be important for them to replace each other, given that human disturbance is not a factor. The two types may replace each other depending on disturbance factors. This combination of vegetation types may also reach into the dry valleys and foothills of the western Andes. It receives up to 1,600 mm of precipitation per year (Gentry 1995b) but has a long and pronounced dry season from May to December or January, when most trees lose their leaves and the grassy stratum becomes dry. One of the most striking features of the dry forest can be its massive flowering of trees and vines at the beginning of the rainy season.

The dominant trees in this open community are all Bombacaceae, especially *Ceiba trischistandra* with its green stems and branches and strikingly grotesque growth form. Other Bombacaceae trees include *Cavanillesia platanifolia, Ceiba pentandra, Eriotheca ruizii, Pseudobombax millei,* and *P. guayasense.* Other common trees and shrubs include *Achatocarpus pubescens* (Achatocarpaceae), *Loxopterygium huasango* (Anacardiaceae), *Tecoma castanifolia* and *Tabebuia chrysantha* (Bignoniaceae), *Cordia* (Boraginaceae), *Bursera* (Burseraceae), *Capparis* (Capparidaceae), *Carica parviflora* (Caricaceae), *Cochlospermum vitifolium* (Cochlospermaceae), *Terminalia* (Combretaceae), *Erythroxylum* (Erythroxylaceae), *Acalypha* and *Croton rivinifolius* (Euphorbiaceae), *Casearia* and *Muntingia calabura* (Flacourtiaceae), *Acacia, Caesalpinia, Cassia, Cercidium praecox, Erythrina velutina, Piptadenia, Pithecellobium, Prosopis juliflora, Mimosa, Parkinsonia aculeata* (Leguminosae, many thorny), *Ficus* (Moraceae), *Coccoloba* (Polygonaceae), *Ziziphus* (Rhamnaceae), *Randia* (Rubiaceae), *Guazuma ulmifolia* (Sterculiaceae), and *Clavija pungens* and *Jacquinia* (Theophrastaceae). There are numerous climbing and erect columnar cacti, such as *Armatocereus cartwrightianus, Hylocereus peruvianus, Neoraimondia,* and *Opuntia* spp. Many climbing vines are apparent after the rains, including *Ipomoea carnea* (Convolvulaceae) and *Apodanthera, Sicyos, Momordica,* and *Luffa* (Cucurbitaceae). Epiphytic bromeliads such as *Tillandsia* and lianas in the Bignoniaceae are common. Annuals include Acanthaceae, Amaranthaceae, Asteraceae, Cucurbitaceae, Leguminosae, Poaceae, and Solanaceae. Common grasses include *Andropogon bicornis, Aristida adsensionis, Chloris virgata, Panicum, Paspalum,* and *Pennisetum purpureum* and *P. occidentale.* Gallery forests may be composed of *Croton* (Euphorbiaceae), *Clusia* (Hypericaceae), *Acacia* (Leguminosae), *Syzygium* (Myrtaceae), *Randia* (Rubiaceae), *Zanthoxylum* (Rutaceae), and *Pouteria* (Sapotaceae).

The most speciose families overall are Asteraceae, Cactaceae, Convolvulaceae, Euphorbiaceae, Leguminosae, Malvaceae, Poaceae, and Solanaceae, also with many Acanthaceae, Amaranthaceae, Cucurbitaceae, Nyctaginaceae, and Scrophulariaceae (Harling 1979; Kessler 1992; Gentry 1993; CDC-Peru 1997; Young and Reynel 1997). For all plants (trees, shrubs, herbs), Neotropical dry forests (as in moist forests) are dominated by two families: Leguminosae (arborescent) and Bignoniaceae (lianas); remaining dry forest dominants include Rubiaceae, Sapindaceae, Euphorbiaceae, Flacourtiaceae, Capparidaceae, Myrtaceae, Apocynaceae, and Nyctaginaceae. In more desert situations, Cactaceae are also an important dry forest family (Gentry 1995b).

In summary, in regard to Neotropical dry forest, Gentry stated that "Neotropical dry forests are generally less diverse than moist forests" and that "With remarkably few exceptions like Cactaceae, Capparidaceae, and Zygophyllaceae, the dry forest flora is a relatively depauperate selection of the same families that constitute moist and wet forest plant communities." Furthermore, "Dry forests generally have fewer epiphytes and more vines than moist or wet forests. They are also distinctive in a higher percentage of trees and lianas with wind-dispersed seeds and conspicuous flowers," and finally, "If entire floras are compared, Leguminosae is always by far the most prevalent family, followed by Euphorbiaceae, Poaceae and Asteraceae, all better represented in dry than in moist forest" (Gentry 1995b). Dodson and Gentry (1991) estimated that dry forest originally occupied 35 percent of western Ecuador, and that less than 1 percent still exists.

MESIC FOREST

For purposes of this discussion, mesic forest includes both the moist and wet forest formations located in the northern and central coastal plain of Ecuador and east to the Andean foothills. These forest types do not seem to reach into coastal Peru. They receive 2,000–6,000 mm of annual precipitation, and elevations range from close to sea level to 900 m. The forest canopy averages 40 m in height (with a few emergents reaching to 60 m) and is mostly continuous with only a few gaps that allow direct sunlight to reach the floor. Neill (1997) feels that these mesic forests have fewer gaps and larger trees than the lowland forests of Amazonian Ecuador. Large forest dominants include *Huberodendron patinoi* (Bombacaceae), *Dacryodes* (Burseraceae), *Humiriastrum procerum* (Humiriaceae), *Carapa guianensis* and *Guarea kunthiana* (Meliaceae), *Brosimum utile* and *Ficus dugandii* (Moraceae), *Virola dixonii* (Myristicaceae), and *Pouteria capacifolia* (Sapotaceae). The rest of the vegetation, including epiphytes and understory plants, is quite diverse, with many species of Araceae, Arecaceae (especially *Geonoma*), Calycanthaceae, Cucurbitaceae, Moraceae, Piperaceae, and Rubiaceae, plus a plethora of ferns. Many endemics, especially Gesneriaceae, are found beneath the canopy. Even though the canopy is closed and appears lush, the diversity of tree and shrub species is thought to be relatively low (Neill 1997).

A number of economic plant species are found in the moist forest. Hardwoods exploited by the timber industry, mostly in plywood manufacture, include *Brosimum utile, Dacryodes* spp., and *Virola dixonii*. Unfortunately, most of the mesic forests on the Pacific coast of Ecuador have been cleared for timber and agricultural purposes. No one knows the exact extent of the mesic forests, but they are thought to have been much more extensive and perhaps continuously distributed throughout the coastal plain. These mesic forests have now been reduced to a few remnants by the action of humans. Dodson and Gentry (1991) estimated that moist forest originally occupied 40 percent of western Ecuador, with less than 4 percent currently remaining. They also estimated that wet forest occupied about 15 percent of western Ecuador, with less than 0.8 percent surviving.

The general floristics of mesic forests is poorly known; therefore, this section summarizes data from the following well-studied sites in western Ecuador: Río Palenque (Dodson and Gentry 1978), Jauneche (Dodson et al. 1985), and two sites belonging to ENDESA (Jørgensen and Ulloa 1989; Neill 1997). The most important families are Apocynaceae, Araceae, Arecaceae, Asteraceae, Bignoniaceae, Bromeliaceae, Cucurbitaceae, Cyclanthaceae, Euphorbiaceae, Gesneriaceae, Leguminosae, Melastomataceae, Moraceae, Orchidaceae, Piperaceae, Poaceae, Rubiaceae, Sapindaceae, Solanaceae, and various families of ferns. The most important trees are Bombacacae (*Ceiba*), Burseraceae (*Protium*), Lauraceae (*Nectandra, Ocotea,* and *Persea*), Leguminosae (*Pithecellobium*), Meliaceae (*Carapa* and *Guarea*), Moraceae (*Brosimum, Coussapoa, Ficus, Pseudolmedia,* and *Pterocarpus*), Myristicaceae (*Dialyanthera* and *Virola*), Sapotaceae (*Pouteria*), and Hypericaceae (*Clusia*). The most important lianas belong to Bignoniaceae (*Amphilophium, Anemopaegma, Arrabidaea,* and *Lundia*) and Leguminosae (*Cassia, Dioclea, Entada,* and *Mucuna*). Important small understory trees include Annonaceae (*Duguetia*), Arecaceae (*Astrocaryum, Catoblastus, Iriartea, Socratea,* and *Wettinia*), Bombacaceae (*Matisia*), Boraginaceae (*Cordia*), Flacourtiaceae (*Casearia*), Hippocrateaceae (*Cheiloclinium*), Leguminosae (*Brownea, Erythrina, Inga, Matisia,* and *Swartzia*), Lecythidaceae (*Gustavia*), Moraceae (*Cecropia* and *Ficus*), Melastomataceae (*Miconia*), Olacaceae (*Heisteria*), Piperaceae (*Piper*), Rubiaceae (*Coutarea,*

Faramea, Psychotria), and Solanaceae (*Solanum*). Important families of shrubs include Acanthaceae, Arecaceae (geonomoid palms), Hypericaceae, Melastomataceae, Piperaceae, and Rubiaceae. Common forest floor herbs are found in the Araceae, Commelinaceae, Cyclanthaceae, Gesnericaceae, Marantaceae, Zingiberaceae, and fern families. Epiphytes and hemiepiphytes are found in the Araceae, Bromeliaceae, Cyclanthaceae, Ericaceae, Gesneriaceae, Orchidaceae, Piperaceae, and ferns.

DESERT VEGETATION

Pacific coastal deserts occur only at the tip of the Santa Elena Peninsula in western Ecuador (near the border with Peru) and then again along the entire coast from northern Peru to Chile. Due to the extreme arid conditions along the coast desert, vegetation is very sparse and entirely lacking over much of the area. One source of water, however, comes from the Andean snowmelt, which fills the rivers that periodically wind their way across the parched landscape. The other source of moisture is derived from the cold, north-flowing Humboldt current, especially as it arrives during the months between April to November along the western Peruvian and northwestern Chilean coasts (and to a lesser extent in Ecuador) in the form of dense fog banks (called *garúas* in Peru and *camanchacas* in Chile). This moisture supports small pockets of mesophytic vegetation, called *lomas*, in an otherwise prominently desert area. Outside the river valleys and the lomas, the Pacific deserts are among the driest places on earth, with some areas averaging only 0.6 mm of precipitation per year (Rundel et al. 1991). According to Dillon (1997), the families of vascular plants containing the greatest diversity throughout the Peruvian and Atacama deserts include Apiaceae, Asteraceae, Boraginaceae, Cactaceae, Leguminosae, Malvaceae, Nolanaceae, and Solanaceae. More detailed studies of the dry coastal vegetation of this region may be found in Weberbauer 1945; Svenson 1946; Acosta-Solís 1970; Ferreyra 1957, 1961, 1979, 1983; Rundel and Mahu 1976; Rauh 1985; Rundel et al. 1991; Dillon 1997; and Dillon and Hoffmann 1997.

RIPARIAN VEGETATION

Snowmelt and rain from the Andes cross the deserts westward to the Pacific Ocean in the form of 52 rivers (only 10 with an uninterrupted flow). Here communities of plants have become established in small but, especially in human terms, important riparian habitats. In these oases along the water's edge, communities of small woodlands or thickets occur, including *Schinus molle* (Anacardiaceae), *Vallesia glabra* (Apocynaceae), *Baccharis, Mikania,* and *Tessaria integrifolia* (Asteraceae), *Cordia* (Boraginaceae), *Muntingia calabura* (Flacourtiaceae), *Acacia macracantha, Caesalpinia tinctoria, Inga, Prosopis chilensis,* and *Vigna* (Leguminosae), *Myrsine* (Myrsinaceae), *Arundo donax, Gynerium sagittatum, Phragmites communis* (Poaceae), *Salix* (Salicaceae), *Sapindus saponaria* (Sapindaceae), *Acnistus, Cestrum* and *Dunalia* (Solanaceae), and a variety of woody and semiwoody shrubs. Where water is more abundant or standing, *Equisetum* (Equisetaceae), *Ludwigia* (Onagraceae), and *Typha* (Typhaceae) also may be found (Weberbauer 1911, 1936; Ferreyra 1983; ONERN 1971).

Many of these species are phreatophytes, with long roots that reach deep into the soil for subterranean water. Much of the land adjacent to the westward-flowing rivers has been used for irrigated agriculture for millennia. These river valleys, with their narrow bands of accompanying riparian vegetation, were attractive to early Precolumbian settlers (and are still heavily populated today) due to the stable water supply and multiple ecotones. Prehistoric coastal farmers produced corn, beans, manioc, squash, peanuts, cotton, and fruit in these oases along the river banks (Brush 1982).

LOMAS

The isolated, islandlike lomas that are seasonally covered by luxuriant vegetation are scattered in a line along the Peruvian and northern Chilean desert coasts. Between April and November, as the warm coast is cooled by the cold Humboldt current, the wet sea fog drifts landward and releases its moisture on the hills and slopes there. Less than 80 mm per year of precipitation is released, but the effect is enhanced by cloud cover. The lomas formations of these coastal deserts are characterized by very strong local endemism (42 percent overall within the Peruvian lomas, fide Müller 1985), especially in the two speciose near-endemic families Nolanaceae and Malesherbiaceae, and in genera such as *Eremocharis* (Apiaceae), *Ambrosia* (Asteraceae), *Argylia radiata* (Bignoniaceae), *Tiquilia* (Boraginaceae), *Astragalus* (Leguminosae), *Dinemandra* (Malpighiaceae), *Cristaria* and *Palaua* (Malvaceae), *Nolana* (Nolanaceae), and *Calceolaria* (Scrophulariaceae) showing significant lomas radiations (Rundel et al. 1991; Dillon 1997). Other characteristic woody species include *Capparis* spp. (Capparidaceae), *Acacia macracantha*, *Caesalpinia spinosa*, and *Prosopis* spp. (Leguminosae). The families Bromeliaceae (e.g., *Tillandsia*, *Puya*) and Cactaceae (e.g., *Haageocereus*, *Melocactus*, *Neoraimondia*) are also dominants in the lomas vegetation, while Aizoaceae, Portulacaceae, and Poaceae also show extremely high endemism in the isolated coastal vegetation.

The lomas formations around Lachay in central Peru have been described by Ferreyra (1953) as having two distinct fog bank zones. Below 100 m there is sparse vegetation, with *Tillandsia* spp., some cacti (e.g., *Espostoa melanoatele*, *Haageocereus decumbens*, and *Melocactus trujilloensis*), prostrate shrubs, and small trees. The lower zone, from 100 to 300 m, is dominated by cryptogams, e.g., *Nostoc commune*, abundant fruticose and foliose lichens, and some vascular plants, both perennial and annual. Above 300 m the vegetation becomes much more diverse. Populations of free-standing bromeliads (*Tillandsia latifolia* and *Puya ferruginea*) and lichens appear on rocky slopes. In the valleys and canyons of this zone are stands of small trees, viz., *Capparis prisca* (Capparidaceae), *Carica candicans* (Caricaceae), and *Caesalpinia spinosa* and *Senna birostris* (Leguminosae). Often they are laden with a variety of epiphytes, such as mosses, ferns, lichens, *Begonia geranifolia*, *Calceolaria pinnata*, and *Peperomia hilli*. Above the 500–600 m line, the woody plants drop out, with herbaceous perennials once again becoming the dominant plants. There is no vegetation above 800 m.

According to Dillon (1997), "Archaeologists have evidence that indigenous peoples have used the lomas formations as a temporal resource for over 5,000 years. . . . in their original state, the lomas provided essential sites for foraging and the periodic cultivation of crops. No sustainable agriculture is currently conducted within the lomas formations."

ATACAMA DESERT

The Atacama Desert extends from the border between Chile and Peru (18°24'S) south to La Serena, Chile (29°55'S). Elevations range from sea level to about 1,500 m on the western slopes of the Andes (Börgel 1973). Rundel and Mahu (1976) considered the Atacama vegetation in the vicinity of Paposo, Chile, a result of influences of elevation and fog bank moisture. The lowest elevations of the coastal plain, below 300 m, receive essentially no moisture from the fog bank, and the vegetation consists largely of scattered stands of the cactus *Copiapoa cinerea* var. *haseltoniana* and infrequent shrubs such as *Chuquiraga ulicina* and *Gypothamnium pinifolium* (Asteraceae). The center of the fog zone (350–650 m) is dominated by *Euphorbia lactiflua* (Euphorbiaceae) and *Eulychnia iquiquensis* (Cactaceae) with other shrubs, viz., *Proustia cuneifolia* (Asteraceae), *Tillandsia geissei* (Bromeliaceae), *Echinopsis coquimbana* (Cactaceae), *Croton chilensis* (Euphorbiaceae), *Balbisia peduncularis* (Geraniaceae), *Oxalis gigantea* (Oxalidaceae), and *Lycium stenophyllum* (Solanacee). Annuals in the same zone include *Alstroemeria graminea* (Alstroemeriaceae), *Chaetanthera glabiata* (Asteraceae), *Cruckshanksia pumila* (Rubiaceae), and *Viola* spp. (Violaceae). Along the lower border of the fog zone, around 300 m, the bromeliads *Puya boliviensis* and *Deuterocohnia chrysantha* predominate. Above the fog zone (850 m) the cactus *Copiapoa* once again becomes dominant, along with *Polyachyrus cinereus* (Asteraceae), *Nolana* spp. (Nolanaceae), and *Oxalis caesia* (Oxalidaceae). Above 900 m, vegetation of any kind vanishes from the landscape (Dillon and Hoffmann 1997).

ANDEAN DIVERSITY

The tropical Andes represent, per unit area, one of the most species-rich plant regions of the tropics and possibly the world (Gentry 1982; Henderson et al. 1991). The Northern Andes, for example, is approximately one-eighth the size of the Amazon Basin, but it contains an equivalent number of flowering plants (ca. 40,000 vs. 30,000 species). Both ferns and mosses (Churchill et al. 1995b) contain about four to five times the number of species in the tropical Andes as in the Amazon Basin. Various animal groups also exhibit high diversity in the Andes, for example, birds (Renjifo et al. 1998). Even groups previously thought to be diverse in lowland tropical forests, such as frogs (cf. Duellman 1979), are now known to be significantly more species-rich in the Andes (Lynch et al. 1997). Ecuador, for example, contains an impressive 15,306 species of vascular plants (Jørgensen and León 1999). Per unit area, tiny Ecuador (283,561 km²), slightly smaller than the state of Nevada, is several magnitudes richer than the combined United States and Canada, with an estimated 18,000 species. Recent data from the checklist of Ecuadorean vascular plants provides some insight into diversity (Jørgensen and León 1999). Keeping in mind the surface area, both Amazonian and Andean regions each cover ca. 35 percent of Ecuador. Most of the Amazon Basin is found within the elevation range from 0 to 500 m and contains 3,996 species. The Andean region at the elevational boundary of 1,500–2,000 m contains 4,071 species and between 2,500 m and 3,000 m there are 3,911 species (of which 1,528 species are shared). Among Neotropical montane forests as a whole, the tropical Andes are richer than the highland forests encountered in either Central America or Mexico (Gentry 1995a). Lower montane forests (mostly at or below 1,500 m) are similar to Amazonian lowland forests both in terms of floristic components and diversity and even, to a certain degree, structure. Plant families found at midelevations (1,500–2,500 m) are represented first, in terms of diversity, by the Lauraceae, followed by the Melastomataceae, Rubiaceae, and Moraceae (Gentry 1995a). The Rubiaceae, including such genera as *Cinchona, Elaeagaia, Palicourea,* and *Psychotria,* are slightly more diverse at lower to midelevations, while Melastomataceae including *Axinaea, Merriania,* and *Miconia* tend to be more diverse at slightly higher midelevations. Additional families include the Araceae, Arecaceae (Palmae), Euphorbiaceae, Hypericaceae (including Clusiaceae and Guttiferae), and Leguminosae. The Lauraceae is apparently first in the upper montane (ca. 2,500–3,000 m), followed by the Melastomataceae and Asteraceae. Additional families include the Aquifoliaceae, Araliaceae, Myrsinaceae, and Solanaceae. The Asteraceae, like the latter four families, increase notably above 2,000 m. Near the transitional montane forest and páramo-puna, ca. 3,000 m, the Asteraceae is the most species-rich family. Additional families important in this zone include Ericaceae, Lauraceae, Melastomataceae, and Myrsinaceae. Genera of the Rosaceae, including *Hesperomeles, Polylepis,* and *Prunus,* for example, are virtually confined to this zone.

The driving force behind this immense diversity of plant life and communities must be attributed in part to the physical evolution of the landscape. On a broad geological scale, tectonic events leading to the dramatic uplift of the Andes (Taylor 1995) provided the template for plant evolution. Short-term geological events such as landslides, related to oversteepened slopes, volcanic activity, and earthquakes at the local level throughout the Andes, provided a continued disturbance that formed a catalyst for evolution.

CONCLUSION

El Dorado was and is immensely rich, unsurpassed, beyond the imagination of those who first encountered it. Even today, new riches in the form of undescribed species are continuously being presented to the scientific community. Future generations will still be challenged by its diversity, both in terms of species and communities of plants and animals that compose the great green El Dorado of the Andes.

The tropical Andes region is one of amazing diversity: from the lower montane forests at the base of the mountains facing the Amazon Basin, to the arid zones of the Atacama Desert along the Pacific coast, with the soaring Andes mountains in between. This truly is a region with a broad span of available resources. Precolumbian inhabitants seemed to prefer the drier zones, such as the upland puna and the low river valleys, as habitation sites. It is in these zones that the Precolumbians developed their agricultural and pastoral systems and where the human impression on the landscape has been most lasting.

As is true all over the Andes, population has grown, bringing with it increased road building and agricultural settlement. Gentry (1997) made several observations that are important to think about and restate here:

> Overall, 96% of the forests of coastal [i.e., western] Ecuador have been destroyed, in an area of high regional local endemics; only 1% of the coastal dry forest remains.

> Deforestation is the major contributor to the loss of biodiversity in the Pacific coastal region of South America.

> Some entire forest types, like the *Podocarpus* forests that used to cover significant parts of the Andes, have already all but disappeared.

> In Ecuador almost nothing is left of the natural forests of the central valley.

> Although there are still areas of relatively intact forest on the eastern slopes of the Andes of Ecuador, Peru and Bolivia [Colombia, Venezuela], all three countries have active road-building programs and rampant deforestation in this region.

Finally, Henderson et al. (1991) suggest that less than 10 percent of the Andean forest remains intact. Therefore, when one considers that reconstruction of Precolumbian Andean vegetation must be based on current (therefore potential) conditions, it is easy to realize that much more needs to be done and that many more isolated or relic forest remnants need to be sampled before we can truly understand what the tropical Andes were like in the past.

REFERENCES

Acosta-Solís, M. 1966. Las divisiones fitogeográficas y las formaciones geobotánicas del Ecuador. *Revista Acad. Colombian Ciencias Exact.* 12: 401–447.

——. 1968. *Divisiones Fitogeográficas y Formaciones Geobotánicas del Ecuador.* Quito: Casa de la Cultura Ecuatoriana.

——. 1970. *Geografía y Ecología de las Tierras Aridas del Ecuador.* Contrib. 72. Quito: Instituto Ecuatoriana de Ciencias Naturales.

——. 1977. *Ecología y Fitoecología.* Quito: Casa de la Cultura Ecuatoriana.

Balslev, H., and J. L. Luteyn, eds. 1992. *Pá-* *ramo: An Andean Ecosystem under Human Influence.* London: Academic Press.

Beard, J. S. 1944. Climax vegetation in tropical America. *Ecology* 25: 127–158.

Berry, P. E. 1982. The systematics and evolution of *Fuchsia* sect. *Fuchsia* (Onagraceae). *Annals of the Missouri Botanical Garden* 69: 1–237.

Bono, G. 1996. *Flora y vegetación del Estado Táchira, Venezuela.* Monografie 20. Turin, Italy: Museo Regionale di Scienze Naturali.

Börgel, R. 1973. The coastal desert of Chile. In D. H. K. Amiran and A. W. Wilson, eds., *Coastal Deserts: Their Natural and Human*

Environments, pp. 111–114. Tucson: University of Arizona Press.

Brockmann, C. E. 1978. *Mapa de Cobertura y Uso Actual de la Tierra, Bolivia*. Memoria explicativa. La Paz: Servicio Geológico de Bolivia, Programa ERTS.

Brush, S. B. 1982. The natural and human environment of the Central Andes. *Mountain Research and Development* 2: 19–38.

Cabrera, A. L. 1957 [1958]. La vegetación de la puna argentina. *Revista Invest. Agric.* (Buenos Aires) 11: 317–412, plates 1–17.

———. 1968. Ecologia vegetal de la puna. In C. Troll, ed., *Colloquium Geographicum, Band 9: Geoecology of the Mountainous Regions of the Tropical Americas*, pp. 91–116. UNESCO.

Cañadas Cruz, L. 1983. *El mapa bioclimático y ecológico del Ecuador*. Quito: Banco Central del Ecuador.

Cañadas Cruz, L., and W. Estrada. 1978. *Ecuador Mapa Ecológico*. Quito: PRONAREG-Ecuador, Ministerio de Agricultura y Ganaderia.

CDC-Peru, Fundación Peruana para la Conservación de la Naturaleza, and O. Herrera-MacBryde. 1997. Cerros de Amotape National Park region (north-western Peru). In S. D. Davis, V. H. Heywood, O. Herrera-MacBryde, J. L. Villa-Lobos, and A. C. Hamilton, eds., *Centres of Plant Diversity: A Guide and Strategy for Their Conservation, 3: The Americas*, pp. 513–518. Cambridge: WWF and IUCN.

Churchill, S. P., H. Balslev, E. Forero, and J. L. Luteyn, eds. 1995a. *Biodiversity and Conservation of Neotropical Montane Forests*. Bronx: New York Botanical Garden.

Churchill, S. P., D. Griffin III, and M. Lewis. 1995b. Moss diversity of the tropical Andes. In S. P. Churchill, H. Balslev, E. Forero, and J. L. Luteyn, eds., *Biodiversity and Conservation of Neotropical Montane Forests*, pp. 335–346. Bronx: New York Botanical Garden.

Cleef, A. M. 1981. The vegetation of the páramos of the Colombian Cordillera Oriental. *Diss. Bot.* 41: 1–320.

———. 1997a. Sierra Nevada del Cocuy-Guantiva (Colombia). In S. D. Davis, V. H. Heywood, O. Herrera-MacBryde, J. L. Villa-Lobos, and A. C. Hamilton, eds., *Centres of Plant Diversity: A Guide and Strategy for Their Conservation, 3: The Americas*, pp. 431–436. Cambridge: WWF and IUCN.

———. 1997b. Páramo de Sumapaz region (Colombia). In S. D. Davis, V. H. Heywood, O. Herrera-MacBryde, J. L. Villa-Lobos, and A. C. Hamilton, eds., *Centres of Plant Diversity: A Guide and Strategy for Their Conservation, 3: The Americas*, pp. 437–441. Cambridge: WWF and IUCN.

Cuatrecasas, J. 1957. A sketch of the vegetation of the North-Andean Province. *Proceedings of the 8th Pacific Scientific Congress* 4: 167–173.

———. 1958. Aspectos de la vegetación natural de Colombia. *Revista Acad. Colombian Ciencias Exact.* 10(40): 221–264. Reprinted in *Perez-Arbelaezia* 11 (1989): 155–283.

———. 1986. Speciation and radiation of the Espeletiinae in the Andes. In F. Vuilleumier and M. Monasterio, eds., *High Altitude Tropical Biogeography*, pp. 267–303. New York: Oxford University Press.

Davis, S. D., V. H. Heywood, O. Herrera-MacBryde, J. L. Villa-Lobos, and A. C. Hamilton, eds. 1997. *Centres of Plant Diversity: A Guide and Strategy for Their Conservation, 3: The Americas*. Cambridge: WWF and IUCN.

Diels, L. 1937. Beiträge zur Kenntnis der Vegetation und Flora von Ecuador. *Bibl. Bot.* 29: 1–190.

Dillon, M. O. 1997. Lomas formations (Peru). In S. D. Davis, V. H. Heywood, O. Herrera-MacBryde, J. L. Villa-Lobos, and A. C. Hamilton, eds., *Centres of Plant Diversity: A Guide and Strategy for Their Conservation, 3: The Americas*, pp. 519–527. Cambridge: WWF and IUCN.

Dillon, M. O., and A. E. Hoffmann-J. 1997. Lomas formations of the Atacama Desert (northern Chile). In S. D. Davis, V. H. Heywood, O. Herrera-MacBryde, J. L. Villa-Lobos, and A. C. Hamilton, eds., *Centres of Plant Diversity: A Guide and Strategy for Their Conservation, 3: The Americas*, pp. 528–535. Cambridge: WWF and IUCN.

Dillon, M. O., A. Sagástegui A., I. Sánchez

Vega, S. Llatas Quiroz, and N. Hensold. 1995. Floristic inventory and biogeographic analysis of montane forests in northwestern Peru. In S. P. Churchill, H. Balslev, E. Forero, and J. L. Luteyn, eds., *Biodiversity and Conservation of Neotropical Montane Forests,* pp. 251–270. Bronx: New York Botanical Garden.

Dodson C. H., and A. H. Gentry. 1978. Flora of Río Palenque. *Selbyana* 4: 1–628.

———. 1991. Biological extinction in western Ecuador. *Annals of the Missouri Botanical Garden* 78: 273–295.

Dodson C. H., A. H. Gentry, and F. M. Valverde. 1985. Flora of Jauneche. *Selbyana* 8: 1–512.

Duellman, W. E. 1979. The herpetofauna of the Andes: Patterns of distribution, origin, differentiation, and present communities. *Museum of Natural History, University of Kansas Monographs* 7: 371–459.

ERTS. 1978. *Mapa de Cubertura y Uso Actual de la Tierra, Bolivia.* La Paz: Servicio Geológico de Bolivia, Programa ERTS.

Espinal, S., and E. Montenegro. 1963. *Formaciones Vegetales de Colombia.* Bogotá: Instituto Geográfico Agustín Codazzi.

Espinosa, R. 1997. *Estudios Botánicos en el Sur del Ecuador, 1 y 2.* 2nd ed. Loja: Herbario Loja, Reinaldo Espinosa (Universidad Nacional de Loja y Universidad de Aarhus).

Ewel, J. J., A. Madríz, and J. A. Tosi Jr. 1976. *Zonas de Vida de Venezuela.* Caracas: Ministerio de Agricultura y Cría, Fondo Nacional de Investigaciones Agropecuarias.

Ferreyra, R. 1953. Comunidades de vegetales de algunas lomas costañeras del Perú. *Estac. Exp. Agrícola La Molina, Bol.* 53: 1–88.

———. 1957. Contribución al conocimiento de la flora costañera del norte peruano (Departamento de Tumbes). *Boletin de la Sociedad Argentina de Botánico* 6: 194–206.

———. 1961. Las lomas costañeras del extremo sur del Peru. *Boletin de la Sociedad Argentina de Botánico* 9: 87–120.

———. 1979. El algarrobal y manglare de la costa norte del Perú. *Bol. Lima* 1: 12–18.

———. 1983. Los tipos de vegetación de la costa peruana. *Anales Jardín Botánico Madrid* 40: 241–256.

Forero, E. 1989. Colombia. In D. G. Campbell and H. D. Hammond, eds., *Floristic Inventory of Tropical Countries,* pp. 353–361. Bronx: New York Botanical Garden.

Forero, E., and A. Gentry. 1989. Lista anotada de las plantas del departamento del Chocó. *Bibliot. José Jerónimo Triana* 10: 1–142.

Gentry, A. 1980. The Flora of Peru: A conspectus. *Fieldiana, Botany* n.s. 5: 1–11.

———. 1982. Neotropical floristic diversity: Phytogeographical connections between Central and South America, Pleistocene climatic fluctuations, or an accident of the Andean orogeny? *Annals of the Missouri Botanical Garden* 69: 557–593.

———. 1986. Species richness and floristic composition of the Chocó region plant communities. *Caldasia* 15: 71–79.

———. 1989. Northwest South America (Colombia, Ecuador, and Peru). In D. G. Campbell and H. D. Hammond, eds., *Floristic Inventory of Tropical Countries,* pp. 391–400. Bronx: New York Botanical Garden.

———. 1993. Overview of the Peruvian flora. In L. Brako and J. L. Zarucchi, eds., *Catalogue of the Flowering Plants and Gymnosperms of Peru (Catálogo de las Angiospermas y Gimnospermas del Perú),* pp. xxix–xl. St. Louis: Missouri Botanical Garden.

———. 1995a. Patterns of diversity and floristic composition in Neotropical montane forests. In S. P. Churchill, H. Balslev, E. Forero, and J. L. Luteyn, eds., *Biodiversity and Conservation of Neotropical Montane Forests,* pp. 103–126. Bronx: New York Botanical Garden.

———. 1995b. Diversity and floristic composition of Neotropical dry forests. In S. H. Bullock, H. A. Mooney, and E. Medina, eds., *Seasonally Dry Tropical Forests,* pp. 146–194. Cambridge: Cambridge University Press.

———. 1997. Regional overview: South America. In S. D. Davis, V. H. Heywood, O. Herrera-MacBryde, J. L. Villa-Lobos, and A. C. Hamilton, eds., *Centres of Plant Diversity: A Guide and Strategy for Their Conservation, 3: The Americas,* pp. 269–307. Cambridge: WWF and IUCN.

Grubb, P. J. 1974. Factors controlling the distribution of forest-types on tropical mountains: New facts and a new perspective. In J. R. Flenley, ed., *Altitudinal Zonation in Malesia*, pp. 13–46. Hull, England: Department of Geography, University of Hull.

Hamilton, L. S., J. O. Juvik, and F. N. Scatena, eds. 1995. *Tropical Montane Cloud Forests*. Ecological Studies 110. New York: Springer-Verlag.

Harling, G. 1979. The vegetation types of Ecuador: A brief survey. In K. Larsen and L. B. Holm-Nielsen, eds., *Tropical Botany*, pp. 165–174. London: Academic Press.

Hastenrath, S. 1968. Certain aspects of the three-dimensional distribution of climate and vegetation belts in the mountains of Central America and southern Mexico. In C. Troll, ed., *Colloquium Geographicum, Band 9: Geoecology of the Mountainous Regions of the Tropical Americas*, pp. 122–130. UNESCO.

Henderson, A., S. P. Churchill, and J. L. Luteyn. 1991. Neotropical plant diversity. *Nature* 351: 21–22.

Herzog, T. 1923. Die Pflanzenwelt der bolivischen Anden und ihres östlichen Vorlandes. In A. Engler and D. Drude, eds., *Die Vegetation der Erde* 15: 1–259.

Holdridge, L. R. 1967. *Life Zone Ecology*. San José, Costa Rica: Tropical Science Center.

Holdridge, L. R., W. C. Grenke, W. H. Hatheway, T. Liang, and J. A. Tosi Jr. 1971. *Forest Environments in Tropical Life Zones: A Pilot Study*. Oxford: Pergamon Press.

Huber, O., and D. Frame. 1989. Venezuela. In D. G. Campbell and H. D. Hammond, eds., *Floristic Inventory of Tropical Countries*, pp. 362–374. Bronx: New York Botanical Garden.

Huber, O. and R. Riina, eds. 1997. *Glosario Fitoecológico de las Américas, vol. 1: América del Sur: Países Hispanoparlantes*. Caracas, Venezuela: UNESCO, Fundación Instituto Botánico de Venezuela.

Hueck, K. 1960. *Mapa de la Vegetación de la República de la Venezuela*. Mérida, Venezuela: Instituto Forestal Latinoamericano de Investigación y Capacitación.

Instituto Geográfico Agustín Codazzi. 1977. *Zonas de Vida o Formaciones Vegetales de Colombia: Memoria Explicativa sobre el Mapa Ecológico*. Bogotá.

Jørgensen, P. M., and S. León, eds. 1999. Catalogue of the vascular plants of Ecuador. *Monographs in Systematic Botany, Missouri Botanical Garden* 75: viii, 1–1181.

Jørgensen, P. M., C. Ulloa Ulloa, J. E. Madsen, and R. Valencia R. 1995. A floristic analysis of the high Andes of Ecuador. In S. P. Churchill, H. Balslev, E. Forero, and J. L. Luteyn, eds., *Biodiversity and Conservation of Neotropical Montane Forests*, pp. 221–237. Bronx: New York Botanical Garden.

Kessler, M. 1992. The vegetation of southwest Ecuador. In B. J. Best, ed., *The Threatened Forests of South-West Ecuador*, pp. 79–100. Leeds, U.K.: Biosphere Publications. [Cited in Gentry 1997].

———. 1995. Present and potential distribution of *Polylepis* (Rosaceae) forests in Bolivia. In S. P. Churchill, H. Balslev, E. Forero, and J. L. Luteyn, eds., *Biodiversity and Conservation of Neotropical Montane Forests*, pp. 281–294. Bronx: New York Botanical Garden.

Kress, W. J., ed. 1986. A symposium: The biology of tropical epiphytes. *Selbyana* 9: 1–270.

Laegaard, S. 1992. Influence of fire in the grass páramo vegetation of Ecuador. In H. Balslev and J. L. Luteyn, eds., *Páramo: An Andean Ecosystem under Human Influence*, pp. 151–170. London: Academic Press.

Leigh, E. G. 1975. Structure and climate in tropical rain forest. *Annual Rev. Ecol. Syst.* 6: 67–86.

Little, M. A. 1981. Human populations of the Andes: The human science basis for research planning. *Mountain Research and Development* 1: 145–170.

Lojtnant, B., and U. Molau. 1982. Analysis of a virgin páramo plant community on Volcán Sumaco, Ecuador. *Nordic J. Botany* 2: 567–574.

Luteyn, J. L. 1992. Páramos: Why study them? In H. Balslev and J. L. Luteyn, eds., *Páramo: An Andean Ecosystem under Human Influence*, pp. 1–14. London: Academic Press.

———. 1999. *Páramos: A Checklist of Plant Diversity, Geographical Distribution, and*

Botanical Literature. Mem. New York Botany Gard. 84: 1–278. Bronx: New York Botanical Garden Press.

Luteyn, J. L., A. M. Cleef, and O. Rangel Ch. 1992. Plant diversity in páramo: Towards a checklist of páramo plants and a generic flora. In H. Balslev and J. L. Luteyn, eds., *Páramo: An Andean Ecosystem under Human Influence,* pp. 71–84. London: Academic Press.

Lynch, J. D., P. M. Ruiz-C., and M. C. Ardila-R. 1997. Biogeographic patterns of Colombian frogs and toads. *Revista Acad. Colombian Ciencias* 21: 237–248.

Madison, M. 1977. Vascular epiphytes: Their systematic occurrence and salient features. *Selbyana* 2: 1–13.

Mena, P., C. Ulloa U., and O. Herrera-MacBryde. 1997. Páramos and Andean forests of Sangay National Park. In S. D. Davis, V. H. Heywood, O. Herrera-MacBryde, J. L. Villa-Lobos, and A. C. Hamilton, eds., *Centres of Plant Diversity: A Guide and Strategy for Their Conservation, 3: The Americas,* pp. 458–464. Cambridge: WWF and IUCN.

Molau, U. 1988. Scrophulariaceae, Part 1: Calceolarieae. *Flora Neotropica Monograph* 47: 1–326.

Molina, E. G., and A. V. Little. 1981. Geoecology of the Andes: The natural science basis for research planning. *Mountain Research and Development* 1: 115–144.

Müller, G. K. 1985. Zur floristischen Analyse der peruanischen Loma-Vegetation. *Flora* 176: 153–165.

Neill, D. A. 1997. Ecuadorian Pacific Coast mesic forests (Ecuador). In S. D. Davis, V. H. Heywood, O. Herrera-MacBryde, J. L. Villa-Lobos, and A. C. Hamilton, eds., *Centres of Plant Diversity: A Guide and Strategy for Their Conservation, 3: The Americas,* pp. 508–512. Cambridge: WWF and IUCN.

Neill, D. A., and W. A. Palacios. 1997. Gran Sumaco and Upper Napo River region (Ecuador). In S. D. Davis, V. H. Heywood, O. Herrera-MacBryde, J. L. Villa-Lobos, and A. C. Hamilton, eds., *Centres of Plant Diversity: A Guide and Strategy for Their Conservation, 3: The Americas,* pp. 496–500. Cambridge: WWF and IUCN.

ONERN. 1971. *Inventario, evaluación, y uso racional de los recursos naturales de la costa: Cuenca del Río Grande (Nazca).* Lima: Oficina Nacional de Evaluación de Recursos Naturales.

Parker III, T. A, and J. L. Carr, eds. 1992. *Status of Forest Remnants in the Cordillera de la Costa and Adjacent Areas of Southwestern Ecuador.* RAP Working Papers 2. Washington, D.C.: Conservation International.

Pittier, H., and L. Williams. 1945. A review of the flora of Venezuela. In F. Verdoorn, ed., *Plants and Plant Science in Latin America,* pp. 102–105. Waltham, Mass.: Chronica Botanica.

Prance, G. T. 1989. American tropical forests. In H. Lieth and M. J. A. Werger, eds., *Tropical Rain Forest Ecosystems,* pp. 99–132. Amsterdam: Elsevier.

Rangel-Ch., J. O., and A. Garzón-C. 1997a. Sierra Nevada de Santa Marta (Colombia). In S. D. Davis, V. H. Heywood, O. Herrera-MacBryde, J. L. Villa-Lobos, and A. C. Hamilton, eds., *Centres of Plant Diversity: A Guide and Strategy for Their Conservation, 3: The Americas,* pp. 426–430. Cambridge: WWF and IUCN.

———. 1997b. Region of Los Nevados Natural National Park (Colombia). In S. D. Davis, V. H. Heywood, O. Herrera-MacBryde, J. L. Villa-Lobos, and A. C. Hamilton, eds., *Centres of Plant Diversity: A Guide and Strategy for Their Conservation, 3: The Americas,* pp. 442–447. Cambridge: WWF and IUCN.

———. 1997c. Colombian Central Massif (Colombia). In S. D. Davis, V. H. Heywood, O. Herrera-MacBryde, J. L. Villa-Lobos, and A. C. Hamilton, eds., *Centres of Plant Diversity: A Guide and Strategy for Their Conservation, 3: The Americas,* pp. 448–452. Cambridge: WWF and IUCN.

———. 1997d. Volcanoes of Nariñense Plateau (Colombia and Ecuador). In S. D. Davis, V. H. Heywood, O. Herrera-DacBryde, J. L. Villa-Lobos, and A. C. Hamilton, eds., *Centres of Plant Diversity: A Guide and Strategy for Their Conservation, vol. 3: The Americas,* pp. 453–457. Cambridge: WWF and IUCN.

Rauh, W. 1985. The Peruvian-Chilean deserts.

In M. Evenari, I. Noy-Meir, and D. W. Goodall, eds., *Hot Deserts and Arid Shrublands, A: Ecosystems of the World, 12A,* pp. 239–267. Amsterdam: Elsevier.

Renjifo, L. M., G. P. Servat, J. M. Goerck, B. A. Loiselle, and J. G. Blake. 1998. Patterns of species composition and endemism in the northern Neotropics: A case for conservation of montane avifaunas. In J. V. Remsen, ed., *Natural History and Conservation of Neotropical Birds,* Ted Parker Memorial Volume, pp. 577–594. Ornithological Monographs 48. Washington, D.C.: American Ornithologist's Union.

Richards, P. W. 1996. *The Tropical Rain Forest.* 2nd ed. Cambridge: Cambridge University Press.

Rundel, P. W., M. O. Dillon, B. Palma, H. A. Mooney, S. L. Gulmon, and J. R. Ehleringer. 1991. The phytogeography and ecology of the coastal Atacama and Peruvian deserts. *Aliso* 13: 1–49.

Rundel, P. W., and M. Mahu. 1976. Community structure and diversity of a coastal fog zone in northern Chile. *Flora* 165: 493–505.

Rundel, P. W., A. P. Smith, and F. C. Meinzer, eds. 1994. *Tropical Alpine Environments: Plant Form and Function.* Cambridge: Cambridge University Press.

Schjellerup, I. 1989. *Children of the Stones.* Publ. 7. Copenhagen: Royal Danish Academy of Sciences and Letters.

Seibert, P. 1983. Human impact on landscape and vegetation in the Central High Andes. In W. Holzner, M. J. A. Werger, and I. Ikusima, eds., *Man's Impact on Vegetation.* The Hague: Dr. W. Junk Publishers.

Smith, A. C., and I. M. Johnston. 1945. A phytogeographic sketch of Latin America. In F. Verdoorn, ed., *Plants and Plant Science in Latin America,* pp. 11–18. Waltham, Mass.: Chronica Botanica.

Smith, J. M. B., and A. M. Cleef. 1988. Composition and origins of the world's tropic-alpine floras. *J. Biogeography* 15: 631–645.

Sodiro, A. 1874. *Apuntes Sobre la Vegetación Ecuatoriana.* Quito.

Solomon, J. C. 1989. Bolivia. In D. G. Campbell and H. D. Hammond, eds., *Floristic Inventory of Tropical Countries,* pp. 457–463. Bronx: New York Botanical Garden.

Stadtmüller, T. 1987. *Cloud Forests in the Humid Tropics: A Bibliographic Review.* (English translation of *Los Bosques Nublados en el Trópico Húmedo: Una Revisión Bibliográfica* by Noël D. Payne.) Tokyo: United Nations University.

Svenson, H. K. 1946. Vegetation of the coast of Ecuador and Peru and its relation to the Galápagos Islands. *American J. Botany* 33: 394–498.

Tamayo, F. 1958. Notas explicativas del ensayo del mapa fitogeográfico de Venezuela (1955). *Revista Forest. Venezuela* 1(1): 7–31.

———. 1979. Vegetación. In *Atlas de Venezuela,* 2nd ed., pp. 184–187. Caracas, Venezuela: Ministerio del Ambiente y de los Recursos Naturales Renovables, Dirección General de Información e Investigación del Ambiente, Dirección de Cartografía Nacional.

Taylor, D. W. 1995. Cretaceous to Tertiary geologic and angiosperm paleobiogeographic history of the Andes. In S. P. Churchill, H. Balslev, E. Forero, and J. L. Luteyn, eds., *Biodiversity and Conservation of Neotropical Montane Forests,* pp. 3–9. Bronx: New York Botanical Garden.

Tosi, J., O. Unzueta, L. Holdridge, and A. González. 1975. *Mapa Ecológico de Bolivia.* La Paz: Ministerio de Asuntos Campesinos y Agropecuarios.

Troll, C., ed. 1968a. *Colloquium Geographicum, Band 9: Geoecology of the Mountainous Regions of the Tropical Americas.* UNESCO.

———. 1968b. The cordilleras of the tropical Americas. Aspects of climate, phytogeographical, and agrarian ecology. In C. Troll, ed., *Colloquium Geographicum, Band 9: Geoecology of the Mountainous Regions of the Tropical Americas,* pp. 15–56. UNESCO.

Unzueta, O. 1975. *Memoria Explicativa: Mapa Ecológico de Bolivia.* La Paz: Ministerio de Asuntos Campesinos y Agropecuarios.

Valencia, R., and P. M. Jørgensen. 1992. Composition and structure of a humid montane forest on the Pasochoa volcano, Ecuador. *Nordic J. Botany* 12: 239–247.

Valverde, F. M., G. de Tazán, and C. García R.

1979. Cubierta vegetal de la peninsula de Santa Elena. *Fac. Cienc. Nat. Publ.* 2: 1–236. [Cited in Dodson and Gentry 1991.]

Vuilleumier, F. 1969. Pleistocene speciation in birds living in the high Andes. *Nature* 223: 1179–1180.

Vuilleumier, F., and M. Monasterio, eds. 1986. *High Altitude Tropical Biogeography*. New York: Oxford University Press.

Walter, H. 1973. *Vegetation of the Earth*. (Translated from the second German edition by Joy Wieser.) New York: English Universities Press. London: Springer-Verlag.

Weberbauer, A. 1911. *Die Pflanzenwelt der Peruanischen Anden*. Vegetation der Erde 12. Leipzig: Englemann.

———. 1936. Phytogeography of the Peruvian Andes. *Publ. Field Mus. Nat. Hist., Bot. Ser.* 13(1): 13–81.

———. 1945. *El Mundo Vegetal de los Andes Peruanos*. Lima: Ministerio de Agricultura.

Webster, G. L. 1995. The panorama of Neotropical cloud forests. In S. P. Churchill, H. Balslev, E. Forero, and J. L. Luteyn, eds., *Biodiversity and Conservation of Neotropical Montane Forests*, pp. 53–77. Bronx: New York Botanical Garden.

Whitmore, T. C. 1998. *An Introduction to Tropical Rain Forests*. 2nd ed. Oxford: Clarendon Press.

Wolf, T. 1892. *Geografía y Geología del Ecuador*. Leipzig.

Young, K. R., and B. León. 1997. Eastern slopes of Peruvian Andes (Peru). In S. D. Davis, V. H. Heywood, O. Herrera-MacBryde, J. L. Villa-Lobos, and A. C. Hamilton, eds., *Centres of Plant Diversity: A Guide and Strategy for Their Conservation, 3: The Americas*, pp. 490–495. Cambridge: WWF and IUCN.

Young, K. R., and C. Reynel. 1997. Huancabamba region (Peru and Ecuador). In S. D. Davis, V. H. Heywood, O. Herrera-MacBryde, J. L. Villa-Lobos, and A. C. Hamilton, eds., *Centres of Plant Diversity: A Guide and Strategy for Their Conservation, 3: The Americas*, pp. 465–469. Cambridge: WWF and IUCN.

Young, K. R., B. León, A. Cano, and O. Herrera-MacBryde. 1997. Peruvian puna (Peru). In S. D. Davis, V. H. Heywood, O. Herrera-MacBryde, J. L. Villa-Lobos, and A. C. Hamilton, eds., *Centres of Plant Diversity: A Guide and Strategy for Their Conservation, 3: The Americas*, pp. 470–476. Cambridge: WWF and IUCN.

CLARK L. ERICKSON

12 | THE LAKE TITICACA BASIN
A Precolumbian Built Landscape

*Any understanding
of contemporary
biodiversity change in
the Americas is likely to
be uninformative and
misleading if it employs
a prehistoric baseline
imbued with pristine
characteristics.*
— STAHL(1996:105)

The landscapes of the Americas hold a material record of a long and complex history of human transformation of the environment. As William Denevan (1992) has pointed out, "the myth of the pristine environment" has long dominated the literature on the environments of the Americas. The human impact on the land before the arrival of Europeans was so profound and at such a massive scale that it could be argued few, if any, of the environments of the Americas occupied by humans past and present could be considered natural or pristine. Humans have cut, cleared, and burned forests for agriculture and settlement, maintained savannas through annual burning, determined plant and animal species composition through direct and indirect selection, transformed hill slopes and wetlands into productive farmland, and made deserts bloom through irrigation agriculture (e.g., Denevan 1992, 2000; Treacy 1994; Stahl 1996; Balée and Posey 1989; Doolittle 1992).

One of the best examples of an anthropogenic or human-built landscape is the Lake Titicaca Basin of present-day Bolivia and Peru (figure 12.1). The high-altitude basin, located between the eastern and western Andean cordilleras, covers an area of 57,000 km². The area adjacent to Lake Titicaca has long been a major center of agricultural production and dense human populations and the home of several important Precolumbian civilizations. Over the past 8,000 years, the environment of the basin has been transformed into a highly patterned, artificial landscape (figure 12.2). The construction of raised fields (*waru waru, suka kollas*), stone-faced terraces (*andenes*), sunken gardens (*q'ochas*), irrigated pasture (*bofedales*), and a multitude

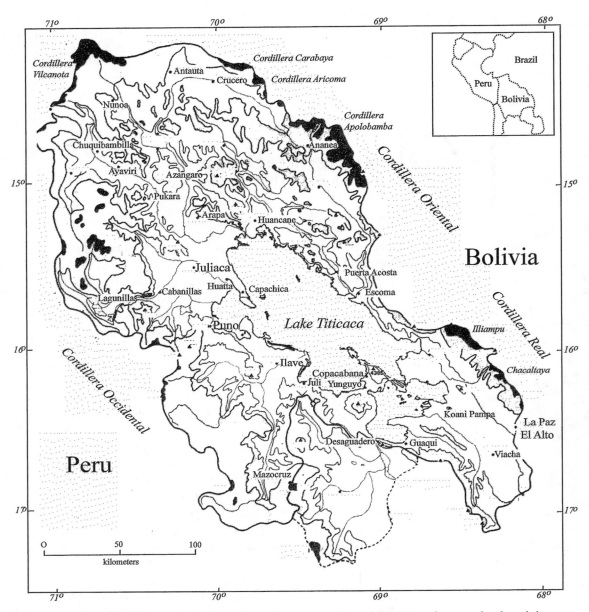

FIGURE 12.1. The Lake Titicaca Basin. The lake surface is approximately 3,810 m above sea level, and the cordilleras bordering the basin are more than 5,000 m above sea level. *Source:* After Boulange and Aquize 1981.

FIGURE 12.2. Patterned raised field landscape near Pomata, Puno, Peru. The platforms (dark linear features) are approximately 20 m wide and 30–70 m long.

of features related to the infrastructure of agriculture and settlement are essential elements of this anthropogenic landscape. Although Precolumbian states may have been responsible for some of the transformation (e.g., Kolata 1993; Stanish 1994), I believe that the majority of the cultural landscape was constructed piecemeal by rural farming peoples through their daily activities. These people, and their descendants, the Aymara and Quechua, are what Netting (1993) describes as "smallholders," farmers who practice small-scale intensive agriculture, making physical improvements to their lands, which are inherited by their descendants. The built environment represents the landscape capital of hundreds of generations of farmers and herders and reflects a rich indigenous knowledge system (Erickson 1993, 1996; Morlon 1996; Denevan 2000; Zimmerer 1996).

In this essay, I will address the issue of the long-term relationship of people and the environment of the Lake Titicaca Basin. Since the end of the Pleistocene, there is little in the landscape that could be considered natural or pristine. This is a truly anthropogenic landscape, one in which humans played a central, active role in shaping the past and present environment. I will discuss the archaeological record for this transformation in terms of a human-centric perspective.

HUMANS AND THE ANDEAN ENVIRONMENT

Our understanding of the relationship between humans and the Andean environment draws upon ecology, evolutionary ecology, cultural ecology, cultural materialism, human ecology, agroecology, geoecology, landscape ecology, political economy, foraging theory, farming systems, system theory, political ecology, and historical ecology. Although these approaches share basic theoretical assumptions about the relationship between humans and the environment, they often differ in important areas such as causality (whether they attribute conditions and change of those conditions to human or natural causes), temporal scale (whether they emphasize the short term or long term), and static-dynamics (whether they stress continuity or change). These approaches can be reduced to four broad and somewhat overlapping categories: (1) the nature-centric perspective, (2) the human adaptation perspective, (3) the environmental determinism perspective, and (4) the human-centric perspective.

THE NATURE-CENTRIC PERSPECTIVE: AN ANDES WITHOUT PEOPLE

The nature-centric perspective (geoecology, landscape ecology) treats the Andes as natural history, a given and often assumed to be constant (at least since the end of the Pleistocene). The Holdridge classification applied by Tosi (1960) to the Andean region is based on the assumption that latitude, altitude, and rainfall determine climax communities of vegetation in tropical montane environments. These "natural" communities are expected to take the form of vertically stacked ecological tiers (e.g., Tosi 1960; ONERN-CORPUNO 1965; Troll 1968; Ellenberg 1979; Dollfus 1982). Environmental change in the Holocene is generally attributed to continental and/or global climatic change (Abbott et al. 1997; Wirrmann et al. 1992; Ybert 1992). Human influence on the natural environment is downplayed or factored out in an attempt to define a "pristine," "original," or "climax" Andean environment (e.g., Seibert 1983; Ellenberg 1979). Precolumbian peoples are often characterized as having (1) had little or no impact on the environment or (2) achieved a harmonious state of equilibrium with the Andean environment through sustainable and ecologically sound agropastoral practices adapted to local conditions (usually attributed to the Inka). In the versions that accept the human factor, the environment is considered "fragile," in that humans can disrupt the mature or climax state of the natural environment (e.g., Glaser and Celecia 1981; Gomez-Molina and Little 1981; Seibert 1983). Environmental change and degradation (removal of natural vegetation through agriculture, overgrazing, fuel collection, burning, soil erosion and exhaustion, and desertification) is often attributed to Colonial policies and modern world systems (ONERN-CORPUNO 1965; Dollfus 1982; Winterhalder and Thomas 1978; Seibert 1983; Gade 1992).

HUMAN ADAPTATION PERSPECTIVE: THE CULTURE OF ECOLOGY

In the human adaptation perspective, humans "adapt to," "interact with," "impact," and "influence" the Andean natural environment (e.g., Seibert 1983; Dollfus 1982; Troll 1968; Knapp 1991; Tosi 1960; Winterhalder and Thomas 1978; Kuznar 1993; Aldenderfer 1998). Humans adapt to the Andean environment through rational and efficient practices of energy use and manage resources through verticality or ecological complementarity (field scattering, sectorial fallow systems, the ideology of reciprocity, scheduling of seasonal activities, high crop diversity, food storage technology, and land races appropriate for specific local conditions). In these functionalist and neofunctionalist interpretations, Andean cultural institutions or strategies attempt to reach a state of equilibrium or homeostasis with local environments. This approach has been incorporated into contemporary schemes for "sustainable development" and "appropriate technology" (e.g., Browder 1989; Morlon 1996). Most human adaptation studies are synchronic and ahistorical, treating the environment and cultural adaptation to it as static and given.

In the revisionist human adaptation perspective, people are given a causal and active role in shaping human-environment relations (e.g., Knapp 1991; Ellenberg 1979; Treacy 1994; Allan et al. 1988; Brush 1976). The environment is an open system capable of change. Based on systemic regularities of the physical environment and rational human decision making, the "adaptative dynamics" of humans to the environment can be predicted or retrodicted (Knapp 1991). The human adaptation perspective and its variants dominate archaeological interpretations of the Andean past.

THE NEOENVIRONMENTAL DETERMINISM PERSPECTIVE: HUMANS AT THE MERCY OF CLIMATE CHANGE

There is an increasing recognition that the environments of the southern Andes have not been stable during the Holocene (Cardich 1985; Binford et al. 1997; Abbott et al.

1997; Thompson et al. 1988; Kolata 1993, 1996; Chepstow-Lusty et al. 1998; Shimada et al. 1991). This perspective acknowledges the long-term historical dimension of human-environment relations in the Andes. Most paleoenvironmental reconstructions for the basin and surrounding regions are based on analyses of sediment and glacial cores (analysis of sediments, ice accumulation, pollen, unstable isotopes, and radiocarbon dating). Climate is assumed to fluctuate around some norm or benchmark (often averages based on historical records of lake level, precipitation, and temperature). Changes in archaeological settlement patterns, agropastoral strategies, sociopolitical organization, and environmental deterioration are causally linked to major "abnormal" climate change (mega El Niños, Little Ice Age, and "chronic droughts") (e.g., Paulsen 1976; Shimada et al. 1991; Binford et al. 1997; Kolata 1993, 1996). In this perspective, humans are passive and assumed helpless in the face of extreme environmental perturbations such as long-term droughts and floods. Anthropogenic processes are secondary to large-scale and long-term natural processes. Human activities are rarely considered as possible explanations of perturbations and discontinuities recorded in sediment and ice cores (e.g., Chepstow-Lusty et al. 1998).

THE HUMAN-CENTRIC PERSPECTIVE:
THE ANTHROPOGENIC ENVIRONMENT

There is an increasing recognition that humans play an active and important role in modifying, creating, transforming, and maintaining the environments in which they live. The human-centric perspective incorporates elements of historical ecology (Crumley 1994; Balée and Posey 1989; Kirch and Hunt 1997), the archaeology of landscapes (Yarmin and Metheny 1996; Tilley 1994; Erickson n.d.), the new ecology (Botkin 1990; Stahl 1996; Zimmerer 1994), and historical geography (Denevan 1992, 2000; Siemens 1998; Zimmerer 1996). This perspective emphasizes the cultural, anthropogenic, or built environment, in this case human modification, transformation, and creation of the landscapes over the long term. The concern is to understand how and why human actors consciously and unconsciously modified and created the cultural landscape for economic, political, social, and religious purposes (Bender 1998; Tilley 1994; Deetz 1990). The patterning of landscape features (pathways, roads, causeways, monuments, walls, gardens and fields and their boundaries, astronomical and calendrical sight lines, shrines, and sacred places) is examined in terms of the "social logic" that can provide insights into indigenous structures such as measurement systems, land tenure, social organization, cosmology, calendrics, astronomy, sacred geography, cognition, and ritual practices (e.g., Miller and Gleason 1994; Aveni 1990; Erickson 1993; Siemens 1998). The perspective considers human land use at multiple spatial and geographic scales (Crumley 1994). The approach also assumes that environments are dynamic and have complex, and often chaotic, histories (Botkin 1990; Zimmerer 1994; Stahl 1996).

I adopt a human-centric perspective for my discussion of the role of Andean peoples, past and present, in creating the landscapes of the Lake Titicaca Basin. I will argue that human agency over the long term must be central to any understanding of past and present environments in the Andes. Humans did not adapt to local environments, but rather they transformed and built the landscape in which they lived. Humans have so altered the natural environment that it no longer exists and probably has not existed for thousands of years. Humans have also played an important role in maintaining and increasing biodiversity of natural resources of the basin. The long-term development of the Andean landscapes is discussed in terms of archaeological, historical, and ethnographic evidence. This record shows that rural farming peoples survived, even thrived during periods of climatic perturbation and environmental change. Much of what has been interpreted in the paleoclimate studies as climate change may in fact be anthropogenic perturbation of the regional environment.

THE NATURAL ENVIRONMENT OF THE LAKE TITICACA BASIN

What is the natural environment of the Lake Titicaca Basin? What is the undisturbed climax ecosystem for the region? What was the pristine or original environment before humans colonized and transformed the region? Is there an environmental benchmark that can be used to compare and contrast the changes imposed by humans on the landscape over time? Most discussions begin with a survey of the natural environment followed by a discussion of how humans have adapted to or adjusted to the "harsh," "hostile," or "marginal" environments of the Lake Titicaca region. The assumption is that the basin environment is a given and stable within a range of cyclical variation or climatic fluctuation.

The transition from the Pleistocene to the Holocene in the Lake Titicaca Basin is believed to have occurred between 12,000 and 10,000 B.P. The reconstruction of the early Holocene environment is based on paleoclimatic evidence derived from dated glacial and lake sediment cores (Wirrmann et al. 1992; Ybert 1992; Binford et al. 1996; Abbott et al. 1997; Binford and Kolata 1996). This evidence suggests that during the Holocene, the climate fluctuated between periods of warm and cold and wetter and drier periods. Unfortunately, these reconstructions are imprecise and ambiguous. For example, one core shows little change over the past 6,000 years, suggesting an environment strikingly similar to that of today (Binford et al. 1996:106; Binford and Kolata 1996:36). The variations in sedimentation rates in other cores covering the same period are interpreted as dramatic climatic variation and lake level change (Binford and Kolata 1996; Abbott et al. 1997; Binford et al. 1997). On the basis of selected cores, Kolata and colleagues conclude that sustained agriculture could not have been practiced in the basin before 3500 B.P. due to an extended period of drought and low lake level (Binford and Kolata 1996; Binford et al. 1996).

More reliable reconstructions of climate and environment are available for later prehistory. By this time full-scale intensive agriculture so altered the landscape that there is no pristine baseline for comparison. To Kolata and colleagues, "normal climatic fluctuation" for the basin is based on averages of the short-term historical record of lake levels, annual rainfall, and temperatures. The cycles of instability and fluctuation interpreted from the cores are compared to these norms (Binford et al. 1997; Abbott et al.

1997; Binford and Kolata 1996). An alternative perspective is provided in the new ecology (Botkin 1990; Zimmerer 1994) and historical ecology (Crumley 1994), which considers the environment to be dynamic. Change, at times chaotic, is natural and expected. Environmental change is historically contingent rather than cyclical, varying around some norm. Thus there is no original pristine or climax baseline environment for comparison.

The present-day environment of the Lake Titicaca Basin has been described in many publications (Winterhalder and Thomas 1978; Kolata 1996; Dejoux and Iltis 1992; Morlon 1996; Luteyn and Churchill this volume). I will present an abbreviated scheme for the Lake Titicaca Basin that is based on Prehispanic, historical, and contemporary land use or production zones (for details, see Erickson n.d.; Kolata 1996; Morlon 1996; Winterhalder and Thomas 1978). The Quechua and Aymara of the Lake Titicaca Basin classify the landscape into the following categories: (1) *lago*, or the lake and permanent and semipermanent wetlands, (2) *pampa*, or the seasonally inundated lake plain, (3) *cerro*, or lower hill slopes near the lake (including islands and peninsulas), and (4) *puna*, or high-altitude grasslands.

LAGO. Lake Titicaca is located at 3,812 m above sea level and covers an area of approximately 8,100 km². Vast areas of seasonal and permanent wetland are found at the lake edge and along the rivers that feed Lake Titicaca. Dense communities of aquatic plants (*Schoenoplectus tatora, Myriophyllum, Elodea, Potamogeton*) dominate the wetlands and shorelines. The annual production of dry biomass in these wetlands has been calculated at 8 MT/ha (Vacher et al. 1991). The most important resource for humans is the totora reed (*Schoenoplectus tatora*), which provides material for roofing, mats, and boats, forage for livestock, and food for humans (the starchy roots). Fish and aquatic birds that thrive here are important in local diets. The annual harvest of fish is estimated to be 12,000 MT (Richarson 1991). Agriculture is more productive near the lake due to higher annual rainfall, warmer temperature, longer growing season, and richer soils (Vacher et al. 1991). Population densities are highest near the lake, in particular the zones with extensive wetland resources, and have been high for thousands of years.

Lake levels are dynamic and have fluctuated 6.5 m in the past century. This, combined with the flat topography surrounding the lake, has important consequences for wetland and farmland distribution. A lake level change of 1 m can either inundate or expose 120,000 ha of land surface. Most human occupation, past and present, is within or adjacent to the area most affected by changes in lake level. Humans play an important role in the creation and management of the lake's resources. Totora is cultivated in areas affected by fluctuating lake levels where stands have been overexploited. The appearance of totora pollen in cores at 3500 B.P. has been interpreted as evidence of cultivation of totora (Binford and Kolata 1996). Most of the wetlands have archaeological evidence of intensive farming and occupation (raised fields, canals, and occupation mounds).

PAMPA. Large expanses of pampa or grassland plains are found adjacent to Lake Titicaca and the major rivers that feed the lake. These flat low-lying areas are subject to frequent flooding by the lake and rivers. Agriculture is considered risky in these areas because of the high water table, frequent frosts, and heavy soils. The pampa currently is used for grazing introduced sheep and cattle. Precolumbian farmers built raised fields and q'ochas and occupied the tens of thousands of artificial mounds distributed across the pampa.

CERRO. The most heavily farmed lands today are the slopes adjacent to the lake (3,800 to 4,200 m). Crops cultivated on lower slopes near the lake and on peninsulas and islands are less affected by waterlogging, frost, and the short growing season. The indigenous crops include potato, oca, ullucu, isañu, tarwi, cañihua, maize, and quinoa. In the immediate vicinity of the lake, stone-faced terracing covers the slopes from valley or lake edge to hilltop, and eroded terraced fields can be found many kilometers from the lake. Because of continuous farming, all natural vegetation has long been removed.

PUNA. The puna is the cold, high-altitude grasslands between 4,100 m to the base of the mountain glaciers (4,600 m and above). The puna of Lake Titicaca is classified as humid puna (800–1,200 mm of annual rainfall). The vegetation is predominantly low mats of herbaceous vegetation, tussock grasses, *Distichia* moors, and remnant groves of *Polylepis* spp. and *Buddleja* spp. This is also the habitat of guanaco, vicuña, deer, and viscacha.

Although the puna appears natural and devoid of human activity, humans have played an important role in shaping this landscape. Ephemeral sites representing 8,000 years of seasonal and permanent occupation are densely distributed throughout the puna (Klink and Aldenderfer 1996). Frost-resistant potatoes can be grown under certain conditions as high as 4,500 m, and herding settlements are found as high as 5,210 m (Bowman 1916:52, figure 24). The remains of agropastoral infrastructure such as residences, walls, corrals, and irrigation canals and large special-purpose sites (cemeteries and pukaras, or forts) are common west of the lake (Stanish et al. 1997; Flores 1979:45–50). Vegetation is annually burned off to improve grazing. T'ola shrubs (*Baccharis* or *Lepidophyllum*) and yareta (*Azorella*), cushion plants rich in resin, have long been harvested for fuel (West 1987; Wickens 1995). Vast wetlands, or bofedales, either anthropogenic or artificial, are found throughout the puna.

EARLY HUMAN MANIPULATION OF THE ENVIRONMENT

The human impact on the landscapes of the Lake Titicaca Basin began with the arrival of the first peoples to the area around 8,000 years ago (table 12.1). Hunter-gatherer settlements dating to 8000–9000 B.P. have been found in the Moquegua drainage southwest of the basin (Aldenderfer 1998). A total of 240 sites dating to the Preceramic period (8000–3500 B.P.) were located in a recent survey of the Ilave River drainage on the western side of Lake Titicaca (Klink and Aldenderfer 1996). Preceramic sites are rare in the immediate vicinity of the lake (Stanish et al. 1997; Albarracín 1996:78–79). By 4000–3500 B.P., hunters and pastoralists were living in permanent settlements in the Ilave River drainage (Klink and Aldenderfer 1996).

Based on elevation and latitude, the Lake Titicaca Basin is classified as moist mon-

TABLE 12.1. CHRONOLOGY OF THE LAKE TITICACA BASIN

Time Scale	Culture	Period/Horizon
1750		
1500		
	Inka	Late Horizon
1250	*Aymara Kingdoms*	Late Intermediate Period
1000	*Tiwanaku V*	
750	*Tiwanaku IV*	Middle Horizon
500	*Tiwanaku III*	
250		Early Intermediate Period
AD/BC	*Pukara*	
250		
500		Early Horizon
750	*Chiripa* *Sillamocco*	
1000	*Qaluya* *Wankarani*	
1250		Initial Period
1500		
1750		
2000		
		Preceramic

tane forest. The stands of native trees (*Polylepis* spp.) are considered to be remnants of a once forested landscape (Ellenberg 1979; Budowski 1968; Winterhalder and Thomas 1978:77; Seibert 1983:265). Deforestation is attributed to long-term climate change (Morlon 1996; Cardich 1985), Precolumbian human degradation (Ellenberg 1979), and Colonial or modern degradation (Morlon 1996; West 1987; Seibert 1983). If the basin was indeed once forested, deforestation by humans for construction material, tool handles, and fuel probably began early and continued throughout prehistory. The early occupants probably burned trees and grasses to improve hunting and collecting. This tradition continues to the present. (During the Festival of San Juan, Aymara and Quechua farmers living near the lake systematically burn all t'ola shrubs and bunch-grasses near their communities.) Sediment cores show continuous presence of charcoal (presumably anthropogenic) from 6000 B.P. to the present (Binford et al. 1997). The trees found in the basin today are cultivated managed stands, not relics or remnants of natural forests (Gade 1981).

Vicuñas are territorial animals, and early hunters would have quickly realized the potential of improving natural wetlands or constructing artificial wetlands (bofedales) to increase populations. The fauna and flora near Preceramic base camps and settlements would have been gradually transformed by human disturbance and daily activities. The coevolution of agricultural and pastoral economies may have its roots in the early transformations of the puna by hunters and gatherers (Piperno and Pearsall 1998; Kuznar 1993). The domestication of camelids is dated to 4000–5000 B.P. in the neighboring upper Moquegua Valley (Aldenderfer 1998:295), although manipulation of wild herds extends further back in time. The impact of large herds of llamas and alpacas on the puna has been substantial since the Late Preceramic period.

Preceramic sites are rare in the wetlands and slopes adjacent to Lake Titicaca (Albarracín 1996; Steadman 1995; Klink and Aldenderfer 1996). Kolata and colleagues argue that the area was uninhabitable before 3500 B.P. due to "chronic drought." Preceramic wetland sites, analogous to the floating island settlements of the ethnographic Uru, would have low archaeological visibility and may be deeply buried under sediments or later occupation. Early hunter-fisher-gatherer populations would have been drawn to the rich resources of the wetlands (Erickson 1996, n.d.). Regular burning and the cultivation of totora may have been early forms of wetland management.

FARMING AND THE CREATION OF AN ANTHROPOGENIC ENVIRONMENT

At the time of Spanish conquest of the southern Andes, most of the Lake Titicaca landscape had long been converted into farm or grazing land. Detailed historical documents of the Lupaca kingdom located in seven major towns on the western shore report that political leaders owned huge herds of llamas and alpacas and exercised control over tens of thousands of human subjects. The Lupaca political organization was based on a hierarchy of leadership, towns divided into upper and lower moieties, dense rural populations, and colonies established in distant lands to exploit nonlocal resources (Murra 1968; Stanish et al. 1997; Graffam 1992). The Lupaca and their neighbors had inherited a rich landscape capital and sophisticated knowledge system from millennia of earlier farming peoples.

Cultivated plants appeared at 8000 B.P. or earlier in South America, and domestication and agriculture soon followed (Piperno and Pearsall 1998). The Lake Titicaca Basin and the south-central Andes have long been considered a "noncenter of domestication" (dispersed center) based on the distribution of wild and weedy species of important highland domesticates (Piperno and Pearsall 1998). Although the Preceramic of the Lake Titicaca Basin is poorly documented, the roots of early agriculture and herding are based in the gradual transformation of local and regional environments by hunter-gatherer-fisher peoples (Piperno and Pearsall 1998). By the time cultigens such as potatoes, quinoa, cañihua, and totora and domesticated animals appear in the archaeological record at 1500 B.C., the agropastoral economies are already well developed and widespread (Browman 1987; Erickson 1976, 1996; Hastorf et al. 1997). Many early farming settlements are located close to wetland resources, and the ethnobotanical records show that these resources played an important role in subsistence (Erickson 1976; K. Moore et al. 1999).

The adoption of farming and herding lifeways between 8,000 and 3,000 years ago must have had a profound and permanent impact on the landscapes of the Lake Titicaca Basin. The removal of vegetation cover increases erosion and evapotranspiration rates. The thick "homogeneous gyttija" in sediment cores, assumed to represent severe drought (Binford and Kolata 1996:106; Binford et al. 1997; Abbott et al. 1997), may actually be the signature of anthropogenic transformation of the puna and cerro above the lake. The grazing of large herds, increased burning, and the harvest of shrubs and trees for building material and fuel permanently altered the puna and upper slopes. By 1000 B.C. farmers began constructing raised fields in wetlands and the pampa of the northern basin (Erickson 1993, 1996). The first evidence of terracing dates to the Early Intermediate period (200 B.C.–A.D. 600), and terracing is well established by the Middle Horizon (A.D. 600–1000) (Albarracín 1996). The initial construction of terraces may have actually increased soil erosion through the removal of slope vegetation and stones (Donkin 1979:131). Dust peaks in the Quelccaya ice core dated to A.D. 920 and A.D. 600 and interpreted as evidence of intensification of raised field agriculture (Thompson et al. 1988) may represent a wider spectrum of anthropogenic activities.

THE TECHNOLOGY OF TRANSFORMATION

The traditional agricultural technology of farming communities in the Lake Titicaca Basin has been described as "hardware poor" (Donkin 1979; Denevan 2000). It is truly remarkable that the massive transformation and building of the landscape was done using simple manual tools without the help of animal traction or metal implements. Metal artifacts of late prehistory were rarely employed for mundane agriculture. The basic tool set for preparing, weeding, and harvesting fields includes the *chakitaqlla*, or Andean footplow, the *rawkana*, or hoe, and the *waqtana*, or clodbuster (Quechua terms) (figure 12.3). These tools have been used by Andean farmers for thousands of years. The modern distribution of the chakitaqlla closely maps Precolumbian terraces and raised fields and certainly played an important role in their evolution (Gade and Rios 1972; Donkin 1979:13).

The chakitaqlla is a simple yet remarkably efficient tool (Gade and Rios 1972; Morlon 1996; Donkin 1979). The footplow is very portable and can be used to turn tough sod of the lake plain and the rocky soils found on steep slopes. Fields are commonly prepared by teams of farmers; two men cut and raise a sod block with their chakitaqllas and a woman flips the sod over. Groups of farmers often line up in formation to turn the soil in community fields. The tool is primarily used today in the Lake Titicaca Basin to create wachos, 0.3 to 1.0 m wide sod lazy beds for planting tubers. The shape and size of the tool varies throughout the central Andes, and distinct forms are often linked to specific regions. A basic chakitaqlla consists of a long straight or curved shaft (1 to

FIGURE 12.3. Tools used to transform the landscapes of the Lake Titicaca Basin: rawkana (left), chakitaqlla (center), and waqtana (right). In the past, the blades would have been made of wood or ground stone.

2 m), with a cutting edge at the end of the shaft, a wooden footpeg, and an optional curved wooden handhold, all lashed together with leather bindings. Today, the blade is cut from a leaf spring of a truck. One end is hammered into a flat blade used for cutting sod, and the other end is hammered into a pointed blade for use in rocky soils. The wood comes from native trees specifically cultivated for tool parts. Traditionally, chakitaqlla blades were made of ground and chipped stone and hardened wood (Gade and Rios 1972; Morlon 1996; Donkin 1979).

The rawkana is a small hoe consisting of a metal blade and a short wooden handle hafted with leather bindings. In the past, the blades were made of wood, bone scapulae, ground and chipped stone, or tabular basalt. The all-purpose tool is used for preparing, planting, and weeding fields, banking tubers, and harvesting crops. The waqtana, a heavy malletlike tool, is used for breaking up clods of soil. It is usually made of a single piece of wood with a dense tree knot serving as the head. Others are made by hafting an oval stone or metal weight to the handle. In addition to these basic tools, wool carrying cloths are often used to transport soil and manure, and long wooden or metal levers are used to pry up and roll stones.

THE ARCHAEOLOGICAL SIGNATURE OF CULTURAL PRACTICES USED TO TRANSFORM THE LANDSCAPE

The cultural practices used by Andean farmers have been well documented in the historical and ethnographic record. Archaeologists often assume that traditional practices of Andean peoples recorded in historical and ethnographic accounts were the same used by their Prehispanic ancestors to shape, transform, and manage the landscape. The naive projection of historical and contemporary practices back into the Prehispanic period recently has been criticized. As Isbell (1997a) has pointed out, archaeologists should not assume continuities (or change, for that matter) in cultural practices, but rather should demonstrate them in the archaeological record. In this section, I will briefly discuss the archaeological signatures of some important Andean cultural practices that are linked to the anthropogenic landscape of the Lake Titicaca Basin.

AGROPASTORAL PRODUCTION ZONES

The rough mountainous terrain of the Andes is often characterized in the literature as stacked vertical ecological tiers determined by altitude and other natural factors, such as soils, rainfall, aspect, slope, elevation, temperature (Holdridge 1947; Tosi 1960; Troll 1968; Ellenberg 1979; Dollfus 1982). In this natural history perspective, environmental vertical zonation is taken as a given, something that humans adapt to. In contrast, Andean communities developed culturally defined land use categories or production zones for their holdings. According to Mayer (1985), production zones provide the structure or rules for the allotment of irrigation water, distribution of communal and individual land, regulation of land use, the scheduling of agricultural activities, definition of crop types, and the cycle of rotational fallow. Another strategy was the development and maintenance of high biodiversity in crops suited for a wide range of environments and culinary purposes (Zimmerer 1996). Prime examples are the land races of bitter potatoes (*Solanum juzepczukii*) and cañihua (*Chenopodium pallidicaule*) cultivated in the cold puna up to 4,450 m (Winterhalder and Thomas 1978:57). Altiplano maize grown on the islands and peninsulas of Lake Titicaca is another example of extending the range of a warm valley crop to a high-altitude environment. When the genetic limits of crop plasticity were reached, Precolumbian farmers often turned to technology to extend the limits of cultivation further. As Zimmerer (1996) has pointed out, Andean farming strategies stress flexibility, not microenvironmental specialization. Rather than adapting to specific ecological conditions, farmers transform nature through their settlement systems, production zones, and farming techniques.

Culturally defined production zones can be identified archaeologically (cf. Hastorf 1993; Goland 1991:514). The abandonment of farmland, and even entire agricultural strategies such as raised field agriculture, limits the direct projection of contemporary production zones back into the remote past. However, as Zimmerer (1996:20) has pointed out (following Murra and Mayer), the use of land is to assert political control over territory. Community territories and, by association, production zones are often defined by physical infrastructure (mojones, or boundary markers, walls, irrigation canals, dispersed seasonal residences, pathways, cemeteries). The major Precolumbian landscape transformations discussed below tend to map roughly onto contemporary production zones despite dramatic changes in the use and intensity of these zones (e.g., raised fields correspond to pampa and lakeshore wetlands, terraces to cerro or slopes, q'ochas to pampa and river valley plains).

THE ARCHAEOLOGY OF AGRICULTURAL LABOR

The basis of the transformation and creation of the cultural landscape of the Lake Titicaca region is raw human labor and its organization in time and space. Much has been written about traditional forms of mobilization and organization of labor at the level of the household (e.g., Orlove 1977; Morlon 1996; Aldenderfer 1993; Golte 1980), the community (e.g., Urton 1990; Winterhalder and Thomas 1978; Erickson n.d.; Golte 1980), the region (e.g., Zimmerer 1996; Masuda et al. 1985; Mayer 1979; Goland 1991), and the state (Albarracín 1996; Hastorf 1993; Kolata 1996). Andean labor institutions, from household to state, are rooted in the ideology of reciprocity. The delayed reciprocity of equal labor exchange between farmers (*ayni*) mobilizes labor for tight schedules demanded by intensive agriculture. Larger work parties, commonly for public works, are mobilized through the practice of *minka*, where a sponsor pays for labor in food and drink, or *faena*, where participants work communally for the good of the community. In later prehistory, Andean peoples paid a labor tax to the state under the *mit'a* system. Laborers working on public projects were housed, fed, and given gifts by the state. The ability to mobilize huge amounts of labor for transforming the land through agriculture and the built environment is the hallmark of Andean civilizations. The archaeological correlates of Andean labor and social organization have long been important issues in Andean prehistory (cf. Moseley 1992; Burger 1992; Isbell 1997a).

Settlement pattern is one index of reciprocal labor used in intensive agriculture. Stone notes that as agriculture becomes more intensive and labor demanding, "residences are 'pulled' towards the plot" and individual households are dispersed (1996: 43). At the same time, the requirements of reciprocal labor for intensive agriculture encourage "gravitation" of households (ibid.:122). Linear landholdings provide the most efficient tenure system in the circumstances, which fits the general Andean pattern of agricultural land divisions, suyus and chutta. Another alternative to increase the labor pool involves multiple family residences, which may have been more common in the past (Isbell 1997b). The dispersed nature of rural settlement patterns in the Lake Titicaca Basin throughout prehistory suggests that farming households and communities attempted to locate residences near their fields (Albarracín and Mathews 1990; Stanish et al. 1997; Albarracín 1996). Because of the practice of field scattering, farmers often maintain multiple residences (Goland 1991; Erickson n.d.). The importance of labor reciprocity can be seen in the clustering of mound settlements on the lake plain (Erickson 1993, n.d.) and hamlets in the lower river valleys throughout late prehistory (e.g., Albarracín 1996). Experimental rehabilitation of terraces and raised fields also has provided valuable indices of labor and social units required for the construction of Precolumbian field systems (Ramos 1986; Treacy 1994; Erickson and Candler 1989; Erickson 1993). The physical patterning of field design reflects these labor and social units. The physical link of land and residence becomes stronger through time as agricultural improvements are accrued and inherited.

PREHISPANIC RURAL SOCIAL ORGANIZATION

Rural farming communities (*pueblos,* Spanish; *llaqta,* Quechua; *marka,* Aymara) of the Lake Titicaca Basin are traditionally divided into upper and lower halves or moieties (*saya* in Quechua and Aymara). The relations between moieties and their subdivisions are hierarchical (Urton 1990). These dual divisions often have spatial and physical components and thus can be identified in the archaeological record (e.g., J. D. Moore 1995; Hyslop 1990). These in turn are divided into ayllus, or local landholding groups that often have spatial integrity (Urton 1990; Wachtel 1990; Carter and Mamani 1982). Ayllus are made of numerous households with individual and communal landholdings dispersed across local territories that are often marked by physical structures on the landscape (figure 12.4; Urton 1990; Wachtel 1990). Traditional ayllus of the Aymara were segmental and hierarchical, ranging from multiple family groupings at the local level to macroayllus at the regional level (Albarracín 1996).

The household, ayllu, and community organization also have archaeological correlates. The rural settlement pattern, past and present, in the Lake Titicaca Basin is highly dispersed. Throughout the archaeological record, regional settlement has fluctuated between weakly and strongly hierarchical, dispersed and agglutinated (Stanish et al. 1997; Albarracín 1996; McAndrews et al. 1997; Albarracín and Mathews 1990). Precolumbian households have been identified in the archaeological record for the basin (Aldenderfer 1993; Bermann 1994; Janusek 1994). The plot where households maintain their primary residence (a house compound often around a patio) and infields (*sayaña,* Aymara) tends to have many physical improvements such as stone walls, corrals, infield gardens, and canals for irrigation and drainage. Each year, individual households were assigned a topo (or *tupu*) of communal land of the ayllu, enough farmland to support a family for a year. The size of a topo would vary according to the productive potential of the land. I have argued that the modular patterning of the smallest units of raised fields reflect topo divisions (Erickson 1993). "Compressed verticality" (Brush 1977) and "field scattering" (Goland 1991) within different environments are household, ayllu, and community strategies to reduce risk. This also is an important factor in determining Prehispanic rural settlement patterning. Multiple residences for individual households or multiple agglutinated settlements for single communities throughout production zones make it difficult to define precisely all components of the individual households through traditional site survey. Local labor units may be reflected in house compounds (Isbell 1997b; Bermann 1994; Janusek 1994),

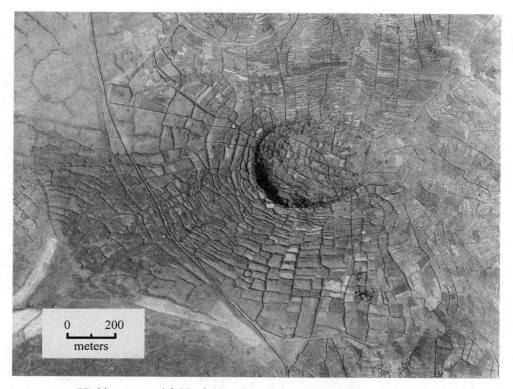

FIGURE 12.4. Highly patterned fields, field walls, pathways, roads, and dispersed farmsteads on gentle slopes between Hatuncolla and Lake Umayo, Puno, Peru. The walls of a community's sectorial fallow system radiate from the top of a hill (left of center).

settlement patterning in relation to intensity of agriculture (e.g., Stone 1996), residence location in reference to fields (Stanish et al. 1997; Erickson n.d.), or the formal patterning of Precolumbian fields (Erickson 1993).

The community and ayllu levels of social organization have a much clearer archaeological signature than individual households and household infields. Traditional community and ayllu lands within production zones are often organized under rotational or sectorial fallow systems (Orlove and Godoy 1986; Wachtel 1990; Goland 1991; Morlon 1996; Mayer 1979, 1985; Carter and Mamani 1982). In ideal cases, communal land is divided into spatially discrete segments, often long linear strips or pie-shaped wedges, called *chutta, suyu, laymi, muyuy,* and *manda* in Quechua and *aynoqa* in Aymara (figure 12.5). Each year, a number of segments are designated to be left in fallow, and the rest are assigned a specific crop. A common cycle is two to three years of cropping followed by two to twelve years of fallow (Winterhalder and Thomas 1978:68). A community's rotational fallow system is often physically defined by stone walls radiating from ridges or tops of hills. Irrigation canals feeding terraced fields often map and bound the social structure of communities (e.g., Treacy 1994; Zuidema

FIGURE 12.5. Massive terraced hillside near Pomata, Puno, Peru. Vertical walls dividing terraces into long strips run from the hill crest (upper left) to the pampa (lower right) several hundred meters below. Faint traces of raised fields can be seen between dispersed farmsteads in the lower right.

1985). Terrace infrastructure (facing walls, pathways, and walls running up and down slopes) often reflects land tenure at the household, ayllu, and community levels. Some grazing lands, raised fields, and q'ochas were also managed by rotational fallow systems and incorporate physical structures such as boundary markers, canals, and walls (Palacios 1977; Erickson n.d.; Flores 1987; Rosas 1986). Ayllus often maintain shrines and "altars" on hilltops overlooking their territories (Candler 1993; Wachtel 1990). Isbell (1997a) has recently argued that the ayllu can be identified in aboveground multiple burial towers (chullpas) that are distributed throughout the Lake Titicaca landscape, often on high ground overlooking community fields. Even natural aquatic resources such as totora reed swamps and fishing grounds are physically marked by ditches or stone cairns to define ayllu and community territories (Levieil and Orlove 1990; Nuñez 1984; Erickson n.d.).

The origins of these cultural institutions remain elusive and debated (Moseley 1992; Kolata 1993; Isbell 1997a). Social and technological institutions are dynamic, and these institutions certainly evolved and transformed over time. Some of the material correlates from these institutions are identifiable in the landscape and built environment by 1000 B.C. Through the practice of everyday life over generations, these structures have been expanded, enhanced, and formalized (Erickson 1993, n.d.). The cultural construction, modification, and improvement of land over long periods of time has been called "landscape capital" (Blaikie and Brookfield 1987b; Netting 1993). The agricultural resources at any point in time represent a history of labor invested by previous generations of farmers. Farmers inherit the improvements (and environmental degradation) of preceding generations, which tends to tie each new generation more tightly to the land. The thousands of years of efforts to demarcate individual and community lands physically through the built environment are now part of the permanent landscape record.

TERRACING

The most striking aspect of the Lake Titicaca Basin for the visitor is the patterned landscape of terraces (*terrazas* or *andenes* in Spanish; *pata*, Quechua; *takha, takhana*, Aymara) on nearly all of the hill slopes (see figure 12.5). Although raised fields (waru waru, suka kollas, or camellones) and sunken gardens (q'ochas) have received the most research attention, the agricultural terraces of the Lake Titicaca Basin are much more impressive in terms of overall labor input and areal extent. Massive conversion of slopes, some quite steep, into productive platforms for agriculture was done at a monumental scale (figure 12.6). There is a nearly continuous distribution of terracing on slopes of both the northeast and southwest shores of Lake Titicaca from the Tiwanaku Valley in the south to the towns of Pukara, Azángaro, and Ayaviri in the north and on slopes rising from all of the river valleys of the basin. Terracing continues unbroken on the adjacent eastern Amazonian watershed of southern Peru and Bolivia (Donkin 1979; Denevan 2000; Goland 1991). I estimate that Precolumbian terracing in the immediate vicinity of the lake and the major river valleys of the basin alone covers 500,000 ha.

Archaeologists, agronomists, soil scientists, and geographers have made detailed studies of Precolumbian and contemporary terrace agriculture in the central Andes (Donkin 1979; Denevan 2000; Treacy and Denevan 1994; Goland 1991; Torre and Burga 1986; Zimmerer 1996) and more specifically in the Lake Titicaca Basin (Morlon 1996; Ramos 1986; Coolman 1986). The functions of terracing have been summarized in numerous publications (Treacy 1994; Treacy and Denevan 1994; Donkin 1979; Ramos 1986; Coolman 1986; Morlon 1996; Torre and Burga 1986).

Treacy and Denevan (1994:93–96) conveniently summarize the functions of terracing as:

• Soil deepening: The soils of Andean slopes tend to be thin and full of stone. Improving the depth of soil increases the retention of water in addition to providing a deeper medium for crop growth (Donkin 1979:131).

FIGURE 12.6. Agricultural terraces, linear land boundaries, pathways and dispersed settlement near Chisi, Copacabana, Bolivia. Precolumbian terrace walls and platforms are still used but are rarely maintained.

- Erosion control: The control of erosion is believed to have been a secondary function of terracing. Andean soils are highly susceptible to erosion, much of which is due to human activities. Informal terracing (lynchets, cross-channel walls) may have been an early response to loss of topsoil. As I will argue below, soils used to fill terraces were often removed from farther up the slopes during construction or through intentional erosion.

- Microclimatic control: The local topography created by terrace walls and platforms provides microclimates that are more favorable for crops. Frost damage can be reduced as terraces interfere with the flow of cold air down slopes and create turbulence, protecting crops. Terraces also modify slope aspect and sun angle for improved growth conditions and reduction of radiant heat loss at night (Donkin 1979:131; Morlon 1996).

- Moisture control: The primary function of Andean terraces may be to control water by artificially flattening surfaces. Irrigation was often combined with terracing in the south-central Andes. Precolumbian terraces in the Lake Titicaca Basin appear to be primarily rainfed, not irrigated. Whatever the source of moisture, terracing improved moisture retention by reducing runoff and providing a deeper soil medium to store moisture.

Most, if not all, slopes adjacent to the lake or lake plains in the Lake Titicaca Basin are terraced. The exception would be rock outcrops with no soil or extremely steep faces. The slopes of all the islands and peninsulas of the lake are completely covered with terracing from the top of the hills to the lake edge. These terraces, categorized as bench terraces, are also the most formal in design and construction (and the most costly in terms of labor). In some cases, the wall height is equal to or greater than the width of the cultivation platform created.

Precolumbian terraces are used by contemporary farmers, but little effort is devoted to their maintenance. Most terraces are in a poor state of preservation, and many walls have been removed to enlarge the fields. With the exception of a few rural development projects, new terraces are not being constructed. Terrace fields are generally farmed using the traditional fallow system of two to three years of cropping followed by two to twelve years of fallow, often combined with grazing (Winterhalder and Thomas 1978:68; Morlon 1996). Because only a portion of the land is in cultivation in any year, much of the landscape covered with terraces appears abandoned (Donkin 1979: 121–122).

Although diverse in form and size, the majority of the terraces of the Lake Titicaca Basin can be classified as bench terraces, contour terraces, and valley floor terraces (Treacy 1994; Treacy and Denevan 1994; Donkin 1979). Walls are constructed of local stone, most often found on the surface of the slope or encountered during construction of the walls and platforms (figure 12.7). Simple terraces without stone walls or facing are found on upper slopes and at the base of slopes. The more formal, stone-faced terraces 5 to 100 m long and 2 to 20 m wide are most common (Ramos 1986; Donkin 1979:120–122; Morlon 1996). Stone retaining walls of 0.5 to 3.5 m tall are either dry-laid or cemented with mud, and most have foundations set in a shallow trench

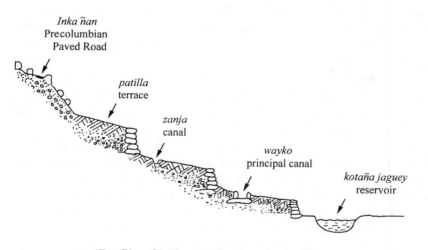

Profile of a Terrace System in Asillo

FIGURE 12.7. Profile of terrace infrastructure. *Source:* After Ramos 1984.

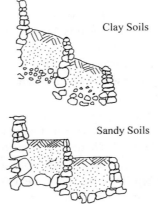

Clay Soils

Sandy Soils

FIGURE 12.8. Profiles of terraces in different soil types. *Source:* After Ramos 1984.   Profiles of Terraces in Asillo

(Donkin 1979:120; Ramos 1986; Coolman 1986). The lower interior fill of walls is generally composed of small stones that provide drainage and prevent blowouts of terrace walls during heavy rains (figure 12.8). The upper fill of terraces is topsoil. Terraces generally receive treatments of dung fertilizer during the cultivation of potatoes.

Terrace construction is labor intensive. Ramos (1986) calculates 225–2,270 person-days/ha (an average of 600 person-days/ha) for terrace rehabilitation in Asillo in the northern basin. Coolman (1986) reports figures of 2,500 person-days/ha for new terrace construction in Puno. The labor for terrace construction was certainly spread out over many thousands of years, and terraces grew through accretion as farmers made improvements on the land.

Terrace design tends to be highly patterned. Discrete blocks of identical terraces are often delineated by vertical walls, canals, and pathways. Many of these features run from the top of the hill to the valley floor (see figures 12.5 and 12.6). A block of stacked terraces is often linked by stairs of stones projecting from the terrace face and vertical and lateral channels. In the Lake Titicaca Basin, these walls and channels serve to (1) control and distribute runoff within and between field platforms and provide drainage for excess water, (2) organize sectorial fallow cycles, and (3) mark individual, ayllu, and community field boundaries (Ramos 1986).

Terracing is often attributed to the Inka, who incorporated the region into their empire in the mid-1400s. The most elaborate terraces are associated with maize production for the Inka located on the Copacabana Peninsula, Isla del Sol, and the Isla de la Luna (Donkin 1979). The Inka were responsible for major landscape engineering feats, which included long causeways, ritual baths and fountains, monumental carving in living rock, and ceremonial centers, in addition to impressive blocks of terracing near state installations (Stanish et al. 1997; Hyslop 1990). Many of these terraces appear overconstructed or overengineered and may have had ritual functions that served to promote the power of the Inka state.

Settlements within and adjacent to the terraced zones of the lake show continuous occupation from 1800 B.C. to the present. The earliest direct evidence of terracing in the region is the massive stone-faced platforms at the site of Pukara (200 B.C.–A.D. 600). These structures provided bases and retaining walls for the monumental buildings at the site. Stone-faced terraces used for agriculture and occupation in the lower Tiwanaku Valley have been securely dated to A.D. 600–1000 (Albarracín 1996). The earliest terraces were probably lynchets, simple barriers of earth, stone, and vegetation constructed to trap soil eroding from slopes. These ephemeral structures would have been erased as more formal, stone-faced terraces were constructed in later prehistory.

RAISED FIELDS

Raised fields (*camellones* in Spanish; *waru waru*, Quechua; *suka kollas*, Aymara) are the best studied of the major technologies of landscape transformation (Smith et al. 1968; Lennon 1983; Erickson 1988, 1993, 1996, 1999; Kolata 1993, 1996). Raised fields are large elevated planting platforms constructed in areas of waterlogged soils or soils prone to annual flooding (figure 12.9). The platforms are accompanied by canals or ditches on one, two, or all sides that were created during the process of raising the field (figure 12.10). Raised fields are highly variable in size and shape. Platforms range from 4 to 10 m wide, 10 to 100 m long, and 0.5 to 3 m tall. Canal size is generally

FIGURE 12.9. Prehispanic raised fields (waru waru, suka kollas) near Huatta, Puno, Peru.

FIGURE 12.10. Rehabilitated raised field platforms planted in potatoes alongside water-filled canal (center) in the community of Viscachani Pampa, Huatta, Puno, Peru.

in proportion to the size of the platform. Bundles of fields are organized in regular patterns, possibly reflecting the social organization of agricultural labor and land tenure, specific functions or crops, or stylistic preferences (figures 12.2 and 12.11; Erickson 1996). Abandoned raised fields are found in most of the seasonally inundated plains and river valleys surrounding Lake Titicaca. A conservative estimate of area of Prehispanic raised field agriculture is 120,000 ha (Erickson n.d.).

The functions of raised field agriculture have been determined through the rehabilitation of Precolumbian fields, ethnographic analogy, and agronomic experiments (Smith et al. 1968; Erickson 1988, 1996; Garaycochea 1986b; Ramos 1990; Kolata 1996). The functions include:

- Soil improvement: The construction of raised field platforms involves increasing the depth of topsoil, aeration, and drainage of heavy and often waterlogged soils.
- Water management: The canals adjacent to the platforms receive water from runoff, lake and river flooding, and the rise in water table during the rainy season. The ability of the cultivated soil to absorb moisture is improved through construction of the platforms. Canals and spillways permit some control over water levels within canals and the water table within fields.
- Capture, production, and recycling of soil nutrients: Soil and nutrients eroding from raised field platforms or carried by floodwaters are captured as sediments in the canals. Mature canals have higher levels of organic matter and nitrogen than nonraised field contexts. Organic matter from harvest stubble and aquatic vegetation growing in the

canals can be incorporated as green manure or muck to renew soil fertility of field platforms for sustained production.

- Improved microclimate: The water in the canals, functioning as a solar heat sink, reduces the daily fluctuation in temperature in the canals and the risk of frost damage
  by increasing local temperatures. The topography of platform-canal decreases heat
  loss during frosts by blocking and reflecting radiation back to the fields.

- Aquaculture: The construction of canals and fields substantially expanded the area of
  wetland conditions. Important wetland resources such as fish, totora, and aquatic birds
  were enhanced and possibly controlled through raised field agriculture.

Like the terraces discussed above, the construction of raised fields caused a major
transformation of the Lake Titicaca Basin landscape. Raised field agriculture involved
reworking the soil profile to a depth of 1–2 m. Biodiversity and carrying capacity of
pampa and wetland ecosystems are improved by artificially expanding the terrestrial-
aquatic interface or ecotone. The microtopography of fields and canals and the standing water in canals may have increased the overall temperature of the basin during the
growing season.

Most of the raised fields were abandoned at the time of or before the arrival of the
Spanish, although some may have remained in cultivation until the last century (Erickson 1993, n.d.). Raised field agriculture probably began as early as the Initial period

FIGURE 12.11. Raised field platforms (light linear features) and canals (dark linear features) on
the edge of Lake Titicaca (lower right) near Huatta, Puno, Peru. Note the long straight canals
that subdivide the raised field landscape into wedges or strips.

(1800–900 B.C.) or Early Horizon (900–200 B.C.) along the lake edge (Erickson 1993). By the Early Intermediate period (200 B.C.–A.D. 600), raised fields were being farmed throughout the basin. Many fields were buried under larger raised fields that were constructed during the Middle Horizon (A.D. 600–1000) and Late Intermediate (A.D. 1000–1475) (Erickson 1993, n.d.; Seddon 1994; Binford et al. 1997; Kolata 1996). Climate change, in particular a long-term drought, has been proposed as the cause of raised field abandonment (Kolata 1996; Kolata and Ortloff 1996a; Binford et al. 1997). I have argued that raised field construction and use continued and actually flourished during the period of presumed drought conditions (Erickson 1993, 1996, n.d.).

Raised field agriculture denotes a substantial modification of the land surface, and thus considerable amounts of labor were invested in construction and maintenance. Based on our experiments, a single farmer can construct 1 m^3 of field/hour or 5 m^3/day (a 5-hour workday). The total of labor dedicated to raised field agriculture is impressive. I estimate that 75.8 million person-days were required to construct the 120,000 ha of raised fields of the Lake Titicaca Basin (Erickson n.d.). Archaeological (Lennon 1983; Seddon 1994; Erickson 1996) and experimental research (Garaycochea 1986b; Erickson 1996) indicates that raised fields were constructed over 2,000 or more years, and thus construction costs were spread out over a long period. Experimental construction suggests that raised fields were built at the beginning and end of the rainy season when conditions are optimal. Initial construction involved the removal of the A horizon of the canals to provide fill for the field platforms. As canals matured, a new organic-rich A horizon formed over time and sediments accumulated in the canals. This was periodically removed during field maintenance and canal cleaning. These rebuilding episodes are clearly recorded in stratigraphic profiles of excavated Prehispanic fields (Seddon 1994; Erickson 1996). A sequence of small fields being replaced by larger fields has been documented in excavations of raised fields. Many generations of farmers were responsible for this growth through accretion.

Experiments showed that major highland Andean crops (potatoes, ocas, ullucus, isañus, quinoa, cañihua, tarwi, and altiplano maize) can be successfully grown on raised fields. Potato yields on experimental raised fields ranged from 5 to 20 MT/ha; traditional agriculture on the slopes yields 2–5 MT/ha (Erickson 1996; Kolata et al. 1996). Based on potato production between 1981 and 1986, I calculate a carrying capacity of 37.5 persons/ha of raised field cultivation platform or 2.25 million inhabitants for the entire raised field system. Of course, the raised fields were not all constructed or all in use at the same time.

Based on the hierarchical settlement patterns and agricultural infrastructure (river canalization, dikes, and aqueducts), Kolata and colleagues (Kolata 1993, 1996) have argued that raised fields on a regional scale could have been constructed only under the direction of a state society, in this case Tiwanaku of the Middle Horizon. Based on the experimental construction of raised fields and archaeological evidence for the patterning of rural settlement and associated field systems, I have argued that farming communities were capable of constructing and maintaining the raised fields of the Lake Titicaca Basin (Erickson 1993, 1996, n.d.).

The networks of canals and platforms increase both the area of wetlands and the rich ecotone or interface between terrestrial and lacustrine habitats (see figure 12.11; Erickson n.d.). These earthworks also function to capture topsoil and important nutrients eroded from the slopes (Carney et al. 1996).

SUNKEN GARDENS

Sunken gardens, or q'ochas (Quechua for "container of water"), *q'otanas, cotaña, cota* (Aymara), *chacras hundidas, pozas, ojos de agua* (Spanish), are the third major element of landscape transformation in the Lake Titicaca Basin (figure 12.12; Flores 1987; Rosas 1986). Q'ochas were first defined in the densely populated pampa of the northern basin between the Río Azángaro on the east and the Río Ayaviri (Pukara) on the west at an elevation of 3,850–3,900 m. Q'ochas and raised fields coexist in the lower Tiwanaku Valley and the Huatta pampa (Albarracín 1996; Erickson n.d.). The total number and areal extent of these features in the Lake Titicaca Basin is unknown, but Flores estimates that the complex of q'ochas in the northern basin covers 530 km² (Flores 1987). A density of over 100 q'ochas per km² has been reported for Mataro Grande (figure 12.13). In the 256 km² area of functioning q'ochas, Flores estimates there are more than 20,240 structures (ibid.:284). Q'ochas are still being maintained and cultivated by the Quechua farmers in the northern basin. The following discussion

FIGURE 12.12. A contemporary sunken garden (q'ocha) in fallow cycle near Llallahua, Puno, Peru. Note the patterns of eroded lazy beds (wachos) in the depression.

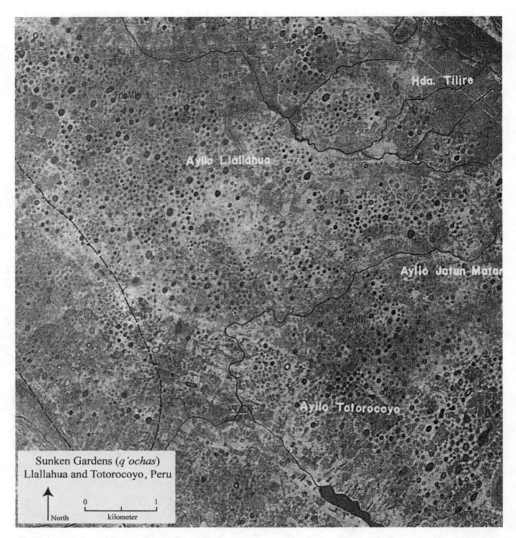

FIGURE 12.13. Aerial photograph of the sunken gardens (dark circles) near the communities of Llallahua, Jatun Mataro, and Totorocoyo, Puno, Peru. *Source:* After ONERN 1965: Hoja 4a.

is based on the well-studied q'ochas of the northern basin that are still in use (Flores 1987; Rosas 1986).

Q'ochas are large shallow depressions ranging from 0.1 to 4 ha and 1.5 to 6 m deep (figure 12.14). They are nearly always located in areas of poor drainage. The forms are highly standardized; round and oval shapes are the most common. The structures are ranked according to size and shape: (1) large and round are most common (*muyu q'ocha*), (2) medium-sized and oval (*suyt'u q'ocha*), and (3) small and rectangular (*chunta q'ocha*) (figure 12.15). The elaborate canal networks that link q'ochas are clearly artificial. It is still not clear whether the depressions themselves are completely artificial or natural formations. The lack of spoil piles of earth at the edges of the northern q'ochas suggests natural formation. Even so, q'ochas show considerable artificial enhancement of their shape, in the formal symmetry of the depression and in the patterning of canals and lazy beds (wachos). A central canal (yani) divides the q'ocha in half and extends beyond the borders of the depression. These canals (0.5–1 m wide and up to 5 m deep) provide a means of capturing runoff to fill the q'ochas and of draining excess water into the river. Other canals encircle the depression to distribute

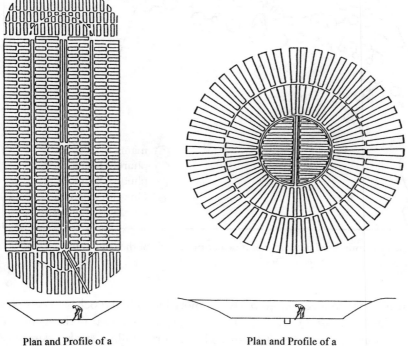

Plan and Profile of a   Plan and Profile of a
Rectangular Q'ocha   Circular Q'ocha

FIGURE 12.14. Plans and profiles of two forms of q'ochas north of Lake Titicaca, Puno, Peru. *Source:* After Flores 1987.

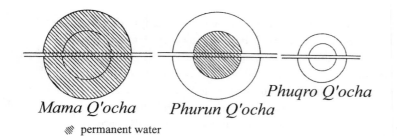

Mama Q'ocha   *Phurun Q'ocha*   *Phuqro Q'ocha*

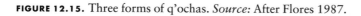 permanent water

FIGURE 12.15. Three forms of q'ochas. *Source:* After Flores 1987.

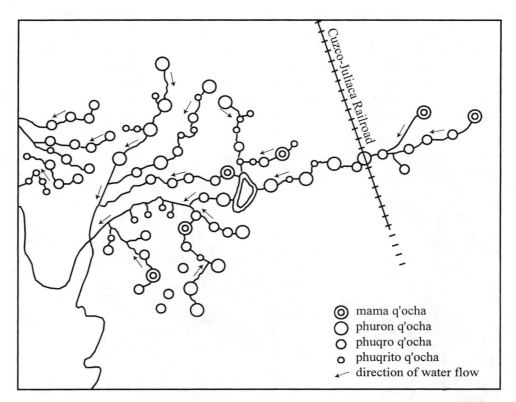

FIGURE 12.16. The interconnected networks of q'ochas and canals of the community of Llalla-hua, Puno, Peru. *Source:* After Angeles 1987.

water. The central canal connects up to twelve or more individual q'ochas in complex hydraulic webs (figure 12.16). Q'ochas are farmed using a three-year cycle of cultivation followed by four to five years of fallow. The sectorial fallow system described above is used to organize the rotation of potatoes, quinoa or cañihua, and barley or wheat (one year each). The fallowed fields are allowed to fill with water and aquatic vegetation. During the years of fallow, animals are allowed to graze on vegetation growing in the depression. The functions of q'ochas include:

- Water management: The depressions function as microcatchment basins to collect and store surface runoff year round. The q'ochas near the lake plain also provide access to the water table during the dry season and droughts.
- Nutrient capture, production, and recycling: The depressions capture organic sediments and topsoil eroded from higher ground. During the fallow stage, the flooded field accumulates organic matter from decomposing aquatic plants.
- Production of forage for domestic animals: During the fallow period native grasses and aquatic plants flourish in the depressions. Domestic animals are allowed to graze on this forage during the dry season.

- Aquaculture: During the fallow stage, the flooded depressions support edible fish and aquatic vegetation such as totora and llachu used as forage during the dry season. Totora is often cultivated for forage, matting, and roofing material in q'ochas on the Huatta pampa.
- Sod construction material: The q'ochas of the Huatta pampa produce thick mats of sod that are cut into blocks and used for the construction of walls, corrals, and temporary shelters on the pampa.
- Source of drinking water: When in the fallow phase, q'ochas are an important source of drinking water for humans and their domestic animals during the dry season. Q'ochas also provide water for making adobes. '
- Improved microclimate: The humid depressions improve the climate conditions for crops and reduce the risk of frost damage.
- Preparation of freeze-dried tubers: In Huatta, q'ochas are excellent locations to freeze-dry potatoes and other tubers during the dry season. The base of the depression is colder than the surrounding pampa during nights of frost.

Q'ochas are owned by individual families today. They can be sold only to other members of the community. Each family controls an average of six to seven q'ochas. Compared to nonq'ocha fields in the northern pampas, crop yields from q'ochas are higher and more consistent. During the droughts of 1982–1983, the q'ochas produced 10 MT/ha of potatoes and 3,600 kg/ha of cañihua (Angeles 1987:69–70, in Morlon 1996:253).

The origins and history of q'ocha cultivation are unknown. References to q'ochas as valuable land appear in early Colonial documents involving land disputes (Flores 1987). The largest concentration of q'ochas is adjacent to the large Early Intermediate (200 B.C.–A.D. 600) site of Pukara. Association with Pukara culture has been suggested, but no q'ocha has been directly dated. The q'ochas of the Huatta pampa (Erickson n.d.) and those of the lower Tiwanaku Valley (Albarracín 1996) are associated with multicomponent occupation sites dating from the Initial period (1800–900 B.C.) to the present. It is quite probable that construction and use were contemporaneous where the distribution of q'ochas and raised fields overlap.

IRRIGATED PASTURE

The high-altitude grasslands, or puna (4,000–4,800 m), have been shaped by human activities over many millennia, first by hunter-gatherers and later by the herders of the native camelids. The llama can survive on the dry tough grasses of the puna, but the alpaca (highly valued for its wool and meat) requires forage that is more succulent. Pasture for herds of alpacas during the dry season can be found only in the *Distichia* moors or bofedales (*oqho* in Aymara). Natural bofedales are not sufficient to support large herds of alpacas. Herders have improved natural pasture and constructed vast artificial bofedales through irrigation (figure 12.17; Palacios 1977, 1984). A hectare of irrigated pasture can support 3 alpacas during the dry season, thus large bofedales are

FIGURE 12.17. Alpacas grazing in a large irrigated pasture (bofedal) near Sandia, Puno, Peru (photograph courtesy of Lisa Markowitz).

necessary to support the 30,000 head of alpaca owned by herders in communities such as Chinchillapi, Peru (Palacios 1984:49). Water, tapped from rivers, streams, and springs, is often brought from long distances in two large feeder canals (*hach'a irpa*) of up to 2 m wide and 0.8 m deep and 17 km long (figure 12.18). Canals are reinforced with sod blocks that take root, forming living walls. These, in turn, supply smaller networks of canals (*hisk'a irpa*) used to create the bofedales. One of the larger bofedales covers 2,200 ha and can support 3,000–4,000 head. It generally takes herders many years of hard work to create mature bofedales capable of supporting large alpaca herds. The sustained addition of nutrients from organic sediments carried by the irrigation water and camelid dung would greatly improve the potential carrying capacity and value of these features.

Bofedales are fragile and require regular maintenance. If allowed to dry out, it can take years to bring them back into production. Canals are cleaned and repaired once a year after the rains. Families are responsible for the canals that cross their property. Each of the four sections of the community of Chinchillapi control separate bofedales. Ownership of a residence near an irrigated bofedal gives a herder grazing rights. When canals pass property lines, owners can tap into the canal to create their own bofedal. Inheritance of access to bofedales is patrilineal, and residences tend to be clusters of related families. Bofedales are foci of subcommunity solidarity and identification. Conflicts are usually resolved at the local level.

Bofedales are mentioned in the early Colonial documents for Puno. The technology was probably an important element in herding economies of the Precolumbian period and apparently has a long history. Vast extensions of bofedales, many more than are in contemporary use, would have been necessary to support the large camelid herds documented in the early Colonial tax records for the Lupaca who occupied the same region in the mid-sixteenth century. Based on these documents, Graffam (1992:889) estimates that the Lupaca controlled a total population of 1.9 million llamas and alpacas. By late prehistory, the nonpuna landscape was dedicated almost entirely to crop production; thus, pastoralism was restricted to the puna (with the exception of grazing in fallowed fields and the lake-edge wetlands). Numerous Late Preceramic (4000–3500 B.P.) occupation sites, possibly associated with herding, have been found near bofedales in the Ilave and upper Moquegua River drainages (Klink and Aldenderfer 1996; Aldenderfer 1998). Because of the physical infrastructure of bofedales (canal

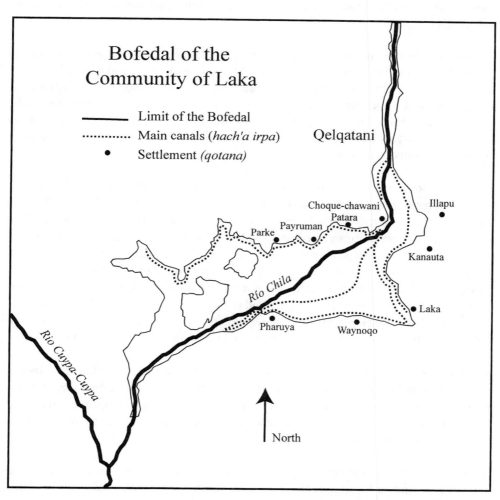

FIGURE 12.18. The irrigated artificial pasture of the community of Laka, Chinchillapi, Peru. *Source:* After Palacios 1977.

networks) and associated residences, it should be possible archaeologically to investigate the social organization of Prehispanic bofedales and the origins and evolution of the technology (e.g., Aldenderfer 1998).

MISCELLANEOUS ARTIFICIAL LANDSCAPE FEATURES

Thus far, I have discussed the major Precolumbian agricultural technologies that were used to transform the Lake Titicaca Basin into a cultural landscape. There are a number of other physical features created by humans that, although less grand in geographical scale, approximate the labor and environmental impact of the major technologies when considered in their entirety. These include modified rivers and streams, artificial canals, improved springs, roads and pathways, causeways, ponds, reservoirs, stone piles, corrals, cemeteries, burial towers, temples and shrines, residences, and settlements.

MODIFIED RIVERS AND STREAMS. Artificially straightened and canalized river and stream channels are found throughout the region. The rerouting of natural channels and the construction of sod walls or dikes on natural levees to reduce flooding are apparently old practices. A 14 km section of the Río Catari was straightened and reinforced by dikes, possibly during the Middle Horizon (Kolata 1993). In recent years, farmers have constructed massive sod walls of many kilometers for flood control along the Illpa, Azángaro, Ramis, and Coata rivers.

ARTIFICIAL CANALS. In addition to the water channels associated with terraces, q'ochas, and raised fields, there are numerous artificial canals in the pampas of the Lake Titicaca Basin. Kolata and Ortloff (1996b) report on the canals of Koani Pampa and the Tiwanaku Valley that were constructed to provide irrigation and drainage for raised fields. In Huatta, artificial canals of up to 5 km long are still used today for poling reed boats through the lake shallows and pampa. One canal network radiates from a large occupation site dated to the Early Intermediate period (figure 12.10; Erickson 1993). Parallel and radial canal networks are used today by communities and individual farmers to mark the boundaries of their wetland resources (Erickson n.d.). These features, combined with raised fields, substantially expanded the wetland ecosystem in the basin.

IMPROVED SPRINGS. Most springs in the Lake Titicaca Basin show modification by humans. These include deepening and enlarging the source into collection tanks. Water is distributed to dispersed farmsteads and villages in networks of open canals. Near springs and along streams, thousands of artificial ponds are dedicated to tunta production, tubers that are water leached and freeze dried.

CAUSEWAYS. Precolumbian causeways of earth, sod, and stone were constructed to cross low-lying pampa and wetlands (Kolata 1993; Smith et al. 1968; Hyslop 1990; Albarracín 1996; Julien 1988). Causeways were often segments of interregional road networks that may date to the Middle Horizon and Late Horizon (Kolata 1996; Hyslop

1990). In addition to transportation, these features may have had a flood control function (Smith et al. 1968; Kolata 1996). Causeways may have also served as aqueducts in the southern basin (Albarracín 1996; Kolata 1996).

ROADS AND PATHS. Elaborate networks of roads and paths cover the landscape of the basin. Formal roadways connect Precolumbian urban centers, administrative sites, and cemeteries (Hyslop 1984; Julien 1988). Many are stone paved and include features such as drains, bridges, gates, curbs, and parallel walls. The Inka constructed a northern and southern road around Lake Titicaca to connect regional centers to Cuzco (Hyslop 1984; Julien 1988; Stanish 1997).

Networks of pathways wind across the agricultural landscape, connecting dispersed fields, farms, villages, and urban centers (see figures 12.4 and 12.5). Paths in densely populated areas show improvements such as stone paving, staircases, and walls to prevent animals from entering fields. Paths often define boundaries between fields. Although difficult to date precisely, paths are associated with some of the earliest agricultural settlements of the basin.

PONDS AND RESERVOIRS. Most contemporary and archaeological settlements in the pampa are associated with artificial ponds. In addition to providing water for humans and livestock, ponds are used to raise fish and totora. Many of these ponds were formed in the process of making adobes and cutting sod for construction of residences and walls.

WALLS. Stone, sod, and adobe walls are ubiquitous features of the built environment (see figure 12.4). Walls tend to be elaborate near farmsteads and within villages. Walls keep domestic animals in or out of fields, depending on the stage of the fallow cycle. Walls are important markers for individual, ayllu, and community lands. Straight walls radiating from the peaks of hills mark Precolumbian land tenure and possibly the practice of sectorial fallow (see figures 12.4, 12.5, and 12.6). Maintenance of walls is often enforced at the level of the community. Walls protect crops and animals from the elements and often shelter important medicinal and wild food plants.

STONE PILES. Stones encountered during agricultural activity are tossed onto piles (a formation process spanning thousands of years). These stone piles and walls may play a significant role in mitigating frosts. The heat absorbed by the stones is released at night, raising the local temperature and reducing the daily fluctuation of temperature (Morlon 1996).

CORRALS AND GARDENS. Contemporary and Precolumbian corrals are ubiquitous features in the basin. Herders and farmers bring their animals into corrals at night. The dung that accumulates is collected as fuel and fertilizer (Winterhalder and Thomas 1978). Corrals are periodically turned into house gardens. Corral walls help protect bitter potatoes and quinoa from frost at higher elevations. The symbiosis between herding and farming is old, and corrals may have played a role in the domestication of Andean crops (Kuznar 1993).

CEMETERIES, BURIAL TOWERS, TEMPLES, AND SHRINES. Cemeteries were often located far from settlements on the tops of hills, mesas, and crests of ridges (Stanish et al. 1997; Hyslop 1977). During late prehistory, cemeteries enclosed by high walls covered hundreds of hectares (Stanish et al. 1997). Stone and adobe burial towers, or chullpas, are prominent features of the landscape (Hyslop 1977, 1984; Stanish et al. 1997). Most date to the Late Intermediate (A.D. 1000–1475) or Late Horizon (A.D. 1475–1532). Chullpas are linked to ancestor worship and the rights of community and ayllu to farmland. The tall towers physically dominate the landscape as symbolic markers of the lands controlled by ayllus and communities (Isbell 1997a). Stone-walled semisubterranean courts dating to the Early Horizon (1800–900 B.C.) through the Middle Horizon (A.D. 600–1000) are found on natural and artificial promontories (Chávez 1988). Landscape shrines, or huacas, are identified by stone altars, hearths, and offerings of fine potsherds, shell, and exotic stones. Most shrines have evidence of long continuous use to the present. Today, these locations are often marked by wooden crosses or stone platforms used in Catholic and traditional rituals (Candler 1993).

RESIDENCES AND SETTLEMENTS. The Lake Titicaca Basin was densely occupied at the time of European contact and has probably been so since the Early Intermediate period (200 B.C.–A.D. 600). Intensive surveys located close to 500 sites within 360 km² of Juli-Pomata (Stanish et al. 1997) and more than 1,000 sites within an area of 400 km² in the Tiwanaku Valley (Albarracín and Mathews 1990). The cycles of population aggregation and dispersion in settlement have been attributed to changes in total population size, political economy, the fortunes of polities, agricultural intensity, and climate (e.g., Stanish et al. 1997; Albarracín 1996; Binford et al. 1997; McAndrews et al. 1997). The largest site, Tiwanaku, covers an estimated 8 km² (Kolata 1993). Although centralized states and urban centers were present at certain times in prehistory, the predominant rural settlement pattern in the basin could be characterized as dispersed.

Precolumbian houses were constructed of adobe, sod, stone, and thatch (Bermann 1994; Janusek 1994). Walled house compounds with sleeping, kitchen, storage, and workshop structures were often arranged around a patio. As structures age and deteriorate, they are often leveled for new construction. Continuous occupation of certain locations in the wetlands and pampa produced huge mounds that dominate the flat landscape. These settlements are surrounded by thousands of smaller mounds that represent small towns, hamlets, and individual family residences. Farmers still choose these locations for settlement. Archaeological excavations of mounds document intense and continuous occupation beginning as early as 1500 B.C. (Stanish and Steadman 1994; Erickson 1988).

THE ANTHROPOGENIC LANDSCAPE LEGACY: LONG-TERM HUMAN-ENVIRONMENT RELATIONSHIPS AS HISTORICAL CONTINGENCY

Traditional perspectives on the relationship of humans and the environment in the Andes assume a separation of culture and nature. According to the nature-centric perspective, the Andean environment since the Pleistocene is viewed as a fragile ahistorical ecosystem in equilibrium, as long as it remains undisturbed by humans. Humans are viewed as forces impacting on and degrading the natural environment. In the human adaptation perspective, people adapt to, respond to, and map onto the structured and ordered Andean environment perceived as discrete ecological tiers. In the neoenvironmental determinism perspective, humans adapt to a limited range of climatic variation but are helpless in the face of extreme climatic fluctuation that surpasses the limits of human adaptation, such as long-term drought (Binford et al. 1997). All three perspectives are limited in their ability to understand the long-term transformation of Andean landscapes.

In contrast, a human-centric perspective assumes that the Andean environment is dynamic, historically contingent, and at times chaotic. Andean peoples did not simply adapt to vertically stacked ecological tiers and influence or impact that natural given environment. Nor were humans helpless in the face of climatic variation, as argued by the neoenvironmental determinists. Humans were and are active agents who shaped, transformed, and created the Andean environment. The Lake Titicaca Basin is a built environment, and the evidence of human presence is continuously distributed across the landscape. The temporal dimension of the human transformation is documented in a palimpsest of material correlates to human activity. I have argued that it would be difficult or impossible to define a wild, natural, pristine environment or climax ecosystem in the Lake Titicaca Basin because of the long-term effect of anthropogenic processes.

The role of native peoples in creating the Andean environment should not be underestimated. Much of the early transformation involved deforestation and burning of grasslands, and the replacement of natural fauna and flora with domesticated crops and animals, weedy species, and cultivated trees. The environment was recreated as a highly patterned built environment of fields, walls, settlements, roads, paths, and canals. Soil structure was reworked to a depth of 2 m or more on a massive regional scale during the construction of raised fields, terraces, and q'ochas. Incidental and induced erosion of soils and nutrients from overgrazing, burning, and deforestation in the upper basin was captured behind terrace walls or in raised field canals below (Carney et al. 1996; Erickson n.d.). Hydrology and temperatures were also modified on a regional scale through construction of raised fields, terraces, and q'ochas. Raised fields permanently transformed and artificially expanded the natural wetlands, increasing biodiversity and overall biomass (Erickson 1996).

ANTHROPOGENIC VERSUS NATURAL CAUSATION

Does a human-centric perspective overemphasize the anthropogenic processes over natural processes of environmental formation and change? Can natural disturbance be distinguished from anthropogenic disturbance (Bray 1995; Stahl 1996)? Proponents of the other perspectives argue that both short- and long-term variations of climate (mega–El Niños, cataclysmic drought, global climate change) have a greater role than humans in determining the past and environment of the Andes. The dynamics of the

Andean environment and Precolumbian societies are attributed to climate fluctuation or climate change. The processes of adoption, expansion, and abandonment of specific farming systems and the rise, flourishing, and collapse of civilizations are reduced to responses to favorable and unfavorable climatic conditions. In the human-centric perspective, climatic fluctuation and change are expected and become a background to human activities. The best evidence for cultural resilience and flexibility in the face of short- and long-term climatic change is the continuity and longevity of rural households and settlements distributed across the landscape and the continued expansion of agricultural production and human population throughout prehistory. The series of short-lived states and urban centers in late prehistory are the best evidence of clear discontinuities in the archaeological record.

DEGRADATION VERSUS ENHANCEMENT
OF THE ENVIRONMENT OF THE LAKE TITICACA BASIN

Accepting for a moment the idea that the environment of the Lake Titicaca Basin is dynamic, historically contingent, and a result of human activities over the long term, does this human landscape represent environmental degradation or environmental enhancement? We would also have to ask, at what temporal and spatial scale? I have argued that there is no pristine or original benchmark for comparison. To argue whether human activities are environmental enhancements, sustainable land use, or environmental degradation requires subjective value judgments (Stahl 1996:118–119; Denevan 1992:381; Kirch and Hunt 1997). Quantitative measures of biodiversity through time are not available for the basin. Contemporary studies show that biodiversity and biomass are high in the wetlands of Lake Titicaca despite the high altitude and harsh climatic conditions (Dejoux and Iltis 1992; Levieil and Orlove 1990). I have argued that these wetlands have been maintained and enhanced by humans for thousands of years. Another measure of the health of a landscape could be the sustained human carrying capacity over the long term. Archaeological surveys show that the Lake Titicaca Basin was densely populated during the 2,000–3,000 years before the arrival of the Spanish (Stanish et al. 1997; Albarracín 1996; Albarracín and Mathews 1990). The basin is still one of the most densely populated agrarian landscapes in the Andes.

Numerous scholars suggest that Prehispanic peoples of the Lake Titicaca Basin caused environmental degradation (Winterhalder and Thomas 1978:77–79; Richarson 1991; Seibert 1983; Ellenberg 1979; ONERN 1965; Budowski 1968). Environmental degradation is difficult to define, document, and measure (Blaikie and Brookfield 1987a; Richarson 1991; Botkin 1990; Stahl 1996). As pointed out by Spriggs (1997), environmental degradation of one part of the landscape can actually be landscape enhancement, improvement, management, and sustainable development in another part. Prehispanic deforestation and topsoil erosion at higher elevations reduced the fertility and moisture-holding ability of soils on slopes to the point that the land lost its ability to produce crops, support camelids, and sustain natural faunal and floral communities. At the same time, the earth, organic matter, and nutrients eroding from these slopes were trapped and deposited on river and lake plains to be recycled by farmers using raised fields and q'ochas for intensive crop production, which in turn sustained large urban populations and state-level societies. Increased nutrient input to the lake may have enhanced the overall biomass of wetlands resources. The tensions and conflicts between competing and often spatially overlapping pastoral and farming activities have a long history (Browman 1987; Winterhalder and Thomas 1978). Terracing, raised fields, q'ochas, and bofedales, considered to be ecologically sound and sustainable strategies, involved massive disturbances of soils, hydrology, and vegetation, replacing nature with artificial ecosystems. From the perspective of smallholders, who pass down improvements made to the land and management techniques (fallowing, crop rotation, soil management, and agricultural infrastructure) these disturbances could be considered long-term conservation strategies (Netting 1993; Blaikie and Brookfield 1987a; Hames 1996).

ARCHAEOLOGY AND THE STUDY OF THE ANDEAN LANDCAPE

Environmental change brought about by introduction of Old World species of domestic plants and animals and farming practices, Colonial economic policies, and contemporary urban and agricultural development has occurred at a scale and intensity far greater than at any comparable span of time in prehistory (Denevan 1992; Zimmerer 1996; Gade 1992; Donkin 1979; Mayer 1979). Despite these transformations, the Andes have not suffered the tragedy of the commons to the degree experienced in other parts of the Americas (Denevan 1992:376; Guillet 1981). Guillet (1981:149–150) believes this is due to the intensification of agriculture, sectorial fallowing, risk management, and strong communal ethic still found in Andean communities. Andean farmers have done a good job of adjusting to changing environmental, economic, political, and social conditions through flexibility, diversification, maintenance of heterogeneity in land races, and adoption of techniques and crops that they find useful (Gade 1992; Zimmerer 1996; Morlon 1996). Zimmerer (1996) has argued persuasively that crop biodiversity and indigenous knowledge systems can best be encouraged and maintained at the scale of regional landscapes.

Many of the Precolumbian technologies, landscapes, and indigenous knowledge systems discussed here are abandoned, underutilized, or forgotten. Their physical imprint on the landscape and built environment is enduring. Despite dramatic changes in land tenure, demography, social organization, and economic systems during the past 500 years, the Precolumbian structures of everyday life (fields, pathways, walls, canals, and other features of the built environment and landscape) still shape contemporary rural life in the region. This valuable record of Andean environmental history is now at risk as terrace walls are removed for pasture and raised fields are erased by mechanized plowing and urban expansion (Garaycochea 1986a; Erickson and Candler 1989).

Archaeology can contribute to our understanding of the relationship between humans and the environment and the long-term dynamics that created the Andean landscape. Many of us have argued that what we learn from the material record of human activities on the land could provide viable models for contemporary land use (Morlon 1996; Denevan 2000; Erickson 1998; Kolata et al. 1996). Significant issues that need to be addressed include justice for all segments of society, land reform, adequate wages, access to capital and markets, and fair prices for agricultural produce if there is any hope for a truly sustainable rural development. The environment of the Lake Titicaca Basin has been shaped by human activities, intervention, and management over many millennia. Botkin (1990:93–201) points out that the future of the environment will depend on similar human input. This future must be informed by a long-term history of the environment that seriously considers the role of humans.

REFERENCES

Abbott, M. G., M. W. Binford, M. Brenner, J. H. Curtis, and K. R. Kelts. 1997. A 3,500 ^{14}C yr high-resolution sediment record of lake level changes in Lake Titicaca, Bolivia/Peru. *Quaternary Research* 47(2): 169–180.

Albarracín-Jordan, J. 1996. *Tiwanaku: Arqueología Regional y Dinámica Segmentaria*. La Paz: Editores Plural.

Albarracín-Jordan, J., and J. E. Mathews. 1990. *Asentamientos Prehispánicos del Valle de Tiwanaku*. Tomo 1. La Paz: Producciones CIMA.

Aldenderfer, M., ed. 1993. *Domestic Architecture, Ethnicity, and Complementarity in the South-Central Andes*. Iowa City: University of Iowa Press.

———. 1998. *Montane Foragers: Asana and the South-Central Andean Archaic*. Iowa City: University of Iowa Press.

Allan, N. J. R., G. W. Knapp, C. Stadel, eds. 1988. *Human Impact on Mountains*. Totowa, N.J.: Rowman & Littlefield.

Angeles, F. V. 1987. *Sistema Technologico Andino en Q'ocha y Organization Campesina*. Thesis in Sociology, Universidad Tecnica del Altiplano, Puno, Peru.

Aveni, A., ed. 1990. *The Nasca Lines*. Philadelphia: American Philosophical Society.

Balée, W., and D. Posey, eds. 1989. *Natural Resource Management by Indigenous and Folk Societies in Amazonia*. Advances in Economic Botany 7. New York: New York Botanical Garden.

Bender, B. 1998. *Stonehenge: Making Space*. Oxford: Berg.

Bermann, M. 1994. *Lukurmata: Household Archaeology in Prehispanic Bolivia*. Princeton: Princeton University Press.

Binford, M. W., M. Brenner, and B. Leyden. 1996. Paleoecology and Tiwanaku agroecosystems. In A. Kolata, ed., *Tiwanaku and Its Hinterlands: Archaeology and Paleoecology of an Andean Civilization 1, Agroecology*, pp. 89–108. Washington, D.C.: Smithsonian Institution Press.

Binford, M. W., and A. L. Kolata. 1996. The natural and human setting. In A. Kolata, ed., *Tiwanaku and Its Hinterlands: Archaeology and Paleoecology of an Andean Civilization 1, Agroecology*, pp. 23–56. Washington, D.C.: Smithsonian Institution Press.

Binford, M. W., A. Kolata, M. Brenner, J. Janusek, M. Seddon, M. Abbott, and J. Curtis. 1997. Climate variation and the rise and fall of an Andean civilization. *Quaternary Research* 47: 235–248.

Blaikie, P., and H. C. Brookfield, eds. 1987a. *Land Degradation and Society*. London: Methuen.

———. 1987b. Defining and debating the problem. In P. Blaikie and H. Brookfield, eds., *Land Degradation and Society*, pp. 1–26. London: Methuen.

Botkin, D. 1990. *Discordant Harmonies: A New Ecology for the Twenty-first Century*. New York: Oxford University.

Boulange, B., and E. Aquize Jaen. 1981. La sédimentation actuelle dans le lac Titicaca et de son bassin versant. *Rev. Hydrobiol. Trop.* 14(4): 269–287.

Bowman, I. 1916. *The Andes of Southern Peru.* New York: American Geographical Society.

Bray, W. 1995. Searching for environmental stress: Climatic and anthropogenic influences on the landscape of Colombia. In P. Stahl, ed., *Archaeology in the Lowland American Neotropics,* pp. 96–112. Cambridge: Cambridge University Press.

Browder, J., ed. 1989. *Fragile Lands of Latin America: Strategies for Sustainable Development.* Boulder, Colo.: Westview Press.

Browman, D. L., ed. 1987. *Arid Land Use Strategies and Risk Management in the Andes: A Regional Anthropological Perspective.* Boulder, Colo.: Westview Press.

Brush, S. B. 1976. Man's use of an Andean ecosystem. *Human Ecology* 4: 147–166.

———. 1977. *Mountain, Field, and Family.* Philadelphia: University of Pennsylvania Press.

Budowski, G. 1968. La influencia humana en la vegetación natural de montañas tropicales Americanas. In C. Troll, ed., *Geoecology of the Mountainous Regions of the Tropical Americas,* pp. 157–162. Bonn: Ferd. Dummlers Verlag.

Burger, R. 1992. *Chavin and the Origins of Andean Civilization.* New York: Thames & Hudson.

Candler, K. L. 1993. *Place and Thought in a Quechua Household Ritual.* Dissertation, Department of Anthropology, University of Illinois, Urbana-Champaign.

Cardich, A. 1985. The fluctuating upper limits of cultivation in the central Andes and their impact on Peruvian prehistory. *Advances in World Archaeology* 4: 293–333.

Carney, H. J., M. W. Binford, and A. L. Kolata. 1996. Nutrient fluxes and retention in Andean raised-field agriculture: Implications for long-term sustainability. In A. Kolata, ed., *Tiwanaku and Its Hinterlands: Archaeology and Paleoecology of an Andean Civilization 1, Agroecology,* pp. 169–180. Washington, D.C.: Smithsonian Institution Press.

Carter, W., and M. Mamani. 1982. *Irpa Chica: Individuo y Comunidad en la Cultura Andina.* La Paz: Editorial Juventud.

Chávez, K. 1988. The significance of Chiripa in Lake Titicaca developments. *Expedition* 30(3): 17–26.

Chepstow-Lusty, A. J., K. D. Bennet, J. Fjeldsa, B. Kendall, W. Galiano, and A. Tupayachi Herrera. 1998. Tracing 4,000 years of environmental history in the Cuzco area, Peru, from the pollen record. *Mountain Research and Development* 18(2): 159–172.

Coolman, B. 1986. Problemática de la recuperación de Andenes: El caso de la comunidad de Pusalaya (Puno). In C. de la Torre and M. Burga, eds., *Andenes y Camellones en el Peru Andino: Historia Presente y Futuro,* pp. 217–224. Lima: Consejo Nacional de Ciéncia y Tecnología.

Crumley, C. L., ed. 1994. *Historical Ecology: Cultural Knowledge and Changing Landscapes.* Santa Fe, N.Mex.: School of American Research.

Deetz, J. 1990. Landscapes as cultural statements. In W. M. Kelso and R. Most, eds., *Earth Patterns: Essays in Landscape Archaeology,* pp. 1–4. Charlottesville: University Press of Virginia.

Dejoux, C., and A. Iltis, eds. 1992. *Lake Titicaca: A Synthesis of Limnological Knowledge.* Dordrecht, Neth.: Kluwer Academic.

Denevan, W. M. 1992. The pristine myth: The landscape of the Americas in 1492. *Annals of the Association of American Geographers* 82: 369–385.

Denevan, W. M. 2000. *Triumph over the Soil.* Oxford: Oxford University.

Dollfus, O. 1982. Development of land-use patterns in the central Andes. *Mountain Research and Development* 2: 39–48.

Doolittle, W. E. 1992. Agriculture in North America on the eve of Contact: A reassessment. *Annals of the Association of American Geographers* 82: 386–401.

Donkin, R. 1979. *Agricultural Terracing in the Aboriginal New World.* Viking Fund Publications in Anthropology 56. New York: Wenner-Gren Foundation for Anthropological Research.

Ellenberg, H. 1979. Man's influence on tropical mountain ecosystems in South America. *J. Ecology* 67: 401–416.

Erickson, C. L. 1976. *Chiripa Ethnobotanical Report: Flotation Recovered Archaeological Remains from an Early Settled Village on the Altiplano of Bolivia.* Senior thesis, Department of Anthropology, Washington University, St. Louis, Mo.

———. 1988. Raised field agriculture in the Lake Titicaca Basin: Putting ancient Andean agriculture back to work. *Expedition* 30(3): 8–16.

———. 1993. The social organization of Prehispanic raised field agriculture in the Lake Titicaca Basin. In V. Scarborough and B. Isaac, eds., *Economic Aspects of Water Management in the Prehispanic New World,* pp. 369–426. Research in Economic Anthropology Supplement 7. Greenwich, Conn.: JAI Press.

———. 1996. *Investigación arqueológica del sistema agrícola de los camellones en la Cuenca del Lago Titicaca del Perú.* La Paz: PIWA and El Centro de Información para el Desarollo.

———. 1998. Applied archaeology and rural development: Archaeology's potential contribution to the future. In M. Whiteford and S. Whiteford, eds., *Crossing Currents: Continuity and Change in Latin America,* pp. 34–45. Upper Saddle, N.J.: Prentice-Hall.

———. 1999. Neo-environmental determinism and agrarian "collapse" in Andean prehistory. *Antiquity* 73: 634–642.

———. n.d. *Waru Waru: Ancient Andean Agriculture.* Cambridge: Cambridge University Press.

Erickson, C. L., and K. L. Candler 1989. Raised fields and sustainable agriculture in the Lake Titicaca Basin. In J. Browder, ed., *Fragile Lands of Latin America: Strategies for Sustainable Development,* pp. 230–248. Boulder, Colo.: Westview Press.

Flores Ochoa, J. A. 1979. *Pastoralists of the Andes.* Philadelphia: Institute for the Study of Human Issues.

———. 1987. Cultivation in the qocha of the South Andean puna. In D. Browman, ed., *Arid Land Use Strategies and Risk Management in the Andes: A Regional Anthropological Perspective,* pp. 271–296. Colorado Springs: Westview Press.

Gade, D. W. 1981. Some research themes in the cultural geography of the central Andean highlands. In P. Baker and C. Jest, eds., *L'Homme et son Environment à Haute Altitude.* Paris: CNRS.

———. 1992. Landscape, system, and identity in the post-Conquest Andes. *Annals of the Association of American Geographers* 82(3): 460–477.

Gade, D. W., and R. Rios. 1972. Chaquitaclla: The native footplough and its persistence in central Andean agriculture. *Tools and Tillage* 2(1): 3–15.

Garaycochea, I. 1986a. Destrucción y conservación de camellones en el departamento de Puno. *Problemática Sur Andina* 9.

———. 1986b. Potencial agrícola de los camellones en el altiplano puneño. In C. de la Torre and M. Burga, eds., *Andenes y Camellones en el Peru Andino: Historia, Presente, y Futuro,* pp. 241–251. Lima: Consejo Nacional de Ciéncia y Tecnología.

Glaser, G., and J. Celecia. 1981. Guidelines for integrated ecological research in the Andean region. *Mountain Research and Development* 1(2): 171–186.

Goland, C. A. 1991. *Cultivating Diversity: Field Scattering as Agricultural Risk Management in Cuyo Cuyo, Department of Puno, Peru, vol. 1 and 2.* Dissertation, Department of Anthropology, University of Michigan, Ann Arbor.

Golte, J. 1980. *La Racionalidad de la Organización Andina.* Lima: Instituto de Estudios Peruanos.

Gomez-Molina, E., and A. V. Little. 1981. Geoecology of the Andes: The natural science basis for research planning. *Mountain Research and Development* 1(2): 115–144.

Graffam, G. 1992. Beyond state collapse: Rural history, raised fields, and pastoralism in the south Andes. *American Anthropologist* 94(4): 882–904.

Guillet, D. 1981. Land tenure, ecological zone, and agricultural regime in the central Andes. *American Ethnologist* 8: 139–156.

Hames, R. 1996. Comments on Algard. *Current Anthropology* 36(5): 804–805.

Hastorf, C. A. 1993. *Agriculture and the Onset of Political Inequality before the Inka.* New York: Cambridge University Press.

Hastorf, C. A., and others. 1997. *Taraco Ar-*

chaeological Project: 1996 Excavations at Chiripa. Manuscript, Department of Anthropology, University of California, Berkeley.

Holdridge, L. R. 1947. *Life Zone Ecology*. San Jose, Costa Rica: Tropical Science Center.

Hyslop, J. 1977. Chullpas of the lupaca zone of the Peruvian High Plateau. *J. Field Archaeology* 4(2): 149–170.

———. 1984. *The Inka Road System*. New York: Academic Press.

———. 1990. *Inka Settlement Planning*. Austin: University of Texas Press.

Isbell, W. 1997a. *Mummies and Mortuary Monuments: A Postprocessual Prehistory of Central Andean Social Organization*. Austin: University of Texas Press.

———. 1997b. Household and ayni in the Andean past. *J. Steward Anthropological Society* 24(1–2).

Janusek, J. 1994. State and local power in a Prehispanic Andean polity: Changing patterns of urban residence in Tiwanaku and Lukurmata, Bolivia. Ph.D. dissertation, Department of Anthropology, University of Chicago.

Julien, C. 1988. The Squire Causeway at Lago Umayo. *Expedition* 30(3): 46–55.

Kirch, P., and T. Hunt, eds. 1997. *Historical Ecology in the Pacific Islands: Prehistoric Environmental and Landscape Change*. New Haven: Yale University Press.

Klink, C., and M. Aldenderfer. 1996. Archaic period settlement on the altiplano: Comparison of two recent surveys in the southwestern Lake Titicaca Basin. Paper presented at the 24th Midwest Conference on Andean and Amazonian Archaeology and Ethnohistory, Beloit, Wisconsin.

Knapp, G. 1991. *Andean Ecology: Adaptive Dynamics in Ecuador*. Boulder, Colo.: Westview Press.

Kolata, A. L. 1993. *The Tiwanaku*. Oxford: Basil Blackwell.

———, ed. 1996. *Tiwanaku and Its Hinterlands: Archaeology and Paleoecology of an Andean Civilization 1, Agroecology*. Washington, D.C.: Smithsonian Institution Press.

Kolata, A. L., and C. Ortloff. 1996a. Agroecological perspectives on the decline of the Tiwanaku state, In A. Kolata, ed., *Tiwanaku and Its Hinterlands: Archaeology and Paleoecology of an Andean Civilization 1, Agroecology*, pp: 181–202. Washington, D.C.: Smithsonian Institution Press.

———. 1996b. Tiwanaku raised-field agriculture in the Lake Titicaca Basin of Bolivia. In A. Kolata, ed., *Tiwanaku and Its Hinterlands: Archaeology and Paleoecology of an Andean Civilization 1, Agroecology*, pp. 109–152. Washington, D.C.: Smithsonian Institution Press.

Kolata, A. L., O. Rivera, J. C. Ramírez, and E. Gemio. 1996. Rehabilitating raised-field agriculture in the southern Lake Titicaca Basin of Bolivia: Theory, practice, and results. In A. Kolata, ed., *Tiwanaku and Its Hinterlands: Archaeology and Paleoecology of an Andean Civilization 1, Agroecology*, pp. 203–230. Washington, D.C.: Smithsonian Institution Press.

Kuznar, L. A. 1993. Mutalism between *Chenopodium*, herd animals, and herders in the south central Andes. *Mountain Research and Development* 13(3): 257–265.

Lennon, T. 1983. Pattern analysis of Prehispanic raised fields of Lake Titicaca, Peru. In J. P. Darch, ed., *Drained Fields of the Americas*, pp. 183–200. International Series 189. Oxford: British Archaeological Reports.

Levieil, D. P., and B. Orlove. 1990. Local control of aquatic resources: Community and ecology in Lake Titicaca, Peru. *American Anthropologist* 92: 362–382.

Masuda, S., I. Shimada, and C. Morris. 1985. *Andean Ecology and Civilization*. Tokyo: University of Tokyo Press.

Mayer, E. 1979. *Land Use in the Andes: Ecology and Agriculture in the Mantaro Valley of Peru with Special Reference to Potatoes*. Lima: Centro Internacional de la Papa.

———. 1985. Production zones. In S. Masuda, I. Shimada, and C. Morris, eds., *Andean Ecology and Civilization*, pp. 45–84. Tokyo: University of Tokyo Press.

McAndrews, T., J. Albarracín-Jordan, and M. Bermann. 1997. Regional settlement patterns in the Tiwanaku Valley of Bolivia. *J. Field Archaeology* 24: 67–83.

Miller, N., and K. Gleason, eds. 1994. *The Ar-

chaeology of Garden and Field. Philadelphia: University of Pennsylvania Press.

Monheim, F. 1963. *Contribucción a la Climatología e Hidrología de la Cuenca del Lago Titicaca*. Puno, Peru: Universidad Técnica del Altiplano.

Moore, J. D. 1995. The archaeology of dual organization in Andean South America: A theoretical review and case study. *Latin American Antiquity* 6(2): 165–181.

Moore, Katherine M., David Steadman, and Susan deFrance. 1999. Herds, fish, and fowl in the domestic and ritual economy of Formative Chiripa. In Christine Hastorf, ed., *Early Settlement at Chiripa, Bolivia: Research of the Taraco Archaeological Project*, pp. 105–116. Contributions 57. Berkeley: University of California Archaeological Research Facility.

Morlon, P., ed. 1996. *Comprender la Agricultura Campesina en Los Andes Centrales: Perú-Bolivia*. Lima: Instituto Francés de Estudios Andinos.

Moseley, M. 1992. *The Incas and Their Ancestors*. New York: Thames & Hudson.

Murra, J. 1968. An Aymara kingdom in 1567. *Ethnohistory* 15(2): 115–151.

Netting, R. M. 1993. *Smallholders, Householders: Farm Families and the Ecology of Intensive, Sustainable Agriculture*. Stanford, Calif.: Stanford University Press.

Nuñez, M. 1984. Manejo y control de totorales en el Titicaca. *Boletín del Instituto de Estudios Aymaras* (Chucuito, Peru) Serie 2, 19 (Abril): 4–19.

ONERN-CORPUNO. 1965. *Inventario y Evaluación Semidetallada de Los Recursos Naturales de Suelos, Uso Actual de la Tierra e Hidrología de la Microregión Puno (Sectores Puno-Manazo)*. Lima: ONERN.

Orlove, B. 1977. Inequality among peasants: The forms and uses of reciprocal exchange in Andean Peru. In R. Halperin and J. Dow, eds., *Peasant Livelihood: Studies in Economic Anthropology and Cultural Ecology*, pp. 201–226. New York: St. Martin's Press.

Orlove, B., and R. Godoy. 1986. Sectoral fallowing systems in the central Andes. *J. Ethnobiology* 6(1): 169–204.

Palacios Rios, F. 1977. Pastizales de regadio para alpacas. In J. Flores Ochoa, ed., *Pastores de Puno: Uywamichiq Punarunak-una*, pp. 155–170. Lima: Instituto de Estudios Peruanos.

———. 1984. Tecnología del pastoreo. *Boletín del Instituto de Estudios Aymaras* (Chucuito, Peru) Serie 2, 18 (Diciembre): 38–53.

Paulsen, A. C. 1976. Environment and empire: Climatic factors in prehistoric Andean culture change. *World Archaeology* 8: 121–229.

Piperno, D. R., and D. M. Pearsall. 1998. *The Origins of Agriculture in the Lowland Neotropics*. New York: Academic Press.

Ramos Vera, C. 1984. Tecnología de la reconstruccion, refaccion, y manejo de andenes y terrazas en el distrito de Asillo. Manuscript, Centro Artesanal Jose Maruri, Asillo.

———. 1986. Reconstruccion, refracción, y manejo de andenes en Asillo (Puno). In C. de la Torre and M. Burga, eds., *Andenes y Camellones en el Peru Andino: Historia, Presente, y Futuro*, pp. 225–239. Lima: Consejo Nacional de Ciéncia y Tecnología.

———. 1990. *Rehabilitación, Uso, y Manejo de Camellones: Propuesta Técnica*. Lima: Programa de Rehabilitación y Uso de Waru Waru, CECI-CCAEP, fondo General de Contravalor.

Richarson, P. J. 1991. Humans as a component of the Lake Titicaca ecosystem: A model system for the study of environmental deterioration. *Proceedings of the Cary Conference* 4.

Rosas, A. 1986. El sístema de cultivo de qocha. In C. de la Torre and M. Burga, eds., *Andenes y Camellones en el Peru Andino: Historia, Presente, y Futuro*, pp. 107–126. Lima: Consejo Nacional de Ciéncia y Tecnología.

Seddon, M. T. 1994. Excavations in raised fields of the Rio Catari sub-basin, Bolivia. Master's thesis, Department of Anthropology, University of Chicago.

Seibert, P. 1983. Human impact on landscape and vegetation in the central High Andes. In W. Holzner, M. J. A. Werger, and I. Ikusinma, eds., *Man's Impact on Vegatation*, pp. 261–276. The Hague: Dr. W. Junk Publishers.

Shimada, I., C. B. Schaaf, L. G. Thompson, and E. Moseley-Thompson. 1991. Cultural impacts of severe droughts in the prehistoric Andes: Applications of a 1,500-year ice core precipitation record. *World Archaeology* 22: 247–270.

Siemens, A. H. 1998. *A Favored Place: San Juan River Wetlands, Central Veracruz,* A.D. *500 to the Present*. Austin: University of Texas Press.

Smith, C. T., W. Denevan, and P. Hamilton. 1968. Ancient ridged fields in the region of Lake Titicaca. *Geographical J.* 134: 353–367.

Spriggs, M. 1997. Landscape catastrophe and landscape enhancement: Are either or both true in the Pacific. In P. Kirch and T. Hunt, eds., *Historical Ecology in the Pacific Islands: Prehistoric Environmental and Landscape Change*, pp. 80–104. New Haven: Yale University Press.

Stahl, P. 1996. Holocene biodiversity: An archaeological perspective from the Americas. *Annual Review of Anthropology* 25: 105–126.

Stanish, C. 1994. The hydraulic hypothesis revisited: Lake Titicaca basin raised fields in theoretical perspective. *Latin American Antiquity* 5: 312–332.

Stanish, C., and L. Steadman. 1994. *Archaeological Research at Tumatumani, Juli, Peru*. Fieldiana Anthropology n.s. 23. Chicago: Field Museum of Natural History.

Stanish, C., E. de la Vega, L. Steadman, C. Chávez-Justo, K. L. Frye, L. Onofre-Mamani, M. Seddon, P. Calisaya-Chuquimia. 1997. *Archaeological Survey in the Juli-Desaguadero Region of Lake Titicaca Basin, Southern Peru*. Fieldiana Anthropology n.s. 29. Chicago: Field Museum of Natural History.

Steadman, L. 1995. Excavations at Camata: An early ceramic chronology for the western Titicaca Basin, Peru. Ph. D. dissertation, Department of Anthropology, University of California, Berkeley.

Stone, G. D. 1996. *Settlement Ecology: The Social and Spatial Organization of Kofyar Agriculture*. Tucson: University of Arizona Press.

Thompson, L. G., M. E. Davis, E. Moseley-Thompson, and K. Liu. 1988. Pre-Incan agricultural activity recorded in dust layers in two tropical ice cores. *Nature* 336: 763–765.

Tilley, C. 1994. *A Phenomenology of Land-scape: Places, Paths, and Monuments*. Providence, R.I.: Berg.

Torre, C. de la, and M. Burga, eds. 1986. *Andenes y Camellones en el Peru Andino: Historia, Presente, y Futuro*. Lima: Consejo Nacional de Ciéncia y Tecnología.

Tosi, J. A. 1960. *Zonas de Vida Natural en el Perú: Memoria Explicativa Sobre el Mapa Ecológico del Perú*. Lima: Instituto Interamericano de Ciencias Agricolas de la OEA, Zona Andina.

Treacy, J. M. 1994. *Las Chacras de Coporaque: Andenería y Riego en el Valle del Colca*. Lima: Instituto de Estudios Peruanos.

Treacy, J. M., and W. Denevan. 1994. The creation of cultivable land through terracing. In N. Miller and K. Gleason, eds., *The Archaeology of Garden and Field*, pp. 91–110. Philadelphia: University of Pennsylvania Press.

Troll, C. 1968. The cordilleras of the tropical Americas: Aspects of climatic, phytogeographical, and agrarian ecology. In C. Troll, ed., *Geo-Ecology of the Mountainous Regions of the Tropical Americas*, pp. 15–56. Bonn: Ferd Dummlers Verlag.

Urton, G. 1990. *The Origin of a Myth: Pacariqtambo and the Origin of the Inkas*. Austin: University of Texas Press.

Vacher, J. J., E. Brasier de Thuy, and M. Liberman. 1991. Influence of the lake on littoral agriculture. In C. DeJoux and A. Iltis, eds., *Lake Titicaca: A Synthesis of Limnological Knowledge*, pp. 517–530. Dordrecht, Neth.: Kluwer Academic.

Wachtel, N. 1990. *Le Retour des Ancêtres: Les Indiens Urus de Bolivie XXe–XVIe siècle. Essai d'Histoire Régressive*. Paris: Bibliothèque des Sciences Humaines, Gallimard.

West, T. L. 1987. The burning bush: Exploitation of native shrubs for fuel in Bolivia. In D. Browman, ed., *Arid Land Use Strategies and Risk Management in the Andes: A Regional Anthropological Perspective*, pp. 151–170. Colorado Springs: Westview Press.

Wickens, G. E. 1995. Llareta (*Azorella compacta*, Umbelliferae): A review. *Economic Botany* 49(2): 207–212.

Winterhalder, B., and R. B. Thomas. 1978. *Geoecology of Southern Highland Peru:*

A Human Adaptation Perspective. Occasional Paper 27. Boulder, Colo.: Institute of Arctic and Alpine Research.

Wirrmann, D., J. P. Ybert, and P. Mourguiart. 1992. In C. DeJoux and A. Iltis, eds., *Lake Titicaca: A Synthesis of Limnological Knowledge,* pp. 40–48. Dordrecht, Neth.: Kluwer Academic.

Yarmin, R., and K. Bescherer Metheny, eds. 1996. *Landscape Archaeology: Reading and Interpreting the American Historical Landscape.* Knoxville: University of Tennessee Press.

Ybert, J. P. 1992. Ancient lake environments as deduced from pollen analysis. In C. DeJoux and A. Iltis, eds., *Lake Titicaca: A Synthesis of Limnological Knowledge,* pp. 49–62. Dordrecht, Neth.: Kluwer Academic.

Zimmerer, K. S. 1994. Human geography and the new ecology: The prospect and promise of integration. *Annals of the American Association of Geographers* 84(1): 108–125.

———. 1996. *Changing Fortunes: Biodiversity and Peasant Livelihood in the Peruvian Andes.* Berkeley: University of California Press.

Zuidema, R. T. 1985. Inka dynasty and irrigation: Another look at Andean concepts of history. In J. Murra, N. Wachtel, and J. Revel, eds., *Anthropological History of Andean Polities,* pp. 177–200. Cambridge: Cambridge University Press.

TERENCE N. D'ALTROY

13 | ANDEAN LAND USE AT THE CUSP OF HISTORY

Western South America is a land of astonishing natural diversity and striking beauty. The region's climate and topography combine to give rise to both the world's driest desert and permanent glaciers on the tallest peaks in the Americas, compressed within a narrow strip along the continent's margin (figure 13.1). Its environmental complexity results from the juxtaposition of the Andes mountains and the Pacific's frigid Humboldt current, coupled with cyclical patterns in the equatorial trade winds. We can gain a sense of the natural diversity, described in greater detail by Luteyn and Churchill (this volume), by envisioning a transect from the shoreline eastward across central Peru. On that path lie 20 of the world's 34 major life zones, in a tightly compacted pattern of horizontal bands that is not duplicated anywhere on the planet (Burger 1992:12). Even where agricultural land is relatively abundant, such as in Peru's coastal valleys, irrigation is often an absolute necessity because rain may not fall for decades on end. The land quickly ascends to two mountain ranges whose highest peaks soar 6,000 m above sea level, amid which lie productive intermontane valleys. Above the treeline lies the puna, a tundra that is the habitat for the Andean camelids, while in the highlands south of Peru is the Bolivian altiplano, a cold, windswept plain with barely a tree to call its own. To the east, the mountains fall precipitously to the tropical rain forest in the north and great plains in the south. In the southern Andes, the lands range upward from longitudinal, desertic valleys to cordilleras of permanent snowcaps.

Since entering South America some 12,000 or more years ago, human populations developed an array of adaptive

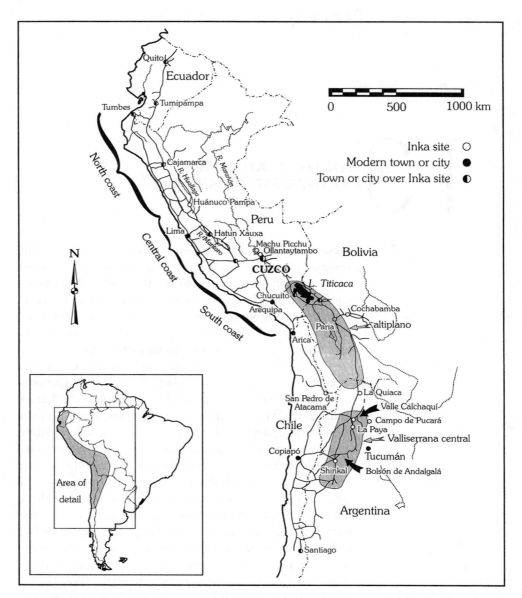

FIGURE 13.1. The Andean region encompassed by the Inka empire, including the main land use regions referred to in the text, the Inka highways, and some of the principal sites. *Source:* After Hyslop 1984.

strategies that provided the foundation for precociously complex societies. The subsistence bases of Andean societies developed over the millennia, as staple foods, such as potatoes (*Solanum tuberosum*), other tubers, and quinoa (*Chenopodium quinoa*), and industrial crops, such as cotton (*Gossypium barbadense*) and gourd, were domesticated in the region between 8000 and 3000 B.C. (Pearsall 1992). Maize, a staple along the coast and in the lower montane reaches, was domesticated by ca. 3500 B.C. in lands north of the present study region but was not a significant element of the Andean diet until ca. 1000 B.C. The marine fisheries and littoral molluscan beds were exceptionally rich until overexploited this century. Together with terrestrial foraging and incipient floodwater farming, they sustained the precocious complex societies that arose along the coastline after 2500 B.C. Equally important for highland populations was the domestication of the camelids by about the fifth millennium B.C. Since then, llama and alpaca herding has become both a fruitful adaptive strategy and a source of wealth for Andean peoples. As a consequence, rather than eking out a living at the constant edge of survival, human societies managed to carve out a highly successful series of adaptations to the demanding natural conditions.

Despite their successes, we should not exaggerate the ease of life, for earthquakes, floods, unpredictable climatic cycles, drought, frost, and other natural forces periodically brought disaster to subsistence systems. Major droughts occurred in the central Andes in 1942–1946, 1956–1957, 1964–1966, and 1982–1983, for example; in the last episode, highland crop production fell 60–70 percent from the norm (Winterhalder 1993:26). Brown (1987, cited in Winterhalder 1993) has noted comparable decrements for the altiplano west of Lake Titicaca, caused by irregular monthly distribution of rainfall, abetted by hard frost. The unreliability of traditional agricultural production is encapsulated by Polo's (1916) 1571 observation that crops failed in the highlands one year out of three and Cobo's (1979) 1653 comment that crops fail on the altiplano twice that often.

The relatively predictable annual cycles are intermittently disrupted by a climatic phenomenon called El Niño Southern Oscillation (ENSO), or more commonly El Niño (the Child), because it peaks around Christmas (Vallis 1986; Diaz and Markgraf 1992). Although the phenomenon is cyclical, events occur at intervals of 2–11 years, making them essentially unpredictable for prehistoric peoples. When an event occurs, warm equatorial waters flow south along the coastline, upwelling of cold southern waters is suppressed, and surface temperatures can reach tropical levels along Peru's north coast. In strong events, such as that of 1997–1998, the ecological effects are catastrophic. Torrential rains fall on the coast, damaging crops and washing away canal systems, roads, and unadvisedly placed settlements. Fish and birds die in massive numbers or migrate out of the range in which traditional methods of capture work effectively. In a landscape in which minor changes in precipitation and temperature can have a profound effect on agriculture, El Niño has been a significant factor in human life over the last 5,000 years.

To speak of a single kind of Andean resource use in such conditions would be to gloss over great variations in the ways that people organized their lives. Even so, it is often noted that many of the societies who inhabited the lands drawn into the Inka empire shared cultural and economic features, some of which were unusual in the premodern world and strikingly novel to the Europeans (see Murra 1975, 1980a). In those regions for which there is reliable documentation, most productive resources were held in common by corporate kin groups, which apportioned access to their members. According to testimony taken down in the early Colonial era, the products of the peasantry's property were at least theoretically inalienable, even if elites and states appropriated productive resources and labor for their own benefit. Except for parts of the north Peruvian coast and Ecuador, market systems and money were unknown in the Andes. Instead, exchange was most often carried out via social relations, barter, or shared access to resources. Specialized food production, artisanry, and services, which are often linked to money and markets, were also more limited or took different forms in the Inka heartland than in many other major civilizations. Ceremonial hospitality, replete with ritualized exchanges of cloth and other fine crafts, provided the context for political relations, while elaborate rituals were integral to activities ranging from tilling the soil to shearing wool (e.g., Polo 1940:24–32; Garcilaso de la Vega 1966: 243–245; Cobo 1990:143–144; see Murra 1962). Together, those qualities point to patterns of resource use shaped by distinctive cultural conventions.

Despite the unusual nature of some Andean cultural features, ecological conditions set inescapable constraints on the agropastoral practices that could maintain human life. Climate, elevation, topography, and the availability of water all restricted the kinds of crops that could be grown and the order in which they could be worked. Over the millennia, Andean peoples pushed the ecological boundaries through ingenuity and sheer hard work, but many choices about land use and labor allocation also followed a logic that still makes sense in a modern economy (see Mitchell and Guillet 1993). Accordingly, in order to understand the main elements of late Precolumbian cultural ecology in the Andes, we need to consider both cultural choices and the kinds of practices that could have been successful under the demanding physical conditions. In addition, we have to be aware that land use practices varied in important ways depending on the interests of the parties involved and the labor at their disposal.

This essay describes the land use strategies of a cross-section of Andean societies at the transition from indigenous to Spanish rule (ca. A.D. 1532). Considering the array of circumstances that existed at the time, only a few examples can be examined, but they provide a sense of the myriad ways in which Andean peoples tailored their lives to the environment and the landscape to their needs. The organizations of the Inka state and aristocracy, which built upon and modified traditional arrangements, are compared to those approaches.

LAND USE CLASSIFICATION

Over the years, geographers have classified the natural environments of the Andes in many ways (see Tosi 1960; Troll 1968; Flannery et al. 1989). Some schemes, such as the widely used formulation devised by J. Pulgar Vidal (1987) for Peru, combine native terminology with modern biotic classification. Because each zone features distinctive climatic, biotic, and topographic properties, it requires different kinds of adaptive strategies by human societies. From Tumbes to southern Chile, the coast (*chala* or *costa*) is a barren desert, punctuated only by drainages that cut ribbons of green through a pastel landscape of drifting sands and jagged rock formations. Along the north and central Peruvian coast, most agriculture is practiced below 150 m elevation,

FIGURE 13.2. Mouth of Nepeña River valley on Peru's north coast, showing the sharp contrast between lush irrigated lands and naturally dry landscape.

which normally receives no rainfall. Where irrigation is practicable, the valley bottoms are lush croplands that supported the principal indigenous crops maize (*Zea mays*), lúcuma (*Lucuma bifera*), cucurbits, gourd, and cotton (figure 13.2); in this century, sugarcane has become a major export crop. The coast presents a variety of other useful microenvironments, including the marine littoral and riparian bands, which are productive in shellfish, fish, and foraged plants. In some areas, seasonal lomas (hills) vegetation supported by heavy fogs provided important additions of plant and animal foods in prehistory.

Above the coastal plain is the yungas zone (300–2,300 m). Warmer than the coastal valleys, the yungas produce maize and tropical fruits, such as cherimoya (*Annona cherimola*), guayabo (*Psidium guajava*), avocado (*Persea americana*), and lúcuma. Even so, the most valued lands in this region have traditionally been the coca (*Erythroxylum coca*) and ají (*Capsicum* spp.) fields. Together with maize, coca was symbolically the most important crop to native Andean peoples. It was used extensively in gift exchanges and sacrifices, and the Inkas gave it in enormous quantities to workers and subjects as part of the largess expected of Andean lords. On the eastern side of the Andes above the Amazonian jungles lies another yungas band, which was equally if not more productive in the same products.

The quechua (quishwa, kichwa) zone (3,100–3,500 m) is the most productive highland ecozone. Because the temperate climate on its valley bottoms is suitable for the cultivation of frost-sensitive crops, traditional dry farming produced maize, beans (e.g.,

FIGURE 13.3. The llama (*Lama glama*), largest of the Andean camelids (photo by author).

Phaseolus vulgaris, P. lunatus, Cannavalia), garden vegetables, quinoa, ulluco (*Ullucus tuberosus*), oca (*Oxalis tuberosus*), mashwa (*Tropaeolum tuberosum*), and talwi (*Lupinus mutabilis*). Above about 3,500 m, maize is subject to killing frosts, so small valleys, quebradas, and rolling uplands are cultivated with chenopods, grains, legumes, and tubers. Small-scale irrigation was common prehistorically in the quechua zone, drawing from springs and streams. Along the soggy margins of springs and lakes, farmers also dug drained-field systems to reclaim productive soils. The suni zone (up to 4,000 m) is characterized by cold hills, ridges, and deep valleys, which yield well in quinoa and talwi.

The puna (up to 4,800 m) is an alpine tundra that is the natural habitat for the Andean camelids (figure 13.3). The weather is usually cold and damp, with heavy fogs and violent storms that roll over the ground during the wet season. The most important human activities on the puna were the herding of llamas and alpacas and, to a lesser extent, the hunting of guanacos and vicuñas. Although the puna is marginal for most kinds of agriculture, some hardy indigenous and European crops can be cultivated for human and animal consumption. Among them are frost-resistant tubers, such as the bitter varieties that provide a main source of chuño, or freeze-dried potatoes.

Above the puna is the janca zone, which features jagged permanent snowcaps and glacial lakes. Despite its brutally cold and oxygen-deprived atmosphere, this highest of the central Andean environmental zones has been exploited for centuries for its abundant mineral wealth.

Down the eastern Peruvian slopes are the warm, wet upper part of the Amazonian jungles, called the montaña, and the lower jungle, called the selva (or rupa-rupa). The upper edge of the Amazonian forest itself is known as the *ceja de selva,* or "eyebrow of the jungle." The montaña is renowned for its steep, forested slopes above the flood-plain of the rain forest proper. In that ecozone, maize, coca, fruits, and a host of other warm-weather crops can be produced both in valley bottoms and in precipitous terrace systems. Below, in the Amazonian rain forests, the verdant forest contains both fertile floodplains (várzea) and slightly raised interfluvial plains that are used to cultivate manioc (*Manihot esculenta*).

The environments to the north and south of Peru differ somewhat from Pulgar Vidal's classification. Along Ecuador's coast, the desert gives way to mangrove swamps in the equatorial latitudes, whereas the grassy highlands are called páramo. The upland valleys are generally smaller and more dissected than the major intermontane valleys of Peru, but the temperate climate makes them one of the world's most pleasant regions to live in. To the south, the main highland environmental feature is the altiplano, which covers most of upland Bolivia and runs into the desiccated northwest corner of Argentina (see Tomka 1994:73–103 for a review of environmental conditions). In the southern part of the Inka domain, the principal region of human occupation was the valliserrana, a high zone of long desertic valleys, shrub forest, and puna that sustained a surprising density of communities in pockets along the main and lateral watercourses.

Researchers working among the traditional societies of the mountains note that these kinds of classifications conform only partially to indigenous conceptions of land use zones (e.g., Murra 1980a; Flores-Ochoa 1977). Flannery et al. (1989:21–24) emphasize a point often made by students of Andean cultures—the landscape is charged with cultural and symbolic significance. The Ayacucho agropastoralists, for example, divide their exploitable landscape into the agricultural zone, called kichwa, and the sallqa, which is the tundra. The kichwa is subdivided primarily into the parts that are made up of human communities and those that are not; seasonal contrasts there are made on the basis of agricultural cycles. The concept of sallqa contains both the ecological features of the puna described above and the view that the region is wild or uncivilized. Above that zone lies the urqu, or glaciated and snowcapped mountains, home to the mountain spirits. Among the late prehistoric and traditional societies of the Colonial era, the mountain peaks were often seen both as the sources of water and as the paqarina, or origin places of the ancestors. As such, they were locations of power and veneration.

TRADITIONAL RESOURCE USE IN THE CENTRAL HIGHLANDS

AGRICULTURE

COMMUNITY PRODUCTION

Traditional highland subsistence strategies, including those in use today and in late prehistory, were often generalized so that any community's holdings incorporated as wide a range of resources as was practicable. As Murra (e.g., 1972, 1975, 1980a) has detailed in a series of insightful studies, the compact verticality of the landscape means that an array of production zones has often been accessible to highland communities. The ideal, both past and present, has been to make the members of individual communities collectively self-sufficient and thus avoid a need for exchange to obtain the basics of life.

S. Brush (1977) has described three ways that modern societies, following traditional land use patterns, have arrayed themselves across highland Peru to achieve this goal. The *compressed* settlement pattern is found in areas characterized by steep verticality, such as the Uchucmarca region. Communities in that kind of landscape attempt to be self-sufficient locally. Most of the population is settled in the upper elevations, often at the ecotone between maize and tuber lands. Smaller subsidiary settlements are founded in areas where particular resources can be procured or tended, such as in the puna for herding or in the more tropical lowlands for fruit and coca cultivation. A second strategy, found in the upper Marañon and Huallaga valleys, for example, is called the *archipelago* settlement pattern. In this approach, there is a wide separation between some of the resource zones, sometimes requiring trips of four to eight days from the parent community. The treks to the distant parcels may involve traversing different ethnic areas. Today, as in prehistory, cultivation of coca is a central factor in this settlement pattern. The third variant, called *extended* settlement, occurs where the topographic gradients are gentler, such as in the Vilcanota-Urubamba Valley near Cuzco. There, the population is more evenly dispersed to take advantage of broader areas of arable land.

The primary ecological determinants of cultivation order in the dry sierra include the type of crops sown, the location of the fields in the compressed zonation, and the particular agricultural cycle being followed (see Mitchell 1980; Hastorf 1993). The planting sequence is especially related to altitude, which has notable effects on temperature, sunshine, moisture, the onset of rains and frost, cloud cover, and evapotranspiration rates. Generally speaking, higher elevations are cloudier and have lower evapotranspiration rates, whereas lower elevations have more sunshine and greater water loss rates from soil and plants. Equally significant, the higher the elevation, the more unpredictable are both rainfall and temperature (Winterhalder 1993).

As an example, we may look briefly at the Quinua region of southern Peru, where multiple ecological zones lie close by one another (Mitchell 1980). Like those described by Brush, the main settlements are frequently situated at ecotones between the maize and tuber zones, although some communities are focused in valley bottoms. That settlement pattern dispersed social groups into several temporally occupied residential locations, e.g., in the puna, in field houses, and in the main settlement. Quinua has two agricultural cycles. The dry-season cycle, called michka, is restricted to the valley bottom and to a small portion of the upper savanna, because it requires irrigation. Two quick-maturing crops are used in succession, one planted in August and the second in November and harvested in April. The rainy-season cycle, called hatun tarpuy, is found in all ecozones. The highest zones are planted in November; planting in the middle and lower zones occurs over time, slowly moving downhill, with the last crops planted in December, with the onset of rains. Cultivation of the frost-sensitive maize, which typically matures over about six months, can be accelerated by irrigation before the inception of the rains, so that it can be harvested before the onset of the killing frosts.

Quinua households normally have fields in several ecozones, which maximizes crop production and spreads risk over zones and crops. The localized climatic variations mean that agricultural failure in one location does not imply failure elsewhere. Mitchell emphasizes that the use of a staggered planting season also provides distinct advantages in the organization of labor, because it spreads out the planting and harvesting workloads over time. Labor is shared among households, but within the community, according to the stage in the sequence of the agricultural cycle.

There is considerable documentary and archaeological evidence that these approaches to settlement and land use were employed by indigenous communities in late prehistory. Murra (1980a) has shown how the Inka state economy itself was initially modeled after and built upon the economies of the Peruvian sierra. The ayllu, a corporate kin group with a common ancestor, was the basic resource-holding unit. With populations ranging from just a few to hundreds of households, the ayllu allocated member households access to resources through usufruct. Like their modern descendants, ayllus and communities often attempted to distribute their members among several complementary ecological zones, so that the products obtained could be pooled and economic independence maintained (Murra 1972; Ortiz de Zúñiga 1967, 1972). Within a family and community framework, household and corporate decision-making were focused around manipulation of customary relationships within stable social structures, all constrained by the ecological conditions.

EXCHANGE NETWORKS

Because not all households and communities could produce all they needed, traditional societies often maintained networks of exchange that integrated or traversed ecozones. Traditional societies organized many of their exchange relationships through reciprocity and redistribution (Alberti and Mayer 1974). The use of these terms by scholars working in the Andes differs from that found in Polanyi's (1957) initial formulation and from that used in much of the general anthropological literature. For example, labor tribute to local elites is often subsumed under *asymmetrical reciprocity* in the Andes, whereas it might be called *mobilization* elsewhere. Similarly, *redistribution* is used to refer to relations more often seen as extraction elsewhere.

Reciprocity, in the terminology employed by Andeanists, takes two forms. Balanced reciprocity, or waje waje, occurs when households of equal status exchange services in expectation of a return of equivalent value. For example, households may share labor in agricultural, herding, or construction tasks. Asymmetrical reciprocity (minka) consists, for instance, of services provided to in-laws or contributions of agricultural labor to upper-status households by lower-status households who then expect to share in the produce. In minka, inequality between the parties is the key to defining the nature of the exchange (Fonseca Martel 1974).

The exchange often called redistribution among Andeanists consisted of two central elements. The first is the elite's provisioning of certain kinds of material goods and edibles, especially cloth and chicha (a fermented beverage made from a variety of plants, most prominently maize), to the subordinate populace as part of the elite's obligations to his group (e.g., Murra 1960; Wachtel 1977; Netherly 1977; Rostworowski 1988). The second consists of the elite's allocation of particular specialized products to the general populace; coca and capsicum peppers have long been among the key goods provided in this manner (Murra 1972). The specialized products could be procured from locations hundreds of kilometers from the home communities, often through maintenance of colonies. At the time of the Spanish Conquest, some colonies contained members of several distinct ethnic groups, who maintained their affiliation with their homelands. In each case, the goods were produced by specialists working directly for the elites and were distributed along sociopolitical lines. This kind of redistribution was not a substitute for basic subsistence production or a market system. Instead, it bonded sociopolitical groups, reinforced unequal statuses, and provided the general populace with access to goods that might otherwise be difficult to obtain. In the volatile pre-Inka era, the distributions also attracted followers to the more powerful elites, thus providing a way to reform political relationships (Toledo 1940).

PASTORALISM

The complement to the agricultural aspect of highland land use was camelid herding. The conquistadores, who saw indigenous pastoralism before the collapse of the Inka state, were struck by the scale of the state herds. Upon capturing Atawallpa, the victorious aspirant to the throne, Francisco Pizarro ordered that the accompanying caravans of some 15,000 llamas be dispersed because they were overrunning the royal camp. The greatest flocks of llamas and alpacas were found on the altiplano, however, where the herds collectively numbered in the hundreds of thousands.

The natural conditions that permitted this scale of herding included the vast altiplano and puna, which provided the habitat for the two domesticated camelids, llama (*Lama glama*) and alpaca (*Lama pacos*), and the two wild species, guanaco (*Lama guanicoe*) and vicuña (*Lama vicugna*) (figure 13.4). The distributions of the domesticates overlap but are not identical. Gade (1977:116) reports that llamas are best suited to elevations above 3,000 m, with the core of their range lying in an area from about 400 km north of Lake Titicaca to 27°S, in Argentina (although see Miller and Gill 1990). Alpacas are naturally adapted to higher elevations, above 4,200 m; their core range lies in a region from about 11°S to 21°S latitude. This relatively restricted natural distribution meant that the appearance of the domesticated camelids on the coast or in many mountainous regions was a direct consequence of human intervention. It is generally thought that the widespread distribution of the camelids in most of the páramo grasslands of northern Peru and Ecuador was a consequence of Inka state activity (see Salomon 1986 for a discussion of the páramo; Troll 1968).

Traditional herding practices in the central sierra have been the subject of concerted research in recent decades, for both cultural and economic reasons (e.g., Flores-Ochoa 1977; Flannery et al. 1989). In the Ayacucho region, the heartland of Peruvian herding, the tundra contains a bewildering array of microenvironmental variations, among them turf, tallgrass, and moor puna (Flannery et al. 1989:16). There are nine to twelve months of frost per year, and the annual temperatures range from a low of 3.2°C to a mean high of 7.2°C. In the puna's lower reaches, talwi and the hardy Andean tubers can be cultivated, especially for chuño, a delicacy that truly exemplifies culturally specific notions of cuisine.

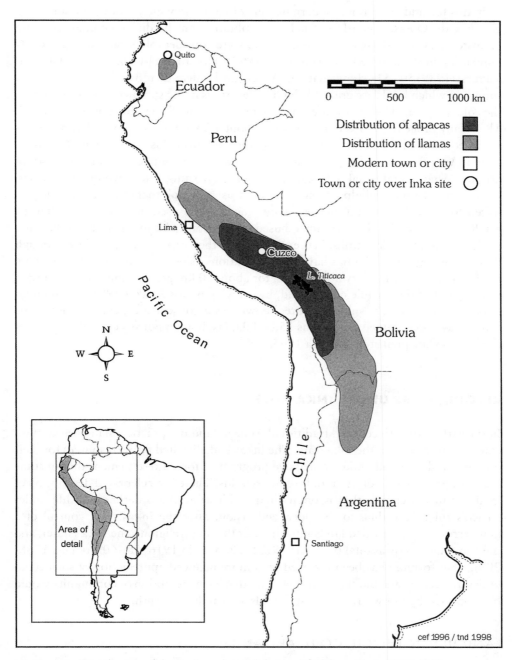

FIGURE 13.4. Distribution of domesticated camelids (modified from Gade 1977:115).

In this forbidding terrain, modern herders often live in widely scattered units called *kanchas,* the Quechua word for "enclosure" (ibid.). Many herding communities of late prehistory, such as those on the Junín puna, were considerably larger, with residents numbering in the hundreds (Parsons et al. 1998). The studies by Flannery and his colleagues (1989:50–51) show that the Ayacucho herders of today follow a seasonal round, in which herds are grazed at higher elevations in the summer and lower ones in the colder season. Their herds are small, with a range of 18–35 and a mean size of 25. Their daily range is restricted, having only about a 2 km radius. In part that pattern is a consequence of the herders' maintaining their potato fields for the manufacture of chuño. As is the case with agricultural activities, herding implies a strong network of group assistance and ritual in herding, butchering, feasting, and ceremonial activities.

Although llamas and alpacas both produce useful wool, that of the alpacas is preferred for its lightness and fineness, while the llama is used more for meat than its smaller relative and is the sole camelid used for transport. Although hardy in the sense of the temperature and altitudinal extremes that they can tolerate, and even though they subsist on forage, llamas have their limitations as beasts of burden. For the long haul, a male buck can carry about 30 kg for about 20 km per day and is usually rested one day out of three while on the road, by carrying no load (West 1981). Even under good care, animals are subject to breakdown. As a consequence, even though some state caravans contained as many as 15,000 llamas, human porters carried the bulk of the loads in late prehistory (Murra 1965, 1980b).

RESOURCE USE UNDER INKA RULE

In the early sixteenth century, traditional Andean land use practices had been complicated by the advent of the Inka state. The Inkas had alienated prime resources for their own use and had undertaken a massive program of forced resettlement that in some regions uprooted tens of thousands of households. There is a celebrated description of land use in the empire that is often cited to illustrate the orderliness of life under Cuzco's rule. According to that idealized vision, after the Inkas took control of a new territory, they divided lands into those of the Sun (religion), the Inka (state), and the communities (peasants) (Cobo 1979:211, 215; Polo 1916:58–60; Arriaga 1968: 209). That formula has been replaced by a more nuanced appreciation of socioeconomics in recent years, but the central premise that each institution or community would be supported by its own resources still holds a great deal of truth.

LAND USE IN THE CUZCO HEARTLAND

THE CUZCO VALLEY

Nowhere in the Andes have scholars studied the relationships that linked the land, irrigation, and society of late prehistory more thoroughly than in the homeland of the Inkas. Examining land and water use around Cuzco provides direct insights into the ways in which the Inkas interacted with the natural world. Cuzco itself lies at 3,450 m above sea level, in a relatively inhospitable location for agriculture. The lands in the immediate vicinity are mostly above the altitude at which maize can be grown, and the distribution of monthly rainfall from year to year is irregular (Winterhalder 1993:35). With rainfall agriculture an uncertain business, the Inkas constructed a complex irrigation network in the valley and sought warmer lands during their early expansions down the Vilcanota-Urubamba drainage (Rowe 1946), where maize could be grown with some degree of security. The latter area quickly became the preferred location for the royal estates and the site of some of the most spectacular land modifications in the world. At the same time that they transformed the landscape, the Inkas venerated the powers that resided in springs, mountain peaks, and stones by building shrines and carving stones and by observing a complicated ritual calendar that linked Cuzco's social hierarchy to water sources, sacred locations, and astronomical cycles (see Farrington 1992). It is no exaggeration to say that the Inkas considered themselves inextricably bound to the generative powers of the earth and cosmos.

The agrosocial organization of Cuzco defies easy description, not simply because it was complex but also because scholars have been forced to piece together its essential features from incomplete records taken down in the early Colonial era, after the upper strata of the social hierarchy had collapsed. The imperial capital itself housed about 50,000 people, mostly royalty, aristocracy, and their servants, while a similar number lived within a radius of about 18 km. The city contained upper and lower divisions, called Hanan Cuzco and Hurin Cuzco, respectively; the former stood higher in both its topographic position and its status. In 1532, Hurin Cuzco comprised the panaqas (descendant corporate kin groups) of the first five Inka rulers in a standardized list, while Hanan Cuzco consisted of the panaqas of the next five. The panaqas held a special role in Inka life, for they were charged with venerating the animate mummies of their deceased ancestors, who were regularly feted and consulted about matters of state. The city, and by extension the empire, was further divided into four parts, called Chinchaysuyu, Antisuyu, Kollasuyu, and Cuntisuyu, radiating out from the center. Thus the empire gained its name Tawantinsuyu, meaning loosely Land of the Four Parts. Outside the city core lay individual aristocratic estates, the lands of the elite kin groups, and the residences of provincials resettled to live in a layout that replicated their location in the empire.

Cuzco's organization was closely tied to both the landscape and the cosmos through an array of 41 or 42 conceptual lines, called zeq'es, which radiated outward from the Temple of the Sun (Qorikancha). Upon those lines lay at least 332 sacred places called wak'a, many of them springs, stones, and mountain peaks. J. Sherbondy (1993:75) has identified over a third of the wak'a as water sources or other features associated with hydraulic works. During a year-long ritual cycle, the panaqas and affiliated ayllus took their appointed turns venerating the powers of the shrines with which they were associated. Hanan Cuzco controlled the water sources and canal systems of the upper half of the valley and Hurin Cuzco the lower (ibid.). Each water source for the major irrigation canals was a shrine on a line assigned to the panaqas, whereas the less important networks received their water at shrines assigned to the lesser kin groups (ibid.:

74). In a pattern that was repeated among many societies in the highlands and on the coast, particular canal segments were cared for by specific kin groups. For the most part, the waters that ran through each canal system were used to irrigate the lands belonging to the associated group. The landscape, its production, and the social order of the valley's inhabitants were thus integrated through a ritual and agricultural cycle.

THE ROYAL AND ARISTOCRATIC ESTATES

As Tawantinsuyu expanded, its elites claimed some of the Andes' finest resources and the labor to work them. Every province was supposed to reserve lands for each ruler, but the most elegant estates lay in the heartland, concentrated in a stretch of the Urubamba between Pisac and Machu Picchu often called the Sacred Valley of the Inkas (figure 13.5). Although the eighth ruler, Pachakuti, was usually credited with the founding of the royal estates, every ruler from his father, Wiraqocha, onward owned estates, and even earlier monarchs may have also had their manors (Sarmiento 1960: 224; Cobo 1979:125, 129; Rowe 1967:68).

The largest estates encompassed thousands of hectares and could boast 4,000 or more families to work them (Rostworowski 1966:32; Villanueva 1971:94, 98, 136, 139; Rowe 1982:100; Niles 1987:13–15). To create their properties, rulers claimed some lands that were already developed, accepted gifts from subjects, confiscated lands after succession conflicts, and even played games of chance with the Sun himself (Albornoz 1989:175, 182; Cabello Valboa 1951:360; Cobo 1979:149; Julien 1993: 184, 209, 233). Some estates were enhanced through ingenious engineering. Wayna Qhapaq's holdings in Yucay, for example, were partially developed by draining swamps and by shifting the river from one side of the valley to the other (Rostworowski 1962:134–135; Villanueva 1971; Niles 1987:13; Farrington 1992, 1995). Like the ayllus' lands, the estates were spread across ecozones to provide access to a wide range of resources. Wayna Qhapaq's manor, for example, contained maize and potato fields, pastures, settlements, forests, parks, a pond and marsh, a hunting range, and salt fields (Villanueva 1971). The gradual creation of the estates in the heterogeneous landscape ultimately produced a patchwork quilt of parcels belonging to rulers, aristocrats, and local communities distributed over the terrain. One result of this process is illustrated in figure 13.5, drawn from the work of A. Kendall (1994) and her colleagues, which shows how the fine terraces of Pachakuti's estate at Cusichaca were intermixed among steep fields, pre-Inka, and intermittent terraces.

The archaeological remains of the estates provide mute, striking testimony to their resources and artisanry. Among the better-known sites are such popular tourist attractions as Machu Picchu, Pisaq, and Ollantaytambo. The Inkas' aptitude for melding landforms and structures is one of the sites' most distinctive design features. In studying these sites, the architect J.-P. Protzen (1993:271) noted an especially neat correspondence between royal estates and the most elegant terracing around Cuzco. Usually situated on rocky promontories, all of the manors exhibit finely crafted waterworks and stonework. Their stones, springs, and peaks were homes to the powerful spirits called apu, and both witnesses and architecture attest to the importance of ceremonial activity at the estates (MacLean 1986; Hyslop 1990:102–145).

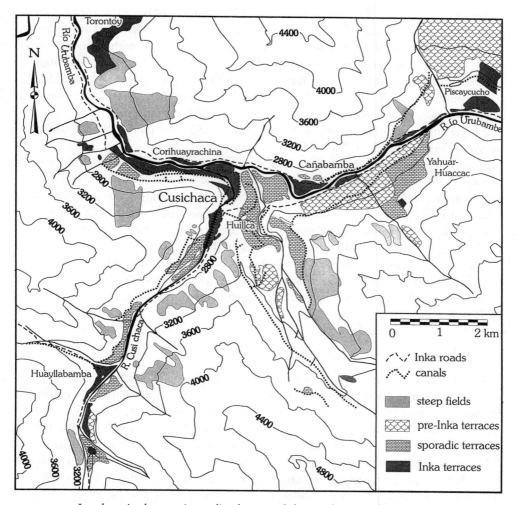

FIGURE 13.5. Land use in the area immediately around the royal estate of Pachakuti, at Cusichaca, about 80 km below Cuzco (modified from figure 3 in Kendall 1994).

From an ecological perspective, the terraces' heat- and moisture-retaining capacities provided a special boon to the estates' residents. The main terraces at Ollantaytambo seem to have raised the ambient temperature by as much as 3°C (Protzen 1993). Similarly, the cultivation of warm-weather crops at Yucay, including coca, ají, cotton, and peanuts (*Arachis hypogaea*), suggests that the estate's designers were able to devise a minimontaña near Cuzco for the emperor's pleasure (Farrington 1983). Their interest in experimenting with the natural ecology is further affirmed by their modest success in growing maize in a microclimate at Copacabana, Bolivia, at an elevation of about 3,900 m.

One of the most intriguing aspects of land use around Cuzco was its political dynamism. For all of the seemingly permanent links between social groups and the land, the system was mutable. Pachakuti was revered as the designer of both the imperial social system and greater Cuzco, as well as the founder of the estate system. However, early chroniclers observed that royal society may have been reformed with the ascension of each ruler in the imperial era. Most important, a new panaqa had to be allocated resources and integrated into the social structure while the principle of balanced duality was maintained (see Rowe 1985). Determining how to put that into practice must have been a thought-provoking challenge and source of much contention.

Interestingly, when the Spaniards took Cuzco in 1533, the panaqa (Tumipampa) of Wayna Qhapaq (ruled ca. 1493–1527) still had not been integrated into the social hierarchy of Cuzco, and thus lands in the zeq'e system may not have been allocated to them. On the other hand, members of etnías (ethnic groups) that had been incorporated into the empire during its last years were installed in greater Cuzco and had to be allocated lands from a scarce supply. Waskhar, the last ruler vested in Cuzco, was so outraged by some of his near relatives during the dynastic war of 1527–1532 and by the lack of unclaimed lands suitable for estates that he threatened to dismantle the whole system of royal entitlements. Whether he would have followed through is anyone's guess, given the political complexities, but the various threads of evidence remind us that only wak'a and canals were carved in stone, not relationships between people and land.

STATE RESOURCES: THE FARMS

One of the essential features of Inka statecraft was the designation of separate farmlands and pastures for the state and the Sun. By and large, those farms were cultivated by mit'ayuqkuna (corvée laborers) as part of their labor service, although colonists were increasingly called upon for this end in the empire's latter stages (Murra 1980b, 1982). The farms were often located near provincial centers, but some were also established in especially favorable agricultural locales. The most prominent was an immense farm that lay in the temperate Cochabamba Valley, Bolivia, where Wayna Qhapaq ordered the western part of the valley vacated to make way for 14,000 agricultural workers, both permanent colonists (mitmaqkuna) and corvée laborers mostly from the adjacent highlands (Wachtel 1982). The maize grown on the lands was reportedly destined for the Inka's armies, as was the produce from other farms in Arica, Arequipa, and Abancay (Urbano 1965:338; Spurling 1982:14). In each case, the mitmaqkuna were given usufruct rights on lands that they used to support themselves.

State farms have also been identified archaeologically in several locations, for example in the Upper Mantaro Valley, Peru (described in the next section) and various locales in northwest Argentina. Although the southern Andes are usually considered marginal to Inka interests, there is a great deal of evidence for intensification that can be attributed to state intervention. At Coctaca-Rodero, Argentina, an immense terraced field system covers about 6 km² on the alluvial fans and piedmont (3,700 m) just below the fringe of the altiplano (Albeck and Scattolin 1991; Albeck 1992–1993; Nielsen 1996). Because many terraces were abandoned before completion, it seems likely that the farm was being developed very late in the empire's run. Large farms may have also lain at the Campo de Pucará, in the Lerma Valley, which reportedly contained 1,700 storehouses (Boman 1908; Fock 1961; González 1983). State farming in the adjacent Calchaquí Valley and farther south in the Bolsón de Andalgalá was more limited than in the other locations, but in each case, state managers intensified production next to Inka installations by irrigating newly created fields (D'Altroy 1994; Williams 1996).

W. Mitchell has observed that the ideology of working the state farms may not have meshed neatly with the requirements of the Andean environment. A number of chroniclers stated that lands in Tawantinsuyu were farmed in a specified sequence. The church's crops came first, followed by the state fields, after which the workers could turn to their own plots (e.g., Cobo 1979; see Mitchell 1980:140–141). Garcilaso (1966:150–152; Mitchell 1980) described an alternate sequence that is not widely cited today: fields of the religion; widows, orphans, the incapacitated, and soldiers on duty; the peasantry; the lords; and the emperor. Farming for the state was a festive occasion when the Inkas displayed their generosity by plying the workers with food and drink. The close link among farming, reciprocal obligations, and ritual can be seen in two definitions of the Aymara term *haymatha,* which is translated as both "to go and work in the fields which are planted communally, like those of the lord . . . or the poor" and "to dance in the ancient way particularly when they go to the fields of their leaders" (Bertonio, cited in Murra 1968:134).

Mitchell points out the rigid labor formulas reported by the chroniclers run afoul of ecological conditions that require that crops be planted, cultivated, and harvested in a staggered sequence over the year (Mitchell 1980; Hastorf 1993). Where it can be grown, maize is the first crop planted, because it takes the longest to mature, but it would have been agriculturally impracticable for farmers to work all state lands across a vertical array before turning their attention to their own fields. A plausible explanation is that state lands were concentrated in a limited set of ecozones near state centers, or that state lands in any given niche were worked first. In any event, Mitchell infers that the strict labor sequences reported in the chronicles reflected elite ideologies of hierarchy and power more than realistic crop scheduling practices.

Even so, the state's reliance on corvée shifted agricultural conditions among the subject populace by tapping into their labor at crucial times during the year. Conversely, as Murra (1982) has emphasized, the creation of colonist farms would have relieved many of the burdens of competitive scheduling. By establishing farms staffed by dedicated personnel, whose lands lay amid the state fields, the Inkas could have partially reduced the problems inherent in using corvée as the principal labor source.

LAND USE BY LOCAL SOCIETIES UNDER INKA RULE: THE UPPER MANTARO VALLEY, CENTRAL PERU

For an exemplary case of land use by local residents, we can turn to Peru's Upper Mantaro Valley, home of the populous Wanka and Xauxa etnías. The 50 km upper drainage is the major intermontane valley of the central highlands (figure 13.6). C. Hastorf's (e.g., 1990, 1993) research there has shown that agricultural lands in its main valley,

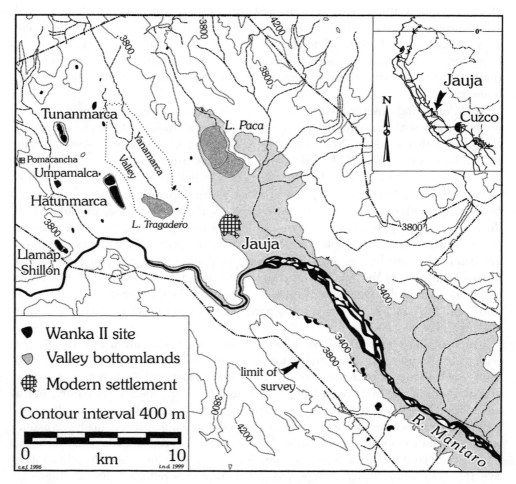

FIGURE 13.6a. Land use zones in the Upper Mantaro Valley, central Peru, during the late Pre-Inka (Wanka II: A.D. 1350–1450) era. *Source:* After Hastorf 1993:151.

adjacent tributary valleys, hillslopes, and rolling uplands were farmed in a fairly standard set of highland crops according to their elevation and microenvironments. Land improvements, such as lynchets, terraces, drained fields, and irrigation, were more common in the upper elevations than in the main valley.

During the truculent period just before the Inka conquest (Wanka II: A.D. 1350–1450), most people lived in defensive settlements at 3,800 m and above. The largest town, called Tunanmarca, contained more than 5,000 dwellings. Hastorf's (1990) studies of excavated botanical remains showed that the foods that were processed and consumed in the residential households matched a local mix of crops fairly closely, which emphasized tubers and quinoa. The main exception lay in maize, for which nearby lands were scarce and whose remains were concentrated in the elite households. The faunal remains also indicated that the people who ate in elite households, whether

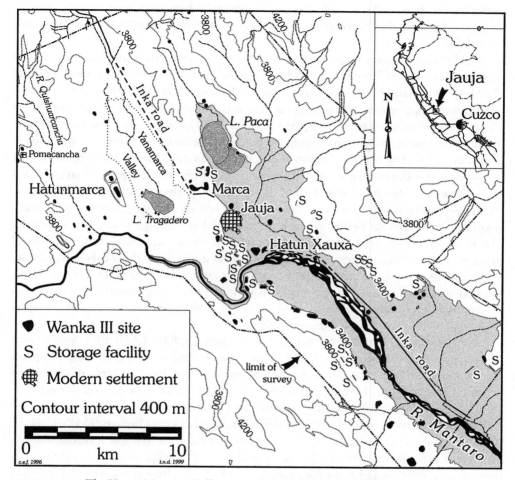

FIGURE 13.6b. The Upper Mantaro Valley, central Peru, during the Inka (A.D. 1450–1533) era; the shaded area is the valley bottom, below 3,400 masl.

the residents or their guests, got a higher proportion of the finer cuts of camelid meats (Sandefur 1988). Surprisingly, some of the strongest evidence for land improvements dates to this period. In the upper Yanamarca Valley, where much of the populace was concentrated, local communities built canal systems as much as 15 km long. Hastorf and Earle (1985) infer that the intensified production served political ends, rather than subsistence, for large tracts of land in the main valley remained untouched during the era.

With the advent of Inka rule, much of the populace moved out of its fortified towns into smaller villages and hamlets strung along the piedmont of the main valley and its tributaries (3,400–3,600 m). Most settlements' immediate catchment thus shifted more in favor of maize-complex crops, a situation that was reflected in the botanical remains recovered from residential households (Hastorf 1990). Stable isotope analysis from human bone collagen indicates that adult males benefited most, largely, Hastorf suggests, because they were given maize beer by the state as partial recompense for their labor.

The Inka occupation transformed use of the region's resources in other ways as well (e.g., Earle et al. 1987; Sandefur 1988; Costin and Earle 1989; D'Altroy 1992; Hastorf 1993; D'Altroy and Hastorf in press). The settlement pattern and distribution of state storehouses suggest that managers set aside lands within about a 5 km radius of Hatun Xauxa, the provincial center, and perhaps as far as 10 km down the main valley. Much of that area had been off-limits during the preceding period, so few farmers would have lost their fields. The produce was stored in more than 2,000 storehouses (qollqa) that lined the valley slopes in more than 30 separate facilities, with a capacity that exceeded 123,000 m³. A little more than half of the qollqa lay within a kilometer of the center, where supplies could be drawn upon to support permanent state personnel, temporary workers, and itinerant personnel such as soldiers. Even though maize was culturally the most important crop, the limited botanical evidence available from state storehouses in central Peru suggests that Inka overseers understood the local environments and cultivated a mix of crops that was appropriate to the immediate vicinity. Thus, quinoa was the crop recovered most frequently from half a dozen Mantaro storehouses, while potatoes held sway in storehouses at the puna site of Huánuco Pampa (Morris 1967; D'Altroy and Hastorf 1984).

Above the main Mantaro farmlands lies the puna (3,900–4,650 m), whose grassy, rolling pastures sustained great flocks of llamas and alpacas. Studies of the Junín puna to the north by Parsons et al. (n.d.) have shown that the local patterns of settlement location and land use changed under Inka rule but that there was an intensive occupation of the region up to and after the Spanish Conquest. The scale of herding can be inferred from early Colonial lawsuits filed in the royal court in Lima in 1558–1561. In 1532, having found themselves on the losing side in the great dynastic war, the Wankas and Xauxas threw their lot in with the Spaniards. In their depositions, they claimed to have turned over 74,521 head for Spanish use in the first four years after the invasion (Espinoza Soriano 1971). What portion of the region's flocks that constituted unfortunately remains unclear.

THE ALTIPLANO

Central Peru was renowned for its herds, but the most productive pastoral region in the Andes was the Bolivian altiplano. Despite its demanding climate, the high plain was heavily populated during late prehistory, especially in the Lake Titicaca basin. Human adaptation to altiplano was actually a complex mix of agropastoral pursuits. In terms of the labor invested and calories produced, agriculture was probably more important than pastoralism, while the latter was more significant for cultural values and as a source of wealth. Tubers, especially potatoes, were the dominant crop, but chenopods were also a major source of food; the latter were particularly valuable because of their high protein content (about 20 percent).

The peoples of the altiplano spent more than a millennium in intensive agricultural pursuits around the margins of Lake Titicaca before Inka rule, where many tens of square kilometers of lands were converted into irrigated or drained fields (Kolata 1991; Erickson 1993; see also Flores-Ochoa and Paz 1984; Erickson this volume). Recently rehabilitated fields are highly productive, in part because their physical characteristics moderate temperatures and thus reduce the chance of killing frosts. Much of the intensified farming occurred during the heyday of Tiwanaku, ca. A.D. 300–750, but an important occupation among the drained fields continued into the Late Intermediate period (Graffam 1992). By the time the Inkas conquered the region in the fifteenth century, the Qolla and Lupaqa polities were among the most powerful of the Andes. Under Inka rule, people continued to concentrate around the margins of the lake, where the societies mixed intensive and extensive agriculture with pastoralism (Stanish and Steadman 1994).

Many of the details of late prehistoric herding practices on the altiplano are inaccessible from our modern vantage point, but we can get a sense of their scope and socio-economic relations from an inspection (*visita*) that Garci Diez de San Miguel conducted for the Spanish crown in Chucuito in 1567 (Diez de San Miguel 1964:40; see Murra 1968). The scale of herding was so vast that one witness commented that the pastures of the altiplano were sometimes insufficient for the great flocks of the Inka era. Individual herds could also be impressive, as one witness testified, perhaps with exaggeration, that a local lord had more than 50,000 animals in his flocks in 1567 (Diez de San Miguel 1964:50; Murra 1968:120).

The visita also provides insight into the allocation of labor for the Aymara lords' benefit (Murra 1968:129). In addition to general farming service claimed in several communities, separate tabulations were kept for the lords' specialist herders, farmers, coastal colonists, household servants, and wood and grass collectors, along with "Indians of service" and yanakuna, or lifelong service personnel. Between them, the two paramount lords laid claim to 27 herders and 26 farmers, out of 107 service personnel enumerated in the capital community of Chucuito. As was typical for lords of the time, the numbers of individuals dedicated to working for the lord were spread more or less evenly across social divisions, called hatha. These figures suggest that although the herds were a main source of wealth, the proportion of labor dedicated to them was less than that dedicated to other activities. This situation repeats a pattern seen in other Andean contexts, in which cultural importance and labor investment were only occasionally in proportion.

NORTH AND CENTRAL COASTS OF PERU

For the last comparative case of Andean land use, we turn to the central and northern Peruvian coast, which differed in important ways from the Inkas' sierra heartland. The most obvious distinctions lay in the ecological setting, but social and economic features also differed in crucial ways. As noted earlier, the Pacific coast is a desert transected by narrow drainages, not all of which are capable of sustaining human life

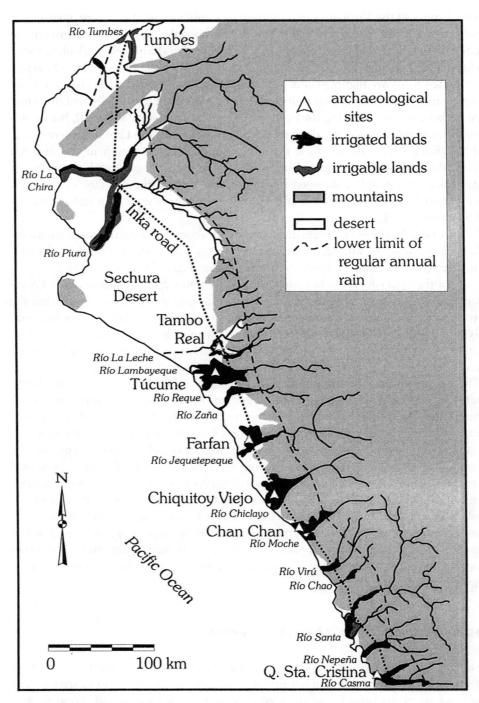

FIGURE 13.7. The north coast of Peru, showing irrigated lands. *Source:* After figure 3.4 in Shimada 1994:41.

(figure 13.7). The catchment basins of the productive north Peruvian valleys range from about 800 km² in the Lacramarca Valley to about 11,000 km² in the Santa Valley (figure 13.7; Kroeber 1930; Shimada 1994:42). The steep terrain and hydrological regimes (e.g., position of the aquifer, permanence of the main watercourses, estuarine encroachment) limit the amount of land that can actually be brought under cultivation to about 7–17 percent of most of the valley's catchments. In the most exaggerated case, the Santa Valley's steep, elongated upper drainage basin and meandering lower river course permit just 86 km², or 0.8 percent, of its total area to be irrigated. Kroeber's data from 1930, as analyzed by Shimada (1994:42), suggest that only 0.3–3.5 percent of a series of major valleys was actually under cultivation at that time. Once they had been planned and built, the coastal irrigation systems—and, as a consequence, social systems—were subject to all manner of disruptions caused by coastal uplift, ENSO-related flooding, river channel downcutting, salinization of coastal aquifers, and aeolian sand movements (e.g., Moseley et al. 1992).

Building and maintaining agricultural systems was a daunting challenge under the circumstances. Irrigation agriculture was nonetheless well ensconced by about 1800 B.C., and entire coastal basins irrigated by the early first millennium A.D., if not earlier. Ultimately, the most expansive canal networks in the Americas lay in the Lambayeque–La Leche region, as several adjacent valleys were linked with interdrainage canal systems. The leadership of the Chimu state, which governed the coast from Chan Chan and was the last major power to fall to the Inkas, favored enormous investments of its public labor in agricultural projects, rather than in monumental architecture (Moseley and Cordy-Collins 1990). A particularly striking example of a state-organized land reclamation project is found in the drained fields of Quebrada Santa Cristina (figure 13.8; Moore 1991).

As in the highlands, the sociopolitical and economic organizations of the north and central coast polities were intimately linked. Entire communities or local sociopolitical groups specialized—as potters, weavers, farmers, fishers, traders, and sandal-makers, for example—exchanging their products for those made by others (Rostworowski 1975, 1977, 1978, 1983; see also Netherly 1977). Despite the specialization, social groups often sought to control as many resources as possible by organizing their holdings longitudinally within valleys (Netherly 1984). Access to resources outside the immediate lands of a given community could also be enhanced as lords up and down the river exchanged use rights for cultivation in maize and coca lands (Rostworowski 1983). Interestingly, the basis of measuring resource rights seems to have been conceived differently on the coast and in the mountains. Coastal farmers measured both field size and standing crops by the number of feeder canals that watered them (Netherly 1984:239). In contrast, highland farmers typically measured their lands in tupus, which referred to the flexible amount of land that a family needed to support itself, but was sometimes taken in the Colonial era to mean the amount of seed to be sown or, by European edict, a fixed area (Murra 1980b:280–281; see Haggard and McLean 1941).

Understanding the ecological arrangements among coastal societies is complicated by the demise of both the population and the political organization in the decades

FIGURE 13.8. Aerial photo of drained field systems in the Quebrada Santa Cristina, Casma Valley, Peru (courtesy Servicio Aerofotográfico Nacional, Peru, and G. Russell).

before detailed records were taken. By 1567, the population had declined to about 10–20 percent of its size in 1532, the Chimu state had been dismantled over a century in part by Inka design, and the Inka empire had collapsed, to be replaced by Spanish authority (Cook 1981; Netherly 1984:230). As in the Cuzco area, however, the middle-level social units had survived and managed the existing irrigation networks.

The engineering skills and labor needed to construct and maintain the irrigation systems were remarkable, but scholars differ over the degree of state intervention involved in their regular use. Netherly (1977:279; 1984) recognizes three levels of agro-social complexity for the region's systems. First were the small canals that were maintained and used by an individual social entity (*parcialidad*, or part of a whole) to water its own lands. Second were the canals shared by more than one parcialidad, among them the canals on the south banks of Pacasmayo river in the Jequetepeque Valley and the Chancay River in the Lambayeque Valley. At a higher level were such intervalley canals as the Taymi and Ynalche of the Lambayeque Valley and the Talambo Canal of the Jequetepeque Valley. At least in the latter half of the sixteenth century, most rights and duties for canal construction and maintenance, as well as the allocation of water, lay in the hands of local groups (Netherly 1977, 1984). Netherly uses the documentary sources to infer that even some of the intervalley irrigation systems were built and maintained in defined segments by local social groups in the pre-Spanish era. She observes that other intervalley canals, such as the Chicama-Moche system, may belong to a separate organizational category, because they were explicitly designed to water Chimu state fields. Under these varied arrangements, labor forces of different scales could be mustered as needed by moving up or down a hierarchy of sociopolitical divisions (Netherly 1984:233). From this perspective, the upper levels of the systems came into play only when state projects were involved, the lower levels failed in their obligations, or local authorities could not resolve disputes.

An alternative argument, for more centralized management of north coast irrigation, is based on different lines of evidence. Farrington, Moseley, Ortloff, and others argue that the scale of some of the canals, the specialized engineering knowledge entailed in their design and construction, and their direct association with state installations indicate that at least the major irrigation systems were managed by a state water bureaucracy during Chimu times and perhaps earlier (e.g., Farrington 1974; Moseley 1983, 1992; Ortloff 1993). Of crucial importance to this position are the observations that it was not unusual for canals to cross through the territories pertaining to multiple named sociopolitical groups without watering their lands (see Netherly 1984:239) and that the segmentary construction implies the presence of central planning. From this viewpoint, the major canal systems could have not operated without state management, even if much of the water allocation and local maintenance was carried out under the auspices of lower-level lords. The irrigation management of the 1560s and after would simply represent the remnants of a more complex system whose upper levels were first dismantled by the Inkas and then collapsed from the disastrous consequences of the Spanish occupation.

This author is not in a position to resolve this disagreement, but two points do need to be emphasized. The first is that late prehistoric Andean states, whether Tiwanaku, the Chimu, or the Inka, were clearly involved in large-scale farming enterprises, including irrigation, land reclamation, and terracing. The second is that local social groups were fully capable of building and maintaining intensified agricultural systems of an expansive nature. My own estimate is that at least on the north coast, Netherly's view may hold a little more water. That is, the state's involvement may have been limited mostly to managing its own resources and intervening when local arrangements proved inadequate or when environmental catastrophe struck.

Typically, water in coastal canals was allocated to farmers from the tail to the head of canals, thus ensuring that downslope users were not cut out of the crucial resource (Netherly 1977:287). In contrast, highland systems often allocated water to users from the head to the tail (W. Mitchell, pers. com. 1998), perhaps to ensure that upper fields received water earlier. No matter the order of irrigation, it is not surprising that people of higher social status had preferred access to the finer lands, especially those toward the canal intakes, which could be multicropped. For the Chincha Valley, Castro and Ortega Morejón (1974) repeat the same kind of formula for irrigation order under Inka rule that we saw for highland farming. First came the lands of the Inka (i.e., royal or state fields), then those of the state religion and the regional lords, and finally the lands of the "poor," by which we can infer the general peasantry (Netherly 1977:284). How the Inkas got their lands on the coast is a matter that remains largely unresolved. For Chincha, Castro and Ortega Morejón state that every 1,000 households ceded 10 hanegadas of (irrigated) lands to the Inka. Roughly speaking, that would have translated into 190 ha (30,000 households and 1.59 acre/hanegada; Haggard and McLean 1941: 77), not a small amount, but certainly nothing approaching the state farms of the highlands.

The relocation of territorial boundaries under Inka rule provides particular insight into their approach to the management of productive resources as part of provincial policy. Canal intakes, which typically lie between 200 and 300 m elevation, were left in the hands of the coastal societies. In contrast, coca lands from 300 m upward in the Chillón, Moche, and Chicama valleys were shifted to the control of the neighboring highlanders, which triggered no end of lawsuits in the early Colonial era (Netherly 1977; Rostworowski 1990). That shift epitomizes key elements of the Inka approach to socioeconomic relations with subject groups. That is, the coastal polities, once subordinated, were left to their own devices for subsistence. Because they were considered rebellious by the Inkas, however, the prestigious and ritually crucial crop was assigned to sierra peoples, thus handicapping the lowlanders and rewarding the serranos (Rostworowski 1970).

In sum, several coastal features diverged markedly from practices in the adjacent highlands. One key distinction is that the scale of the agricultural enterprises among the coastal populace required a level of management or cooperation that was usually unnecessary in the highlands, except for state projects. Second, the arrangements on the coast involved more specialization and exchange than found among highland groups, which more typically sought to control as many resources within their own territories as possible. Third, the conception of agricultural rights was different, measured as it was in terms of feeder canals rather than productive lands. Finally, protein in the coastal diet was enhanced primarily by the consumption of shellfish and fish, rather than camelid meat, which was the principal source of faunal protein in the highlands.

REFLECTIONS ON CULTURAL ECOLOGY AND CULTURAL IDEOLOGY

This discussion has highlighted only a few of the many points that can be developed about Andean land use around the time of the Spanish invasion. The most outstanding characteristic of the natural environment is that its wide array of opportunities for agropastoral exploitation, set in a vertically compressed topography, was matched by the formidable challenges it set for those who wished to take advantage. Human adaptations to the conditions were as diverse as the environment. There was no one Andean adaptation, nor did all peoples living in the same region exploit it in an equivalent fashion. There were, however, a number of common threads to ways in which societies organized themselves. Most notably, many local corporate groups went to considerable lengths to gain access to as many resource zones as they could, in the interests of self-sufficiency. Even though many households did gain access to a wide range of resources, the ideal was tempered by social status differences, and differential access was mediated by political and exchange relationships.

Agriculture and pastoralism are often rightfully viewed as complementary aspects of verticality in the highlands. As Thomas (1973) has shown, the complementarity is not simply one of cultural preference, but the food bank provided by the herds forms an effective means of risk reduction in a difficult environment. Murra's research has defined two principal kinds of verticality: household or community dispersal and the archipelago model. As important a principle as verticality has been to Andean adaptations and to modern interpretations of human ecology, it must be noted that such an approach to landscape use is not unique to the Andes. Throughout the world, societies that inhabit lands with steep gradients often arrange their territories longitudinally to ensure that corporate social groups control a variety of ecozones. Examples may be found among the Ainu of Japan, the Polynesians of Hawaii, the Kwakiutl of northwestern North America, traditional societies of Nepal, the Owens Valley Paiute in California, and traditional Swiss farmers.

What may be the most interesting aspects of Andean vertical integration, then, are the unusual distances involved and how the approaches worked in practice. For this purpose, we may look briefly at resource ownership across some of the lands discussed earlier, specifically patterns of land use that cross-cut the altiplano, the Peruvian highlands, and the coast. How and when the altiplano polities held resources on either side of the cordilleras has long been the subject of speculation, but only recently the subject of concentrated research. The results to date are illuminating. Working in the Moquegua region, members of the Proyecto Condesuyo have shown that the great highland urban powers of Wari and Tiwanaku established colonies in the warm coastal lowlands during the Middle Horizon (e.g., Watanabe et al. 1990; Moseley et al. 1992; Owen 1993). Similarly, Tiwanaku appears to have founded colonies in or set up exchange relationships with the Cochabamba Valley, where maize could be grown for transport to the highlands (Céspedes Paz 1982; although see Higueras-Hare 1996). In contrast to Murra's (1968, 1972) suggestion that the altiplano polities held resources on the south coast in the Late Intermediate period, however, archaeological research by Bürgi (1993) suggests that they may have regained rights under Inka rule that they had lost during the Late Intermediate, while the Inkas took the Cochabamba Valley for themselves (Wachtel 1982). Even when the archipelago model was in place, the purpose seems to have been largely one of providing the highland lords with prestigious and exotic commodities that they could use to sustain sociopolitical status. That is, the archipelago was not part of a basic subsistence system but was central to the political economies of highland states.

At the local level, proximity to resources, status, and family structure each seem to have contributed to resource access in the highlands. A useful example may be found in the 1591 inspection of the Yanque-Collaguas, a subgroup of the larger Collagua ethnic group (Pease 1978). There, ayllus living in the same reducción (a settlement in which indigenous people were resettled by the Spaniards) tended to exploit the same resources. Most households had access to two adjacent production zones out of the pastures, tuber lands, quinoa fields, and maize lands but had to call on exchange relationships to gain access to the full range of resources (Tomka 1987). Elite households, however, typically held lands in more ecozones than did commoners (Deustua 1978).

Tomka (1987:24) also notes that households with complete nuclear or extended families had direct access to a wider range of agricultural crops than did households formed of old, single, or orphaned individuals, because they could call on more useful labor and had a higher proportion of producers to dependents (see also Harris 1978:57). At least in early Colonial Collaguas then, the practice of land use was somewhat removed from and more complex than the Andean ideal of generalized self-sufficiency among households. Multiple factors impinged on the ideal and required households to be flexible in their arrangements.

It is equally important to recognize how the cultural priorities of Andean societies reciprocally affected the natural environment. Rivers were moved, the hydrological patterns of entire valleys were modified, wetlands drained, hillsides terraced, and plants and animals domesticated. The development of the lands around Cuzco exemplifies how landscapes were modified and prime resources converted into corporate domains nominally held in a public trust for the reverence of the emperors. We often think of land improvements as though they are designed to increase production. While that was true for the Inkas, the main goals at the royal estates had as much to with defining symbolic space, status display, and perhaps aesthetics as with subsistence intensification (Farrington 1983; see also Protzen 1993). Even while they were revering and staking a claim to nature, however, they improved production by increasing the arable land area, reducing erosion, providing irrigation water, and creating favorable micro-climates.

In closing, we may return to the topic of traditional Andean and modern perceptions of relationships between human society and natural forces. It is clear that peoples living in the Andes in 1532 thought about natural forces and land use in different ways from modern agronomists. Even so, indigenous practices often illustrated that farmers and herders had a finely tuned sense of the environmental impacts of long-term use of many natural resources. It is in this light that we have to take the representations of land use that found their way into the chronicles. Those principles, while containing much truth, were also often generalizations or elite views of the way things ought to have been. Only with meticulous analysis of local documents and archaeological research have scholars been able to show how the specifics of practice met the ideals of culture. One observation here in this vein was that despite a ceremonial emphasis on maize, archaeological evidence suggests that both the Inka state and local societies tended to cultivate a mix of crops that permitted a sustainable rotation locally. When and where possible, they intensified maize production, but potatoes, other tubers, and chenopods remained staple crops and may well have been more important in terms of land use, energy investment, and diet. Similarly, even in the heartland of pastoralism, agriculture seems to have been more important in terms of both labor investment and diet. Those judgments do not alter the significance of maize or camelids for Andean societies but simply indicate that we need to keep our analytical contexts straight. Ultimately, it will only be in the intersection of several elements—environmental conditions, human modifications, decision-making strategies, and cultural conceptions—that Andean land use can be understood most fully.

REFERENCES

Albeck, M. E. 1992–1993. Areas agrícolas y densidad de ocupación prehispánica en la Quebrada de Humahuaca. *Avances en Arqueología* (Instituto Interdisciplinario Tilcara, Jujuy, Argentina) 2: 56–77.

Albeck, M. E., and M. C. Scattolin. 1991. Cálculo fotogramétrico de superficies de cultivo en Coctaca y Rodero. Quebrada de Humahuaca. *Avances en Arqueología* (Instituto Interdisciplinario Tilcara, Jujuy, Argentina) 1: 43–58.

Alberti, G., and E. Mayer, eds. 1974. *Reciprocidad e Intercambio en Los Andes Peruanos*. Lima: Instituto de Estudios Peruanos.

Albornoz, C. de. 1989 [1584]. *Instrucción Para Descubrir Todas Las Guacas del Piru y sus Camayos y Haziendas,* H. Urbano and P. Duviols, eds., pp. 161–198. Crónicas de America 48. Madrid.

Arriaga, Father P. J. de. 1968 [1621]. *Extirpación de la Idolotría del Pirú*. Biblioteca de Autores Españoles 209, pp. 191–275. Madrid: Ediciones Atlas.

Boman, E. 1908. *Antiquités de la Région Andine de la République Argentine et du Désert D'Atacama*. Paris: Imprimerie Nationale.

Brown, P. F. 1987. Economy, ecology, and population: Recent changes in Peruvian Aymara land use patterns. In D. Browman, ed., *Arid Land Use Strategies and Risk Management in the Andes,* pp. 99–120. Boulder, Colo.: Westview Press.

Brush, S. 1977. *Mountain, Field, and Family: The Economy and Human Ecology of an Andean Valley*. Philadelphia: University of Pennsylvania Press.

Burger, R. L. 1992. *Chavín and the Origins of Andean Civilization*. London: Thames & Hudson.

Bürgi, P. 1993. *The Inka Empire's Expansion into the Coastal Sierran Region West of Lake Titicaca*. Ph.D. dissertation, University of Chicago. Ann Arbor, Mich.: University Microfilms.

Cabello Valboa, M. 1951 [1586]. *Miscelanea Antártica*. Lima: Universidad Nacional Mayor de San Marcos.

Castro, C. de, and D. de Ortega Morejón. 1974 [1558]. Relación y declaración del modo que este valle de Chincha y sus comarcanos se governaven antes que oviese Ingas y despues q(ue) los vuo hasta q(ue) los Cristianos entraron en esta tierra. Introduction by J. C. Crespo. *Historia y Cultura* (Lima) 8: 91–104.

Céspedes Paz, R. 1982. *La Arqueología del Area de Pocona,* Cuadernos de Investigación, Serie Arqueología 1: 89–99, Instituto de Investigaciones Antropológicas (Museo Arqueológico), Universidad Mayor de San Simon, Cochabamba.

Cobo, B. 1979 [1653]. *History of the Inca Empire*. R. Hamilton, trans. Austin: University of Texas Press.

———. 1990 [1653] *Inca Religion and Customs*. Roland Hamilton, trans. and ed. Austin: University of Texas.

Cook, N. D. 1981. *Demographic Collapse: Indian Peru, 1520–1620*. Cambridge: Cambridge University Press.

Costin, C., and T. Earle. 1989. Status distinction and legitimation of power as reflected in changing patterns of consumption in late prehispanic Peru. *American Antiquity* 54: 691–714.

D'Altroy, T. N. 1992. *Provincial Power in the Inka Empire*. Washington, D.C.: Smithsonian.

———. 1994. Public and private economy in the Inka empire. In E. Brumfiel, ed., *The Economic Anthropology of the State,* pp. 171–222. Society for Economic Anthropology Monograph 11. Lanham, Md.: University Press of America.

D'Altroy, T. N., and C. A. Hastorf. 1984. The distribution and contents of Inca state storehouses in the Xauxa region of Peru. *American Antiquity* 49: 334–349.

———, eds. In press. *Empire and Domestic Economy*. New York: Plenum.

Deustua, J. 1978. Acceso a recursos en Yanque-Collaguas 1591: Una experiencia estadística. In M. Koth de Paredes and A. Castelli, comp., *Etnohistoria y Antropología Andina,* pp. 41–51. Lima: Primera

Jornada del Museo Nacional de Historia (1974).

Diaz, H. F., and V. Markgraf, eds. 1992. *El Niño: Historical and Paleoclimatic Aspects of the Southern Oscillation.* New York: Cambridge University Press.

Diez de San Miguel, G. 1964 [1567]. *Visita Hecha a la Provincia de Chucuito por Garci Diez de San Miguel en el Año 1567.* Lima: Casa de Cultura.

Earle, T. K., T. N. D'Altroy, C. A. Hastorf, C. J. Scott, C. L. Costin, G. S. Russell, and E. Sandefur. 1987. *Archaeological Field Research in the Upper Mantaro, Peru, 1982–1983: Investigations of Inka Expansion and Exchange.* Monograph 28. Los Angeles: Institute of Archaeology, University of California.

Erickson, C. L. 1993. The social organization of Prehispanic raised field agriculture in the Lake Titicaca basin. In V. L. Scarborough and B. L. Isaac, eds., *Economic Aspects of Water Management in the Prehispanic New World,* pp. 369–426. Research in Economic Anthropology Supplement 7. Greenwich, Conn.: JAI Press.

Espinoza Soriano, W. 1971. Los Huancas, aliados de la conquista: Tres informaciones inéditas sobre la participación indígena en la conquista del Perú, 1558–1560–1561. *Anales Científicos de la Universidad de Centro del Perú* (Huancayo) 1: 3–407.

Farrington, I. S. 1974. Irrigation and settlement pattern: Preliminary research results from the north coast of Peru. In T. E. Downing and M. Gibson, eds., *Irrigation's Impact on Society,* pp. 83–94. Anthropological Papers of the University of Arizona 25. Tucson.

———. 1983. Prehistoric intensive agriculture: Preliminary notes on river canalization in the Sacred Valley of the Incas. In J. Darch, ed., *Drained Field Agriculture in Central and South America,* pp. 221–235. International Series 189. Oxford: British Archaeological Reports.

———. 1992. Ritual geography, settlement patterns, and the characterization of the provinces of the Inka heartland. *World Archaeology* 23: 368–385.

———. 1995. The mummy, estate, and palace of Inka Huayna Capac at Quispeguanca.

Tawantinsuyu (Canberra, Australia) 1: 55–65.

Flannery, K. V., J. Marcus, and R. G. Reynolds. 1989. *The Flocks of the Wamani.* San Diego, Calif.: Academic Press.

Flores-Ochoa, J. A. 1977. *Pastores de Puna, Uywamichiq Punarunakuna.* Lima: Instituto de Estudios Peruanos.

Flores-Ochoa, J. A., and P. Paz Florez. 1984. El cultivo en Qocha en la Puna sur Andina. In S. Masuda, ed., *Contribuciones a los Estudios de los Andes Centrales,* pp. 59–100. Tokyo: University of Tokyo Press.

Fock, N. 1961. Inka imperialism in northwest Argentina and Chaco burial forms. *Folk* (Copenhagen) 3: 67–90.

Fonseca Martel, C. 1974. Modalidades de la minka. In *Reciprocidad e intercambio en los Andes Peruanos,* pp. 89–109. Lima: Instituto de Estudios Peruanos.

Gade, D. W. 1977. Llama, alpaca, y vicuña: Ficción y realidad. In J. Flores-Ochoa, comp., *Pastores de Puna, Uywamichiq Punarunakuna,* pp. 113–120. Lima: Instituto de Estudios Peruanos.

Garcilaso de la Vega, El Inca. 1966 [1609]. *Comentarios reales de los Incas.* Biblioteca de Autores Españoles 133–135. Madrid: Ediciones Atlas.

González, A. R. 1983. Inca settlement patterns in a marginal province of the empire: Sociocultural implications. In Evon Z. Vogt and R. M. Leventhal, eds., *Prehistoric Settlement Patterns: Essays in Honor of Gordon R. Willey,* pp. 337–360. Cambridge: Harvard University.

Graffam, G. 1992. Beyond state collapse: Rural history, raised fields, and pastoralism in the South Andes. *American Anthropologist* 94: 882–904.

Haggard, J. V., and M. D. McLean. 1941. *Handbook for Translators of Spanish Historical Documents.* Archives Collections, University of Texas. Oklahoma City: Semco Color Press.

Harris, O. 1978. El parentesco y la economía vertical en el ayllu Laymi (norte de Potosí). *Avances* (La Paz) 1.

Hastorf, C. A. 1990. The effect of the Inka state on Sausa agricultural production and crop consumption. *American Antiquity* 55: 262–290.

————. 1993. *Agriculture and the Onset of Political Inequality before the Inka.* Cambridge: Cambridge University Press.

Hastorf, C. A., and T. K. Earle. 1985. Intensive agriculture and the geography of political change in the Upper Mantaro region of central Peru. In I. Farrington, ed., *Prehistoric Intensive Agriculture in the Tropics,* pp. 569–595. International Series 232. Oxford: British Archaeological Reports.

Higueras-Hare, A. 1996. *Prehispanic Settlement and Land Use in Cochabamba, Bolivia.* Ph.D. dissertation, University of Pittsburgh. Ann Arbor, Mich.: University Microfilms.

Hyslop, J. 1984. *The Inka Road System.* New York: Academic Press.

————. 1990. *Inka Settlement Planning.* Austin: University of Texas Press.

Julien, C. J. 1993. Finding a fit: Archaeology and ethnohistory of the Incas. In M. Malpass, ed., *Provincial Inca: Archaeological and Ethnohistorical Assessment of the Impact of the Inca State,* pp. 177–233. Iowa City: University of Iowa Press.

Kendall, A. 1994. *Proyecto Arqueológico Cuschicaca, Cusco: Investigaciones arqueológicas y de rehabilitación agricola, 1.* Lima, Southern Peru Copper Corp.

Kolata, A. L. 1991. The technology and organization of agricultural production in the Tiwanaku state. *Latin American Antiquity* 2: 99–125.

Kroeber, A. L. 1930. *Archaeological Exploration in Peru, 2: The Northern Coast.* Anthropology Memoirs 2(2). Chicago: Field Museum of Natural History.

MacLean, M. G. 1986. *Sacred Land, Sacred Water: Inca Landscape Planning in the Cuzco Area.* Ph.D. dissertation, Department of Anthropology, University of California, Berkeley. Ann Arbor, Mich.: University Microfilms.

Miller, G. R., and A. L. Gill. 1990. Zooarchaeology at Pirincay, a Formative Period site in highland Ecuador. *J. Field Archaeology* 17: 49–68.

Mitchell, W. 1980. Local ecology and the state: Implications of contemporary Quechua land use for the Inca sequence of agricultural work. In E. B. Ross, ed., *Beyond the Myths of Culture: Essays in Cultural Materialism,* pp.139–154. New York: Academic Press.

Mitchell, W. P., and D. Guillet. 1993. *Irrigation at High Altitudes: The Social Organization of Water Control Systems in the Andes.* Society for Latin American Anthropology Publication Series 12. Washington, D.C.: American Anthropological Association.

Moore, J. D. 1991. Cultural responses to environmental catastrophes: Post–El Niño subsistence on the prehistoric north coast of Peru. *Latin American Antiquity* 2: 27–47.

Morris, C. 1967. *Storage in Tawantinsuyu.* Ph.D. dissertation, Department of Anthropology, University of Chicago.

Moseley, M. E. 1983. The good old days were better: Agrarian collapse and tectonics. *American Anthropologist* 85: 773–799.

————. 1992. *The Incas and Their Ancestors.* London: Thames & Hudson.

Moseley, M. E., and A. Cordy-Collins, eds. 1990. *The Northern Dynasties: Kingship and Statecraft in Chimor.* Washington, D.C.: Dumbarton Oaks.

Moseley, M. E., D. Wagner, and J. B. Richardson III. 1992. Space shuttle imagery of recent catastrophic climate change along the arid Andean coast. In L. L. Johnson, ed., *Paleoshorelines and Prehistory: An Investigation of Method,* pp. 215–235. Boca Raton, Fla.: CRC Press.

Murra, J. V. 1960. Rite and crop in the Inca state. In S. Diamond, ed., *Culture in History,* pp. 393–407. New York: Columbia University Press.

————. 1962. Cloth and its functions in the Inca state. *American Anthropologist* 64: 710–728.

————. 1965. Herds and herders in the Inca state. In A. Leeds and A. P. Vayda, eds., *Man, Culture, and Animals,* pp. 185–215. Publication 78. Washington, D.C.: American Association for the Advancement of Science.

————. 1968. An Aymara kingdom in 1567. *Ethnohistory* 15(2): 115–151.

————. 1972. El control vertical de un máximo de pisos ecológicos en la economía de las sociedades andinas. In J. V. Murra, ed., *Visita de la Provincia de Leon de Huá-*

nuco, tomo 2, pp. 429–476. Huánuco, Peru: Universidad Hermilio Valdizan.

———. 1975. *Formaciones Económicas y Políticas del Mundo Andino.* Lima: Instituto de Estudios Peruanos.

———. 1980a [1956]. *The Economic Organization of the Inka State.* Greenwich, Conn.: JAI Press.

———. 1980b. Derechos a las tierras en el Tawantinsuyu. *Revista de la Universidad Complutense* (Madrid) 28(117): 273–287.

———. 1982. The mit'a obligations of ethnic groups in the Inka state. In G. A. Collier, R. I. Rosaldo, and J. D. Wirth, eds., *The Inca and Aztec States, 1400–1800,* pp. 237–262. New York: Academic Press.

Netherly, P. 1977. *Local Level Lords on the North Coast of Peru.* Ph.D. dissertation, Department of Anthropology, Cornell University, Ithaca, N.Y.

———. 1984. The management of late Andean irrigation systems on the north coast of Peru. *American Antiquity* 49: 2: 227–254.

Nielsen, A. 1996. Demografía y cambio social en Quebrada de Humahuaca (Jujuy, Argentina) 700–1535 D.C. *Relaciones de la Sociedad Argentina de Antropología* (Buenos Aires) 21: 307–385.

Niles, S. 1987. *Callachaca.* Iowa City: University of Iowa Press.

Ortiz de Zúñiga, I. 1967 [1562]. *Visita de la Provincia de León de Huánuco en 1562, Iñigo Ortiz de Zúñiga, visitador, vol. 1.* J. V. Murra, ed. Huánuco, Peru: Universidad Nacional Hermilio Valdizán.

———. 1972 [1562]. *Visita de la Provincia de León de Huánuco en 1562, Iñigo Ortiz de Zúñiga, visitador, vol. 2.* J. V. Murra, ed. Huánuco, Peru: Universidad Nacional Hermilio Valdizán.

Ortloff, C. R. 1993. Chimu hydraulics technology and statecraft on the north coast of Peru, A.D. 1000–1470. In V. L. Scarborough and B. L. Isaac, eds., *Economic Aspects of Water Management in the Prehispanic New World,* pp. 327–367. Research in Economic Anthropology Supplement 7. Greenwich, Conn.: JAI Press.

Owen, B. D. 1993. A model of multiethnicity: State collapse, competition, and social complexity from Tiwanaku to Chiribaya in the Osmore Valley, Peru. Ph.D. dissertation, University of California, Los Angeles. Ann Arbor, Mich.: University Microfilms.

Parsons, J. R., C. M. Hastings, and R. Matos Mendieta. 1998. Rebuilding the state in highlands Peru: Herder-cultivator interaction during the Late Intermediate Period in the Tarama-Chinchaycocha region. *Latin American Antiquity* 8: 4: 317–341.

———. n.d. *Settlement Patterns in the Tarma-Junín region, Peru,* manuscript.

Pearsall, D. M. 1992. The origins of plant cultivation in South America. In C. W. Cowan and P. J. Watson, eds., *The Origins of Agriculture,* pp. 173–206. Washington, D.C.: Smithsonian Institution Press.

Pease, G. Y. 1978. *Del Tawantinsuyu a la Historia del Perú.* Lima: Instituto de Estudios Peruanos.

Polanyi, K. 1957. The economy as instituted process. In K. Polanyi, C. Arensberg, and H. Pearson, eds., *Trade and Market in the Early Empires,* pp. 243–270. New York: Free Press.

Polo de Ondegardo, J. 1916 [1571]. Relación de los fundamentos acerca del notable daño que resulta de no guardar a los indios sus fueros. In H. H. Urteaga, ed., *Colección de Libros y Documentos Referentes a la Historia del Perú, tomo 3,* pp. 45–188. Lima: Sanmartí.

———. 1940 [1561]. Informe del Licenciado Juan Polo de Ondegardo al Licenciado Briviesca de Muñatones sobre la perpetuidad de las encomiendas en el Perú. *Revista Histórica* (Lima) 13: 128–196.

Protzen, J.-P. 1993. *Inca Architecture and Construction at Ollantaytambo.* Oxford: Oxford University Press.

Pulgar Vidal, J. 1987. *Geografía del Perú.* 9th ed. Lima: PEISA.

Rostworowski de Diez Canseco, M. 1962. Nuevos datos sobre la tenencia de tierras en el incario. *Revista del Museo Nacional* (Lima) 31: 130–164.

———. 1970. Mercaderes del valle de Chincha en la época prehispánica: Un documento y unos comentarios. *Revista Española de Antropología Americana* 5: 135–178.

———. 1975. Pescadores, artesanos, y mercadores costeños prehispánicos. *Revista del Museo Nacional* (Lima) 41.

———. 1977. *Etnía y Sociedad Costa Peruana prehispánica.* Lima: Instituto de Estudios Peruanos.

———. 1978. *Señorios Indígenas de Lima y Canta.* Lima: Instituto de Estudios Peruanos.

———. 1983. *Estructuras Andinas del Poder.* Lima: Instituto de Estudios Peruanos.

———. 1988. *Historia del Tahuantinsuyu.* 2d ed. Lima: Instituto de Estudios Peruanos.

———. 1990. *Conflicts over Coca Fields in Sixteenth-Century Peru.* Memoirs of the Museum of Anthropology 21. Ann Arbor: University of Michigan.

Rowe, J. H. 1946. Inca culture at the time of the Spanish Conquest. In J. Steward, ed., *Handbook of South American Indians, 2,* pp. 183–330. Bulletin 143. Washington, D.C.: Bureau of American Ethnology.

———. 1967. What kind of settlement was Inca Cuzco? *Ñawpa Pacha* (Berkeley) 5: 59–76.

———. 1982. Inca policies and institutions relating to the cultural unification of the empire. In G. A. Collier, R. I. Rosaldo, and J. D. Wirth, eds., *The Inca and Aztec States, 1400–1800,* pp. 93–118. New York: Academic Press.

———. 1985. La constitución inca del Cuzco. *Histórica* (Lima) 9(1): 35–73.

Salomon, F. 1986. *Native Lords of Quito in the Age of the Incas.* Cambridge: Cambridge University Press.

Sandefur, E. C. 1988. *Andean Zooarchaeology: Animal Use and the Inka Conquest of the Upper Mantaro Valley.* Ph.D. dissertation, Archaeology Program, University of California. Ann Arbor, Mich.: University Microfilms.

Sarmiento de Gamboa, P. 1960 [1572]. *Historia Indica.* Biblioteca de Autores Españoles 135: 193–279. Madrid: Ediciones Atlas.

Sherbondy, J. 1993. Water and power: The role of irrigation districts in the transition from Inca to Spanish Cuzco. In W. P. Mitchell and D. Guillet, eds., *Irrigation at High Altitudes: The Social Organization of Water Control Systems in the Andes,* pp. 69–97. Society for Latin American

Anthropology Publication Series 12. Washington, D.C.: American Anthropological Association.

Shimada, I. 1994. *Pampa Grande and the Mochica Culture.* Austin: University of Texas Press.

Spurling, G. 1982. *Inka militarism,* manuscript.

Stanish, C., and L. Steadman. 1994. *Archaeological Research at the Site of Tumatumani, Juli, Peru.* Fieldiana, Anthropology 23. Chicago: Field Museum of Natural History.

Thomas, R. B. 1973. *Human Adaptation to a High Andean Energy Flow System.* Occasional Papers in Anthropology 7. University Park: Department of Anthropology, Pennsylvania State University.

Toledo, F. de. 1940 [1570]. Informacion hecha por orden de Don Francisco de Toledo en su visita de las Provincias del Peru. In R. Levillier, ed., *Don Francisco del Toledo, Supremo Organizador del Peru, Su Vida, Su Obra 1515–1582,* vol. 2, pp. 14–37. Buenos Aires: Espasa-Calpe.

Tomka, S. A. 1987. Resource ownership and utilization patterns among the Yanque-Collaguas as manifested in the visita de Yanque-Collaguas, 1591. *Andean Perspective Newsletter* 5: 15–24. Institute of Latin American Studies, University of Texas, Austin.

———. 1994. *Quinua and Camelids on the Bolivian Altiplano: An Ethnoarchaeological Approach to Agro-Pastoral Subsistence with an Emphasis on Agro-Pastoral Transhumance.* Ph.D. dissertation, University of Texas, Austin. Ann Arbor, Mich.: University Microfilms.

Tosi, J. A. Jr. 1960. *Zonas de Vida Natural en el Perú: Memoria Explicativa Sobre el Mapa Ecológico del Perú.* Boletín Técnico 5. Instituto Interamericano de Ciencias Agrícolas de la OEA, Zona Andina.

Troll, C. 1968. The cordilleras of the tropical Americas. In C. Troll, ed., *Geo-ecology of the Mountainous Regions of the Tropical Americas,* pp. 13–56. Proceedings of the UNESCO Mexico Symposium, 1966. Bonn: Ferd. Dümmlers Verlag.

Urbano Martínez Carreras, J. 1965 [1557–1586]. *Relaciones Geográficas de Indias.*

Tomos 183–185. Biblioteca de Autores Españoles. Madrid: Ediciones Atlas.

Vallis, G. K. 1986. El Niño: A chaotic dynamical system? *Science* 232: 243–5.

Villanueva Urteaga, H. 1971. Documentos sobre Yucay, siglo 16. *Revista del Archivo Histórico de Cuzco* 13: 1–148.

Wachtel, N. 1977. *The Vision of the Vanquished*. Ben and Sian Reynolds, trans. New York: Barnes & Noble.

———. 1982. The *mitimas* of the Cochabamba Valley: The colonization policy of Huayna Capac. In G. A. Collier, R. I. Rosaldo, and J. D. Wirth, eds., *The Inca and Aztec States, 1400–1800: Anthropology and History*, pp. 199–235. New York: Academic Press.

Watanabe, L. K., M. E. Moseley, and F. Cabieses, comp. 1990. *Trabajos Arqueológicos en Moquegua, Perú*. Programa Conti-

suyu del Museo Peruano de Ciencias de la Salud, Southern Peru Copper Corporation. Lima: Editorial Escuela Nueva.

West, T. 1981. Llama caravans of the Andes. *Natural History* 90: 12: 62–73.

Williams, V. I. 1996. *La Ocupación Inka en la Región Central de Catamarca (República Argentina)*. Ph.D. dissertation, Universidad Nacional de La Plata, Argentina.

Winterhalder, B. 1993. The ecological basis of water management in the central Andes: Rainfall and temperature in southern Peru. In W. P. Mitchell and D. Guillet, eds., *Irrigation at High Altitudes: The Social Organization of Water Control Systems in the Andes*, pp. 21–67. Society for Latin American Anthropology Publication Series 12. Washington, D.C.: American Anthropological Association.

DOUGLAS C. DALY
JOHN D. MITCHELL

14 | LOWLAND VEGETATION OF TROPICAL SOUTH AMERICA
An Overview

Tropical lowland South America boasts a diversity of vegetation cover as impressive, and often as bewildering, as its diversity of plant species. In this essay, we attempt to describe the major types of vegetation cover in this vast region as they occurred in Precolumbian times and outline the conditions that support them. Examining the large-scale phytogeographic regions characterized by each major cover type (figure 14.1), we provide basic information on geology, geological history, topography, and climate; describe variants of physiognomy (vegetation structure) and geography; discuss transitions; and examine some floristic patterns and affinities within and among these regions. We mention some of the better-known useful plant species native to each region.

Preceding the primarily descriptive body of the text are brief discussions on vegetation classification and on folk classifications of vegetation. For humid forests, we briefly explore trends and patterns of floristic diversity. Throughout the essay, we cite relevant references from both classical and recent literature.

The map in figure 14.1 reflects the classification of vegetation cover on a broad scale as an integration of floristics and physiognomy; one vegetation cover type (or a particular mosaic) predominates in each region. The figure is modified from Prance (e.g., 1989), who classified the American tropical forests using the "phytochorion," a concept widely used in studies of African vegetation, which emphasizes species distributions rather than vegetation types while stressing the correlation between the two. White (1979, see

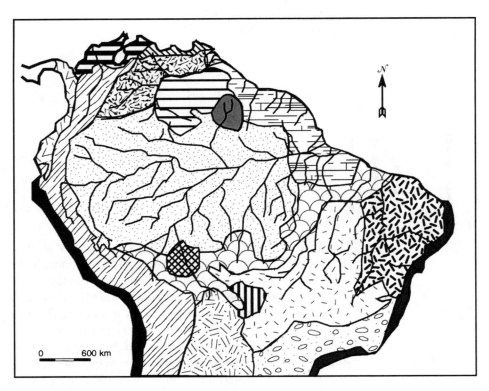

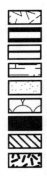 Chocó		Chaco	
Magdalena/NW Caribbean		S. Brazilian Region	
Venezuelan Guayana		Cerrado	
Guianas - E. Amazonia		Llanos of Venezuela & Colombia	
W. and Central Amazonia		Roraima-Rupununi Savannas	
S. Amazonia Transitions		Llanos de Moxos	
Atlantic Forest Complex		Pantanal	
Coastal Cordillera		Andean Region	
Caatinga		S. Pacific Coast	

FIGURE 14.1. The major phytochoria of northern South America. *Source:* After Prance 1989.

applications in White 1983) considered phytochoria to have 50 percent endemism or more than 1,000 endemic species, excepting mosaics and transitions. Historically, the integration of floristics and physiognomy was dealt with by Beard (e.g., 1955) in a hierarchical fashion: the "formations" he defined were physiognomic reduction series, while floristics characterized the next level, "associations."

OBSERVATIONS ON VEGETATION CLASSIFICATION

Continental-scale vegetation maps of South America published to date are largely unsatisfactory (see compilation of Küchler 1980). The UNESCO (1981a) map includes only a half-dozen cover types for all of Amazonia. Veloso's (1966) first classification of vegetation in Brazil relied on very broad physiognomic types with no regional specificity, although the accompanying text contains floristic information. Hueck and Seibert's (1972) map is much more detailed, as is the accompanying text (Hueck 1972), but it still fails to do justice to the diversity of cover types and ultimately to the beta-diversity (essentially the habitat diversity) of the continent.

The question to be posed is whether it is possible to classify vegetation cover adequately on a continental scale. Recent work in Peruvian Amazonia has identified numerous floristic consociations and vegetation cover types even on a rather local scale, based on multiple sampling using two taxonomic groups of plants (Tuomisto and Ruokolainen 1994) and on satellite images (Tuomisto et al. 1994, 1995), respectively. Indeed, the use of pixel (i.e., spectral) diversity on satellite images as a tool for measuring beta-diversity and for sampling biodiversity shows great promise (e.g., Podolsky 1992, 1994).

Such characterization can work on a local up to a landscape or sometimes regional scale but runs into trouble toward the continental scale. One fundamental quality of the lowland vegetation of South America (and the tropics in general) is that similar climatic and edaphic conditions in geographically distant regions can give rise to strikingly similar vegetation physiognomies. In some instances the floristic connections are minor, such as the classical campo rupestre (figure 14.2) of the Serra do Espinhaço in southeast-central Brazil (e.g., Giulietti and Pirani 1988) versus the so-called campos rupestres of the Serra do Cachimbo in southwestern Pará and of the Río Cururu in the Río Tapajós basin (see Pires and Prance 1985 and discussion below).

In other cases, there are strong floristic as well as physiognomic affinities that signal former physical connections and therefore genetic exchange, such as the now interrupted areas of dry forest that form an arc around the southern periphery of Amazonia. Numerous species show distributions around part or all of this arc from Ceará in northeastern Brazil around and up to Acre in Brazil and even to Tarapoto in Peru (Prado and Gibbs 1993).

In general, however, the floristic component of a given physiognomy must be painted with very broad brush strokes between regions of humid tropical forests. For example, Condit (results cited in Cook 1998) examined the composition of humid forests at a number of sites in Costa Rica with comparable climates and topographies and found that the only reliable predictor of community similarity between sites was proximity. Indeed, as Condit (1996) has noted elsewhere, it is unclear how well climatic and structural classifications correlate with species distributions.

This was borne out in a study by Terborgh and Andresen (1998): seeking to address the lack of composition-based vegetation classification systems in Amazonia, they examined the relative densities of tree taxa in 48 forest inventories in several portions of

FIGURE 14.2. *Top:* Campo rupestre sensu lato, summit of Serranía de Santiago, Santa Cruz, Bolivia, August 1983. *Bottom:* Abrupt savanna–gallery forest transition, between Macapá and Porto Grande, Amapá, Amazonian Brazil, November 1986. (Photos by D. Daly.)

the region. They found that ordinations based on species and genera were essentially unusable because site similarity diverged so much over distance. Ultimately they used relative densities of families, and based on this criterion, at least, they concluded that floodplain and terra firme forests in a given area are more similar to each other than to the same forest type in another part of Amazonia. Again, history must play a key role, but this time in the form of historic barriers rather than connections; one must also consider that on a regional scale, floodplain taxa may have given rise to terra firme taxa (as suggested by Goulding 1993) or vice versa.

FOLK CLASSIFICATIONS

Local classifications, reflecting cultural perceptions of the surroundings, may be based on physiognomy or topography-drainage; a sort of folk Braun-Blanquet plant-sociology system is applied to some formations in which a species or set of confamilials is dominant. The varillales (from the word for "pole") on white sand in Peruvian Amazonia are characterized by the densely packed slender trees implicated by the name. *Caatinga* is derived from the Tupi for "white forest," referring to the blanched trunks so common in arid northeastern Brazil. *Mata de cipó,* or liana forest, is self-explanatory. The various popular terms for formations in the Cerrado (e.g., *campo limpo, campo cerrado, cerradão*) describe the relative density of woody plant cover.

Local names for humid forest formations tend to be oriented toward terrain, as are many of those in the scientific literature. Such terms as *várzea, baixio* (*bajio* in Spanish, for low areas), *terraço* (*alto* and *baixo*), and *barranco* (levee) involve the elevation of the terrain relative to river levels and to flooding regimes.

Popular names refer to particular taxa when those taxa are locally dominant in some way. In Peruvian Amazonia, *irapayales* refers to relatively tall forests on sandy soils whose understory is dominated by the palm *Lepidocaryum tenue* (see Whitney and Álvarez 1998), and moenales are forests on young terraces dominated by several Lauraceae. In southwestern Brazilian Amazonia and contiguous Peru, the tabocais or pacales are forests dominated by arborescent bamboos in the genus *Guadua*. The wallaba forests of Guyana are dominated by *Eperua* spp.

This phenomenon is more frequent in drier formations, where dominance by one or a few taxa is more common. The quebrachales of the Chaco are dominated by *Schinopsis* spp. or *Aspidosperma quebracho-blanco*. Also in the Chaco, palo-santales are dominated by *Bulnesia sarmientoi,* and algarrobales are dominated by *Prosopis* spp.

Some terms are not used consistently, or the same term may have different meanings among regions. The caatinga of northeastern Brazil has nothing to do with the Amazonian caatingas of the upper Río Negro described below, except perhaps for the sun-bleached barks of some of their trees. *Igapó* as used by Prance (e.g., 1979) applies to forests seasonally inundated by black-water rivers, whereas most Brazilian botanists (e.g., Pires 1973) have used the term to describe permanently flooded forests.

HUMID FORESTS

Lowland moist to wet forests dominate the landscape of northern South America, although not to the extent often assumed (Mares 1992). The largest single area of these mostly humid forests, the Amazonian hylaea, covers approximately 6 million km², and the remaining areas cover perhaps an additional 500,000 km², but the assemblage of drier lowland formations totals nearly that much: the caatinga vegetation of northeastern Brazil covers 600,000–900,000 km² (Andrade-Lima 1981; Sampaio 1995), the Chaco of Paraguay and Bolivia another 800,000–1 million km² (Hueck 1972; Galera and Ramela 1997), the Cerrado of central Brazil approximately 1.8 million km² (Ab'Sáber 1971; Coutinho 1990), the Llanos of Venezuela and Colombia 500,000 km² (Sarmiento 1984), the Llanos de Moxos savannas 150,000 km² (Beck and Moraes 1997), the Pantanal 150,000–170,000 km² (Prance and Schaller 1982; Dubs 1992a), and the Roraima-Rupununi savannas 54,000 km² (Pires 1973; Eden and McGregor 1992).

The humid forests comprise a complex array of formations; only an exceedingly small portion of these forests correspond to the cathedral-like rain forests usually associated with Amazonia and other parts of northern South America. Many of the formations have canopies that are discontinuous and far from towering, and many have an understory congested with coarse herbs or lianas or slender trees. Similarly, most of the Neotropical lowland forests occur in climates that are markedly or at least somewhat seasonal (see Walsh 1998).

The largest and floristically most significant regions of lowland humid forest in northern South America are the Amazonian hylaea, largely continuous floristically with the Guianas (figure 14.3); Brazil's Atlantic coastal forests, extending from Río Grande do Norte around the horn of Brazil and south to Paraná state; and the Chocó biogeographic region, west of the Andes and extending from southern Panama south to central Ecuador (see figure 14.1). The precise area originally occupied by each of these is obscured by unclear transitions to dry or montane formations and increasingly to anthropically modified or degraded regions. The limits of Amazonia have been variously defined (see review in Daly and Prance 1989); the Amazon rivers drain an area of some 7,050,000 km² (Sioli 1984), but the hydrological basin extends farther than the Amazonian vegetation, whose extent in Brazil has been estimated at between 4 million km² (Nelson 1994) and 6 million km² (Pires 1973).

STRUCTURE OF HUMID FORESTS

Lowland humid tropical forests can be characterized by a disparate set of factors. The canopy is generally high (to 40 m) but often discontinuous, and emergent trees may reach 65 m in height. It is usually not possible to distinguish strata. Palms and lianas occur in varying densities. Cauliflory and buttresses are common. Compared to dry or lower montane forests, the lowland humid forests have greater stand height, height to the first branch, trunk volume, buttress height, and buttress area (see table 295 in

FIGURE 14.3. *Top:* Profile of tropical moist forest on terra firme. Road construction near EMBRAPA reserve, Camaipi, Amapá, Amazonian Brazil, September 1983. *Bottom:* Profile of disjunct campinarana (Amazonian caatinga). São Paulo de Olivença, upper Río Solimões, Amazonas, Brazil, November 1986. (Photos by D. Daly.)

Holdridge 1971). Many tree species have trunks that are not cylindric but rather sulcate (grooved), fluted, or fenestrate (reticulate) (Pires 1973; see definitions in Mori et al. 1997). Unlike many trees in dry forests and particularly savannas, few show adaptations to fire such as very thick bark or underground trunks, and some even have highly combustible sap, such as the resin of many Burseraceae. Hepatics (liverworts) are high in both density and diversity, and epiphytic mosses are common but not so much as in montane forests. Also compared to montane forests, the lowland humid forests are poorer in epiphytes in general and particularly in epiphytic monocots (e.g., Ducke and Black 1953), epiphytic shrubs, and tree ferns (Holdridge 1971). The understory is occupied more by tree regeneration and less by herbs and shrubs than in montane forests. A number of tree and epiphyte species have various morphological adaptations that make them hosts to biting or stinging ants (Ducke and Black 1953). A number of species are monocarpic (Pires 1973).

The diversity of these lowland forests is derived more from trees as well. In an inventory of all plants over 0.5 m in height in 1.8 ha each of upland and floodplain forests near Araracuara in Colombian Amazonia, in both formations trees contributed 65 percent of the diversity, followed by climbers (24 percent), shrubs (8 percent), and terrestrial herbs (3 percent) (Londoño-Vega and Álvarez 1997). On 1 ha plots (100 × 100 m) at Cuyabeno, Amazonian Ecuador, Valencia et al. (1994) found 307 species of trees 10 cm or greater DBH (diameter at breast height), but Poulsen and Balslev (1991) found only 96 species of terrestrial (vascular) herbs.

FORMATIONS AND HOW TO DEFINE THEM

Formations are defined on a purely physiognomic basis. A given phytogeographic subdivision is often characterized by the predominance of a given formation or formation series, while each formation in that region may be characterized by taxa of various ranks. Formations within the broader category of moist and wet forests have been defined using canopy height; canopy continuity; presence and number of strata; stand density; dominance; species richness; presence and density of lianas as well as epiphytes, palms, and herbs; biomass (often expressed in relative terms using basal area); number of deciduous species; leaf type; density of mosses; bark types; and other characters.

The formations or vegetation types of Amazonia account for much of the physiognomic spectrum of humid forests in northern South America and were summarized by Pires (1973) and revised by Pires and Prance (1985), Prance and Brown (1987), and Prance (1989).

Dense forests, or mata pesada (literally, "heavy forest"), show a basal area of 40 m² per hectare or more. The canopy is high and often continuous, the understory relatively open. There are relatively few grasses and sedges, shrubs, herbs, and vines. Prance and Brown (1987) recognized a distinct forest type on undulating terrain, "dense and open hill forest on richer soils."

Open forests have a basal area of 18–24 m². In these forests, there are more shrubs and lianas but fewer epiphytes. Open forests may be associated with a low water table, impermeable soils, poor drainage, poor root penetration, a longer dry season, and/or low relative humidity. Some open forests lack palms almost entirely, while others show a strong presence of palms; in Amazonia some of the more prevalent palms in open forests are *Oenocarpus bacaba* and *O. mapora* (bacaba), *O. (Jessenia) bataua* (patauá), *Euterpe* spp. (açaí), and *Attalea (Maximilliana) maripa* (inajá). Both these types of open forest were considered transitional by Prance (1989), even though they are not confined to the periphery of Amazonia. Open forests occur in many hilly regions where the trees are subject to unstable slopes and especially winds.

Although moist forests on latosols (red, porous clays) occur in the Río Negro region, the most characteristic vegetation of the region is the Amazon caatinga or campina forest (also called campinarana; see figure 14.3; Anderson 1981), particularly in the upper Río Negro, which drains much of northern Brazil, part of southern Amazonas state in Venezuela, and part of Vaupés and Guainía departments in Colombia. They are edaphic formations not separated climatically from the rest of Amazonia. These forests on extremely nutrient-poor podzols (white-sand soils) are still inadequately studied. The most developed of them consist of closely packed slender trees with an often closed canopy to approximately 20 m high, with occasional large emergents. The understory has a thick layer of poorly decomposed litter. Epiphytes are common, but not lianas.

There is actually a continuum from campina forest to open, shrubby campina (called bana in Venezuela); this is a function of the elevation above the water table (Jordan 1989). The area covered by the Amazon caatingas in Brazil has been estimated at 30,000 km², and a similar figure for low campina (Pires and Prance 1985), but these have yet to be confirmed by ground-truthing and interpretation of satellite images.

Above the Río Negro's confluence with the Río Branco, there is a complex mosaic of soil types and correlated moist forests, Amazon caatingas, and shrub communities (Projeto Radambrasil 1975–1981). The great paradox of the upper Río Negro is the occurrence of high species diversity in a region characterized by poor soils (podzols predominant) and often perched water tables (Goulding et al. 1988). There is considerable endemism at the generic level (e.g., Ducke and Black 1953; Pires and Prance 1985), which is not surprising, considering that most of the white-sand soils of Amazonia are concentrated here. Some of the typical genera of the Río Negro are *Clusia, Tovomita, Lissocarpa, Pagamea, Campnosperma, Retiniphyllum, Barcella, Platycarpus,* and *Henriquezia.* In the Venezuelan part of the Río Negro basin, some taller forests little affected by inundation are dominated by *Eperua purpurea* and others by *Monopteryx uacu,* while true Amazon caatingas (forests) there may be dominated by *Eperua leucantha* or *Micrandra sprucei* (O. Huber 1995). Other tree species characteristic of that region are *Lissocarpa benthamii, Hevea pauciflora, Pradosia schomburgkiana, Pagamea coriacea,* and *Sclerolobium dwyeri* (ibid.).

Formations of varying extents that are disjunct but physiognomically similar to the Río Negro vegetation occur elsewhere in Amazonia. Areas containing campina forest include the upper Río Solimões near São Paulo de Olivença, the lower Río Tonantins, and the varillales of Amazonian Peru. Shrubby and open campinas occur sporadically through much of Amazonia; large areas of campina occur south of the upper Río Madeira. Räsänen (1993) interpreted many of the white-sand soils in Amazonia as being derived from the Guayana Shield, from which they were transported to western South America prior to the Andean orogeny when continental drainage flowed east-west.

Arborescent bamboos are the dominant element of a vegetation type covering more than 180,000 km² in the southwest Brazilian Amazon and neighboring Peru and Bolivia (expanded from Nelson 1994). This unusual and until recently unstudied vegetation type was remarked upon by early travelers to the region (e.g., Chandless 1866; J. Huber 1906). Physiognomically, *Guadua* bamboo forests resemble liana forests, and like that formation they were considered transitional by Prance (1989), although they are not transitional to extra-Amazonian vegetation. Adult culms (stems) of these species average 8 cm in diameter and 15 m in height but may reach a height of 30 m, approximately equivalent to the top of the forest canopy. Bamboo culms fill gaps between individual trees and cover their canopy. In the forest understory, young erect culms and fallen branches can produce an almost impenetrable tangle of spiny vegetation. These so-called bamboo forests dominate the lower portions of terra firme landscapes in parts of southwest Amazonia, including valleys and relict river terraces, but not usually on plateaus.

Two aspects of this bamboo's basic ecology have a profound impact on the forest dynamics of the region. One is its ability to spread aggressively by means of rhizomes. There is evidence that the already enormous area covered by this formation is expanding; bamboo forests may be favored by large-scale natural or anthropogenic disturbance. The other conspicuous aspect of their ecology is the synchronous flowering and subsequent die-back that characterize many species of bamboo worldwide, and this appears to involve numerous populations on at least a regional scale. By their cycles of expansion and collapse, to a large degree these forests mediate the rather tumultuous forest dynamics of the region.

Campo rupestre, or "rocky field," is a term that has been applied to several open formations on stony soils that are geographically distant and floristically rather distinct. The classical campo rupestre is that occurring on the quartzite Serra do Espinhaço in southeastern Brazil, where that vegetation occurs between (700) 1,000 and 2,000+ m. This vegetation consists of a more or less continuous herbaceous stratum with evergreen small shrubs and subshrubs. Characteristic taxa include Velloziaceae, Eriocaulaceae, Xyridaceae, Poaceae, Melastomataceae, and Bromeliaceae. (e.g., Giulietti and Pirani 1988, 1997; see discussion below under "Savannas").

Two other major areas of campo rupestre in the broader sense occur toward the southern periphery of the Amazon Basin; the one on part of the Serra do Cachimbo in southwest Pará covers some 16,000 km², while a smaller extent is found near the Río Cururu in the upper Río Tapajós. Despite their physiognomy, some typical cerrado species do not occur in these formations, such as *Curatella americana, Hancornia speciosa, Salvertia convallariodora,* and *Qualea grandiflora.* Certain plant groups important in the Serra do Espinhaço are also prominent there, such as *Vellozia,* Eriocaulaceae, *Byrsonima,* and several genera of terrestrial Bromeliaceae (see Pires and Prance 1985), but otherwise their floras, which are still poorly known, appear to be rather distinct.

INCLUSIONS AND ARCHIPELAGOS

The extent of humid lowland forests is in places vast but not unbroken. There may be continuous mosaics of edaphic conditions defining the beta-diversity of a region. Large regions of humid forests may also contain anomalous formations either covering considerable areas or occurring as irregular islands. The most significant of these are the Amazonian savannas, which may cover 100,000–150,000 km² or about 3–4 percent of Brazilian Amazonia (Pires and Prance 1985). These include a number of nonflooded formations such as the subcoastal savannas in Amapá (Sanaiotti et al. 1997; see figure 14.2), the upland savannas of Marajó Island at the mouth of the Río Amazonas (Bastos 1984), the campos de Ariramba of the upper Río Trombetas in northern Pará (Egler 1960), and those near Santarém in western Pará (Miranda 1993). They also include riverine flooded savannas. Although the upland savannas share some species with the Cerrado, there are important physiognomic differences indicating nonadaptation to fire: the root systems are more superficial, the relative humidity is higher, and fewer woody species are capable of sprouting from roots or stumps. Some of the Amazonian savannas are species-poor, such as those in Amapá, while others, like those of the upper Río Trombetas, are relatively biodiverse (Pires and Prance 1985).

Shrubby campinas on partly exposed white sands (discussed in previous section) occur sporadically throughout much of Amazonia as islands on the scale of hectares or of square kilometers. The campinas have little in common floristically with the surrounding tall forests on clay soils (Guillaumet 1987), and they function as islands, with numerous adaptations to relatively long-distance dispersal (Anderson 1981; Macedo and Prance 1978). They show a high degree of endemism, even including some endemic genera (e.g., Struwe et al. 1997).

Another archipelago of inclusions occurs in the southern part of Venezuela's Guayana Highland (see figure 14.1), where isolated plateaus—the tepuis—are all that remain of an ancient sandstone tableland. The (pre)montane forests on the upper slopes of the tepuis and the vegetation of the bogs and rock outcrops on their summits contrast greatly with the surrounding Amazonian forests (O. Huber 1995). Smaller and lower sandstone plateau outcrops with related floras extend into the Caquetá-Araracuara-Cahuinarí region of Colombia (Sánchez et al. 1997).

Inselbergs are isolated granitic mountains up to 700 m in elevation (Sarthou and Villiers 1998). The inselbergs are restricted to the Guayana and Brazilian shields, principally in French Guiana, Suriname, and Brazil and including the Serra do Mar along the Atlantic coast in the latter. The summit vegetation consists mostly of mats of monocots, characterized by Velloziaceae but also including Bromeliaceae and Orchidaceae, plus some dicots, including *Clusia* and (in Brazil) principally Cactaceae. The inselbergs include Nouragues and Mitaraka in French Guiana (Granville and Sastre 1974; Barthlott et al. 1993) and the Pão de Açucar in Río de Janeiro, Brazil. On the inselbergs in French Guiana, Cactaceae are scarce and Velloziaceae lacking (Sarthou and Villiers 1998).

In some regions it is the humid forests that occur as islands. In most of the northeastern Brazilian states, isolated patches of moist forest occur (or occurred) on the brejos. These are granitic plateaus of up to 1,000 m elevation, found 30–90 km from the coast. Humid microclimates occur at higher elevations on the southern and eastern slopes, which are exposed to coastal winds. These islands of forest have affinities with Amazonia and with the Atlantic coastal forests farther south (e.g., Andrade-Lima 1982).

CONDITIONS THAT SUPPORT MOIST FORESTS

Humid lowland forests occur in a remarkable range of climatic regimes and substrates. Mean temperatures do not vary much within the lowland tropics; in this context, the principal climatic factor determining vegetation is the equilibrium between precipitation and evapotranspiration, and above certain minima, the limiting factor is seasonality of rainfall. Schimper (1903) was among the first to note the importance of this factor in the tropics. A key parameter for the tropics in the classical climatic classification of Köppen (1901) is the number of months with less than 60 mm of precipitation.

Annual precipitation exceeds 9,000 mm in some northern parts of the Chocó region (Herrera-MacBryde et al. 1997), and in most regions of northern South America the lower limit for humid forests is approximately 1,700 mm. The importance of seasonality is underscored, however, in the southern Bahian portion of Brazil's Atlantic forest complex, where species-rich evergreen moist forest thrives under only 1,850 mm of precipitation that is evenly distributed throughout the year (Mori et al. 1982).

Rainfall patterns, particularly in large regions such as Amazonia, are far from uniform. The spectrum of modified Walter-Lieth climate diagrams presented by Walsh (1998) show at least three degrees of seasonality in the region, ranging from superwet to wet to seasonal. It is significant to note that four of the phytogeographic divisions of Amazonia proposed by Ducke and Black (1953) correspond to "eco-climate" subdivisions of Aubréville, which were defined by the timing and number of dry months with under 30 mm rainfall (see Daly and Prance 1989). Annual precipitation often exceeds 3,000 mm in the eastern Andean piedmont, while farther east there is a large swath cutting northwest-southeast across the Río Trombetas and the lower Río Xingu that receives only 1,750 mm of rain. This is known as the transverse dry belt (Pires-O'Brien 1997) or Aw belt, referring to a category in Köppen's climate system in which the driest month has less than 60–65 mm rainfall and results in seasonal formations and savannas. Although consideration of the length and severity of the dry season is a more appropriate measure of climate (see Walsh 1998), in this region Köppen's system accurately predicts the vegetation.

TRENDS AND PATTERNS OF DIVERSITY IN HUMID FORESTS

The plant diversity of moist and wet tropical forests of northern South America is accounted for primarily by trees. Monodominant and oligarchic forests do occur, usually but not always under relatively extreme environmental conditions, but the majority of humid forests show marked dominance only by a changing consortium of multiple species. Conversely, often half the tree species are rare on a scale of 1–50 hectares or more. Among moist forest regions, there are recognizable trends in relative alpha-diversity (species richness per site) related to soil nutrients, total rainfall, seasonality, and latitude. Major changes over distance in composition and often physiognomy usually correspond to existing barriers or transitions of soil or climate, but in some cases they coincide with nonobvious historical geological features. On a landscape scale, species richness and composition correlate with topography and soils, but only up to a point.

One of the hallmarks of humid lowlands is the rapidity with which the forests change over distance, be it from one vegetation type or even one spot to the next (e.g., Campbell et al. 1986). For example, near the confluence of the Río Negro and Río Uaupés in northern Amazonian Brazil, Takeuchi (1962) found remarkable differences between two sites within a high campina formation in terms of humus depth, soil moisture, relative humidity, canopy height, canopy tree species composition, the number of strata, the density and composition of understory, dominance, and the occurrence of palms and epiphytes.

Compared to montane forests, trees are a far more important source of diversity in the lowland forests. In the Río Caquetá of Colombia, for example, most species-rich families and genera tend to be tree groups, although in contrast to other regions much of the diversity is represented by small trees of 2.5 cm DBH or less (ca. 1.3 m), and those less than 10 cm DBH were the most species-rich habit group (Duivenvoorden 1994). Considering margins of error and variability, equally megadiverse forests have been documented in Amazonian Ecuador (Valencia et al. 1994), Amazonian Peru (Gentry 1988), the Chocó region of Colombia (Gentry 1986), and the vicinity of Manaus in Brazil (Oliveira and Mori 1999), all with approximately 300 species of trees and lianas of 10 cm DBH or greater per hectare.

Forests strongly dominated by one or a few species (monodominant and oligarchic forests, respectively) are not rare. For example, mora forests dominated by *Mora excelsa* are frequent in low-lying areas of Guyana and contiguous Delta Amacuro, Venezuela (e.g., Connell and Lowman 1989; ter Steege et al. 1993; O. Huber 1995), and tabocais or pacales dominated by *Guadua weberbaueri* cover large areas of southwestern Amazonia (e.g., Nelson 1994).

Many monodominant and oligarchic forests are associated with extremes of drainage or poor soils. Floodplains and other relative edaphic extremes tend to support forests with relatively lower alpha-diversity and higher dominance, some of them dominated by well-known economic species (cf. Peters 1992); examples are the palm forests of the Orinoco Delta and Río Amazonas that are often dominated by buriti or moriche (*Mauritia flexuosa*) and açaí (*Euterpe oleracea*) palms (O. Huber 1995; Pires and Prance 1985). It has been estimated that almost 10,000 km² of pure stands of the former occur in Peru alone, and the same estimate has been made for pure stands of the latter in the Amazon estuary (reported in Peters 1992).

At one site at Yanamono in Peru, there were only twice as many individuals as species of trees (Gentry 1988). Salomão and Lisboa (1988) reviewed thirteen published

Amazonian forest inventories of 1–3.5 ha, most with a 9.5–10 cm DBH cutoff, and found that in most cases between 36 and 53 percent of the trees censused were single-tons, i.e., represented by only one individual per hectare or more.

In the Neotropics overall, there are fairly consistent trends in alpha-diversity across several gradients. In general, humid lowland forests are more alpha-diverse in areas with higher total rainfall, with more aseasonal climates, and at lower latitudes within the tropics (Gentry 1988). Also, greater rainfall is correlated with poorer soils, higher productivity, higher stem density, and somewhat smaller stature (Huston 1994). Within Amazonia, forests tend to be more alpha-diverse along an east-west gradient.

There are notable exceptions to these trends. For example, one of the highest levels of alpha-diversity in the world for trees of 10 cm DBH or greater has been found in the region of Manaus (Oliveira and Mori 1999), but it is located in central Amazonia, and the rainfall is relatively low and markedly seasonal. One possible explanation for this is the status of Manaus as a biotic crossroads where the distributional limits of numerous taxa of plants and animals coincide; this may be a region of reconvergence for distributions disrupted by one or another factor during glacial periods (Oliveira and Daly 1999).

Up to a point, differences in soils and topography correlate with changes in species diversity, species composition, and forest structure; indeed, they often determine distinct forest types. In the mosaic of habitats one often encounters in regions of lowland humid forests, particularly in western Amazonia, there are generalists in each component, but there are also edaphic specialists (e.g., Young and León 1989), and the compositional differences are often statistically rather clear-cut except when it comes to comparing sites with different clay soils. Guillaumet (1987) compared sites of terra firme (upland) forest, swamp forest, campinarana, and campina in central Amazonia and found that the latter two were closely linked floristically with each other but not to the terra firme or swamp forest. Not one tree species was shared by all four formations.

The work of Tuomisto et al. (1995) in Peruvian Amazonia confirmed and expanded on these results. They found that plant communities were sharply differentiated on sandy versus clayey soils, ridge tops versus valleys, and white-sand formations (campinarana analogues) versus swamp forests, but among different clayey soils they found only weak evidence that tree communities correlate with soil nutrients (data cited in Condit 1996). In an area of clayey soils in French Guiana, Lescure and Boulet (1985) mapped trees greater than 20 cm in diameter onto drainage types (which affected soil structure) and found that most trees could occur in more than one drainage type. In a central French Guianan forest, Mitchell and Mori (1987) found that some species of Lecythidaceae preferred ridgetops, while others were associated either with hill slopes or low-lying areas.

On the middle Caquetá River in Colombian Amazonia, Duivenvoorden (1995) found meaningful patterns of tree species composition related to drainage, flooding, humus forms, and soil nutrients. On well-drained upland sites, species composition depended significantly on soil properties, but it was basically a dichotomous pattern and not the superhigh beta-pattern that might have been expected. Comparing swamp forest and formations on sandy (podzolized) soils versus forests on well-drained soils, in the former he found more dominance in general and greater importance of palms, Clusiaceae, Bombacaceae, and Apocynaceae, and in the latter greater importance of Lauraceae, Chrysobalanaceae, Moraceae, and Lecythidaceae (ibid.).

Working with Amazonian palms, Clark et al. (1995) found there was species sorting by soils and topography, e.g., between infertile upland areas and adjacent young alluvial terraces. In Ecuadorean Amazonia, Poulsen and Balslev (1991) found that spatial distribution of understory herbs may be correlated with topography.

CHANGES WITHIN HUMID FOREST REGIONS

The features and factors that separate one phytogeographically distinct area of humid forest from another are usually evident, but some are more subtle, others appear to be historic, and still others are invisible. Barriers such as the Andes separating the Chocó from western Amazonia, the cordilleras that flank the Magdalena Valley in Colombia, and the Serra do Mar that sets apart the Mata Atlântica of Brazil help to circumscribe phytochoria. Major rivers such as the Río Solimões-Amazonas function as barriers to dispersal and can create and help define biogeographic subdivisions (see Wallace 1849; Ayres and Clutton-Brock 1992).

Other changes in vegetation observable on satellite images coincide with historical geological features such as paleoarches, long zones of ancient uplift barely perceptible except on satellite images; an example occurs on either side of the Fitzcarrald Arch in southwestern Amazonia (R. Kalliola pers. comm., and pers. obs.). It is interesting to note that another paleoarch in western and southwestern Amazonia, the Iquitos Arch, runs perpendicular to the Río Juruá and coincides with a sharp line of genetic divergence in several genera of arboreal rodents between the upper and lower Juruá (Patton et al. 1997). The role of these paleoarches in vicariant events much older than the Pleistocene has been explored only superficially (see Räsänen et al. 1987; Daly and Prance 1989; Patton et al. 1997). Still other changes observable on satellite images do not coincide with any known barrier, past or present. Many of these unpredictable differences may be due to natural disturbances (Nelson 1994), pest outbreaks, or chaotic fluctuations (Condit 1996); given the slow turnover rate in some tropical forests (e.g., Hartshorn 1978) and the limited dispersal capacity of many humid forest tree species, such changes could persist for centuries (cf. Condit 1996).

TRANSITIONS

The shifts from lowland humid forests to other formations may be abrupt, such as the transition from gallery forests to grassy campos in subcoastal Amapá, but much more often they are gradual or patchy and occur along gradients of elevation, soils, and rainfall; Prance (1975) observed that transitions tend to occur in areas with moderate (2,000 mm/year) and highly seasonal rainfall. Moreover, humid forests grade into markedly different formations in different regions.

In Amazonia, each portion of the periphery may show a different kind of transition to the extra-Amazonian floras (see review in Daly and Prance 1989). In general, the northern perimeter is characterized by rather abrupt transitions to savannas, while the southern perimeter usually shows mosaiclike transitions involving semideciduous or liana forests (see figure 14.1). An exception in the north is the mosaic of campina, low caatinga (or campina forest), and high caatinga in the upper Río Negro region. In northwestern Amapá and northern Pará, the middle and upper reaches of the Trombetas and Paru rivers show extensive savannas that reach up to the Serra do Tumucumaque and the borders of French Guiana and Suriname. The transitions to the Roraima savannas are often abrupt, but there are also some patches of poorly studied semideciduous forests.

Semideciduous forests comprise the principal type of transition in Amazonia. These forests, essentially Amazonian but with some endemics and Cerrado elements, frequently occur in patches along the southern periphery of Amazonia, as well as in Roraima, where their floristic composition is different. A number of tree species show largely peri-Amazonian distributions (especially around the southern periphery in what could be considered transition zones), including the balsam of Peru, *Myroxylon balsamum* (see Granville 1992; Prado and Gibbs 1993).

Brazil borders on a part of the Guayana Highland (see figure 14.1) of sandstone plateaus (tepuis), and in places along the Venezuelan border in northwestern Roraima and northeastern Amazonas the sandy Amazonian campinas or caatingas give way to tall forests on the lower slopes of the tepuis, which in turn may grade into montane forests on the upper slopes. The upper Río Negro basin is characterized by the Amazonian caatingas, which extend beyond Brazil into southwestern Venezuela and southeastern Colombia, where a drier climate, slightly higher elevation, and other edaphic factors mark the transition to the savannas of the Llanos region.

The southern Amazonia transition zone was mapped as a distinct transition-type phytochorion by Prance (1989; see also figure 14.1) and is discussed in a separate section at the end of this essay. There are diverse transitions to cerrado vegetation, mostly semideciduous forests but also including liana forests and campo rupestre.

FLOODPLAIN FORESTS

The periodically flooded forests that flank the rivers of the Amazon and the many other river systems of northern South America are significant features of the landscape in many parts of the region. Although a given parcel of floodplain forest will show lower alpha-diversity than its upland counterpart, these forests contribute greatly to overall floristic diversity because both species composition and physiognomy change not only between regions but also locally as a function of small changes in topography (and therefore flooding regimes). The resulting mosaic of habitats is most complex and most finely divided in western Amazonia. The floodplains have ecological importance way out of proportion to their area because of their roles in capturing and cycling nutrients, harboring (and feeding) incomparably rich freshwater fish life and great invertebrate diversity, stabilizing flooded soils and landscapes, requiring remarkable physiological adaptations, and possibly constituting a source over millennia of new taxa that colonize surrounding terra firme.

The greatest extent of floodplain forests occurs in Amazonia, where Goulding (1993) estimated that they cover some 150,000 km². The principal areas of várzea or white-water floodplains are the Río Solimões-Amazonas and large areas along the Madeira and Purus rivers (figures 14.4 and 14.5); on the lower Río Branco there is a large region of igapó or black-water floodplain forest (figure 14.5).

In Amazonia, the river levels can fluctuate 7–13 m/year (Goulding 1993) and several meters more in extreme years. Still, the floodplains are mostly evergreen, even in lower-lying areas where many of the trees are completely under water during annual flooding. Most floodplain trees have no visible adaptations to having at least their

FIGURE 14.4. Várzea (Amazonian white-water floodplain forest), river receding, Riozinho do Andirá (basin of Río Purus), Acre, Brazil, April 1995 (photo by D. Daly).

FIGURE 14.5. *Top:* Early successional bands on inner (convex) curve of meandering tributary in southwestern Amazonia, Río Acre, Acre, Brazil, October 1997. The lower shrubs are *Alchornea castaneifolia,* backed by *Tessaria integrifolia,* then *Gynerium sagittatum,* then *Cecropia* sp. (photo by D. Daly). *Bottom:* Igapó (Amazonian black-water floodplain forest) at high water, Río Cueiras, Amazonas, Brazil, June 1989 (photo by J. D. Mitchell).

root systems submerged for part of the year, although in the Amazon estuary it is more common to find trees with aerial roots or pneumatophores, as in mangroves (Pires 1973).

It appears that plant species in floodplain habitats have different preferences (i.e., tolerances and competitive abilities) depending on the depth and duration of the annual floods, consequently slight topographic differences yield dramatic changes in species composition. On the lower Río Negro, Keel and Prance (1979) found distinct zonation of species along the flooding gradient, although the boundaries were not clear. Downstream from Manaus in a white-water várzea, Junk (1989) found clear zonation along a flood-level gradient of different depths and durations of inundation: (1) a low-lying shrub community at 20.5 m above average sea level, including *Coccoloba ovata*, *Eugenia inundata*, *Ruprechtia ternifolia*, and *Symmeria paniculata;* (2) a midlevel tree community at 22+ m, including *Acosmium nitens*, *Buchenavia macrophylla*, *Cecropia latiloba*, *Crateva benthamii*, *Ficus anthelminthica*, *Machaerium leiophyllum*, *Macrolobium angustifolium*, *Piranhea trifoliata*, *Pseudobombax munguba*, and *Tabebuia barbata;* and (3) a higher-ground tree community at 25+ m, including *Calycophyllum spruceanum*, *Ceiba pentandra*, *Couroupita guianensis*, *Genipa americana*, *Hevea brasiliensis*, *Lecointea amazonica*, *Sterculia elata*, and *Rheedia brasiliensis*. Vegetation zones and their boundaries are often very clear in the successional bands found in the white-water floodplains of western Amazonia, where river courses are continually changing (e.g., Puhakka et al. 1993; figure 14.5).

Floodplain forests in Amazonia do show lower alpha-diversity than adjacent terra firme (Balslev et al. 1987; Campbell et al. 1986; Duivenvoorden 1994), but their distinct composition adds significantly even to local floristic diversity. At Araracuara in Colombian Amazonia, an inventory of all plants over 0.5 m in height in 1.8 ha of terra firme and floodplain found the floristic similarity to be high (67 percent) at the family level and moderate (36 percent) at the generic level, but only 5 percent at the species level (Londoño and Álvarez 1997).

RIVER TYPES

The chemical and optical properties of rivers reflect the soil properties of the regions they drain. Different conditions of parent material, topography, age, and weathering result in rivers with different pH, transparency, sediment and nutrient load, and even temperature. These provide a basis for drawing inferences about floodplain vegetation.

Alfred Russel Wallace (1853) was the first to classify the Amazon and its tributaries into white-, clear-, and black-water rivers. The white-water rivers drain the Andes and Andean forelands; the latter are still weathering and often steep. The waters are rather opaque and neutral or slightly basic and carry a nutrient-rich sediment load, so the floodplains are essentially fertilized each year. The white-water rivers cut through soft sediments, and many are continually changing course; this and river-level fluctuations over geological time result in a tremendously complex mosaic of habitats including levees and the backslopes behind them, seasonal lakes, oxbow lakes, recent and older forested terraces, canals, curves with successional bands on new waves of depositions, and various stages of succession in places where sedimentation has progressed. Peanuts, native in southwestern Amazonia, are cultivated on recently exposed silty beaches. The character of the white waters dominates the main course of the Río Solimões-Amazonas. The floodplain forests of these rivers are called várzea forests, while those of the clear- and black-water rivers are usually referred to as igapó forests (Prance 1979).

The clear-water rivers of Amazonia drain the ancient, highly weathered Brazilian and Guayanan crystalline shields that flank the Amazon Valley. The riverbeds are rocky, and the courses are stable. The waters are transparent and nutrient-poor and have a very low sediment load. Their floodplains are not broad, and their floodplain forests are relatively species-poor.

One of three explanations for the black waters of the Río Negro proposed by the great Amazon explorer Alexandre Rodrigues Ferreira in the diaries of his three years on that river during 1785–1787 was that they are derived from plant extracts. This was echoed several years later by Alexander von Humboldt (see Goulding et al. 1988). The Río Negro drains the usually nutrient-poor, acidic, sandy lowlands of the northwestern Amazon Basin. Litter decomposition is slow and incomplete, and there is no clay to filter the humic acids. The result is tea-colored, cool, acidic, nutrient-poor waters with a low sediment load.

The igapó forests have lower within-site diversity than the várzeas and a lower diversity of habitats as well. Without a rich sediment load, there are no levees. Some of the more common tree species in the Río Negro igapós are *Eugenia inundata, Alchornea castaneifolia, Allantoma lineata, Copaifera martii, Couepia paraensis, Licania apetala, Piranhea trifoliata, Tabebuia barbata,* and *Triplaris surinamensis* (Pires and Prance 1985; Prance 1989).

Largely as a consequence of the river types, three sectors can be distinguished on the main course of the Amazon: the estuary, the lower Amazonas, and the upper Amazonas. The estuary is a labyrinth of islands and canals formed of sediments from slowing rivers. The influence of the tides penetrates far in from the mouth, and this, like the estuaries of the Río Orinoco and of rivers draining western Colombia, is a zone of tidal várzeas that are flooded twice daily by a freshwater backup. A colonizing species on freshly exposed muds is *Montrichardia arborescens* (see figure 14.4). There is a strong presence of palms in Amazonia, including *Euterpe oleracea, Manicaria saccifera, Socratea exorrhiza, Astrocaryum murumuru, Mauritia flexuosa, Raphia taedigera, Oenocarpus bacaba,* and *O. pataua.* In addition to palms, characteristic trees are *Calophyllum brasiliense, Ficus* spp., *Macrolobium acaciifolium, Pachira aquatica, Symphonia globulifera,* and *Triplaris surinamensis* (Pires 1973; Prance 1989).

In contrast, the lower Amazonas várzeas have few palms. In this region there are many grassy meadows and seasonal lakes behind forested levees, the latter being especially large near Monte Alegre in Pará (Pires and Prance 1985). Some of the most common trees of the várzea forests of the lower Amazonas are *Hura crepitans, Triplaris surinamensis, Calycophyllum spruceanum, Pseudobombax munguba,* and *Ceiba pentandra* (Pires 1973).

There are few grassy meadows in the upper Amazonas. Here, the várzea forests show the greatest complexity, diversity, and extent. The area flooded is very large between the Japurá and Içá rivers (e.g., Pires and Prance 1985). Above Parintins one begins to see the tall grass canarana (*caña brava* in Spanish), *Gynerium sagittatum*, which is a prominent species in the early successional bands on recent sediments. In western and southwestern Amazonia, other species associated with the early successional bands are *Alchornea castaneifolia*, *Salix humboldtiana*, and *Tessaria integrifolia*. The understory of shallowly inundated várzeas in this region is rich in *Heliconia* spp., Cyclanthaceae, Marantaceae, and Zingiberaceae; characteristic trees include *Carapa guianensis*, *Caryocar microcarpum*, *Hevea brasiliensis*, and *Eschweilera* spp. (Prance 1989).

OTHER INUNDATED FORESTS

Swamp forests on gleyed soils (deeply buried clayey soils in waterlogged conditions) occur sporadically in the northern South American lowlands. One formation that covers large areas in western Amazonia is the often monodominant aguajales, *Mauritia flexuosa*, or buritizais stands on poorly drained sands.

The mangroves of northern South America are extremely depauperate in species compared to many of their Old World counterparts. This is particularly true on the Caribbean and Atlantic coasts, where the mangroves tend to be more discontinuous and where the dominant trees are mostly *Rhizophora mangle*, *Avicennia tomentosa* and *A. germinans* farther inland, and *Laguncularia racemosa* in brackish, sandy areas. On the Pacific side, the mangroves are relatively more continuous and species-rich down to San Lorenzo, Ecuador, and in the Tumbes Delta in Peru (Prance 1989).

PHYTOCHORIA

Each of the following sections corresponds to one of the phytochoria mapped in figure 14.1, and while each contains a range of vegetation types, humid forests predominate in all, with two exceptions: the Magdalena–northwest Caribbean coastal region, which is rather evenly divided between dry and humid formations, and the southern Amazonia transitions to various drier formations.

CHOCÓ

The Chocó sensu lato (figure 14.6) extends along the Pacific coast of South America from southern Panama south to inland from Guayaquil in central-west Ecuador, where the desiccating Humboldt current from the south is deflected westward into the Pacific. A thin strip of mesic forest inland from the coastal plain extends into southern Ecuador. This topographically diverse region of mostly lowland tropical pluvial to moist forests originally covered approximately 180,000 km² (Herrera-MacBryde et al. 1997; Neill 1997). It is flanked by Andean montane forests to the east, tropical dry forests to the south, and gradually drier conditions to the northwest in Panama. Along the Colombian coast it is interrupted by mangroves, while southward from the Ecuadorean border the progressively drier climate supports coastal dry forests, and the western

FIGURE 14.6. *Top:* Wet tropical forest near base of Cerro Torrá, Chocó, Colombia, February 1977. *Bottom:* Atlantic coastal forest, Estação Ecológica de Juréia, São Paulo, Brazil, January 1994 (photo by D. Daly).

edge of the original mesic forests shifts inland. The Chocó flora was once continuous with that of Amazonia but has been isolated from it since the uplift of the Andes in the Tertiary, resulting in a level of endemism above 20 percent, including many species whose closest relatives are in Amazonia (Gentry 1989).

The northern (mostly Colombian) part of the Chocó experiences aseasonal climate and extremely high rainfall, often 4,000–9,000 mm/year and averaging 11,770 mm in one locality (Herrera-MacBryde et al. 1997), and in places supports pluvial forests that are physiognomically distinct from those with lower and particularly with more seasonal rainfall. The density of small and medium-sized trees is high; there are fewer lianas and more hemiepiphytic climbers; and a large number of species have remarkably large fruits or large leaves (Gentry 1986).

Some Chocó forests show less dominance of Fabaceae (Leguminosae) sensu lato than do most Neotropical forests. There is a prevalence of Clusiaceae, Arecaceae, Myrtaceae, and Melastomataceae, also Rubiaceae, Annonaceae, Sapotaceae, and (on good soils) Moraceae.

Well-known useful plants of this region include the prized endemic fruit-medicinal tree borojó (*Borojoa patinoi*; cf. Ricker et al. 1997), vegetable ivory (tágua, *Palandra [Phytelephas] aequatorialis*), and one of the early sources of rubber, *Castilla elastica*.

MAGDALENA AND NORTHWEST CARIBBEAN COAST

This complex region is bordered to the north by the Caribbean, to the east by the Coastal Cordillera of Venezuela, to the southeast by the Sierra Nevada de Mérida, and to the west by the Chocó region, whose flora is likewise poorly investigated. The natural vegetation consists of mangroves along parts of the coast, dry forests in broad bands flanking Lake Maracaibo and across to near the Gulf of Urabá (cf. Hueck and Seibert 1972), humid forests in parts of the Magdalena Valley and near Lake Maracaibo and the Gulf of Urabá, and, on the mountainous Guajira peninsula, an elevational series ranging from thornscrub to dry forest to montane cloud forest (Forero 1988).

The Magdalena Valley of northern Colombia formerly contained extensive humid forests, with large swampy areas in the northern part (Gentry 1989; Prance 1989); the flora is difficult to reconstruct because it was largely destroyed before it could be adequately inventoried. It is possible to say that the region's forests were a mix principally of Central American and Amazonian elements that reached their southeastern and northwestern limits there, respectively, plus some Chocó elements and a number of endemics. The latter are mostly woody taxa and include the endemic genera *Tetralocularia, Romeroa,* and *Brachycyclix* (Gentry 1989). *Caryocar amygdaliferum* is an example of a Magdalena Valley endemic, and *Cariniana pyriformis* occurs there and in the northern Chocó (Prance 1989).

Also formerly covered by tall humid forests, but even more often overlooked and poorly known, are the lowland regions proximal to the mangroves around the Gulf of Urabá in Colombia (extending into part of Darién in Panama) and around Lake Maracaibo flanking the Colombia-Venezuela border. One species endemic to that region is *Bursera inversa* (Daly 1993). The map in Prance (1989) includes this region in his Magdalena phytochorion.

VENEZUELAN GUAYANA–GUAYANA HIGHLANDS

The widely scattered table mountains, the steep-walled tepuis, that characterize Venezuelan Guayana are ancient sandstones and quartzites overlying the even more ancient and mostly granitic Guayana Shield. More than 90 percent of these table mountains are in Venezuela (Amazonas and Bolívar states); the rest are in northernmost Brazil, eastern Colombia, and northwestern Guyana. Most of the region is drained by the Orinoco River, while the remainder is drained by the Cuyuni and Mazaruni rivers of Guyana and by the Río Branco and Río Negro, which drain into the Amazon (O. Huber 1995).

The circumscription of the region containing the tepuis embraces a vast area of mostly forested lowlands and piedmont areas. The numerous isolated montane and summit regions above 1,300 m, collectively referred to as Pantepui or the Guayana Highlands, comprise a straightforward example of an archipelago-type phytochorion (sensu White 1979). They cover only about 6,000–7,000 km², although the area under their influence due to runoff (of both water and nutrients) and wind patterns is far larger and results in an expanded interpretation of the piedmont (see O. Huber 1995). The broader Venezuelan Guayana region, which includes Delta Amacuro state and so the Orinoco Delta, is about 83 percent forest. The vegetation map prepared by O. Huber (1995) distinguishes 54 forest units—18 of them in the lowlands—that are both physiognomically and floristically defined, reflecting the extensive if incomplete ecological and botanical surveys of the region. Some of them can be summarized as follows:

- On the lower Orinoco there are coastal mangroves and three delta zones distinguished by flooding regimes and species composition. The middle delta forests reach 25 m; they are seasonally flooded and include a number of species characteristic of the Amazon estuarine region, including *Euterpe oleracea, Manicaria saccifera, Mauritia flexuosa, Pachira aquatica,* and *Symphonia globulifera.*
- The riverine forests flanking the middle and lower Orinoco are similar in both physiognomy and composition to the várzeas of Amazonia.
- Much of southeastern Delta Amacuro and northern Bolívar experiences a drier and more seasonal climate than areas farther south. On the gently rolling hills between the Cuyuní and Caroní rivers, tall forests alternate with scrub savannas. Farther north and west one finds semideciduous forests 15–25 m tall. The sandy or rocky hilltops south of the middle and lower Orinoco are covered by low deciduous forests; dominants include *Bursera simaruba, Copaifera pubiflora, Tabebuia capitata,* and *Bourreria cumanensis.* The northwestern piedmont, between the Orinoco and the western highlands, is mostly a mosaic of savannas (disjunct with those of northeastern Bolívar) and semideciduous, medium to low forests dominated in places by *Swartzia laevicarpa, Anadenanthera peregrina, Cassia moschata,* and *Copaifera pubiflora.*
- The most extensive continuous lowland humid forests of the region (and of Venezuela)

occur on the undulating terrain of the lower and middle Caura and Paragua river basins; some of the tallest trees belong to the genera *Anacardium, Calophyllum, Protium, Parkia, Copaifera,* and *Erythrina,* and well-known economic species include tonka beans (*Dipteryx* spp.) and balata (e.g., *Manilkara bidentata* and *Pradosia surinamensis*).

• Most of the southwestern lowlands (essentially Amazonas state) experience an average annual temperature of more than 24°C and rainfall rather evenly distributed throughout the year in excess of 2,000 mm, increasing along north-south and east-west gradients. This region displays a complex mosaic of forest types and associations determined by the climatic gradients plus differences in underlying parent rock and geomorphological processes (alluvial, depositional, and erosional) that have produced a variety of soil conditions.

• Hilly central and southeastern Amazonas state is characterized by tall, dense forests; common trees are *Lecointea amazonica, Clathrotropis glaucophylla, Peltogyne venosa, Erisma uncinatum;* various species in the genera *Ocotea, Nectandra, Licania, Trichilia, Guarea,* and *Toulicia;* and a high density of tall palms in the genera *Oenocarpus, Socratea, Leopoldinia,* and *Bactris.* Much of southwestern Amazonas state is a continuation of the upper Río Negro vegetation and flora, mostly on podzolized white-sand soils, with a reduction series from relatively tall forest to open scrub, depending largely on elevation above the perched water table and the duration of periods of flooding alternating with water stress. Numerous types of scrub formations are found at elevations under 1,000 m in Venezuelan Guayana, including those on level to inclined sandstone in the Caroní-Paragua drainage, the rocky sandstones and deep sands of the Gran Sabana (up to 1,500 m), the plinthic and bauxite substrates on dry hilltops in drier northwest Bolívar, and the dune islands of the often flooded Sipapo, Atabapo, and Guainía lowlands (O. Huber 1995).

The highlands themselves are constantly humid; they receive approximately 2,000–4,000 mm of rain per year and experience no true dry season. The mean annual temperatures range between 8° and 20°C, depending on elevation, with an absolute minimum of near 0°C on the highest summits above 2,800 m.

The vegetation of the Pantepui region is physiognomically diverse—no fewer than fourteen vegetation types have been recognized in the zones above 1,300 m—but it can be broken down into four broad categories: forests, scrub, high-mountain fields, and what some call pioneer formations on rock outcrops and walls. Most of the highland forests occur on the upper slopes of the tepui bases. These include cloud forests and some extremely tall (to 60 m) forests; the most frequent families are Lauraceae, Magnoliaceae, Elaeocarpaceae, Rubiaceae, and Myrtaceae. Some summit areas also support dense, low forests (8–12 m high); some of the typical summit tree genera are *Bonnetia, Schefflera,* and *Stenopadus* (O. Huber 1995).

The scrub formations are the most diverse, both physiognomically and floristically. Taxa frequently found in the summit scrub include *Tepuianthus, Gongylolepis* (and other Asteraceae), *Clusia, Bonnetia, Maguireothamnus, Tyleria, Ilex,* and *Blepharandra.* On the higher summits one finds páramolike formations with colonies of stem-rosette Asteraceae as well as thick-stemmed *Bonnetia* spp. The high-mountain fields are rich in Rapateaceae, Xyridaceae, Cyperaceae, and Eriocaulaceae. The pioneer formations are characterized most by terrestrial rosettes of several genera of Bromeliaceae (O. Huber 1995).

GUIANAS — EASTERN AMAZONIA

East of the transverse dry belt discussed above, the region including Amapá, eastern Pará, and northwestern Maranhão states and extending through the Guianas has been treated in several ways by phytogeographers; much of the following review is taken from Daly and Prance (1989). It has strong floristic affinities with Brazil's Atlantic forests and probably was connected to them via the coast during past interglacials (e.g., Mori et al. 1981; Andrade-Lima 1982). Ducke and Black (1953) considered the region under the influence of the tides, i.e., the estuary of the Río Amazonas and the lower Río Tocantins ("Atlantic Sector"), to be a separate region strongly characterized by tidal várzeas; Marajó, at the mouth of the Río Amazonas, is a still poorly known island the size of Switzerland with vast floodplain forests and flooded campos but also large areas of humid terra firme forest (J. Huber 1898).

Ducke and Black (1953) subdivided the eastern region into the northeastern and southeastern hylaea. Hueck's (1972) homologue of the latter excluded the basin of the Río Xingu and used the low mountains separating Brazil from the Guianas to divide northeastern Amazonia into the Northeast Sector and the Guianas. Prance (1989) eventually considered the entire eastern region (minus a homologue of Hueck's Northeast Sector) to be a phytochorion distinct from the rest of Amazonia, even though it does not meet White's (1979) criteria of 50 percent endemism or 1,000 endemic species. Mori (1991) considered the Guianas and parts of adjacent Brazil and Venezuela to be a distinct floristic province. In contrast, Huber (1994) proposed a more restricted eastern Guayana province consisting essentially of the Guianas plus the Orinoco Delta and so excluding the northeastern Amazon Valley. He argued that this represents a more homogeneous geomorphologic unit, i.e., alluvial coastal plains alternating with hilly terrain on the eastern Guayana Shield, while maintaining high endemism, including the genera *Potarophytum* and *Windsorina* (Rapateaceae), *Elephantomene* (Menispermaceae), and *Lembocarpus* (Gesneriaceae). Clearly, the delimitation of this region is still under debate.

This region is more diverse in climates and topography than western Amazonia, if not biologically. The southern Guianas are bordered by modest east-west oriented mountain ranges plus some high inselbergs, and southeastern Amazonia by a number of lower north-south oriented ranges. In the northeastern portion of the region there is the "hill country" (Ducke and Black 1953) of the middle and upper courses of the Atlantic-draining rivers between the Amazon and Essequibo. Many of the sediments of the region are old and derived from the Barreira formation; one finds mostly nutrient-poor kaolinic soils on level ground and sandy podzols on slopes.

Rainfall is moderate to relatively high and seasonal to markedly so in most places; Mazaruni Station in Guyana has ca. 2,570 mm of rain and no month with less than 100 mm; on the other hand, Belém receives more rain (ca. 2,730 mm) but experiences a brief dry season, and Sipaliwini, Suriname, receives ca. 2,070 mm of rain but experiences a five-month dry season and hydrologic stress for about two months (cf. climate diagrams in Walsh 1998).

It is a region mostly of clear-water rivers draining the ancient, weathered crystalline Guayana and Brazilian shields. The rivers have well-defined, often rocky beds and stable banks, so the terra firme and floodplains are better defined than in western Amazonia; the upper reaches of many have rapids, which make transport difficult but provide excellent habitats for Podostemaceae and some other aquatics adapted to fast-flowing clear waters. The lower reaches of many have broad mouth-bays.

The vegetation in the northeastern part of this region is heterogeneous; it is dominated by humid forests, but there is also a strong presence of seasonal forests. There are also edaphic monodominant and oligarchic forests, notably the aforementioned wallaba (*Eperua* spp.) forests on white-sand soils and the mora (*Mora excelsa*) forests in low-lying and some poorly drained areas, both in Guyana. A species characteristic of dry hills from Monte Alegre to Macapá is *Peltogyne paradoxa* (Ducke and Black 1953).

The southeastern portion is more homogeneous and more thoroughly dominated by humid forests. Prominent elements here are the Brazil nut (*Bertholletia excelsa*), *Cenostigma tocantina*, *Bombax tocantinum*, and *Bauhinia bombaciflora* (Ducke and Black 1953).

Dominant families in much of eastern Amazonia and the Guianas include the Lecythidaceae, Chrysobalanaceae, Burseraceae, Fabaceae sensu lato, Lauraceae, and Sapotaceae. One of many species restricted to the eastern region as a whole is the important leguminous timber tree acapú, *Vouacapoua americana*. There are more narrow endemics in Franch Guiana and Guyana than elsewhere in this phytochorion.

WESTERN AND CENTRAL AMAZONIA

Except for the vicinity of Manaus, the flora of the remainder of Amazonia is considerably more poorly known that that of eastern Amazonia and the Guianas. Although this vast region is mapped here as a single unit, further study of its parts will permit better definition of their boundaries and may reveal that they are as distinct from each other as they are from the eastern phytochorion. Several phytogeographic subdivisions of Amazonia have been proposed since that presented by Ducke and Black in 1953 (see comparison in Daly and Prance 1989), but all have been based on that seminal work. Several of the phytogeographic subdivisions are described briefly here.

Prance's (1977) northwestern sector includes all of the upper Río Negro plus additional Amazon tributaries in Colombia; this unites a region of high rainfall and relatively aseasonal climate, black-water rivers, Amazonian caatingas and campinas-banas on very poor white-sand soils, and outlying sandstone and quartzite plateaus. The western Guayana province of O. Huber (1994) is similarly defined, but he ascribed its affinities to the greater Guayana region. Indeed, most of the region lies on the Guayana Shield, most of the soils are derived from granitic parent rock or weathering of quartzite or sandstone mountains, and several of the typical Guayanan families are well represented there (Rapateaceae, Tepuianthaceae, Theaceae, Humiriaceae, Xyridaceae).

The western hylaea of Ducke and Black (1953) is essentially an immense alluvial plain, with flat or usually gently undulating terrain, on both sides of the Río Solimões-Marañon. It is a region of great beta-diversity (habitat diversity) characterized by instability. The white-water rivers are meandering and often braided, and their courses are continually changing; as a consequence, one quarter of the lowland forests in this region show characteristics of recent erosional or depositional activity, and 12 percent of these forests in Peru are currently in successional stages along rivers (Salo et al. 1986). In this region, the distinction between terra firme and the often broad várzeas (floodplains) is blurred.

The southern hylaea of Ducke and Black (1953) includes the basins of the Tapajós, Madeira (except for its extreme southern and southwestern portions), and possibly lower Purus rivers; Prance (1977) included part of the Xingu Basin. Much of this region is on the Brazilian Shield, so most of the rivers are clear water and have stable courses. Most of the vegetation is humid nonflooded forest, but as much of this region experiences a pronounced dry season, there are some areas of semideciduous forest (notably near Belterra in the Tapajós Basin), and there are large areas of white-sand campinas on the upper Tapajós and particularly the upper Madeira.

The southwestern sector consists of the upper Purus and Juruá river basins (and possibly upper Ucayali) of the state of Acre in Brazil and contiguous portions of Peru, northwestern Bolivia, and western Rondônia state. Its floristic affinities are complex, with the result that it is the Amazon Basin's center of diversity for palms; some 75 species occur in Acre alone (E. Ferreira, unpublished data). In the upper Purus Basin, precipitation is lower and more seasonal, and a number of species represent extensions of Cerrado or Paraguayan elements or of circum- or peri-Amazonian distributions. The upper Juruá receives more rainfall and experiences less of a dry season; it includes outlying Andean foothills such as the Serra do Divisor and therefore some Andean elements. The southwestern sector is also the geographic center of the bamboo-dominated forests of southwestern Amazonia, discussed above.

A number of small plant families are endemic to Amazonia or Amazonia plus the Guianas, although few of them are widespread in the region. They include the Dialypetalanthaceae, Peridiscaceae, Duckeodendraceae, Rhabdodendraceae, Lissocarpaceae, and Polygonanthaceae. Well-known plant resources native to at least part of Amazonia include guaraná (*Paullinia cupana*), manioc (*Manihot esculentum* cvs.), rubber (*Hevea brasiliensis*, etc.), ipecac (*Psychotria [Cephaelis] ipecacuanha*), cocoa (*Theobroma cacao*), several types of curare (e.g., *Chondodendron tomentosum*), rosewood (*Aniba duckei*, etc.), Brazil nut (*Bertholletia excelsa*), and vegetable ivory (*Phytelephas* spp.).

SOUTHERN AMAZONIA TRANSITIONS

Semideciduous forests comprise the principal type of transition in southern Amazonia. These forests, essentially Amazonian but with some endemics and Cerrado elements, frequently occur in patches along the southern periphery of Amazonia.

Moving from west to east, near the border with northeastern Bolivia, the Serra dos Parecis and the Serra dos Pacaás Novos in Rondônia are long narrow strips of high ground running northwest-southeast which include often rocky campos of uncertain affinities; they probably should not be classified under *campo rupestre,* a term that carries strong geological and floristic connotations with the Serra do Espinhaço in Minas Gerais to northern Bahia (discussed below under "Savannas"). In Mato Grosso, the Amazon forest reaches south and up into many of the river's southern tributaries, gradually narrowing to gallery forests as they penetrate into the Cerrado.

The limits of Amazonia in Maranhão are confused by the dissected terrain and the roles of additional vegetation types. The moister northwestern part of the state is already markedly seasonal, and a significant minority of the trees are deciduous. The center is at present a large, highly disturbed area of mostly secondary forests and anthropic savannas, both dominated by the babassu palm (*Attalea speciosa*). To the east, babassu formations, seasonally flooded grasslands, and hardpan savannas intergrade irregularly not into cerrado but into the xerophytic caatinga vegetation of northeastern Brazil, with some Cerrado elements present. In the rolling hills and small plateaus toward the southern periphery, semideciduous forests and liana forests (including babassu forests) form a mosaic with patches of cerrado vegetation.

In southeastern Pará, the boundary of Amazonia angles northeastward, following but not including the Río Araguaia basin, crossing it and cutting across the northern tip of Goiás just south of the confluence with the Río Tocantins. This region shows the greatest extent of liana forests.

The babassu forests and liana forests are not easily distinguished and indeed have been treated either together (e.g., Pires and Prance 1985) or separately (Prance and Brown 1987). Here the former is considered an anthropic version of the latter, and judgment is reserved on the origin of liana forests.

Liana forests display moderate biomass, a highly irregular and generally lower canopy, few large trees, and a rather impenetrable understory of climbing, fallen, and winding lianas. Although their floras are poorly documented, they appear to contain Amazonian elements plus some Cerrado elements and endemics. In Amazonia, the principal liana families are the Fabaceae sensu lato, Bignoniaceae, Malpighiaceae, Dilleniaceae, and Menispermaceae; the babassu palm is of course often common; and tall trees include *Hymenaea parvifolia, Bagassa guianensis, Tetragastris altissima, Astronium graveolens* (syn. *A. gracile*), *Apuleia leiocarpa* var. *molaris, Sapium marmieri, Castilla ulei, Myrocarpus frondosus, Acacia polyphylla,* and two important economic trees: Brazil nut (*Bertholletia excelsa*) and, on more humid sites, mahogany (Pires and Prance 1985; Prance 1989). Pires (1973) estimated that liana forests occupy as much as 100,000 km² in southern Pará state between Cametá on the Río Tocantins and Altamira on the Río Xingu.

Liana forests have been the object of more discussion than research. There are serious disagreements as to whether they are natural or anthropic formations. There is no dispute that they occur in climatic and vegetational transitions between moist forests and cerrado or other savanna–savanna woodland vegetation, including other large areas in the Río Jari basin and in Roraima (Prance 1989), as well as a long inland belt in Brazil's Atlantic forest complex (e.g., Thomas et al. 1998). Pires (1973) observed that many are rich in mineral deposits. Nelson (1994) observed that postburn sites tend to have high densities of lianas and stressed the need to compare the species of burn survivors with those in liana forests. He proposed that larger patches of liana forest could be a result of larger natural fires such as El Niño fires, and small ones could be due to indigenous swidden fires, as proposed by Balée and Campbell (1990).

ATLANTIC FOREST COMPLEX

Brazil's Complexo Mata Atlântica, or Atlantic forest complex (see figure 14.6), comprises a relatively narrow and essentially continuous fringe of moist vegetation between the coast and the drier uplands of the Brazilian Shield. In southern Bahia the width of the fringe is 100–200 km (references in Thomas et al. 1998). The vegetation cover consists principally of humid forests but includes relatively small areas of floodplain forests, species-poor mangroves, transitional liana forests (inland in southern Bahia), piedmont slope forests of the Serra do Mar, and edaphic grassy or thicket-type savannas. The coastal plain supports a complex continuum of vegetation types on sandy soils referred to collectively as restinga, which ranges from sparse coastal scrub (including dune vegetation) to forest (e.g., Thomas et al. 1998; Peixoto and Silva 1997; D. S. Araújo 1997; Mamede et al. 1997). The restinga forests harbor *Allagoptera, Bonnetia*, and, in coastal Bahia, the economically important paxiúba, *Attalea funifera*.

From southern Bahia south to Río de Janeiro state, most of the lowlands do not receive more than 1,500 mm of rain annually, but in many places it is well distributed throughout the year and the relative humidity is high; in these areas the region can support well-developed humid forests with higher densities of epiphytes than in most Amazonian forests. Southward from São Paulo, most of the lowland forests are semideciduous.

In stark contrast with Amazonia, the Atlantic forest complex is the land of Myrtaceae. This family and the legumes are usually the two dominant tree families in the humid forests of southern Bahia, the tabuleiro forests on low flat tablelands in Espírito Santo, and the piedmont slope forests of the Serra da Juréia in São Paulo. At one site in southern Bahia, the Myrtaceae comprised 20–25 percent of the trees over 10 cm in diameter (Mori et al. 1983). Other important tree families include the Sapotaceae and Euphorbiaceae in southern Bahia (Thomas et al. 1997); Sapotaceae, Lauraceae, and Rubiaceae in Espírito Santo (Peixoto and Silva 1997); and Melastomataceae and Annonaceae in Juréia (Mamede et al. 1997).

The flora of the Atlantic forest complex is ancient. Climatic cycles not only affected the width of the fringe, but they apparently caused cyclical interruptions as well (e.g., Daly 1992, and references cited therein). The region shows strong affinities with Amazonia (Rizzini 1963, 1979), with which it shares a number of species; clearly the two floras were linked during more humid interglacials via the coast and gallery forests (Mori et al. 1981). Mixtures of Amazonian and Atlantic complex elements are found on the isolated moist inland granitic plateaus, or brejos, of northeastern Brazil (e.g., Andrade-Lima 1982).

For a continental flora, however ancient, the endemism in the Atlantic forest complex is remarkable and has been estimated from various study sites at approximately 41.6–44.1 percent overall (Thomas et al. 1998), 53 percent for trees (Mori et al. 1983), 74 percent for Bromeliaceae, and 64 percent for palms, plus a stunning

40.9 percent of the genera of bambusoid grasses occurring in the region (see Mamede et al. 1997).

Brazil's Atlantic coastal range forms an interrupted chain 0–25 km inland from the coastal plain in Río de Janeiro, São Paulo, and eastern Minas Gerais states. The Serra da Mantiqueira parallels and is separated from the Serra do Mar by the valley of the Río Paraíba do Sul near the intersection of the three states.

The topography of the mountain ranges is irregular and highly dissected. Several peaks near Teresópolis exceed 2,200 m, and those of Itatiaia in the Serra da Mantiqueira reach nearly 2,800 m. Rainfall varies greatly by location, ranging from ca. 1,000 mm in the lowlands to 2,000–2,500 mm in montane forests (Guedes-Bruni and Lima 1997).

The rounded and dissected hillsides of the piedmont zones have not only topographies but also soils very different from the sandy restingas of the coastal plains, and they support a distinct flora. Trees characteristic of the up to 35 m canopy include *Cariniana estrellensis, Hyeronima alchorneoides, Virola oleifera, Jacaratia spinosa, Eugenia* spp., *Pseudopiptadenia inaequalis, Moldenhawera floribunda, Chrysophyllum imperiale,* and *Aspidosperma parvifolium.*

The montane zones above 800 m support dense, humid forests rich in epiphytes and lianas. Canopy species include numerous Lauraceae, *Eugenia, Tibouchina, Solanum swartzianum, Vernonia arborea, Cabralea canjerana,* and *Symplocos variabilis,* while typical understory taxa are *Hedyosmum brasiliensis, Myrcia* spp., *Psychotria velloziana, Guatteria nigrescens,* and large numbers of *Euterpe edulis* (a source of hearts of palm) and the tree fern *Cyathea delgadii.*

Between 1,400 and 1,800 m are lower and more open forests on shallow soils and large rock outcrops. Tree species include *Miconia* spp., *Rapanea* spp., *Lamanonia speciosa, Weinmannia* spp., and *Drimys brasiliensis.* Finally, in similar substrates at the highest elevations are the campos de altitude ("high fields"), which support an open formation of herbs and shrubs, the former rich in Bromeliaceae, Cyperaceae, Xyridaceae, Eriocaulaceae, and Orchidaceae, and the latter in Melastomataceae, *Fuchsia* spp., Asteraceae, Ericaceae, and the bamboo *Chusquea pinnifolia* (Guedes-Bruni and Lima 1997).

The more important useful plants of this region include an important source of hearts of palm, *Euterpe edulis,* the timber tree caxeta (*Tabebuia cassinoides*), Brazilian rosewood (*Dalbergia nigra*), one of the piassava palms, *Attalea funifera,* and brazilwood (*Caesalpinia echinata*), whose wood is used for dyestuff and fine woodwork and for which the country of Brazil was named (e.g., Mamede et al. 1997; Thomas et al. 1998).

TROPICAL DRY FORESTS

Tropical dry forests are the dominant type of vegetation cover in a significant portion of the Neotropics, including large areas of Mexico, the Greater Antilles, northern Colombia, northern Venezuela, Brazil south of the Amazon Basin, the western

FIGURE 14.7. *Top:* Disturbed tropical dry forest, João Pinheiro, Minas Gerais, July 1984 (photo by J. D. Mitchell). *Bottom:* Tropical dry forest, Tucavaca Valley, Santa Cruz, Bolivia, July 1983 (photo by D. Daly).

Ecuador-Peru frontier region (Centro de Datos et al. 1997), southeastern Bolivia, Paraguay, and northern Argentina. They support a significant percentage of Neotropical biodiversity (figure 14.7; Mares 1992).

These forests generally have an annual preciptation of less than 1,600 mm, whereas tropical moist forests generally receive more than 2,000 mm/year. Some tropical dry forests have two dry seasons per year. Sometimes edaphic factors play a role in producing forests that resemble dry forests growing under humid climatic conditions (e.g., forests growing on limestone, such as the mogotes of Cuba).

Tropical dry forests are often similar physiognomically to tropical humid forests. The major differences are that most of the woody plants are deciduous during the dry season, they are generally shorter in stature, and they are lower in biomass, diversity, density of epiphytes, and density of lianas (see Gentry 1995). Toward the more arid extreme, as in most of the Guajira Peninsula in Colombia and in most of Lara and Falcón states in northwestern Venezuela, tropical dry forest formations generally become even lower in stature and have more open tropical arid scrub formations with an abundance of cacti.

Tropical dry forests are generally much less rich in plant species than tropical humid forests, typically ca. 50–70 species greater than 7.5 cm DBH in 0.1 ha sample plots (Gentry 1995). In some of the most arid tropical dry forest formations, monospecific stands can occur, such as *Loxopterygium huasango* or *Prosopis* sp. woodlands in the region of Tumbes in southwestern Ecuador–northwestern Peru. On the other hand,

FIGURE 14.8. Open caatinga with *Pilosocereus* cf. *gounellii* (Cactaceae), west of Fortaleza, Ceará, northeastern Brazil, January 1990 (photo by D. Daly).

tropical dry forests are often higher in endemism than adjacent humid forests (ibid.). The most diverse tropical dry forests known in the Neotropics occur at higher latitudes; examples are found in Chamela, Jalisco, in Mexico and Chuquimayo in southern Bolivia (ibid.).

An interesting structural feature of the transition between tropical humid forests and tropical dry forests or other seasonal formations such as savannas is the frequent occurrence of a belt of liana forest, i.e., forest with a greater than usual abundance of lianas. Examples of liana forest occur in the transition between the Mata Atlântica and Caatinga and between the southern Amazonian forests and the Cerrado. Their occurrence also has been noted in the Old World, e.g., liana forests in eastern Australia (e.g., Webb and Tracey 1994).

A significant difference between tropical dry and humid forests is the greater importance of seed dispersal by wind in the former (Gentry 1982; Wikander 1984; Armesto 1987). In some dry forests, the canopy is made up almost entirely of wind-dispersed tree species (Killeen et al. 1998). By contrast, in the moist forests of central French Guiana only 9.8 percent of dicotyledonous plants are wind-dispersed (Mori and Brown 1994), and in the wettest of the tropical forests, e.g., the Chocó of the Pacific coast of Colombia, more than 90 percent of the trees are animal-dispersed (Gentry 1982, 1991).

The flowering plant family composition of tropical dry forests is similar to that of humid forests, with a few exceptions, such as the greater richness and abundance in dry forests of the Capparidaceae, Cactaceae, Erythroxylaceae, Zygophyllaceae, Anacardiaceae, Asteraceae, Malvaceae, and Lamiaceae. An example of this pattern is illustrated by the Anacardiaceae. In tropical humid forest plots of 0.1–1+ ha, this family is usually represented by only a few individuals, but it is dominant or codominant in some dry forests, e.g., *Schinopsis*-dominated forests in the Chaco.

Tropical dry forests occur in many of the phytochoria covered by this essay (see figure 14.1). This important type of vegetation cover in tropical South America is introduced to the reader by highlighting the two most significant regions dominated by dry forest, the Caatinga of northeastern Brazil and the Chaco of south-central South America. For an enlightening discussion of the disjunct patches of tropical dry forest that arch across south-central South America, we refer the reader to Prado and Gibbs (1993).

Additional important areas of tropical dry forests occur in the following areas: the Tumbes region of the western Ecuador-Peru frontier; northern Colombia-Venezuela (Catatumbo-Magdalena); the fringes of the llanos region of Venezuela and Colombia; parts of northern Venezuelan Guayana; parts of Roraima, Brazil; the Tarapoto region of central Peru; some dry inter-Andean valleys; the transition zone between Amazonia and the Cerrado; the Cerrado region; the Pantanal; and the diagonal dry belt crossing through Santarém in eastern Amazonia.

COASTAL CORDILLERA OF VENEZUELA

Venezuela's Coastal Cordillera is an ancient complex of approximately 45,000 km². The uplift in this region took place in the lower Tertiary some 60 million B.P., making it almost twice as old as the Andes, with whose northeastern branch it is in contact. The western portion of the Coastal Cordillera consists of two parallel, essentially east-west oriented mountain chains, the Serranía del Litoral and behind it the Serranía del Interior. The eastern portion, separated from the west by the Depression of Unare, is the Turumiquire massif (2,590 m) and the lower cerros Humo and Patao, continuing on to Trinidad as the Northern Range (O. Huber 1997).

The topography ranges from sea level to a number of peaks over 2,000 m, reaching 2,765 m on Pico Naiguatá. The climate varies greatly as a function of elevation and orientation. The lower elevations tend to be highly seasonal and arid: Barcelona (7 m elev.) experiences hydrological stress seven months of the year (climate diagram in UNESCO 1981b), and in places the average annual temperature is over 24°C (O. Huber 1997), whereas above 800 m on the windward slopes and 1,000 m on the leeward slopes there are frequent mists and cooler temperatures (10–20°C).

Forests predominate, but the vegetation types range from coastal mangroves to coastal thornscrub, hill savanna, deciduous forest, montane forests that are semidecid-uous to evergreen, montane cloud forest, upper montane elfin forest, and upper montane scrub (subpáramo). The cloud forests are the most complex in structure, most species-rich, and highest in number of endemics. The floristic affinities of the Coastal Cordillera are with Mesoamerica, the Caribbean, and, to a lesser degree, the Andes (O. Huber 1997).

CAATINGA

The semiarid region of northeastern Brazil commonly known as the Caatinga was probably covered to a large extent in early Holocene times by tropical dry forest or woodland. However, much of the Caatinga vegetation is a shrub steppe, with extensive patches of bare soil or rock. The dominant plant cover today reflects the long history of human occupation of northeastern Brazil. Prado and Gibbs (1993) have postulated that the caatinga is the northeastern extension of a formerly much broader arc of tropical dry forest that covered a large portion of south-central South America, particularly during the period 18,000–12,000 B.P. and probably earlier dry periods as well.

There are two published opinions as to the origin of the Tupi word *caatinga*. The most commonly cited etymology traces it to a term meaning "white forest" (Andrade-Lima 1954), which probably refers to the whitish bark of many of the trees, most apparent during the dry season. Another interpretation is "open forest" or "open vegetation" (Andrade-Lima 1981). This vegetation should not be confused with the caatinga or campinarana of the upper Río Negro region of Brazil and Venezuela.

The caatinga phytogeographical region extends from about 2°54'–17°21'S, and covers an area between 600,000 and 900,000 km² (Andrade-Lima 1981; Sampaio 1995). Caatinga vegetation covers part or all of the Brazilian states of Piauí, Ceará, Río Grande do Norte, Paraíba, Pernambuco, Alagoas, Sergipe, Bahia, and smaller parts of Minas Gerais and Maranhão. Rainfall varies from 300 to 1,000 mm/year and is concentrated in a three- to five-month period. The rainfall pattern is extremely erratic, characterized by years of extreme drought, followed by an occasional year of torrential rains. The rainy season occurs during the Southern Hemisphere summer and autumn

(January–June). The climate is consistently hot, the average temperature ranging between 23° and 27°C. The average relative humidity is about 50 percent, and the rate of evapotranspiration is very high.

The geological substrate of the caatinga is severely eroded crystalline bedrock of the Precambrian Brazilian Shield and Paleozoic and Mesozoic sedimentary basins. The generally level terrain, which lies at approximately 400–500 m elevation, is broken by occasional mesas and isolated mountain ranges, which reach 1,000 m elevation. The soils of the caatinga are highly variable, ranging from shallow and rich in clay to rocky, deep, and sandy. In the southern extension of the caatinga into southern and northern Minas Gerais, soils derived from the Bambuí Group limestone are calcareous. In general, the soils of the caatinga are poor in organic material but richer in nutrients than the soils that support cerrado vegetation (Andrade-Lima 1981).

The boundaries of the caatinga are not always clear-cut. The caatinga grades into cerrado toward the south and west. In many cases the change from caatinga to cerrado follows the topographic gradient: the basins support caatinga vegetation, and the plateaus support cerrado. To the east, particularly in Bahia, the caatinga grades into a transitional sequence of tropical dry forest, tropical semideciduous forest, liana forest, and tropical humid forests (Mata Atlântica). In parts of Río Grande do Norte, caatinga extends virtually to the coast.

The caatinga is essentially a mosaic of vegetation types. The core caatinga vegetation is characterized by trees and shrubs that are often spiny, mostly deciduous during the dry season, and sometimes having small leaves with a waxy cuticle. Only a few woody species retain their leaves during the dry season, for example, *Ziziphus joazeiro* and *Maytenus rigida*. Several of the woody species store water in their swollen trunks, e.g., *Cavanillesia arborea* and *Chorisia glaziovii*. Others, such as *Spondias tuberosa*, *Sterculia striata*, and *Thiloa glaucocarpa*, store water or food in tuberous roots. Succulents such as cacti, bromeliads, and *Euphorbia phosphorea* are components of typical caatinga vegetation. This mosaic also includes gallery forest along the Río São Francisco and its tributaries, cerrado (Castro et al. 1998), humid submontane forest (brejo de altitude), and grasslands that may or may not be floristically part of the Cerrado phytogeographical region.

The core caatinga is divided by some authors into two types based on a moisture gradient: an eastern band of more humid vegetation called agreste and a drier inland type called sertão. This division is somewhat arbitrary. Andrade-Lima (1981) classified the vegetation into six units with twelve subunits, and Eiten (1983) divided it into eight physiognomic categories. A thorough review of their classification schemes is beyond the scope of this summary; a few characteristic types are mentioned here.

- Arboreal caatinga (figure 14.9). Height 8–10 m, on soils deeper and moister relative to other caatinga units; canopy coverage over 60 percent. Characteristic taxa include *Cavanillesia arborea*, *Tabebuia* spp., *Myracrodruon urundeuva* (aroeira), *Schinopsis brasiliensis* (brauna), *Aspidosperma* spp., *Pterogyne nitens*, *Cereus jamacaru*.
- Carrasco. Arboreal to shrubby caatinga with tree cover ranging from 10 to 60 percent (Eiten 1983).

FIGURE 14.9. Cerrado landscape on Chapada dos Guimarães, with patch of gallery forest at Véu da Noiva waterfall, Mato Grosso, Brazil, January 1989 (photo by D. Daly).

- Seridó. A short-grass savanna with deciduous scrub (Eiten 1983).
- Caatinga proper. The most common form of caatinga is a closed-canopy thornscrub 3–5 m in height with an occasional tree 6–8 m tall.

The species richness of individual sites within the caatinga varies markedly. In a survey of quantitative inventories cited by Sampaio et al. (1993), there were 38–195 species of woody plants over 3 cm in diameter at 50 cm above the ground on sites 3–5 ha in area. Empéraire (1989) catalogued 615 species in Serra da Capivara National Park, Piauí. The most speciose flowering plant families are the Fabaceae sensu lato, Euphorbiaceae, and Cactaceae. Among the species with the broadest distributions in the caatinga vegetation are *Commiphora leptophloeos, Myracrodruon urundeuva,* and *Anadenanthera colubrina* var. *cebil* (angico; *A. macrocarpa* in much of the caatinga literature).

In addition to the core caatinga vegetation of tropical dry forest, woodland, or shrub steppe, there are a few other discrete vegetation units. The summits of hills and isolated mountain ranges often have patches of humid submontane forest called brejos or brejos de altitude (e.g., Mayo and Fevereiro 1982). These mostly broad-leaved evergreen forests with trees up to 30–35 m tall have floristic affinities with both the Amazonian and the Atlantic humid forests. Some of the characteristic species of brejos are *Cedrela odorata, Schefflera morototoni, Pilocarpus jaborandi, Dalbergia variabilis, Machaerium amplum, Hymenaea courbaril, Manilkara rufula,* and *Symphonia globulifera* (Andrade-Lima 1981; Lleras 1997).

One distinctive plant community of the caatinga is on the sparsely vegetated rocky outcrops dominated by cacti and bromeliads (see figure 14.8). A plant community often associated with alluvial soils is the *Copernicia prunifera, Licania rigida,* and *Geoffroea spinosa* association (Andrade-Lima 1981). Stands dominated by palms are found in various parts of the caatinga, including stands of carnaúba (*Copernicia prunifera*), babassu (*Attalea speciosa* [*Orbignya phalerata*]), tucum (*Astrocaryum aculeatissimum*), and macaúba (*Acrocomia aculeata*).

The caatinga is recognized by both botanists and zoologists as a biogeographic region and center of endemism (Cabrera and Willink 1973; Andrade-Lima 1981; Cracraft 1985; Prado and Gibbs 1993; Sampaio 1995; Lleras 1997). Its status as a center of endemism for plants is supported by the presence of several endemic genera, such as *Apterokarpos* (Anacardiaceae), *Auxemma* (Boraginaceae), *Neoglaziovia* (Bromeliaceae), and *Fraunhofera* (Celastraceae). Approximately 180 woody plant species are endemic to the caatinga; examples include *Cereus jamacaru, Pilosocereus gounellei, Cyrtocarpa caatingae, Spondias tuberosa,* and *Patagonula bahiana* (figure 14.8; Mitchell and Daly 1991; Lleras 1997). It should be noted that the floristic inventory of this region is still rather incomplete, and modern taxonomic treatments for many of the key genera are not available. Contrary to the opinion of some authors (e.g., Andrade-Lima 1981; Bucher 1982), the floristic link between the Chaco and the caatinga is negligible (Prado and Gibbs 1993).

CHACO

The vast region called the Gran Chaco is one of the few places in the world where the transition between the tropics and the temperate zone is not a desert but a transition of landscapes dominated by dry forests and woodlands. A shortened version of the etymology of *Chaco* translates as "hunting land," from the Quechua word *chaku* (Prado 1993b). The Chaco phytogeographical region falls between about 17°–33°S and 57°–67°W, covering an area estimated at between 800,000 km² (Hueck 1972) and 1,010,000 km² (Galera and Ramela 1997).

Precipitation varies along an east-west gradient from 1,267 mm/year in Formosa City to about 350 mm/year in the southwestern Chaco, both in Argentina. The rainy season occurs in the months of October–April, followed by a harsh dry season in the remaining months of the year. This is the general pattern for the Chaco; however, there is a strong east-west gradient in the duration of the dry season, varying from no dry months in Formosa City to a six-month dry season in Rivadavia, Salta, Argentina.

The climate of the Chaco is strongly continental, with temperature maxima as high as 48°C, and lows to -7.2°C. During the austral winter, no point in the Chaco is free from frost except for a narrow band paralleling the major rivers in the eastern Chaco. The mean annual temperature follows a north-south latitudinal gradient varying from about 26°C in northern Paraguay to 17°C or less in the Sierra Chaco. There is an east-west gradient in lowest absolute temperatures varying from -1.1°C in Corrientes to -7.2°C in Santiago de Estero, both in Argentina.

The Chaco is a vast, virtually unbroken plain 100–500 m above sea level, which is tilted at a very slight angle to the west. It is underlain by a massive accumulation of Quaternary, Tertiary, Mesozoic, and Paleozoic sediments. The unconsolidated aeolian (i.e., formed by wind) and fluvial Quaternary sediments originated from erosion of the Andes. This layer of sediments, approximately 3,000 m thick, presses down on the deeply sunken pre-Cambrian Brazilian Shield. The Chaco plains are interrupted by the emergence of hills and mountains such as the Sierra Pampeanas, Argentina, Cerro León, Paraguay, and the Sierra Chiquitos, Bolivia, which range in elevation from 400 to 2,400 m above the plain. The soils of the Gran Chaco are derived from parent materials consisting of aeolian, fluvial, lacustrine, or marine sediments. In general, the size of the soil particles remains the same along an east-west gradient in the Chaco. The soils are generally fertile, but the main limitations of Chaco soils are poor drainage, high salinity or alkalinity, occasional severe flooding, and some areas of hardpan (Prado 1993a).

The Chaco sensu stricto vegetation is bounded by several different vegetation types, some of which are transitional in floristic composition, particularly along the eastern Chaco boundary. The Chaco grades into the arid Monte vegetation in Argentina (Cabrera 1976). The westernmost type of Chaco, the Sierra Chaco, is bordered by transition forests (e.g., tipa-pacará and palo blanco forests, which will be discussed later), which in some parts separate the Chaco from the Yungas and the Tucumán-Bolivian montane forests (Cabrera 1976; Prado 1993b).

The dominant form of Chaco vegetation is characterized by a forest of trees and shrubs that are often spiny and mostly deciduous in the dry season. The understory is often shrubby, with a variety of herbs and grasses growing in association with bromeliads and cacti, some of which are arborescent. As is the case with the caatinga, the Chaco is a mosaic of vegetation types. This mosaic includes annually flooded gallery forests, the adjacent selva de ribera, wetlands, palm savannas, grasslands, halophytic shrubby steppes, cactus stands, and communities dominated by *Prosopis* spp. (e.g., algarrobales).

A unique aspect of Chaquenian vegetation is the dominance of the genus *Schinopsis* (Anacardiaceae). In the more humid, frequently waterlogged woodlands of the eastern Chaco, the dominant species is the simple-leaved *Schinopsis balansae*. In the semiarid central and western Chaco, the dominant species is the multifoliolate *S. quebracho-colorado,* and on the lower slopes of the western sierras, the dominant species is another multifoliolate species, *S. haenkeana*.

The vernacular name for *Schinopsis* species in the Chaco is *quebracho* (from *quebra hacha,* or "break axe"), which refers to the very hard, tannin-rich wood (Prado 1993a). The quebrachal of *Schinopsis balansae* is a forest with the canopy dominated by *S. balansae* in association with *Aspidosperma quebracho-blanco* and *Caesalpinia paraguariensis.* This semiopen forest grows in seasonally waterlogged, saline soils. Its understory often includes thickets of *Schinus fasciculata,* various Cactaceae, *Capparis* spp., and *Celtis* spp. The branches are often festooned with lichens and *Tillandsia* spp.

In addition to the dry forests and woodlands previously described, the Chaco sensu stricto includes several other vegetation types, such as extensive stands of *Prosopis* spp. and the palm stands and palm savannas of *Copernicia alba* and halophytic grasslands dominated by *Spartina argentinensis* (with or without trees; this is in the same genus of grasses that dominate north-temperate salt marshes).

Occupying a narrow belt between the Chaco and the more humid submontane forests known as the Yungas and the Tucumán-Bolivian forests are the transition forests (Prado 1993b; Cabrera 1976). The two major forest types in this category are the palo blanco forests dominated by *Calycophyllum multiflorum,* which are taller than Chaco, reaching 30 m in height, and the tipa-pacará forest of the legumes *Tipuana tipu* and *Enterolobium contortisiliquum.* The transition forests have virtually disappeared due to human activity; they are floristically related to the arc of tropical dry forest that extends from the Caatinga in the northeast of Brazil to the foothills of the Andes (Prado and Gibbs 1993; Prado 1993b).

Another transitional vegetation type is the austro-Brazilian transitional forest. Its well-developed to intermediate soils support an association of woody plants such as *Myracrodruon balansae, Myrcianthes pungens, Eugenia uniflora, Tabebuia impetiginosa,* and a weird Sapindaceae, *Diplokeleba floribunda.* Sometimes *Schinopsis balansae* is present. Additional transitional vegetation types are discussed in great detail in Prado et al. (1992) and Dubs (1992a).

Some of the most important plant families of the Chaco sensu stricto are the Fabaceae sensu lato (e.g., *Acacia, Prosopis, Caesalpinia, Cercidium,* and *Geoffroea*), the Anacardiaceae (*Schinopsis, Lithrea, Schinus*), Apocynaceae (*Aspidosperma*), Arecaceae (*Copernicia, Trithrinax*), Zygophyllaceae (*Bulnesia*), Rhamnaceae (*Ziziphus*), Cactaceae (*Opuntia, Cereus, Stetsonia, Pereskia, Quiabentia*), Bromeliaceae (*Dyckia, Bromelia, Puya, Ananas*), and Poaceae (*Eleoneuris, Paspalum, Chloris, Spartina*) (Galera and Ramela 1997). The Chaco has several endemic genera such as *Lophocarpinia, Mimozyganthus,* and *Stenodrepanum.* The Chaco is the domain of a few wild relatives of the pineapple (*Ananas* spp.; ibid.).

SOUTHERN BRAZILIAN REGION

The southern Brazilian region (Paranaense Province of Cabrera and Willink 1973) lies in Brazil south of the Cerrado and west of the Atlantic forests; it also includes northeastern Argentina (Misiones) and eastern Paraguay (see figure 14.1). The original vegetation consisted primarily of a mosaic of forests dominated by *Araucaria angustifolia,* semideciduous forests, temperate grasslands (particularly in Río Grande do Sul), and small outliers of cerrado in São Paulo and Paraná. Some taxa typical of this region include *Podocarpus lambertii, Dicksonia sellowiana, Holocalyx balansae, Tabebuia ipe, Myrocarpus frondosus, Balfourodendron riedelianum, Syagrus romansoffiana, Myracrodruon balansae, Cedrela fissilis, Cariniana estrellensis,* and yerba mate (*Ilex paraguariensis*) (Takhtajan 1986; Keel 1997; Leitão-Filho 1997; Hueck and Seibert 1972; Spichiger et al. 1995; Reitz 1965).

SAVANNAS

In South America, savannas cover an area of approximately 2 million km² (Sarmiento 1983). They comprise the second most extensive vegetation type in tropical South America after tropical humid forests. The largest expanse of savanna in South America is the Cerrado phytogeographic province (Eiten 1978) of central Brazil and adjacent areas of Paraguay and Bolivia. The second largest savanna-dominated region is the Llanos of eastern Venezuela and Colombia. Smaller areas of savanna include the Llanos de Moxos of northern Bolivia, islands of savanna surrounded by humid forest in the Amazon Basin, the Gran Sabana of Venezuelan Guayana, the Roraima-Rupununi savannas, the Guianan savannas, and disjunct areas of campo rupestre. The Pantanal region of Brazil and contiguous Bolivia also includes extensive areas of hyperseasonal savanna and cerrado vegetation; this is discussed in a separate section.

The area of savanna in South America has apparently expanded and contracted during cool-dry and warm-wet intervals during the Quaternary. It has been postulated that two or more recent episodes of savanna and other dry vegetation type expansion and concomitant tropical humid forest contraction occurred during glacial periods in the Northern Hemisphere. The most severe episode of cooling and drying probably occurred during the Northern Hemisphere Würm-Wisconsin ice age (Brown and Ab'Sáber 1979; van der Hammen 1982; Dickinson and Virji 1987; Absy et al. 1991).

Savannas are tropical grasslands varying from treeless areas dominated by grasses, sedges, and other herbs to dense woodlands with grass-dominated understories. Sarmiento (1983) divided Neotropical savannas into three basic types. The most widespread type of savanna is the seasonal savanna, which is maintained by a severe dry season and frequent fires (e.g., central Brazilian Cerrado and the Llanos of western Venezuela and contiguous Colombia). The second type of savanna, the hyperseasonal savanna, is a product of excessive drought and fires, with severe flooding during the wet season. These savannas are particularly prominent in the poorly drained bottomlands of the Grão Pantanal, the Llanos de Moxos, the Roraima-Rupununi savannas, and the llanos of Colombia-Venezuela. Hyperseasonal savannas often have large numbers of monocots, represented by the Commelinaceae, Costaceae, Eriocaulaceae, Iridaceae, Rapateaceae, and Xyridaceae. They are also characterized by palm stands, especially *Mauritia flexuosa*. The third type of savanna is the semiseasonal savanna, which occurs as disjunct patches surrounded by tropical humid forest. These occur in a more humid climate with one or two short dry seasons and less frequent fires (e.g., some of the Amazonian and Guianan savannas).

The savanna flora contains a suite of species that are widespread throughout the

FIGURE 14.10. *Top:* Cerrado landscape with esteiro (permanently wet meadow; right foreground) between gallery forest (right background) and cerrado (left foreground), near Gama, Distrito Federal, Brazil, July 1984. *Bottom:* Cerrado landscape with the widespread Neotropical savanna species *Salvertia convallariodora* (Vochysiaceae) in foreground, São Gonçalo, Minas Gerais, Brazil, July 1984. (Photos by J. D. Mitchell.)

Neotropics, such as the woody plants *Byrsonima coccolobifolia, B. crassifolia, B. verbascifolia, Bowdichia virgilioides, Anacardium occidentale* (includes *A. microcarpum*), *Curatella americana, Hancornia speciosa,* and *Salvertia convallariodora* (figure 14.10). Floristic richness in Neotropical savannas varies considerably, from core central Brazilian cerrado plots with more than 100 species of trees and shrubs (not including perennial herbs and subshrubs with woody underground structures) to plots in some Amazonian savannas with only 7. A comparison of data from 98 savanna sites from the Cerrado and Amazonia, which lists 534 species of trees and large shrubs, concludes that the Amazonian savannas are depauperate, disjunct islands of cerrado vegetation (Ratter et al. 1996). In contrast, Pires and Prance (1985) state that savannas of the upper Río Trombetas (Ariramba and Tiriós) and of Roraima-Rupununi are species rich and that the latter show considerable endemism while lacking *Hancornia* and *Salvertia*. Both studies agree that the savannas of Amapá are particularly species-poor.

CERRADO

The Cerrado phytochorion covers an area of about 1.8 million km² (Ab'Sáber 1971; Coutinho 1990). It covers most of central Brazil, from southern Maranhão and Tocantins in the north to southern cerrado outliers in São Paulo and Paraná, and extends west to include parts of Paraguay and eastern Bolivia. Rainfall is generally higher than in the caatinga region to the northeast and less than that of the humid forests to the southeast and northwest. Annual precipitation averages about 1,500 mm and varies from 750 mm at the caatinga ecotone to more than 2,000 mm at the Amazonian transition zone. The dry season lasts about three to five months (May–September) during the austral winter. At its southern limit, cerrado vegetation is occasionally affected by frosts, which can severely damage woody plants. Occasional severe frost may be a factor that sets the southern limits of cerrado vegetation (Sarmiento 1983).

The Cerrado is essentially a landscape of savanna-covered plateaus and tablelands, or chapadas (figure 14.9), that are remnants of a vast mid-Tertiary peneplain separated by broad river valleys with gallery forests. The deep bedrock underlying most of the Cerrado are of the pre-Cambrian Brazilian Shield. Much of the Cerrado in central Brazil occupies terrain ranging from 1,000 to 1,500 m elevation. Cerrado outliers in southern Brazil and the Pantanal of southwestern Brazil and Paraguay occur at much lower elevations (Sarmiento 1983).

Soil fertility and drainage play preeminent roles in determining the large-scale vegetation patterns in the Cerrado. Soils supporting the dominant savanna vegetation are dystrophic, i.e., very poor in nitrogen, phosphorus, calcium, etc., and high in aluminum (Ratter et al. 1996; Oliveira-Filho and Ratter 1995; Coutinho 1990; Goodland 1971; Filgueiras 1997). Mesotrophic soils, which are much richer in calcium and other essential nutrients, support tropical dry forests (deciduous or semideciduous) and sclerophyllous woodlands (e.g., mesotrophic cerradão) under relatively undisturbed conditions. These woodlands and forests are usually in valleys where erosion of the tablelands has cut into nutrient-rich underlying rock (e.g., silts and mudstones), or they are growing on limestone-derived soils (Oliveira-Filho and Ratter 1995; Prado and

Gibbs 1993). Seasonally waterlogged soils are the substrate for hyperseasonal savannas and esteros (sedge- and grass-dominated, often permanently wet campos). Moist soils adjacent to rivers support a dendritic network of gallery forests. Cerrado is frequently separated from gallery forest by a narrow band of esteiro (figure 14.10; Sarmiento 1983).

In Brazil, the core vegetation of cerrado is a complex savanna continuum varying from treeless grassland to dense woodland. This continuum is often broken down into five physiognomic units (e.g., Eiten 1972; Sarmiento 1983): campo limpo ("clear field"), a treeless grassland with herbs, a few small shrubs, and suffrutices (subshrubs); campo sujo ("dirty field"), a grassland with less than 2 percent woody plant cover; campo cerrado, a grassland with 2–15 percent woody plant cover; and cerrado, an open woodland with 15–40 percent woody plant cover; and cerradão, a sclerophyllous woodland with more than 40 percent woody plant cover and 3,000–4,000 trees per ha.

The tortuous and often grotesque habit and thick bark of most cerrado trees and shrubs belies the essential role of fire in the ecology of the Cerrado (Coutinho 1990). Cerrado differs from many other types of dry tropical vegetation in that many of its woody plants have mesophyllous, thickly coriaceous leaves that are retained during the dry season. The more open forms of cerrado vegetation depend on fire for their maintenance. In the past, humans have undoubtedly played an important role in the management and possibly expansion of cerrado vegetation. The numerous and complex adaptations of cerrado plants to fire, however, in addition to the high degree of endemism in the region, suggest a long history of fire ecology predating man by millions of years. Many woody plants escape destruction from fire by having most of their biomass beneath the soil surface. Frequently, these species have woody underground structures called xylopodia or lignotubers. These structures are large and have dormant buds at or below the ground surface that are capable of resprouting shortly after a fire has scorched the surface (ibid.). Sometimes these xylopodia appear to be underground tree trunks; two notable examples of underground trees are *Andira humilis* and the dwarf cashew, *Anacardium humile* (López Naranjo 1977; Mitchell and Mori 1987; Rawitscher and Rachid 1946). In the case of *Anacardium humile* and many other suffrutices with large xylopodia, the only structures that emerge above the ground are the inflorescences and a few orthotropic leafy shoots. These suffrutices often have arborescent relatives in the Cerrado and sister taxa that may be giant trees in the Amazon Basin; the counterpart of *Anacardium humile* in Amazonia is *A. spruceanum*.

It has been assumed that some xylopodia function as water storage organs. A study of the xylopodia of *Isostigma peucedanifolium* and *Lantana montevidensis* revealed that the water content of the xylopodia remained relatively constant during both dry and wet seasons. This study found that after a fire the xylopodia absorbed minerals and other nutrients released into the soil by the burning of vegetation (Coutinho 1990). The diversity and abundance of suffrutices with xylopodia or similar adaptations for resistance to and resprouting after fires comprise one attribute of Cerrado phytogeography that sets it apart from Amazonian and other savanna regions of the Neotropics (Pires and Prance 1985). The only other region of the world with a similarly rich flora of underground trees is the Zambesian region of Africa (White 1976).

Even though the dominant vegetation of the Cerrado is a savanna continuum, a major portion of the region was originally covered by woodlands and forests. There may have been times in the past several thousand years when dry forests were the dominant type of vegetation (Prado and Gibbs 1993; Oliveira-Filho and Ratter 1995). It is very difficult to reconstruct the vegetation history of the Cerrado due to the paucity of palynological studies. A recent palynological study by Ledru (1993) carried out at Salitre, western Minas Gerais, suggests that between 33,000 and 25,000 B.P. that region was cooler and moister than today and was covered by semideciduous forests that included southern Brazilian genera such as *Araucaria, Podocarpus, Symplocos,* and *Drimys.* Remnants of these formerly extensive forests can still be seen today.

The type of forest present is determined by the edaphic gradient. Cerrado grades into two different types of cerradão (sclerophyllous woodland), depending on whether the soils are dystrophic or mesotrophic. In the transition zone between the Amazon forest and the Cerrado, dystrophic cerradão is prevalent. Characteristic species of dystrophic cerradão include *Hirtella glandulosa, Emmotum nitens, Aspidosperma macrocarpon, Vochysia haenkiana,* and *Xylopia sericea.* Two indicator species for mesotrophic cerradão, common in central Brazil and the fringes of the Pantanal, are *Magonia pubescens* and *Callisthene fasciculata.* The vegetation types dystrophic cerradão and mesotrophic cerradão are based on Oliveira-Filho and Ratter's (1995) modification of the Brazilian government vegetation classification scheme (Veloso et al. 1991). Tropical dry forests (including deciduous and semideciduous forests) are widespread and grow in mesotrophic soils. They are floristically linked to the caatinga, the sub-Andean dry forests, and, in the southern areas of the Cerrado province, to the southern Brazilian (Paranaense) forests. Some characteristic species are *Anadenanthera colubrina, Tabebuia impetiginosa, Myracrodruon urundeuva, Dilodendron bipinnatum, Sterculia striata,* and *Enterolobium contortisiliquum.* The extensive networks of gallery forests in the river valleys of the Cerrado comprise a corridor linking the Amazonian forests to the Atlantic coastal forests of Brazil. Many of the woody species in the gallery forests are habitat generalists, e.g., *Tapirira guianensis, T. obtusa, Protium heptaphyllum, Hymenaea courbaril,* and *Virola sebifera.* Some of the gallery forests contain local endemics, viz., *Virola malmei, V. urbaniana,* and *Hirtella hoehnei* (Oliveira-Filho and Ratter 1995). The central and southern gallery forests are floristically linked with the southern Brazilian semideciduous forests. In Mato Grosso, some Amazonian tree genera reach their southern limit in gallery forests, e.g., *Hevea,* a genus considered Guianan-Amazonian.

The Cerrado is recognized by both botanists and zoologists as a major center of biodiversity. For example, Dias (1992) estimated that 160,000 species of plants, animals, and fungi occupy the cerrado biome. It is also an important center of endemism for plants and animals (Cracraft 1985; Heringer et al. 1977; Cabrera and Willink 1973; Filgueiras 1997; Ratter et al. 1996). It is very heterogeneous in floristic composition. In a comparative study of 98 areas of cerrado and Amazonian savannas, 158 (30 percent) of 534 species of trees and shrubs occurred at a single site only. Only 28 species were present at 50 percent or more of the study sites. Local areas of cerrado can be extremely rich in plant species, for example, Heringer (1971) found more than 300 species in one hectare of protected cerrado near Brasília, D.F.

Some of the most important plant families in the Cerrado are the Fabaceae sensu lato, Poaceae, Asteraceae, Orchidaceae, Rubiaceae, Myrtaceae, Melastomataceae, and Apocynaceae. Among the more species-rich genera are *Paspalum, Panicum, Habenaria* (including *Platanthera*), *Vernonia* sensu lato, *Chamaecrista, Senna,* and *Hyptis.* Some of the most widespread species of trees and shrubs, occurring on at least 50 percent of 98 sites sampled throughout the Cerrado, include *Qualea grandiflora, Annona crassiflora, Astronium fraxinifolium, Bowdichia virgilioides, Lafoensia pacari, Kielmeyera coriacea, Hymenaea stigonocarpa, Copaifera langsdorffii, Caryocar brasiliense, Machaerium acutifolium, Tocoyena formosa, Tabebuia aurea,* and *Byrsonima coccolobifolia* (Ratter et al. 1996). Some of the characteristic suffrutices include *Anacardium humile, A. nanum, Andira humilis, Annona pygmaea, Parinari obtusifolia,* and *Stryphnodendron confertum.* Grass genera include *Andropogon, Aristida, Axonopus, Elionurus, Paspalum,* and *Trachypogon,* which are widespread in most Neotropical savannas (Sarmiento 1983). Some of the characteristic palms include *Butia leiospatha, Syagrus acaulus, Attalea exigua, Astrocaryum campestre,* and *Acanthococos emensis* (Sarmiento 1983; Coutinho 1990). The Cerrado is the domain of the important fruit piquí (*Caryocar brasiliensis;* F. D. Araújo 1995; Filgueiras 1997) and wild relatives of the cashew of commerce (*Anacardium* spp.; Mitchell and Mori 1987).

LLANOS OF VENEZUELA AND COLOMBIA

The Llanos, covering some 500,000 km^2 (Sarmiento 1984), is the second most extensive region dominated by savanna in South America after the Cerrado (see figure 14.10). The vegetation of the Llanos and its relationship to the regional physiography were comprehensively reviewed by Sarmiento (1983). This region is a huge plain drained by the Orinoco River and its tributaries, enclosed to the west and north by the Andes and the Coastal Cordillera, respectively. The Guaviare River in central Colombia forms the southern boundary with Amazonia. It is bordered to the east by the Orinoco Delta and the Guayana Highlands. Most of the Llanos is underlain by Quaternary alluvial and aeolian sediments. The major area of tropical dry forests in the Llanos occupies terrain underlain by Tertiary clays and shales which extend south from the Coastal Cordillera. Annual rainfall varies from 1,000 mm in parts of the eastern Llanos to 2,200 mm in the southwest corner at the Guaviare River. The dry season becomes less pronounced across a northeast-southwest gradient, decreasing from five to six months to one to two months.

The Llanos can be divided into four regions:

THE PIEDMONT

This is a narrow strip bordering the Andes and the western portion of the Coastal Cordillera. It is a mosaic of semideciduous forests, savannas, and gallery forests, growing on a complex of alluvial terraces and fans. Savanna woodland usually occupies the oldest Quaternary deposits, and the youngest and lowest alluvial terraces are usually occupied by gallery forests. Hyperseasonal savannas in the piedmont are fairly local in their distribution. The woody plants of the piedmont are mostly widespread species typical of most Neotropical savannas, such as *Bowdichia virgilioides, Byrsonima coc-*

colobifolia, B. crassifolia, Cochlospermum vitifolium, Curatella americana, Genipa americana, and *Xylopia aromatica.*

THE HIGH PLAINS

Mesas are the dominant landform of much of the eastern and southern llanos (a large portion of the Colombian llanos). They are the eroded remnants of a vast tableland of late Pleiocene or early Pleistocene age. The predominant vegetation on the mesas, which are often 200–300 m above sea level, is an open savanna to savanna woodland with widespread tree species such as *Bowdichia virgilioides, Byrsonima crassifolia,* and *Curatella americana.* Treeless, seasonal, and hyperseasonal savannas dominate the wide valleys separating the mesas. Hyperseasonal savannas, with the locally abundant palm *Copernicia tectorum* being the only woody plant present, are common in the valleys of the high plains region. Savanna parkland, consisting of groves of tropical dry forest (including species such as *Copaifera officinalis* and *Vochysia venezuelana*) surrounded by treeless hyperseasonal savanna, is another common vegetation type in this region.

THE ALLUVIAL OVERFLOW PLAINS

A huge depression in the central part of the Llanos, bounded by the piedmont to the west and north and by the high plains to the south and east, is covered by the alluvial overflow plains. Its vegetation varies tremendously along an extremely slight elevational gradient of 1–2 m. It is reminiscent of the well-known wetland ecosystems, the Everglades of Florida and the Okavango Delta of Botswana. It is the northern South American counterpart of the hyperseasonal savannas and wetlands of the Pantanal. Both the alluvial overflow plains and the Pantanal are characterized by enormous flocks of waterfowl, ibises, and storks and large herds of capybaras, the world's largest rodent.

During the rainy season, most of the region is flooded for periods of a few weeks to several months. The flooding is not caused by streams overflowing their banks but rather by the accumulation of rainwater, which drains very slowly on the remarkably flat terrain. The upper portions of the elevational gradient support savannas with scattered trees, savanna woodlands, or groves of trees (matas) in a hyperseasonal savanna parkland landscape. These groves of trees typically include tree species such as *Spondias mombin, Annona jahnii, Platymiscium pinnatum, Vochysia venezuelana,* and *Pterocarpus podocarpus.* The lowest areas in the elevational gradient are characterized by species-poor esteros, commonly dominated by grasses such as *Paspalum fasciculatum* or *Leersia hexandra* and *Hymenachne crassicaulis,* or sometimes extensive herbaceous wetlands. The riverbanks support either savanna or gallery forest, depending on the height of the water table.

THE AEOLIAN PLAINS

A continuous belt characterized by extensive dune fields extends from the upper Meta River in central Colombia to the Cinaruco River in southern Venezuela. These plains are apparently the remnants of an arid landscape whose origin dates back to the

Würm-Wisconsin glacial period (Tricart 1974). A seasonal savanna with a very depauperate tree flora grows on these dunes (often *Byrsonima crassifolia* is the only tree species present). The ground layer is dominated by *Trachypogon ligularis* and *Paspalum carinatum*. Hyperseasonal savannas and esteros occupy depressions between the dunes and the plains. The hyperseasonal savannas are often dominated by grasses of the genus *Mesosetum,* and the esteros often include morichales (stands of *Mauritia flexuosa*).

The Llanos flora includes many of the widespread woody plants of the Cerrado and widespread Amazonian savannas. Its woody plant flora is depauperate in comparison with the Cerrado. Some of the widespread grass and sedge genera that dominate the ground layer of the llanos include *Andropogon, Aristida, Axonopus, Panicum, Paspalum, Mesosetum, Elionurus, Sporobolus, Trachypogon, Bulbostylis, Rhynchospora,* and *Scleria*.

As is the case in the Cerrado, fire plays an extremely important role in maintaining the savanna landscape of the Llanos. According to Sarmiento (1983), the Llanos flora is perhaps one of the best-known savanna floras in the Neotropics.

RORAIMA-RUPUNUNI SAVANNAS

The savannas of Roraima, Brazil, are continuous with the Rupununi savannas of Guyana, and together they cover some 54,000 km² (Pires and Prance 1985). These are not Amazonian savannas in the sense that they occur in a moist climate; the region is the northern end of the roughly northwest-southeast transverse dry belt that crosses the Amazon east of Manaus (see map in Pires-O'Brien 1997), and it experiences a long dry season and relatively low rainfall (Richards 1998). The southern part of the region is flatter, while the terrain of the northern portion is more undulating and the savannas lie between hills (Pires and Prance 1985).

These are mostly well-drained savannas with few trees. As in parts of Amazonia, there are many swampy areas of sands underlain by an impermeable hardpan that are dominated by *Mauritia flexuosa*. Some of the swamps have *Xyris,* Eriocaulaceae, *Cephalostemon, Abolboda, Rapatea,* and some *Drosera* spp. The well-drained savannas share many tree species characteristic of the Cerrado, except for *Hancornia speciosa* and *Salvertia convallariodora,* but there is also appreciable endemism (Pires and Prance 1985).

LLANOS DE MOXOS

The Llanos de Moxos (Mojos in some references) region is a mosaic of forests (ca. 120,000 km²) and savannas (ca. 150,000 km²) and various associated wetlands in northeastern Bolivia. It is the third largest complex of savannas in South America after the Cerrado and the Llanos of Venezuela and Colombia. This region is primarily in Beni Department and extends into other adjacent departments, 11–16°S, 64–69°W. The brief description here of the region and its vegetation is based on the relatively scant literature available (Beck and Moraes 1997; Beck 1984; Haase and Beck 1989).

It is a flat plain, between 130–235 m elevation, in the lowlands northeast of the Andes. The three main rivers that drain the Llanos de Moxos are the Beni, Mamoré, and Guaporé (or Iteñez), which unite in the north to form the major southwestern tributary of the Amazon, the Río Madeira. This huge plain is bounded to the east by the Serra dos Pacaás Novos and Chapada dos Parecis, Brazil (Rondônia and Mato Grosso states). Mean annual rainfall increases along an east-west gradient from 1,300 mm/year in Magdalena to over 2,000 mm/year in Rurrenabaque. The austral summer is characterized by heavy rainfall and a two- to three-month dry season that usually lasts from June to August. The mean annual temperature is ca. 26°C, but during the austral winter, cold fronts (surazos) frequently chill the air. The lowest temperature recorded in the Llanos de Moxos was 6°C at Santa Ana de Yacuma.

The Llanos de Moxos are underlain by extensive Andean-derived alluvial sediments dating from the late Pleistocene to the Quaternary. Because of its low relief, the vegetation of the region varies tremendously along an extremely slight elevational gradient. This is particularly true for the central portion of the region between the Beni and Mamoré rivers, which is reminiscent of the alluvial overflow plains of the Pantanal and the Llanos of Venezuela and Colombia.

This central portion is dominated by esteros and hyperseasonal savannas, with an abundance of associated wetlands, some of which are dominated by sedges and *Thalia geniculata*, others by *Cyperus giganteus*, and palm stands of *Copernicia alba* are common in some areas. Beck (1984) classified more than 30 plant communities with more than 400 species of vascular plants within a small portion of the central Llanos de Moxos. The savannas of the northwestern Llanos de Moxos, in the vicinity of the Río Beni, occupy acidic soils, are depauperate in woody plant diversity, and show termite mounds as a characteristic feature. One type of mostly treeless savanna is characterized by a *Leptocoryphium lanatum–Trachypogon plumosus* association; another type is characterized by the melastomataceous shrub *Macairea scabra*, with a ground layer mostly of *Mesosetum penicillatum* and *Bulbostylis juncoides*. Palm stands of *Mauritia flexuosa* or sometimes of *Mauritiella aculeata* are common. Gallery forests line both black-water and white-water rivers in this region. A detailed study of savanna in an area of ca. 2,000 km² just west of the Río Beni identified ca. 600 species of vascular plants and classified several vegetation types (Haase and Beck 1989). The northwestern savannas are floristically and physiognomically similar to portions of the Llanos of Venezuela and Colombia and other northern South American savannas. The savannas of the northern portion, between the Beni and Iteñez-Guaporé or Mamoré rivers, are predominantly seasonal savannas that are floristically and physiognomically similar to the Cerrado.

The northern portion of the Llanos de Moxos meets the southernmost extension of the Amazon forest. Two of the typical Amazonian species are the Pará rubber tree (*Hevea brasiliensis*) and the Brazil nut (*Bertholletia excelsa*). A more humid forest covers the foothills of the Andes bordering the Llanos de Moxos to the west. Gallery forests and parkland savannas (i.e., with forest islands) are widespread in the savanna-dominated portions of this region. The southern portion of the Llanos de Moxos supports a mostly evergreen forest with a strongly seasonal climate.

The vascular flora of the Llanos de Moxos is estimated to consist of more than 5,000 species, of which ca. 1,500 are found in the savannas. The flora includes widespread Neotropical savanna, cerrado, and Amazonian species, primarily. The number of endemic species is probably low (Beck and Moraes 1997). The flora includes wild relatives of the pineapple (*Ananas comosus*) and numerous cultivars of the peanut (*Arachis hypogaea*).

PANTANAL

The Pantanal is treated as a regional mosaic or transition in Prance's (1989) description of South American phytochoria. It is not a center of endemism; instead, it is a region at the vertex of the Cerrado, Amazonia, Gran Chaco, and southern Brazilian floras. Based on the additional floristic elements it contains, it must also lie in current or historic migration routes of the caatinga of northeastern Brazil (Prado and Gibbs 1993).

The Pantanal is considered by many to be one of the most extensive and significant wetland-savanna complexes in the world. It is an alluvial plain of 150,000–170,000 km² in the upper Paraguay River basin. The Pantanal is situated between 16–22°S and 55–58°W. It is bordered by the Cerrado to the north and east, the Serra de Bodoquena to the south, and the Paraguay River to the west. Most of the Pantanal is in Brazil (Mato Grosso and Mato Grosso do Sul), but a small portion crosses the Paraguay River into southeastern Bolivia. It is a relatively flat plain that lies mostly at 100 m elevation, tilted very slightly in a northeast-to-southwest direction. This very level terrain is broken in a few places by a few emergent pre-Cambrian uplands, such as the Serra de Amolar along the Paraguay River. Annual rainfall of 1,000–1,400 mm falls mostly during the rainy season from November to April. The harsh dry season lasts three to four months during the austral winter. This brief review is based primarily on Dubs (1992a,b), Prance and Schaller (1982), Frey (1995), Sarmiento (1983), and Ratter et al. (1988b).

The Pantanal is underlain by a sediment-filled subsidence zone extending from the Llanos of Venezuela and Colombia south to the Gran Chaco. This subsidence zone was formed at the time of the Andean orogeny. The Pantanal's complex mosaic of wetlands, savannas, and forests is a product of annual flooding and its dynamic effects on a landscape of alluvial and aeolian forms shaped by Quaternary climatic fluctuations. The contemporary landscape and vegetation of this region is very similar to that of the Llanos of Venezuela and Colombia and Bolivia's Llanos de Moxos.

Cerrado vegetation, including the whole continuum from campo limpo to cerradão, dominates in the eastern and northern frontiers and is common on uplands and on sandy interfluves with deep water tables (Dubs 1992b; Sarmiento 1983; Prance and Schaller 1982). Typical cerrado and widespread Neotropical savanna taxa such as *Bowdichia virgilioides*, *Caryocar brasiliense*, *Curatella americana*, *Qualea parviflora*, and *Anacardium humile* occur in the cerrados of the Pantanal.

Mesotrophic cerradão is also widespread in areas unaffected by flooding. The cerradão of the Pantanal often contains a mixture of typical central Brazilian mesotrophic cerradão species such as *Magonia pubescens*, *Luehea paniculata*, *Buchenavia tomen-*

tosa, Astronium fraxinifolium, and some calcicolous species typical of deciduous forests rooted in calcareous soils, such as *Myracrodruon urundeuva, Vitex cymosa, Sterculia striata, Dipteryx alata,* and *Platypodium grandiflorum.* Its canopy is generally 10–14 m high, and the ground layer often contains numerous terrestrial bromeliads (Dubs 1992a).

Tropical dry forests (deciduous and semideciduous) are particularly common in the uplands and areas bordering the Pantanal. In the southern Pantanal, semideciduous forests with a canopy 18–21 m high often contain a dense understory of urucurí palms (*Attalea phalerata*). Typical trees of these forests are *Myracrodruon urundeuva, Astronium fraxinifolium, Talisia esculenta, Hymenaea courbaril, Tabebuia ochracea, T. impetiginosa, T. roseo-alba, Acacia paniculata, Caesalpinia floribunda, Dilodendron bipinnatum, Casearia gossypiosperma,* and *Combretum leprosum* (Dubs 1992a; Ratter et al. 1988b; Prance and Schaller 1982). Some species of the Chaco or Chaco transitions occupy dry uplands, but these are a minor component of the vegetation. Some of these transitional or chaquenian species include *Bulnesia sarmientoi, Pereskia saccharosa,* and *Schinopsis balansae* and transitional forest species such as *Calycophyllum multiflorum* and *Enterolobium contortisiliquum.*

Gallery forests and some upland forests, particularly in the northern frontier of the Pantanal, are evergreen. These evergreen forests include some Amazonian species such as *Guarea macrophylla, Abuta grandifolia, Mouriri guianensis,* and *Rudgea cornifolia.*

The floodplains are a mosaic of narrow to broad rivers, oxbow lakes, ponds, permanent swamps, and hyperseasonal savannas. The wetlands are characterized by mostly pantropical and Neotropical floating and rooted aquatic plants, e.g., *Eichornia crassipes, Salvinia auriculata, Pistia stratiotes, Ludwigia* spp., *Isoetes* spp., *Utricularia* spp., and *Victoria amazonica* (Dubs 1992a; Prance and Shaller 1982; Frey 1995). Dense marshes dominated by *Cyperus giganteus* or *Typha dominguensis* are also characteristic. Hyperseasonal savannas flooded up to several months during the rainy season include treeless grasslands to lightly wooded savannas with *Curatella americana, Byrsonima orbignyana, Bactris glaucescens,* and *Licania parvifolia; Copernicia alba* palm savannas; and savannas with small forest islands associated with termite or ant colonies (sometimes called campos de murundus; Oliveira-Filho 1992a). *Tabebuia aurea* is a characteristic tree of decayed termite mounds. The term *capão* is sometimes used to refer to the kind of larger forest islands in the hyperseasonal savannas of the Pantanal. These capões should not be confused with those of the campo rupestre, which has a completely different flora.

CAMPO RUPESTRE

The complex of montane vegetation commonly referred to as campo rupestre is primarily associated with the discontinuous range of rocky summits called the Serra do Espinhaço. This narrow range, 50–100 km wide, extends about 1,000 km from northern Bahia south to Minas Gerais (a short distance south of Belo Horizonte), 10°–20°35′S. The northern portion of the campo rupestre is nested within the caatinga phytochorion. The southern portion penetrates the Cerrado region. The area covered by campo rupestre is 6,000–7,000 km². The numerous low mountains range in elevation from 900 to 2,107 m. Outliers of campo rupestre vegetation are found in some of the low mountains (serras) in Goiás, Mato Grosso, and southeastern Bolivia (see figure 14.2). A few areas of Amazonian Brazil have vegetation that has been called campo rupestre, such as part of the Serra do Cachimbo in southwestern Pará and at Ariramba on the Río Trombetas in northern Pará (Pires and Prance 1985). The application of the term *campo rupestre* to these areas needs further study, as these are rather different from the core area of campo rupestre associated with the Serra do Espinhaço.

The climate of the Serra do Espinhaço is characterized by a mild summer (average temperature varies between 17.4° and 19.8°C and below 22°C even at the hottest time of the year). The dry season lasts three to four months during the austral winter, but the higher elevation areas remain fairly moist due to dew and rain, especially on the eastern slopes. Pre-Cambrian rocks of the Brazilian Shield underwent extensive folding and subsequent erosion from the Paleozoic until the Tertiary. The rocky substrate for the unique campo rupestre vegetation consists mostly of metamorphosed sedimentary rocks and includes sandstones, quartzites, schists, filites, and dolomites. The soils derived from these rocks are frequently shallow, sandy or stony, and highly acidic. Forests occur in areas of deeper soil accumulation (Giulietti and Pirani 1988, 1997; Harley 1995).

Most of the vegetation is open, with many plants possessing special adaptations to growing on rocks. Some of these adaptations include specialized roots, dense vestiture, and, in some families, persistent leaf bases (e.g., Velloziaceae, Bromeliaceae, and Orchidaceae). Sandy soils are generally covered by diverse grassland dominated by Poaceae, Cyperaceae, Eriocaulaceae, and Xyridaceae. Poorly drained areas often have bogs characterized by Xyridaceae, Eriocaulaceae, Cyperaceae, Gentianaceae, Lentibulariaceae, Droseraceae, and Burmanniaceae (Giulietti and Pirani 1997). Woody plants in the open campo rupestre are represented by scattered small trees, shrubs, or suffrutices, which are usually sclerophyllous and evergreen. As mentioned, campo rupestre is not a homogeneous vegetation type but rather consists of a mosaic of additional plant formations, such as oligotrophic marshes, cerrados, gallery forests, deciduous forests, and capões, which are isolated islands of forest occurring in grasslands at higher elevations.

It has been suggested that the campo rupestre has the highest degree of endemism of any Brazilian vegetation type. More than 70 percent of the world's ca. 250 species of Velloziaceae and more than 60 percent of the world's Eriocaulaceae and Xyridaceae are endemic to the Serra do Espinhaço (Giulietti and Pirani 1988, 1997). Examples of endemic genera include *Burlemarxia* (Velloziaceae), *Pseudotrimezia* (Iridaceae), *Cipocereus* (Cactaceae), *Raylea* (Sterculiaceae), *Morithamnus* (Asteraceae), and *Bishopiella* (Asteraceae).

The core area of campo rupestre is amazingly rich in species; for example, 1,590 species of vascular plants are estimated to occur in ca. 200 km² of the Serra do Cipó. The flora of the core area is estimated to exceed 4,000 species of vascular plants (Giulietti and Pirani 1997). The largest dicot families in the campo rupestre are the Asteraceae, Melastomataceae, Rubiaceae, Fabaceae sensu lato, Myrtaceae, Malpighiaceae, and Euphorbiaceae; the largest monocot families are the Poaceae, Eriocaulaceae, Orchidaceae, Velloziaceae, Xyridaceae, Bromeliaceae, and Cyperaceae (Giulietti and Pirani 1997).

In Minas Gerais, the campos rupestres are surrounded by lower-lying areas dominated by cerrado. In fact, the cerrado flora shares many species and genera with the campo rupestre. Some genera occurring in both floras include *Eremanthus, Qualea, Campomanesia, Hyptis, Kielmeyera, Jacaranda, Diplusodon,* and *Aristida.* In central Bahia, campo rupestre is often separated by lower-lying areas dominated by caatinga vegetation. In contrast with the Cerrado, the campo rupestre and caatinga floras have very little in common floristically (Giulietti and Pirani 1997).

The campo rupestre has floristic connections with the tepuis of Venezuelan Guayana and the restinga vegetation of Atlantic coastal Brazil. Both the tepuis and campos rupestres are humid, higher-elevation islands formed in part from pre-Cambrian shields; the tepui summits contain rocky outcrops and bogs and are rich in Eriocaulaceae and Xyridaceae. There are even a few species shared by and restricted to these two regions.

CONCLUSIONS

A continental-scale introduction to the late Holocene vegetation of tropical South America must of necessity paint the landscape in broad strokes. Behind the simplicity implied by the unifying name for each phytogeographic region or phytochorion is a riot of heterogeneity, of changes across often independently varying gradients and shifts in key environmental factors. Transitions to other phytochoria may occur abruptly or gradually, as continuums or as mosaics of vegetation cover derived from two or three surrounding regions. In addition to transitions, there may be inclusions, often totaling large areas and in some cases occurring as archipelagos of vegetation cover that is either unique or has analogues or homologues elsewhere.

The vegetation types themselves include numerous variations on themes, and even where large areas are under the influence of relatively uniform conditions, nonobvious changes in such factors as drainage can yield dramatically different physiognomies and species compositions. An example is the alluvial overflow plains of the Llanos and the Pantanal.

Still, there are patterns. Each phytochorion is characterized by prevailing climates, geology, topography, vegetation types, principal life forms, adaptations, dominant plant groups, floristic affinities, diversity patterns, and species distribution patterns. Each has a set of endemic taxa.

We noted that similar climatic and edaphic conditions in geographically distant regions can give rise to strikingly similar vegetation physiognomies. In lowland northern South America, some of the same taxa can be found in comparable vegetation types hundreds or thousands of kilometers distant, particularly in savannas (figure 14.2), floodplain forests, and swampy palm forests; particularly interesting examples occur between true campo rupestre in east-central Brazil and the tepuis of Venezuelan Guayana. Given similar vegetation cover for any pair of distant localities, however, most of the species and some of the genera differ, as do family dominance patterns.

It is essential to expand on the measures that have been taken toward a standardized analysis of vegetation cover—including both physiognomic and floristic information—on a finer scale across the continent. There is a bewildering array of vegetation cover terminology, much of it derived from local or regional folk classifications. The compilation of definitions for Spanish South America developed by O. Huber and Riina (1997) lays a partial foundation for making comparisons. In addition to thorough descriptions of physiognomies and the edaphic conditions that give rise to them, these comparisons depend on adequate and accurate floristic surveys, which ultimately depend on updated monographs and revisions, i.e., modern taxonomic classification. There is an urgent need for more analyses of floristic affinities, such as those carried out for the Venezuelan Guayana (Berry et al. 1995) and for the Serranía de Chiribiquete (Cortés and Franco 1997; Cortés et al. 1998) in Colombia. There is also a need for the application of more statistical methods in floristics. Mechanisms are needed for gathering and mapping information from disparate sources on plant distributions. With massive amounts of fieldwork, vastly increased efforts in systematic studies, and improved bioinformatics, the vegetation cover of tropical South America can be understood and mapped.

ACKNOWLEDGMENTS

We thank Scott Mori and Hans ter Steege for their comments on the manuscript and for assistance with the literature survey, Marina Tereza Campos for information on the capão vegetation and the campos rupestres, Marlene Bellengi for preparing the map, and Muriel Weinerman for preparing the photographs from color slides.

REFERENCES

Ab'Sáber, A. N. 1971. Organização natural das paisagens intere subtropicais brasileiras. In M. G. Ferri, ed., *Simpósio Sôbre o Cerrado*, pp. 1–14. São Paulo: Universidade de São Paulo.

Absy, M. L., A. Cleef, M. Fournier, L. Martin, M. Servant, A. Siffedine, M. F. F. Silva, F. Soubiès, K. Suguiu, B. T. Urk, and T. van der Hammen. 1991. Mise en évidence de quatre phases d'ouverture de la forêt dans le sud-est de l'Amazonie au cours des 60,000 dernières années. Première comparaison avec d'autres régions tropicales. *Compt. Rend. Hebd. Séances Acad. Sci.* sér. 2, t. 312: 673–678.

Anderson, A. B. 1981. White-sand vegetation of Brazilian Amazonia. *Biotropica* 13(3): 199–210.

Andrade-Lima, D. 1954. *Contribution to the Study of the Flora of Pernambuco, Brazil*. Monografia 1. Recife: Universidade Rural de Pernambuco.

———. 1981. The caatingas dominium. *Rev. Brasil. Bot.* 4: 149–153.

———. 1982. Present-day forest refuges in northeastern Brazil. In G. T. Prance, ed., *Biological Diversification in the Tropics*, pp. 245–251. New York: Columbia University Press.

Araújo, D. S. Dunn de. 1997. Cabo Frio region: Southeastern Brazil. In S. D. Davis, V. H. Heywood, O. Herrera-MacBryde, J. Villa-Lobos, and A. C. Hamilton, eds., *Centres of Plant Diversity: A Guide and Strategy for Their Conservation, 3: The Americas*, pp. 373–375. Cambridge: WWF and IUCN.

Araújo, F. D. 1995. A review of *Caryocar*

brasiliense (Caryocaraceae), an economically valuable species of the central Brazilian cerrados. *Economic Botany* 49: 40–48.

Armesto, J. J. 1987. Mecanismos de diseminación de semillas en el bosque de Chiloé: Una comparación con otros bosques templados y tropicales. *An. 4 Congreso Latinoamericano de Botánica* 2: 7–12.

Ayres, J. M., and T. H. Clutton-Brock. 1992. River boundaries and species range size in Amazonian primates. *American Naturalist* 140(3): 531–537.

Balée, W., and D. G. Campbell. 1990. Evidence for successional status of liana forest (Xingu River Basin, Amazonian Brazil). *Biotropica* 22(1): 36–47.

Balslev, H., J. Luteyn, B. Ollgaard, and L. B. Holm-Nielsen. 1987. Composition and structure of adjacent unflooded and floodplain forest in Amazonian Ecuador. *Opera Bot.* 92: 37–57.

Barthlott, W., A. Groger, and S. Porembski. 1993. Some remarks on the vegetation of tropical inselbergs: Diversity and ecological differentiation. *Biogéographica* 69: 105–124.

Bastos, M. N. C. 1984. Levantamento florístico dos campos do Estado do Pará: Campo de Joannis (Ilha de Marajó). *Boletim do Museu Paraense, Bot.* 1(1/2): 67–86.

Beard, J. S. 1955. The classification of tropical American vegetation types. *Ecology* 36(1): 89–99.

Beck, S. G. 1984. Comunidades vegetales de las savanas inundadizas en al noreste de Bolivia. *Phytocenologia* 12: 321–350.

Beck, S. G., and M. Moraes-R. 1997. Llanos de Mojos region, Bolivia. In S. D. Davis, V. H. Heywood, O. Herrera-MacBryde, J. Villa-Lobos, and A. C. Hamilton, eds., *Centres of Plant Diversity: A Guide and Strategy for Their Conservation, 3: The Americas,* pp. 421–425. Cambridge: WWF and IUCN.

Berry, P. E., O. Huber, and B. K. Holst. 1995. Floristic analysis and phytogeography. In Berry, P. E., B. K. Holst, and K. Yatskievych, eds., *Flora of the Venezuelan Guayana, 1: Introduction,* pp. 161–191. St. Louis: Missouri Botanical Garden.

Brown, K. S. Jr., and A. N. Ab'Sáber. 1979. Ice-age forest refuges and evolution in the Neotropics: Correlation of paleoclimatological, geomorphological, and pedological data with modern biological endemisms. *Paleoclimas* 5: 1–30.

Bucher, E. H. 1982. Chaco and caatinga: South American arid savannas, woodlands, and thickets. In B. J. Huntley and B. H. Walker, eds., *Ecology of Tropical Savannas,* pp. 48–79. Berlin: Springer-Verlag.

Cabrera, A. 1976. *Regiones Fitogeográficas Argentinas,* 2nd ed. Buenos Aires: Enciclopedia Argentina Agricultura y Jardinería ACME.

Cabrera, A. L., and A. Willink. 1973. *Biogeografía de América Latina.* Washington, D.C.: Programa Regional de Desarrollo Científico y Tecnológico, Departamento Asuntos Científicos, Secretario General de la Organización de los Estados Americanos.

Campbell, D. G., D. C. Daly, G. T. Prance, and U. N. Maciel. 1986. Quantitative ecological inventory of terra firme and varzea tropical forest on the Rio Xingu, Pará, Brazil. *Brittonia* 38: 369–393.

Castro, A. A. J. F., F. R. Martins, and A. G. Fernandes. 1998. The woody flora of cerrado vegetation in the state of Piauí, northeastern Brazil. *Edinburgh J. Botany* 55: 455–472.

Centro de Datos para la Conservacion-Peru, Fundación Peruana para la Conservación de la Naturaleza, and O. Herrera-MacBryde. 1997. Cerros de Amotape National Park region, northwestern Peru. In S. D. Davis, V. H. Heywood, O. Herrera-MacBryde, J. Villa-Lobos, and A. C. Hamilton, eds., *Centres of Plant Diversity: A Guide and Strategy for Their Conservation, 3: The Americas,* pp. 513–518. Cambridge: WWF and IUCN.

Chandless, W. 1866. Ascent of the River Purús. *J. Royal Geographical Society* 36: 86–188 (plus map).

Clark, D. A., D. B. Clark, R. Sandoval M., and M. V. Castro C. 1995. Edaphic and human effects on landscape-scale distributions of tropical rain-forest palms. *Ecology* 76: 2581–2594.

Condit, R. 1996. Defining and mapping vegetation types in megadiverse tropical forests.

Trends in Ecology and Evolution 11(1): 4–5.

Connell, J. H., and M. D. Lowman. 1989. Low-diversity tropical rain forests: Some possible mechanisms for their existence. *American Naturalist* 134: 88–119.

Cook, S. 1998. A diversity of approaches to the study of species richness. *Trends in Ecology and Evolution* 13(9): 340–341.

Cortés-B., R., and P. Franco-R. 1997. Analisis panbiogeográfico de la flora de Chiribiquete, Colombia. *Caldásia* 19: 465–478.

Cortés-B., R., P. Franco-R., and J. O. Rangel-C. 1998. La flora vascular de la Sierra de Chiribiquete, Colombia. *Caldásia* 20: 103–141.

Coutinho, L. M. 1990. Fire in the ecology of the Brazilian cerrado. In J. G. Goldammer, ed., *Fire in the Tropical Biota: Ecosystem Processes and Global Challenges*, pp. 82–105. Berlin: Springer-Verlag.

Cracraft, J. 1985. Historical biogeography and patterns of differentiation within the South American avifauna: Areas of endemism. In P. A. Buckley, M. S. Foster, E. S. Morton, R. S. Ridgely, and F. G. Buckley, eds., *Neotropical Ornithology*, pp. 49–84. Ornithological Monographs 36. Washington, D.C.: American Ornithologists' Union.

Daly, D. C. 1992. Two new taxa of *Protium* from eastern Brazil. Studies in Neotropical Burseraceae 5. *Kew Bulletin* 47: 713–719.

———. 1993. Notes on *Bursera* in South America, including a new species. Studies in Neotropical Burseraceae 7. *Brittonia* 45: 240–246.

Daly, D. C., and G. T. Prance. 1989. Brazilian Amazon. In D. G. Campbell and H. D. Hammond, eds., *Floristic Inventory of Tropical Countries*, pp. 401–426. New York: New York Botanical Garden.

Dias, B. F. S. 1992. Cerrados: Uma caracterização. In B. F. S. Dias, coord., *Alternativas de Desenvolvimento dos cerrados: Manejo e Conservação dos Recursos Naturais Renováveis*, pp. 11–25. Brasília, D.F.: Funatura-IBAMA.

Dickinson, R. E., and H. Virji. 1987. Climate change in the humid tropics, especially Amazonia, over the last twenty thousand years. In R. E. Dickinson, ed., *The Geo-physiology of Amazonia: Vegetation and Climate Interactions*, pp. 91–101. New York: J. Wiley & Sons.

Dubs, B. 1992a. Observations on the differentiation of woodland and wet savanna habitats in the Pantanal of Mato Grosso, Brazil. In P. A. Furley, J. Proctor, and J. A. Ratter, eds., *Nature and Dynamics of Forest-Savanna Boundaries*, pp. 431–449. London: Chapman & Hall.

———. 1992b. *Birds of Southwestern Brazil: Catalogue and Guide to the Birds of the Pantanal of Mato Grosso and Its Border Areas.* Küsnacht, Switz.: Betrona-Verlag.

Ducke, A., and G. A. Black. 1953. Phytogeographical notes on the Brazilian Amazon. *An. Acad. Brasil. Ciências* 25(1): 1–46.

Duivenvoorden, J. F. 1994. Vascular plant species counts in the rain forests of the middle Caquetá area, Colombian Amazonia. *Biodiversity and Conservation* 3: 685–715.

———. 1995. Tree species composition and rain forest-environment relationships in the middle Caquetá area, Colombia, NW Amazonia. *Vegetatio* 120: 91–113.

Eden, M. J., and D. F. M. McGregor. 1992. Dynamics of the forest-savanna boundary in the Rio Branco-Rupununi region of northern Amazonia. In P. A. Furley, J. Proctor, and J. A. Ratter, eds., *Nature and Dynamics of Forest-Savanna Boundaries*, pp. 77–89. London: Chapman & Hall.

Egler, W. A. 1960. Contribuições ao conhecimento dos campos da Amazônia 1: Os Campos de Ariramba. *Boletim do Museu Paraense, Bot.* n.s. 4: 1–36 (plus figs.).

Eiten, G. 1972. The cerrado vegetation of Brazil. *Botanical Review* 38: 201–341.

———. 1978. Delimitation of the cerrado concept. *Vegetatio* 36: 169–178.

———. 1983. *Classificação da Vegetação do Brasil.* Brasilia: CNPQ, Coordenação Editorial.

Empéraire, L. 1989. *Végétation et Gestion des Resources Naturelles dans la Caatinga du Sud-Est du Piauí (Brésil).* Collection Travaux et Documents Microédités 52. Paris: Editions de l'Orstom.

Filgueiras, T. 1997. Distrito Federal, Brazil. In S. D. Davis, V. H. Heywood, O. Herrera-MacBryde, J. Villa-Lobos, and A. C. Hamilton, eds., *Centres of Plant Diversity: A*

Guide and Strategy for Their Conservation, 3: The Americas, pp. 405–410. Cambridge: WWF and IUCN.

Forero, E. 1988. Botanical exploration and phytogeography of Colombia: Past, present, and future. Taxon 37: 561–566.

Frey, R. 1995. Flora and vegetation of Las Piedritas and the margin of Laguna Cáceres, Puerto Suárez, Bolivian Pantanal. Bulletin of the Torrey Botanical Club 122(4): 314–319.

Galera, F. M., and L. Ramela. 1997. Gran Chaco: Argentina, Paraguay, Brazil, Bolivia. In S. D. Davis, V. H. Heywood, O. Herrera-MacBryde, J. Villa-Lobos, and A. C. Hamilton, eds., Centres of Plant Diversity: A Guide and Strategy for Their Conservation, 3: The Americas, pp. 411–415. Cambridge: WWF and IUCN.

Gentry, A. H. 1982. Patterns of Neotropical plant diversity. Evolutionary Biology 15: 1–84.

———. 1986. Species richness and floristic composition of Chocó region plant communities. Caldásia 15(71–75): 1–91.

———. 1988. Tree species richness of upper Amazonian forests. Proceedings of the National Academy of Sciences USA 85: 156–159.

———. 1989. Northwest South America (Colombia, Ecuador, and Peru). In D. G. Campbell and H. D. Hammond, eds., Floristic Inventory of Tropical Countries, pp. 391–400. Bronx: New York Botanical Garden.

———. 1991. The distribution and evolution of climbing plants. In F. E. Putz and H. A. Mooney, eds., The Biology of Vines, pp. 3–42. Cambridge: Cambridge University Press.

———. 1995. Diversity and floristic composition of Neotropical dry forests. In S. H. Bullock, H. A. Mooney, and E. Medina, eds., Seasonally Dry Tropical Forest, pp. 146–194. Cambridge: Cambridge University Press.

Giulietti, A. M., and J. R. Pirani. 1988. Patterns of geographic distribution of some plant species from the Espinhaço Range, Minas Gerais and Bahia, Brazil. In P. E. Vanzolini and W. R. Heyer, eds., Proceedings of Workshop on Neotropical Distri-

bution Patterns Held 12–16 January 1987, pp. 39–69. Rio de Janeiro: Academia Brasileira de Ciências.

———. 1997. Espinhaço Range region, eastern Brazil. In S. D. Davis, V. H. Heywood, O. Herrera-MacBryde, J. Villa-Lobos, and A. C. Hamilton, eds., Centres of Plant Diversity: A Guide and Strategy for Their Conservation, 3: The Americas, pp. 397–404. Cambridge: WWF and IUCN.

Goodland, R. J. A. 1971. Oligotrofismo e alumínio no cerrado. In M. G. Ferri, ed., III Simpósio Sôbre o Cerrado. São Paulo: Universidade de São Paulo.

Goulding, M. 1993. Flooded forests of the Amazon. Scientific American 266(3): 114–120.

Goulding, M., M. Leal Carvalho, and E. G. Ferreira. 1988. Rio Negro: Rich Life in Poor Water. The Hague: SPB Academic.

Granville, J. J. de. 1992. Un cas de distribution particulier: Les espèces forestières peri-Amazoniennes. Compt. Rend. Sommaire Séances Soc. Biogéogr. 68: 1–33.

Granville, J. J. de, and C. Sastre. 1974. Aperçu sur la végétation des inselbergs du sudouest de la Guyana Française. Compt. Rend. Sommaire Séances Soc. Biogéogr. 439: 54–58.

Guedes-Bruni, R. R., and H. C. Lima. 1997. Mountain ranges of Rio de Janeiro, southeastern Brazil. In S. D. Davis, V. H. Heywood, O. Herrera-MacBryde, J. Villa-Lobos, and A. C. Hamilton, eds., Centres of Plant Diversity: A Guide and Strategy for Their Conservation, 3: The Americas, pp. 376–380. Cambridge: WWF and IUCN.

Guillaumet, J.-L. 1987. Some structural and floristic aspects of the forest. Experientia 43: 241–251.

Haase, R., and S. G. Beck. 1989. Structure and composition of savanna vegetation in northern Bolivia: A preliminary report. Brittonia 4(1): 80–100.

Harley, R. M. 1995. Introduction. In Stannard, B. L., ed., Flora of the Pico das Almas: Chapada Diamantina, Bahia, Brazil, pp. 1–42. Kew: Royal Botanic Gardens.

Hartshorn, G. S. 1978. Tree falls and tropical forest dynamics. In P. B. Tomlinson and M. H. Zimmermann, eds., Tropical Trees as Living Systems, pp. 617–638. Cambridge: Cambridge University Press.

Heringer, E. P. 1971. Propagação e sucessão das espécies arbóreas em função do fogo do capim, da capina, e de Aldrin. In M. G. Ferri, ed., III Simpósio Sôbre o Cerrado, pp. 167–179. São Paulo: Universidade de São Paulo.

Heringer, E. P., G. M. Barroso, J. A. Rizzo, and C. T. Rizzini. 1977. A flora do cerrado. In IV Simpósio Sôbre o Cerrado: Bases para Utilização e Agropecuário, pp. 211–232. Belo Horizonte: Editora da Universidade de São Paulo.

Herrera-MacBryde, O., O. Rangel-Ch., M. Aguilar-P., H. Sánchez-C., P. Lowy-C., D. Cuartas-Ch., and A. Garzón-C. 1997. Colombian Pacific coast region (Chocó): Colombia. In S. D. Davis, V. H. Heywood, O. Herrera-MacBryde, J. Villa-Lobos, and A. C. Hamilton, eds., Centres of Plant Diversity: A Guide and Strategy for Their Conservation, 3: The Americas, pp. 501–507. Cambridge: WWF and IUCN.

Holdridge, L. R. 1971. Forest Environments in Tropical Life Zones: A Pilot Study. Oxford: Pergamon.

Huber, J. 1898. Materiaes para a flora Amazônica 1: Lista das plantas colligidas na Ilha de Marajó no anno de 1896. Boletim do Museu Paraense Hist. Nat. 2: 288–321.

———. 1906. Guadua superba Hub. n. sp., a taboca gigante do alto Río Purus. Boletim do Museu Paraense Hist. Nat. 4(1/4): 479–480.

Huber, O. 1994. Recent advances in the phytogeography of the Guayanan region, South America. Mém. Soc. Biogéogr. (3éme sér.) 4: 53–63.

———. 1995. Vegetation. In J. A. Steyermark, P. E. Berry, and B. K. Holst, eds., Flora of the Venezuelan Guayana, 1, pp. 97–160. St. Louis: Missouri Botanical Garden.

———. 1997. Coastal Cordillera, Venezuela. In S. D. Davis, V. H. Heywood, O. Herrera-MacBryde, J. Villa-Lobos, and A. C. Hamilton, eds., Centres of Plant Diversity: A Guide and Strategy for Their Conservation, 3: The Americas, pp. 308–315. Cambridge: WWF and IUCN.

Huber, O., and R. Riina, eds. 1997. Glosário Fitoecológico de las Américas, 1: América

del Sur, Países Hispanoparlantes. Caracas: UNESCO.

Hueck, K. 1972. As florestas da América do Sul. Translation of Hueck 1966. São Paulo: Polígono.

Hueck, K., and P. Seibert. 1972. Vegetationskarte von Südamerika. Stuttgart: G. Fischer-Verlag.

Huston, M. A. 1994. Biological Diversity: The Coexistence of Species on Changing Landscapes. Cambridge: Cambridge University Press.

Jordan, C. F. 1989. An Amazonian Rain Forest: The Structure and Function of a Nutrient-Stressed Ecosystem and the Impact of Slash-and-Burn Agriculture. Man and the Biosphere Series 2. Paris: UNESCO.

Junk, W. J. 1989. Flood tolerance and tree distribution in central Amazonian floodplains. In L. B. Holm-Nielsen, I. C. Nielsen, and H. Balslev, eds., Tropical Forests, Botanical Dynamics, Speciation, and Diversity, pp. 47–64. San Diego, Calif.: Academic Press.

Keel, S. H. K. 1997. Mbaracayu Reserve, Paraguay. In S. D. Davis, V. H. Heywood, O. Herrera-MacBryde, J. Villa-Lobos, and A. C. Hamilton, eds., Centres of Plant Diversity: A Guide and Strategy for Their Conservation, 3: The Americas, pp. 389–392. Cambridge: WWF and IUCN.

Keel, S. H. K., and G. T. Prance. 1979. Studies of the vegetation of a white-sand black water igapó (Rio Negro, Brazil). Acta Amazonica 9: 645–655.

Killeen, T. J., A. Jardim, F. Mamani, and N. Rojas. 1998. Diversity, composition, and structure of a tropical semideciduous forest in the Chiquitanía region of Santa Cruz, Bolivia. J. Tropical Ecology 14: 803–827.

Köppen, W. P. 1901. Versuch einer Klassifikation der Klimate Vorzugweise nach Ihren Beziehungen zur Pflanzenwelt. Leipzig: B. G. Teubner.

Küchler, A. W. 1980. International Bibliography of Vegetation Maps, 1: South America. 2nd ed. Lawrence: University of Kansas Libraries.

Ledru, M. P. 1993. Late Quaternary environmental and climatic changes in central Brazil. Quaternary Research 39: 90–98.

Leitão-Filho, H. F. 1997. Serra do Japi, southeastern Brazil. In S. D. Davis, V. H. Heywood, O. Herrera-MacBryde, J. Villa-Lobos, and A. C. Hamilton, eds., *Centres of Plant Diversity: A Guide and Strategy for Their Conservation, 3: The Americas,* pp. 381–384. Cambridge: WWF and IUCN.

Lescure, J.-P., and R. Boulet. 1985. Relationships between soil and vegetation in a tropical rain forest in French Guiana. *Biotropica* 17: 155–164.

Lleras, E. 1997. Caatinga of northeastern Brazil: Brazil. In S. D. Davis, V. H. Heywood, O. Herrera-MacBryde, J. Villa-Lobos, and A. C. Hamilton, eds., *Centres of Plant Diversity: A Guide and Strategy for Their Conservation, 3: The Americas,* pp. 393–396. Cambridge: WWF and IUCN.

Londoño-Vega, A. C., and E. Alvarez. 1997. Comparación florística de dos bosques (tierra firme y varzea) en la región de Araracuara, Amazonía colombiana. *Caldásia* 19: 431–463.

López Naranjo, H. J. 1977. Hábito de crescimiento y estructura de las yemas de *Anacardium humile* St. Hil., Anacardiaceae. *Rev. Forest. Venezuela* 27: 159–173.

Macedo, M., and G. T. Prance. 1978. Notes on the vegetation of Amazonia 2: The dispersal of plants in Amazonian white sand sampinas: The campinas as functional islands. *Brittonia* 30: 203–215.

Mamede, M. C. H., I. Cordeiro, and L. Rossi. 1997. Juréia-Itatins Ecological Station, southeastern Brazil. In S. D. Davis, V. H. Heywood, O. Herrera-MacBryde, J. Villa-Lobos, and A. C. Hamilton, eds., *Centres of Plant Diversity: A Guide and Strategy for Their Conservation, 3: The Americas,* pp. 385–388. Cambridge: WWF and IUCN.

Mares, M. A. 1992. Neotropical mammals and the myth of Amazonian biodiversity. *Science* 255: 976–979.

Mayo, S. J., and V. P. B. Fevereiro. 1982. *Mata de Pau Ferro: A Pilot Study of the Brejo Forest of Paraíba, Brazil.* London: Royal Botanic Gardens, Kew.

Miranda, I. S. 1993. Estrutura do estrato arbóreo do cerrado amazônico em Alter-do-Chão, Pará, Brasil. *Rev. Brasil. Bot.* 16(2): 143–150.

Mitchell, J. D., and D. C. Daly. 1991. *Cyrtocarpa* Kunth (Anacardiaceae) in South America. *Annals of the Missouri Botanical Garden* 78: 184–189.

Mitchell, J. D., and S. A. Mori. 1987a. Ecology. In S. A. Mori and collaborators, *The Lecythidaceae of a Lowland Neotropical Forest: La Fumée Mountain, French Guiana,* pp. 113–123. Memoirs of the New York Botanical Garden 44.

———. 1987b. *The Cashew and Its Relatives (Anacardium: Anacardiaceae).* Memoirs of the New York Botanical Garden 42.

Mori, S. A. 1991. The Guayana lowland floristic province. *C. R. Soc. Biogéogr.* 67(2): 67–75.

Mori, S. A., and J. L. Brown. 1994. Report on wind dispersal in a lowland moist forest in central French Guiana. *Brittonia* 46: 105–125.

Mori, S. A., B. M. Boom, and G. T. Prance. 1981. Distribution patterns and conservation of eastern Brazilian coastal forest tree species. *Brittonia* 33: 233–245.

Mori, S. A., G. Lisboa, and J. A. Kallunki. 1982. Fenologia de uma mata higrófila baiana. *Revista Theobroma* 12(4): 217–230.

Mori, S. A., B. M. Boom, A. M. de Carvalho, and T. S. dos Santos. 1983. Southern Bahian moist forests. *Botanical Review* 49(2): 155–232.

Mori, S. A., G. Cremers, C. Gracie, J.-J. de Granville, M. Hoff, and J. D. Mitchell. 1997. *Guide to the Vascular Plants of Central French Guiana.* Memoirs of the New York Botanical Garden 76.

Neill, D. A. 1997. Ecuadorian Pacific coast mesic forests: Ecuador. In S. D. Davis, V. H. Heywood, O. Herrera-MacBryde, J. Villa-Lobos, and A. C. Hamilton, eds., *Centres of Plant Diversity: A Guide and Strategy for Their Conservation, 3: The Americas,* pp. 508–512. Cambridge: WWF and IUCN.

Nelson, B. W. 1994. Natural forest disturbance and change in the Brazilian Amazon. *Remote Sensing Reviews* 10: 105–125.

Oliveira, A. A. de, and D. C. Daly. 1999. Geographic distribution of tree species occurring in the region of Manaus, Brazil: Implications for regional diversity and conservation. *Biodiversity and Conservation* 8: 1245–1259.

Oliveira, A. A. de, and S. A. Mori. 1999. A central Amazonian terra firme forest, 1: High tree species richness on poor soils. *Biodiversity and Conservation* 8: 1219–1244.

Oliveira-Filho, A. T. 1992. Floodplain murundus of Central Brazil: Evidence for the termite origin hypothesis. *J. Tropical Ecology* 8(1): 1–19.

Oliveira-Filho, A. T., and J. A. Ratter. 1995. A study of the origin of central Brazilian forests by the analysis of plant species distribution patterns. *Edinburgh J. Botany* 52(2): 141–194.

Patton, J. L., M. N. F. Silva, M. C. Lara, M. A. Mustrangi. 1997. Diversity, differentiation, and the historical biogeography of nonvolant small mammals of the Neotropical forests. In W. F. Laurence and R. O. Bierregaard Jr., eds., *Tropical Forest Remnants: Ecology, Management, and Conservation of Fragmented Communities,* pp. 455–465. Chicago: University of Chicago Press.

Peixoto, A. L., and I. M. Silva. 1997. Tabuleiro forests of northern Espírito Santo: Southeastern Brazil. In S. D. Davis, V. H. Heywood, O. Herrera-MacBryde, J. Villa-Lobos, and A. C. Hamilton, eds., *Centres of Plant Diversity: A Guide and Strategy for Their Conservation, 3: The Americas,* pp. 369–372. Cambridge: WWF and IUCN.

Peters, C. M. 1992. The ecology and economics of oligarchic forests. *Advances in Economic Botany* 9: 15–22.

Pires, J. M. 1973. Tipos de vegetação da Amazonia. *Publ. Avulsas do Museu Paraense Emílio Goeldi* 20: 179–202.

Pires, J. M., and G. T. Prance. 1985. The vegetation types of the Brazilian Amazon. In G. T. Prance and T. E. Lovejoy, eds., *Key Environments: Amazonia,* pp. 109–145. Oxford: Pergamon Press.

Pires-O'Brien, M. J. 1997. Transverse dry belt of Brazil. In S. D. Davis, V. H. Heywood, O. Herrera-MacBryde, J. Villa-Lobos, and A. C. Hamilton, eds., *Centres of Plant Diversity: A Guide and Strategy for Their Conservation, 3: The Americas,* pp. 319–324. Cambridge: WWF and IUCN.

Podolsky, R. 1992. Remote sensing, geographic data, and the conservation of biological resources. *Endangered Species Update* 9(12): 1–4.

———. 1994. Ecological hot spots: A method for estimating biodiversity directly from digital earth imagery. *Earth Observation* June 1994: 30–36.

Poulsen, A. D., and H. Balslev. 1991. Abundance and cover of ground herbs in an Amazonian rain forest. *J. Vegetation Science* 2(3): 315–321.

Prado, D. E. 1993a. What is the Gran Chaco vegetation in South America? 1: A review. Contribution to the study of flora and vegetation of the Chaco 5. *Candollea* 48(1): 145–172.

———. 1993b. What is the Gran Chaco vegetation in South America? 2: A redefinition. Contribution to the study of flora and vegetation of the Chaco 7. *Candollea* 48(2): 615–629.

Prado, D. E., and P. E. Gibbs. 1993. Patterns of species distributions in the dry seasonal forests of South America. *Annals of the Missouri Botanical Garden* 80(4): 902–927.

Prado, D. E., P. E. Gibbs, and V. J. Pott. 1992. The Chaco-Pantanal transition in southern Mato Grosso, Brazil. In P. A. Furley, J. Proctor, and J. A. Ratter, eds., *Nature and Dynamics of Forest-Savanna Boundaries,* pp. 451–470. London: Chapman & Hall.

Prance, G. T. 1975. Estudos sobre a vegetação das Campinas Amazônicas 1. *Acta Amazonica* 5(3): 207–209.

———. 1979. Notes on the vegetation of Amazonia 3: The terminology of Amazonian forest types subject to inundation. *Brittonia* 31(1): 26–38.

———. 1989. American tropical forests. In H. Lieth and M. J. A. Werger, eds., *Tropical Rain Forest Ecosystems, Biogeographical and Ecological Studies: Ecosystems of the World 14,* pp. 99–132. Amsterdam: Elsevier.

Prance, G. T., and K. S. Brown Jr. 1987. The principal vegetation types of the Brazilian Amazon. In T. C. Whitmore and G. T. Prance, eds., *Biogeography and Quaternary History in Tropical America,* fig. 2.4, pp. 30–31. Oxford: Clarendon Press.

Prance, G. T., and G. B. Schaller. 1982. Preliminary study of some vegetation types of the Pantanal, Mato Grosso, Brazil. *Brittonia* 34(2): 228–251.

Projeto Radambrasil. 1975–1981. *Levantamento de Recursos Naturais, 8–18.* Rio de Janeiro: Ministério de Minas e Energia.

Puhakka, M., R. Kalliola, J. Salo, and M. Rajasilta. 1993. La sucesión forestal que sigue a la migración de ríos en la selva baja peruana. In R. Kalliola, M. Puhakka, and W. Danjoy, eds., *Amazonía Peruana: Vegetación Humeda Tropical en el Llano Subandino,* pp. 167–201. Jyväskylä, Finland: Gummerus.

Räsänen, M. E. 1993. La geohistoria y geología de la Amazonía peruana. In R. Kalliola, M. Puhakka, and W. Danjoy, eds., *Amazonía Peruana: Vegetación Humeda Tropical en el Llano Subandino,* pp. 43–67. Jyväskylä, Finland: Gummerus.

Räsänen, M. E., J. Salo, and R. Kalliola. 1987. Fluvial perturbance in the western Amazon Basin: Regulation by long term sub-Andean tectonics. *Science* 238: 1398–1401.

Ratter, J. A., H. de F. L. Filho, G. Argent, P. E. Gibbs, J. Semir, G. Shepherd, and J. Tamashiro. 1988a. Floristic composition and community structure of a southern cerrado area in Brazil. *Notes from the Royal Botanical Garden Edinburgh* 45(1): 137–151.

Ratter, J. A., A. Pott, V. J. Pott, C. N. Cunha, and M. Haridasan. 1988b. Observations on the woody vegetation types in the Pantanal and at Corumbá, Brazil. *Notes from the Royal Botanical Garden Edinburgh* 45: 503–525.

Ratter, J. A., S. Bridgewater, R. Atkinson, and J. F. Ribeiro. 1996. Analysis of the floristic composition of the Brazilian cerrado vegetation, 2: Comparison of the woody vegetation of 98 areas. *Edinburgh J. Botany* 53(2): 153–180.

Rawitscher, F., and M. Rachid. 1946. Troncos subterraneos de plantas brasileras. *An. Acad. Brasil. Ciências* 18: 261–280.

Reitz, P. R., ed. 1965. *Flora Ilustrada Catarinense 1: As Plantas.* Santa Catarina, Brazil: Itajaí.

Richards, P. W. 1998. *The Tropical Rain Forest.* 2nd ed. Cambridge: Cambridge University Press.

Ricker, M., J. H. Jessen, and D. C. Daly. 1997. The case for *Borojoa patinoi* in the Chocó region, Colombia. *Economic Botany* 51(1): 39–48.

Rizzini, C. T. 1963. Nota prévia sobre a fitogeografia do Brasil. *Rev. Brasil. Geogr.* 1 (Ano 25): 1–64.

———. 1979. *Tratado de fitogeografia do Brasil, 2: Aspectos Sociológicos e Florísticos.* São Paulo: Editora Universidade de São Paulo.

Salo, J., R. Kalliola, I. Hakkinen, Y. Makinen, P. Niemela, M. Puhakka, and P. Coley. 1986. River dynamics and the diversity of Amazon lowland forest. *Nature* 322: 254–258.

Salomão, R. P., and P. L. B. Lisboa. 1988. Análise ecológica da vegetação de uma floresta pluvial tropical de terra firme, Rondônia. *Boletim do Museu Paraense Hist. Nat., Bot.* 4: 195–234.

Sampaio, E. V. S. B. 1995. Overview of the Brazilian caatinga. In S. H. Bullock, H. A. Mooney, and E. Medina, eds., *Seasonally Dry Tropical Forests,* pp. 35–63. Cambridge: Cambridge University Press.

Sampaio, E. V. S. B., I. H. Salcedo, and J. B. Kauffman. 1993. Effect of different fire severities on coppicing of caatinga vegetation in Serra Talhada, PE, Brazil. *Biotropica* 25: 452–460.

Sanaiotti, T., S. Bridgewater, and J. A. Ratter. 1997. A floristic study of the savanna vegetation of the state of Amapá, Brazil, and suggestions for its conservation. *Boletim do Museu Paraense Hist. Nat., Bot.* 13: 3–29.

Sánchez-S., M., L. E. Urrego, J. G. Saldarriaga, J. Fuertes, J. Estrada, and J. F. Duivenvoorden. 1997. Chiribiquete-Araracuara-Cahuinarí region, Colombia. In S. D. Davis, V. H. Heywood, O. Herrera-MacBryde, J. Villa-Lobos, and A. C. Hamilton, eds., *Centres of Plant Diversity: A Guide and Strategy for Their Conservation, 3: The Americas,* pp. 338–343. Cambridge: WWF and IUCN.

Sarmiento, G. 1983. The savannas of tropical

America. In F. Bourlière, ed., *Tropical Savannas: Ecosystems of the World, 13,* pp. 245–288. Amsterdam: Elsevier.

———. 1984. *The Ecology of Neotropical Savannas.* Cambridge: Harvard University Press.

Sarthou, C., and J.-F. Villiers. 1998. Epilithic plant communities on inselbergs in French Guiana. *J. Vegetation Science* 9: 847–860.

Schimper, A. 1903. *Plant Geography upon a Physiological Basis.* Oxford: Clarendon Press.

Sioli, H. 1984. The Amazon and its main affluents: Hydrography, morphology of the river courses, and river types. In H. Sioli, ed., *The Amazon: Limnology and Landscape Ecology of a Mighty Tropical River,* pp 127–165. Dordrecht, Neth.: Dr. W. Junk.

Spichiger, R., R. Palese, A. Chautems, and L. Ramella. 1995. Origin, affinities, and diversity hot spots of the Paraguayan dendrofloras. *Candollea* 50: 515–537.

Struwe, L., P. J. M. Maas, and V. A. Albert. 1997. *Aripuana cullmaniorum,* a new genus and species of Gentianaceae from white sands of southeastern Amazonas, Brazil. *Harvard Papers in Botany* 2: 235–254.

Takeuchi, M. 1962. The structure of the Amazonian vegetation 4: High campina forest in the Upper Rio Negro. *J. Fac. Sci. University of Tokyo, Bot.* Sect. 2, 8: 279–288.

Takhtajan, A. 1986. *Floristic Regions of the World.* Berkeley: University of California Press.

Terborgh, J., and E. Andresen. 1998. The composition of Amazonian forests: Patterns at local and regional scales. *J. Tropical Ecology* 14: 45–664.

ter Steege, H., V. G. Jetten, A. M. Polak, and M. J. A. Weger. 1993. Tropical rain forest types and soil factors in a watershed area in Guyana. *J. Vegetation Science* 4(5): 705–716.

Thomas, W. W., A. M. Carvalho, and O. Herrera-MacBryde. 1997. Atlantic moist forest of southern Bahia: Southeastern Brazil. In S. D. Davis, V. H. Heywood, O. Herrera-MacBryde, J. Villa-Lobos, and A. C. Hamilton, eds., *Centres of Plant Diversity: A Guide and Strategy for Their Conservation, 3: The Americas,* pp. 364–358. Cambridge: WWF and IUCN.

Thomas, W. W., A. M. V. de Carvalho, A. M. A. Amorim, J. Garrison, and A. L. Arbeláez. 1998. Plant endemism in two forests in southern Bahia, Brazil. *Biodiversity and Conservation* 7: 311–322.

Tricart, J. 1974. Existence de périodes sèches au Quaternaire en Amazonia et dans régions voisines. *Rev. Géomorphol. Dyn.* 23: 145–158.

Tuomisto, H., and K. Ruokolainen. 1994. Distribution of Pteridophyta and Melastomataceae across an edaphic gradient in an Amazonian rain forest. *J. Vegetation Science* 5: 25–34.

Tuomisto, H., A. Linna, and R. Kalliola. 1994. Use of digitally processed satellite images in studies of tropical rain forest vegetation. *International J. Remote Sensing* 15(8): 1595–1610.

Tuomisto, H., K. Ruokolainen, R. Kalliola, A. Linna, W. Danjoy, and Z. Rodriguez. 1995. Dissecting Amazonian biodiversity. *Science* 269: 63–66.

UNESCO. 1981a. *Vegetation Map of South America.* Paris: UNESCO.

———. 1981b. Vegetation map of South America: Explanatory notes. *Natural Resources Research* 17. Paris: UNESCO.

Valencia, R., H. Balslev, G. Paz y Miño. 1994. High tree alpha-diversity in Amazonian Ecuador. *Biodiversity and Conservation* 3: 21–28.

van der Hammen, T. 1982. Paleoecology of tropical South America. In G. T. Prance, ed., *Biological Diversification in the Tropics,* pp. 60–66. New York: Columbia University Press.

Veloso, H. P. 1966. *Atlas Florestal do Brasil.* Rio de Janeiro: Ministério da Agricultura, Conselho Florestal Federal.

Veloso, H. P., A. L. R. Rangel Filho, and J. C. A. Lima. 1991. *Classificação da Vegetação Brasileira Adaptada a um Sistema Universal.* Rio de Janeiro: IBGE.

Wallace, A. R. 1849. On the monkeys of the Amazon. *Proceedings of the Zoological Society of London* 20: 107–110.

———. 1853. *Narrative of Travels on the Amazon and Río Negro.* London: Reeve & Co.

Walsh, R. P. D. 1998. Climate. In P. W. Richards, *The Tropical Rain Forest,* 2nd ed., pp. 159–205. Cambridge: Cambridge University Press.

Webb, L. J., and J. G. Tracey. 1994. The rainforest of northern Australia. In R. H. Groves, ed., *Australian Vegetation,* 2nd ed., pp. 87–129. Cambridge: Cambridge University Press.

White, F. 1976. The underground forests of Africa: A preliminary review. *Gard. Bull. Straits Settlem.* 29: 57–71.

———. 1979. The Guineo-Congolian region and its relationships to other phytochoria. *Bull. Jard. Bot. Nac. Belg.* 49: 11–55.

———. 1983. *The Vegetation of Africa.* Paris: UNESCO.

Whitney, B. M., and J. Álvarez Alonso. 1998. A new *Herpsilochmus* antwren (Aves: Thamnophilidae) from northern Amazonian Peru and adjacent Ecuador: The role of edaphic heterogeneity of terra firme forest. *Auk* 115(3): 559–576.

Wikander, T. 1984. Mecanismos de dispersion de diasporas de una selva decidua en Venezuela. *Biotropica* 16: 276–283.

Young, K. R., and B. León. 1989. Pteridophyte species diversity in the central Peruvian Amazon: Importance of edaphic specialization. *Brittonia* 41: 388–395.

ANNA C. ROOSEVELT

15 | THE LOWER AMAZON
A Dynamic Human Habitat

The Amazon has been portrayed as a pristine and ancient habitat vulnerable to destruction by modern civilization. Native Amazonians are pictured as Stone Age peoples living in harmony with the ancient forest. Recently, such models have been weakened by evidence of significant changes in environment and human adaptation before the European conquest of the Americas. One implication of the evidence is that ancient Amazonians had considerable impacts on the habitat in areas adjacent to their settlements. Some distinctive forest patterns once thought purely natural now may be seen as having been influenced by past human activities.

In this essay, I will examine the evidence for indigenous human impacts in the Lower Amazon. Following the case-study approach of this book, I have chosen to discuss one of the regions of my recent research: the Tapajós-Amazon confluence at Santarém–Monte Alegre. Since the largest discernible indigenous impacts were from late prehistoric activities, I will focus on the late periods. But since human impacts can be assessed only against a background of environmental history, I will also summarize what is known of habitat change independent of the human occupation. Finally, I will compare what is known of indigenous impacts with what has happened since the European conquest, when large, centralized, and hierarchical human organizations from abroad began influencing the environment.

THEORETICAL BACKGROUND
AND METHODOLOGICAL PROBLEMS

Much research in Amazonia has focused on natural diversity and the limits it is supposed to place on human adaptation and evolution. In the heyday of equilibrium models, Amazonia was assumed to have been stable for much of its existence, and the stability fostered diversity. Interest was focused on ecosystem maintenance and succession (Odum 1971). In the 1980s, alternative understandings emerged. The high-diversity forests were considered to have been biological refuges from savannas that spread during Pleistocene aridity (Absy 1982; Clapperton 1993; Prance 1982; Prance and Lovejoy 1985; Whitmore and Prance 1987; van der Hammen and Absy 1994). The extensive tropical rain forest was seen as a recent formation, a product of Postpleistocene warming.

Anthropologists' models of the human occupation have tracked the environmental models. During the early days of systems theory, societies were considered stable adaptations to their habitats. Research focused on cultural continuity and the sustainability of low-technology indigenous land use and shifting settlement (Hames and Vickers 1983). The closed-canopy rain forests were considered a barrier to the development of civilization, in contrast to the arid, open habitats of the Andes, where civilization flourished (Meggers 1954, 1971; Steward and Faron 1959). Elaborate mound-building cultures that turned up in the Amazon were explained as short-lived Andean invasions (Meggers and Evans 1957). During refugium-theory days, archaeologists envisioned temperate savannas inhabited by Paleoindian hunter-gatherers, for whom the game-poor tropical rain forests were believed to form a barrier. The Holocene human occupation was seen as unstable, a succession of tropical forest cultures that invaded from the outside during periodic droughts (Meggers 1975, 1977, 1979, 1988).

All along, the various theories were adopted by consensus before empirical evidence to evaluate them had been collected, and they have proved difficult to revise in the face of new data. In a circle of reasoning, data theorized to be effects of hypothetical past climatic events were used, in turn, as evidence for the climatic events. In refugium theory, modern Amazon biogeography and undated geomorphological features were explained as the product of Pleistocene drying and then became the evidence for the late Pleistocene drying, in the absence of Pleistocene biological information (Whitmore and Prance 1987). The only pollen profiles at the time were dated to the Holocene (Absy 1979, 1982). When intact Pleistocene pollen cores and paleobotanical specimens were finally collected, analyzed for carbon isotopes, and published, they held no evidence for the replacement of tropical rain forest by Pleistocene savannas, but this has not deterred refugium enthusiasts (Clapperton 1993; Piperno and Pearsall 1998; Roosevelt 1995a).

In anthropology, too, theories have been pursued regardless of the relevant data. The original settlement chronologies of environmental limitation theory were merely guess-dates from sites that had not been systematically excavated, sampled, dated, mapped, or recorded (Evans 1950; Meggers 1952). When radiocarbon dates were published, the supposed Andean centers of expansion turned out to be much younger than the Amazonian sites hypothesized as their offshoots (Roosevelt 1991:313–314). The idea that early foragers and civilized farmers could not live in the tropical rain forest arose before any research at preceramic sites or centers of complex cultures, and it continues today (Meggers 1996; Sales Barbosa 1992) despite the abundant stratigraphic, artifactual, radiometric, isotopic, and paleobotanical evidence for the existence of such cultures (Roosevelt 1989, 1991, 1993, 1995b; Roosevelt et al. 1991, 1996).

This essay confronts the theoretical formulations with relevant bodies of data. The state of the habitat at different times and places is evaluated on the basis of quantified, statistically significant, vouchered, species-level identifications of excavated specimens with published stratigraphic contexts, direct radiocarbon dates, and measured stable carbon isotope ratios. The chronology of cultures is based on consistent series of statistically evaluated radiometric and luminescence dates run directly on cultural objects from stratigraphic contexts. Human population is assessed from detailed maps and counts of sites and excavated structures and strata. Human resource use is inferred from excavated plant and animal remains and/or quantitative stable isotope analyses of the human bones from recorded contexts.

Taken as a whole, the evidence presented in this essay and in the references indicates that late prehistoric humans in the Lower Amazon significantly transformed landscapes in the vicinity of their settlements. Some societies in floodplain regions widely altered topography by building large earth constructions that now support forests where only annual herbs grew previously. Societies in many regions built up extensive garbage dumps that altered soil quality over wide areas. People introduced numerous cultivars or exotic species into the forest and favored or inadvertently attracted certain local species through their activities. In all known occupied regions, people clustered particular species at or near their living areas. Although such activities changed the distribution of species in the affected zones, there is as yet no documented species extinction by a prehistoric Amazonian group. Apparently, the large size of Amazonia and its great physical and biological diversity protected economic species from regional extinction by prehistoric humans. Whether any rare endemic species incidental to human survival died out due to interference or habitat destruction by prehistoric humans we do not yet know; since such species often escape the notice of modern biologists, the hope of finding them in the archaeological record may be unrealistic.

In contrast, there are numerous examples of large-scale, detrimental alterations and local extinctions due to urbanization and industrialized utilization of Amazonian habitats (e.g., Skole et al. 1994). Neither these nor ancient human impacts are adequately considered in explanations of Amazon biodiversity patterns. Biogeographers attribute forest diversity patterns to climate, topography, and soil (Whitmore and Prance 1987), despite a repeated association of some of the diversity with human crops, cultural remains, and anthropic soils. Paleoclimatologists treat rainfall patterns as purely natural conditions (Clapperton 1993; van der Hammen and Absy 1994) without considering the impact extensive human-caused deforestation has on rain levels (Salati 1985; Salati and Marques 1984). In the same vein, ecological research on current forests treats them as virgin a priori (Grace et al. 1995; Phillips and Gentry 1994), without even consulting archival or archaeological records of their history of human use. The lack of recognition of past and present human impacts in such studies impairs the validity of conclusions about the causes of diversity patterns.

Knowledge of the effects of past human activities on Amazonian habitats can inform environmental conservation and development programs about the long-term viability of different land use systems in Amazonia. The compatibility of indigenous occupations and forest conservation has been broadly questioned (Redford 1992), and cultural diversity is rarely considered in conservation projects. Yet, the association of rain forests and indigenous peoples reveals a strong positive correlation between cultural and natural survival (Clay 1988). The history of forest use by prehistoric and living peoples suggests that they are compatible unless machines or outside markets intrude (Balée 1989; Hames and Vickers 1983; Jordan 1987; Piperno and Pearsall 1998; N. Smith 1999). The viability of agriculture in Amazonia has often been questioned, but the prehistory nonetheless shows that shifting, long-fallow cultivation was widely practiced for thousands of years in areas well forested today. Even localized intensive cultivation is viable, to judge from prehistoric evidence. Depending on local soils and methods, it can enhance regionwide sustainability by taking pressure off the lower-quality soils and fragile microenvironments (N. Smith 1999). In environmental conservation, as in paleoecology and human prehistory, practitioners need to evaluate their theoretical assumptions with the empirical evidence and adjust them accordingly.

THE LOWER AMAZON REGION

The Lower Amazon (figure 15.1) is one of the areas in the basin most impacted by human occupation. In this region the main channel runs roughly east-west along the equator from the mouth of the Tapajós River at Santarém and Monte Alegre to the mouth of the Amazon at Belém. The watershed extends south from the Guianas and north from the Brazilian Shield. An abundance of rainfall and high groundwater linked to current high ocean levels create both the rain forest and a dense network of rivers, streams, and lakes. Active floodplains lie near mean sea level; relict, earlier floodplains lie between 3 and 50 m above sea level; and upland plateau and hill ranges can reach from 50 to over 2,000 m elevation.

The quality of natural soils in the region depends largely on subsoil composition. Where there are alkaline volcanic bodies, lateritic crusts (iron-rich formations from tropical weathering of volcanic rock), limestone, or recent alluvium, the soils have good nutrients and moisture-holding capacity. Such areas are extensive at Monte Alegre, at Altamira, in Carajas, and on the large islands and Bragantina uplands at the mouth of the Amazon near Belém (MDME 1976). Here, intensive cultivation can be productive and reforestation vigorous. Sandy or rocky soils on acid crystalline rocks are more vulnerable under permanent cultivation because weathering cannot replace nutrients lost to leaching or erosion. Low-lying flooded land does not leach as readily but can suffer poor drainage and salt concentrations during the seasonal drought.

Even though well-textured, high-nutrient soils are a minority (about 25 percent),

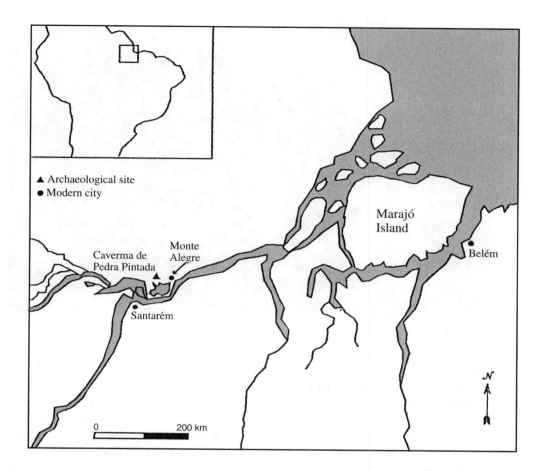

FIGURE 15.1. The eastern Amazon Basin, including Santarém–Monte Alegre area, and Marajó Island, Para, Brazil.

the region is so large that they make up a considerable resource in absolute terms. Even on soils of indifferent quality, slash-and-burn cultivation is quite sustainable if population levels remain low. Contrary to early thinking that soil nutrients were irrevocably lost during cultivation, it has been shown that nutrients locked up by chelation in cultivated soil can be released by mulching and fallowing (Jordan 1987). Even heavy, hard-to-manage hydromorphic soils were cultivated by prehistoric people by building fields up above the water table (Denevan 1966). Only the sourest, stoniest mountaintops or deepest, gluelike swamps in the Lower Amazon can be considered uncultivable by indigenous methods. All the more populous late prehistoric indigenous societies documented so far relied on crops for staple foods.

Although some sources on Amazonian soils consider only natural factors in assessing soils (Sombroek 1966), past human impacts are important factors in soil quality

FIGURE 15.2. Black Indian soil at Alter do Chaõ, near Santarém.

(Falesi 1974), creating both the best and the worst of soils. On the one hand, bad management has left some areas bereft of topsoil. In the hills of Monte Alegre, for example, more than 100 years of repeated cutting and burning for fuel and pasture has exposed the dark, loamy forest topsoils to intense oxidation and erosion, leaving little but infertile, droughty sands over wide areas. On the other hand, the most extensive agricultural soils in the Lower Amazon are in fact anthropic soils (Kern 1994; N. Smith 1980). Locally called Black Indian soils, they form when upland soils are altered by the buildup of human garbage rich in decayed vegetal and faunal matter, shells, bones, finely divided charcoal, and human excrement (figure 15.2). Although zonal upland soils usually must be long-fallowed to recover from cultivation, anthrosols have been successfully cultivated continuously for more than 100 years. Lands on the upper terraces along the Lower Amazon served as productive commercial sources of sugarcane and timber in the nineteenth century (H. H. Smith 1879). They have rapidly regained closed-canopy tropical rain forest since abandonment at the turn of the century. Nineteenth-century illustrations show denuded hills that today bear tall, diverse, mature evergreen tropical rain forest (ibid.:152).

The climate of the Lower Amazon is typically very rainy and warm (Salati 1985; Salati and Marques 1984). Although temperatures are not high under mature tropical rain forests, because of shade and evaporation, the extensive deforestation of the last 100 years in the Lower Amazon has resulted in high ground temperatures. Annual rainfall in the Lower Amazon varies from highs of more than 3,000 mm in the east to less than 2,000 mm in the west. Along the mainstream at the Tapajós confluence recently deforested areas have much lower rainfall than areas of mature forest. Studies of tropical rain forest water budgets suggest that as much as half of rain is recirculated from the vegetation, so rainfall can drop by as much as half when the vegetation is removed (Salati 1985; Salati and Marques 1984). At Taperinha, which reforested early in this century, annual rainfall is nearly 3,000 mm. Just 8 km away, the ravaged landscape at Monte Alegre City gets only 1,700 mm. Progressively increasing deforestation there has been accompanied by statistically significant reductions in rainfall, according to municipal records (Sadeck n.d.). The vast majority of Amazonia has more seasonal rain than most textbooks state, and some trees are briefly deciduous, either in the dry or the wet season.

The current vegetation of the Lower Amazon is a diverse mosaic of tropical evergreen forest, semievergreen forest, woodland, savanna, swamp, and marsh (figure 15.3; Murca Pires and Prance 1985). Because of the huge biomass of these varied forests, they are the main source of food for both humans and animals. A pattern of diverse, widely dispersed species is common, but there also are numerous areas where a few species dominate. The multispecies forests are tall and dense but some lack true three-part stratification and tightly closed canopies. Many trees have starchy or oil-rich fruit pulp and kernels. These include palms (figure 15.4), which also have starchy stems, legumes, and a wide range of other food-bearing trees. Most are angiosperms, but some gymnosperm species (e.g., *Podocarpus*) occur dispersed with the other trees, especially on hilly land or low-lying land where moisture is abundant year-round (Murca Pires and Prance 1985; Lorenzi 1992; Robin Foster pers. comm.). Most forests

FIGURE 15.3. Floodplain grasses, Lago Grande de Vila Franca, near Santarém.

FIGURE 15.4. Understory forest palm, sacurí (*Attalea microcarpa*), at Caverna da Pedra Pintada, Monte Alegre hills.

also have crassulacean acid metabolism (CAM) photosynthesis epiphytes, which shift from low to high stable carbon isotope ratios in the dry season. The monodominant Amazon forests tend to have a single palm or leguminous tree stratum with herbs below. Closed-canopy Amazon forests have more negative stable carbon isotope means (−26 to −28 per mil) compared to temperate forests or open-canopy tropical forests (−22 to −26 per mil) (Tieszen 1991; van der Merwe and Medina 1989, 1991). The long-term history of vegetation and stable carbon isotope ratios in the Lower Amazon, given below, suggest that both the diverse and monotypic upland forests have been chronically disturbed by humans for a long time.

Influences on vegetation patterns include soil nutrients and texture, topography, rainfall, and history of land use (Murca Pires and Prance 1985). The relatively high overall rainfall means that major vegetation patterns are determined less by rainfall than by local drainage. Coarse, low-nutrient sediments and indurated clays may bear low, deciduous, drought-resistant woody or herbaceous vegetation even in high-rain areas, if moisture is seasonally distributed. Where adequate moisture is available year-round, even poor, coarse sediments support evergreen rain forest. Organic-rich coarse soils or alluvial soils with some water-holding clay can support evergreen rain forest even in areas with seasonal rainfall (figures 15.5, 15.6). Where soils are wet year-round, low-diversity swamp forest vegetation of the palm *Mauritia flexuosa* and certain herbs proliferate, regardless of absolute local rainfall. Such vegetation is found in high-rainfall areas such as eastern Marajó Island, medium-rainfall areas such as Monte Alegre (figure 15.7), and low-rainfall areas such as poorly drained areas in the Guayana Shield (Roosevelt 1980:161–181, 1991:7–26).

Certain vegetation patterns can only be explained by ancient and recent human impacts. When several species dominate in terra firme forests, human activity is usually implicated (Anderson 1983; Balée 1989; Balée and Campbell 1990). In numerous published inventories of "virgin" tropical rain forest in the Lower Amazon, certain species are more common than expected (D. G. Campbell et al. 1986). These comprise trees that are introduced, cultivated, or otherwise encouraged by human disturbance: taperebá (*Spondias mombin*), jutaí (*Hymenaea courbaril*), Brazil nut (*Bertholletia excelsa*), silk cotton tree (*Ceiba pentandra*), açaí (*Euterpe oleracea*), tucumã (*Astrocaryum vulgare*), pupunha (*Bactris gasipaes*), babassu (*Attalea speciosa*), citrus (*Citrus* spp.), mango (*Mangifera indica*), and *Cecropia peltata*. When excavated, the soil of

FIGURE 15.5. Igapo forest, at Igarape de Gorgota near Monte Alegre.

FIGURE 15.6. Várzea forest on floodplain between Santarém and Taperinha.

FIGURE 15.7. Moriche (*Mauritia excelsa*) forest in the Lago Grande, between Monte Alegre and Paituna.

such oligarchic forests yields abundant remains of past human activities, such as structure foundations, garbage dumps, or cemeteries. Such remains are surprisingly widespread throughout Amazonia, and the anthropic upland forests that grow on the sites are extensive enough to be detected in remote sensing data due to the distinctive texture and reflectivity of their vegetation (Roosevelt n.d.). Other human-associated forests are the forests on the artificial earth mounds erected in floodplains by past and present humans on Marajó and the coastal plain of the Guianas. These include many of the same dominants in human-influenced terra firme rain forests (Anderson in Roosevelt 1991:171–172). Without the mounds, such floodplain areas bear herbaceous vegetation rich in grasses and sedges.

In disturbed forests, understory palms like tucumã and sacurí spread more rapidly than many other trees (Henderson 1995). Highly desirable species such as açaí can occur as monodominants through the direct involvement of humans. The dense açaí patches common along the lower Amazon floodplain are coincident with present or past human dwellings where açaí was heavily utilized (figure 15.8). Their existence correlates with intensive, long-term açaí seed discard or planting by humans.

Grasses and other herbs are a normal part of the Amazonian vegetation complex. The tall evergreen forests include a wide range of herbs, many of which are good food sources, like the starchy stems and roots of species of Marantaceae. Grasses that grow under the shade of forest are usually C_3 in photosynthetic pattern. Nonwoody vegetation with carbon-4 photosynthetic pattern is a common native Amazonian successional stage that predominates on recent river floodplains and lake edges. Locally, it has a very high harvestable biomass (Junk 1984). The edible grasses among these include C_4 *Leersia hexandra,* as well as some recent introductions, like the C_3 *Oryza sativa.* Which of the edible herbs of the floodplain was the "brown rice" collected by indigenous populations at the mouth of the Amazon is not known (Brochado 1980). In heavily overgrazed areas, like eastern Marajó, some native grasses are extinct locally, and poisonous or inedible herbs have taken their place (Roosevelt 1991: 13–17).

On patches of coarse, droughty, or poorly drained soil within zones of terra firme and floodplain forests, there are vegetation communities structurally similar to wooded and grass savanna. The upper Río Negro has extensive, low-statured, drought-resistant forest on deep, white sands (Murca Pires and Prance 1985). Savannalike vegetation predominates on the heavy, seasonally flooded and desiccated soils of eastern Marajó (Roosevelt 1991:13–19) and on recently denuded rock outcrops and clays in the Monte Alegre hills cleared for pasture or maize cultivation (figure 15.9). Edaphic conditions hinder tree roots from having year-round access to moisture, oxygen, and nutrients, and repeated burning limits the spread of woody vegetation. Accordingly, although these communities have abundant annual rainfall between 2,000 and 3,000 mm, in physiognomy they resemble the low or sparse vegetation of much drier regions. Trees are short, often twisted, and have coarse, leathery leaves. Carbon-4 grasses and other herbs form a dense basal story.

The Amazon savanna patches have been assumed to be composed of species that invaded from the Orinoco plains and central Brazil cerrado during Pleistocene aridity (Whitmore and Prance 1987), but vegetation surveys do not support this idea. Few

FIGURE 15.8. Anthropogenic açaí (*Euterpe oleracea*) grove at Taperinha.

Amazonian savanna animal or plant species are found in the cerrado or Orinoco plains, which have a different set of species (Cardoso 1995; Murca Pires and Prance 1984). The former appear to be a group of long-term endemics of Amazonian microenvironments plus a few Holocene invaders from the adjacent Atlantic littoral and elsewhere (Cavalcante 1991).

Some savanna patches of the Lower Amazon appear to be anthropogenic, not natural climaxes (Sioli 1974), and persistent human clearing and burning can maintain savannalike vegetation on coarse or heavy soils whose climax would ordinarily be tall evergreen forest. Some open, grassy landscapes, such as occur in the sandy plateaus of Monte Alegre, were considered a natural consequence of Pleistocene savannaization (Meggers 1975; Whitmore and Prance 1987), but numerous species identifications and stable carbon isotopes of 56 dated late Pleistocene tree specimens from prehistoric deposits in that area show closed-canopy tropical rain forest without the grasses or savanna trees in late Pleistocene and early Holocene prehistory (figure 15.10; Roosevelt et al. 1996). Similarly, C_4-type stable isotope ratios are absent or rare in Amazonian sites until after 4000 B.C. The historical ecology reveals that the vegetation of these particular savanna patches is not a long-term natural climax but an adaptation to recent human deforestation and soil degradation. For example, native people in the Bolivian Amazon clear forest to open habitat for certain game species (Robin Foster pers. comm.). Such recent human impacts on vegetation and soil are an integral but little recognized part of the causality of microenvironmental patterning in the Amazon today.

Among the fauna of Amazonia, large terrestrial game makes up the smallest part of the biomass. However, even these rare creatures are a useful source of food or other materials for small, relatively mobile human populations. Since large game is a much less abundant and less predictable calorie source than that of other resources, such as oily tree fruits, starchy roots or stems, and aquatic fauna, it invariably is a supplement rather than a staple in Amazonian diets, which are limited primarily by calories and certain vitamins, not by protein availability (Dufour 1994).

Terrestrial game is influenced by both forest and hydrological regimes and can be ambushed at water bodies and groves of fruiting trees. Large, conspicuous, slow-reproducing water mammals, such as the manatee (*Trichechus inunguis*), are particularly vulnerable to hunting pressure. Manatee, forest deer, and some primates become

FIGURE 15.9. Amazonian anthropogenic savannas on denuded rock outcrop, Monte Alegre hills.

FIGURE 15.10. Burned food remains from Pleistocene campsite at Caverna da Pedra Pintada: jutaí (*Hymenaea* cf. *oblongifolia* and *H. parvifolia*) seeds. The left seed gave a stable carbon isotope ratio of −27.0 per mil (normalized to wood value), indicating that it came from a dense, closed-canopy tropical rain forest.

locally extinct under modern hunting pressure. Among fauna, invertebrates such as beetles, ants, and termites are the majority of the biomasss by weight, and the more digestible larval stages can be gathered from breeding locations. However, the largest and most accessible, reliable faunal food sources for humans are the smaller aquatic species: fish and shellfish. Shellfish are available year-round in localized beds, but fish tend to be seasonally available. Dry-season fishing can take a large proportion of the breeding population, but wet-season flooding limits fishing effectiveness and allows recolonization of depleted areas through the flooded forests. Seasonal abundance can be smoked, dried, and kept for several months. Isotopic studies show that the food base of fish is primarily litter washed into the water from mature forest, not floodplain herbs, which are food for creatures such as the manatees and capybaras. Fish carbon isotope ratios are usually C_3 due to the dominance of forest products at the base of the food chain. Just as on land, the smallest aquatic faunas are the most abundant in the biomass and the main source of animal protein for large, sedentary human populations.

PALEOENVIRONMENT

Many sources state that the evidence shows that the Amazon region was a temperate savanna with only small forest patches until the Holocene, when the tropical rain forests spread to cover most of the basin (Absy 1979, 1985; Clapperton 1993; Haffer 1969, 1987, 1993; Meggers 1996; van der Hammen and Absy 1994). As mentioned above, however, the evidence originally cited for Pleistocene savannas consisted of modern biogeographic and geomorphological patterning, not paleoclimatic data. Modern biological patterns cannot be proof of paleoclimate, and there is no consensus about the reasons for this patterning, whether collecting bias, alluvial processes, tectonic events, human impacts, or climate change (K. S. Brown 1987; K. S. Brown and G. C. Brown 1992; K. E. Campbell and Frailey 1984; Nelson et al. 1990; Salo et al. 1986). Geomorphological evidence for Pleistocene savannas is a mixed bag of undated features that either are perfectly consistent with rain forest conditions or are Holocene, not Pleistocene. The features' varied characteristics are explained primarily by humid tropical weathering processes and eustatic (sea level) and tectonic changes (Irion 1984a,b; Klammer 1984; Putzer 1984; Sanchez 1976). The supposed desert "red beds" are lateritic soils, formed under humid tropical conditions, not in superarid climates. The "stone lines" are either lateritic formations or gravels cut from bedrock by headwater rivers; they still form today in the tropical rain forests. The tropical sand dunes were in the Orinoco, not in the Amazon, and have Holocene, not Pleistocene, radiocarbon dates (Roa 1980). They form when sand is blown from river beaches exposed during the season of low water (Roosevelt 1980:166–178).

Sparse, poorly dated pollen and macroplant data from early research were at first interpreted as strong support for the refuge theory (Clapperton 1993). Most cores claimed as evidence for "severe" Pleistocene desiccation (Clapperton 1993; Haffer 1969, 1987:13–15; van der Hammen and Absy 1994; Whitmore 1990:87–94) were cores from present-day semiarid areas distant from the Amazon lowlands or from high-elevation or high-latitude regions marginal to the equatorial lowlands. Since high elevations and areas distant from the equator are particularly vulnerable to impacts from global climate change, they cannot be valid proxies for equatorial lowland regions, where hyperhumidity and high temperatures buffer against global climate change. Some regions touted as having evidence of Pleistocene savannas are regions that today have major herbaceous components, such as the Guianas and Brazilian highlands. That they might have had as much herbaceous vegetation along with trees in the Pleistocene is hardly surprising, but it is not evidence for Pleistocene savannas in present-day rain forest areas. Interestingly, even in the peripheral regions, comparisons of present and prehistoric vegetation show a rainier, more wooded landscape for most of the Pleistocene than today, rather than drier and more open than today (Absy et al. 1991; Behling 1995; Chaves and Renault-Miskovsky 1996; Ledru 1993; Wijmstra 1971; Wijmstra and van der Hammen 1966; van der Hammen 1963).

The abundant and well-dated pollen and macroplant data that have recently become available from the forested equatorial lowlands indicate that tropical lowlands were dynamic but did not become savannas. The evidence from securely dated pollen series, numerous dated and taxonomically identified macroscopic floral and faunal specimens (figure 15.10), and extensive stable carbon isotope analyses show the undeniable presence of extensive, heterogeneous, canopied lowland rain forest and river floodplains in the equatorial lowlands of Amazonia in the late Pleistocene and no evidence of extensive grass savannas or savanna woodlands (Absy 1979, 1985; Absy and van der Hammen 1976; Absy et al. 1991; Athens and Ward 1999; Behling 1996; Bush et al. 1989; Colinvaux et al. 1996; Haberle and Maslin 1999; Liu and Colinvaux 1985; Roosevelt et al. 1991, 1996; van der Hammen and Absy 1994; van der Hammen et al. 1992).

The securely dated Pleistocene core assemblages of numerous sites in the Ecuadorean Amazon, the Colombian Amazon, and the Brazilian Amazon document the presence of closed-canopy tropical rain forests, not savanna woodlands or grasslands. All had pollen typical of rain forest sediments (such as Myrtaceae, *Miconia,* and *Alchornea,* etc.) but not the assemblages characteristic of savannas. Percentages of grass pollen range from traces to over 35 percent, percentages similar to Holocene Amazon rain forests but much lower than in savannas, which have over 90 percent (Absy 1979, 1985; Absy et al. 1991; Behling 1995; Ledru 1993; Wijmstra and van der Hammen 1966). The large series of 56 stable carbon isotope ratio assays on trees dating between 11,200 and 10,000 at Monte Alegre in the Brazilian Amazon are typical of closed-canopy tropical rain forest (Roosevelt et al. 1996), as are the ratios from the Ecuadorean Amazon (Athens and Ward 1999: 292, table 1).

One of two cores claimed as evidence for Pleistocene savannas was from a petroleum exploration core in Ecuador (Athens and Ward 1999), but it has tropical rain forest arboreal taxa and only minor percentages (15 percent) of grasses, less than many rain forests (Absy 1979, 1985). No savanna, which by definition has a dominance of grass (Harris 1980), had that little grass. Certain taxa (such as *Croton* and *Byrsonima* and Asteraceae) whose presence is cited as evidence of savanna woodlands are in fact commonly present in Amazon rain forests and their cores today (Absy 1979, 1985). Furthermore, throughout both the Pleistocene and Holocene, the stable carbon isotopes of the Ecuadorean core fall in the negative range typical of dense tropical rain forest (-25.5 to -31.1 per mil, average 28.1 per mil, $N = 7$), not of savanna (-8 to -15 per mil) or of savanna woodland (-15 to -26 per mil) (Tieszen 1991; van der Merwe and Medina 1989, 1991). The date of the Ecuadorean core segment is unknown, because of its discordant and infinite dates (Athens and Ward 1999).

The single existing Amazon core with very high grass percentages and C_4 carbon isotopes (Absy and van der Hammen 1976) still lacks the tree combination considered typical of savanna: *Byrsonima* with *Curatella americana* with no rain forest species (Absy 1979, 1985; Absy et al. 1991; Behling 1995; Wijmstra and van der Hammen 1966). Instead, it resembles recent Amazon floodplain cores, which have abundant C_4 grass pollen with small amounts of pollen of rain forest indicator trees. The age of this pollen assemblage is not known because it has strongly discordant radiocarbon dates and combines Holocene and Pleistocene sediments through a stratigraphic reversal (Absy and van der Hammen 1976; van der Hammen and Absy 1994: 255, fig. 2, stratum C). For example, the same sample location gave the dates 18,500 and 42,000 B.C.

The widely espoused hypothesis of extreme cold in the Pleistocene is not supported by the occurrence of taxa or carbon isotopes in cores, either. During certain times in the late glacial period, humid tropical montane species now common between 1,000 and 2,000 m elevation today in Amazonia increase in some lowland cores, a finding interpreted as evidence for extremely cold, dry nontropical conditions in the lowlands (Clapperton 1993) or for very cold, moist conditions (Behling 1995; Colinvaux et al. 1996; Liu and Colinvaux 1985). However, none of the lowland cores has more than minor percentages (ca. 10 percent or less) of the common montane genus *Podocarpus.* The mere presence of such pollen does not mean a humid tropical forest was replaced by montane forest. The very abundant, buoyant pollen of *Podocarpus, Alnus,* and *Hedyosmum* are strongly dominant where they grow and travel in large quantities long distances in air and water. Their proportions in sediments are greatly overrepresented in comparison with other trees, especially lowland tropical rain forest trees, most of which are fauna pollinated and are rare or absent in sediments. Cores from tropical lowlands below the montane forests of the Andean slope would be expected to have high percentages. *Podocarpus* has the most abundant and buoyant pollen of the three genera (Coetzee 1967:54) and is the genus most common in lowland cores both today and in the past. Two species of *Podocarpus* and several of *Hedyosmum* trees commonly occur dispersed through the Amazon rain forest today. Where many grew near a coring site, their pollen representation would be even higher than in the Pleistocene cores. Like the lowland rain forest indicator species they occur with, podocarps require constant high moisture and cannot be evidence of dry conditions. Modest gymnosperm percentages similar to the Pleistocene ones are routinely found in recent sediments of the Amazon and other lowland equatorial rain forests and are properly considered to be allogenic indicators for the existence of highland vegetation far distant from the lowlands (Absy 1985; Bonnefille 1995; Coetzee 1967:59–60). Conclusions of cooling in the Amazon based on such occurrences are invalid. Furthermore, forests in cold climates have significantly more positive stable isotope ratios than do tropical rain forests (van der Merwe and Medina 1991), so the existence of montane forest in the lowlands is not supported. The negative stable carbon isotope ratios and the association of substantial pollen of humid tropical rain forest tree species with the gymnosperms shows that the rain forest was the dominant woody vegetation, given the overrepresentation of gymnosperms and underrepresentation of lowland rain forest trees in pollen rain.

Certain changes in the above-mentioned offsite sediments and archaeological deposits signal hydrological change linked to eustatic fluctuations and tectonic activity, rather than changes in local rainfall and temperature (Absy 1979, 1985; Irion 1984a). Geophysical studies of the Amazon bed show deep canyon cutting during glacial maxima, when much of the world's water was bound up in ice and sea levels plummeted. In the pollen of the time, species adapted to poorly drained soil, such as moriche, disappeared from wetland cores, and Amazonian forest taxa adapted to well-drained soil microenvironments increased. In contrast, when groundwater levels are higher during early Holocene eustatic highs, sediments reflect higher and less seasonal rivers, and species such as moriche become common in some regional cores. Such changes were formerly interpreted as evidence for glacial Amazonian savannas alternating with rain forests, but the record shows minimal grass and continuous presence of rain forest throughout the sequence.

The pollen and isotope records and current taxonomy indicate that the Amazon forest belt never opened sufficiently during the Pleistocene for savanna species to invade from the north or south. Isotopic and botanical evidence of drying and opening of the lowland equatorial Amazon rain forest is first detected only in the later Holocene, and its timing and biocultural associations implicate human disturbance as an important cause.

LATE HOLOCENE CULTURES OF THE TAPAJÓS CONFLUENCE REGION

Comprehensive knowledge of late prehistoric land use at the Tapajós confluence exists only for the Incised and Punctate horizon, between about 1,950 and 400 radiocarbon years ago (A.D. 1000 and 1550). This widespread horizon occurs both in the Amazon and in northern South America. Its place of origin is uncertain, but possible precursors have been found in eastern South America in the period 1450 to 1950 B.P. (A.D. 500 to 1000) (Roosevelt 1997). Several Santarém region sites of the horizon give a useful picture of local ecology and land use in the eastern Amazon just before the penetration of Europeans. The horizon has been associated with the ethnohistoric Tapajó culture, observed in the same area in early Contact times (Palmatary 1960). The elaborate Santarém material culture; large, complex archaeological sites; and Contact period data on political organization, ritual, and warfare in the area are interpreted as evidence for the existence of a populous, warlike indigenous complex society ruled by paramount chiefs, organized by ranking and labor specialization, and supported by intensive agriculture, collecting, and craft manufacture and trade.

The Santarém Incised and Punctate horizon is remarkable for its elaborate effigy pottery, complex geometric incision, and figurative modeling (Palmatary 1960). The pottery sometimes also bears rectilinear polychrome painting or overall red paint. Diverse ground lithic artifacts range from jade ornaments to axes, projectile points, and probable war clubs. The present city of Santarém may have been one of the culture's centers; the archaeological site there is about 4 km^2 (the Aldeia plus the commercial port). Our excavations and radiocarbon dates indicate that the classic Santarém culture flourished mainly in the prehistoric period, rather than at the time of Contact (Quinn n.d.). There were few Postconquest radiocarbon dates and objects of European origin in undisturbed deposits at the three sites. The society's population was apparently large and permanently settled. There are numerous large archaeological middens on the bluffs and banks overlooking the floodplain. Remains of substantial structures and large black-soil garbage dumps have been found in all the sites studied.

The Tapajó culture is known from accounts by early explorers, colonial administrators, and missionaries between about 400 and 250 calendar years ago (A.D. 1541 to 1700) (Acuna 1891; Bettendorf 1910; Heriarte 1964; Nimuendaju 1949; Rowe 1952). Large villages stretched almost continuously along the floodplains at the time of the first European entry, and the invaders entered a town with ceremonial structures and numerous large residences. The area was well known for its fine, decorated pottery and decorated gourd vessels, which were traded widely. Early observers stated that maize was the staple food, and manioc was not as important as in other regions. All the common indigenous cultigens and foraged foods of present-day Amazonian indigenous groups were known there also. Maize and other crops were grown on the extensive floodplain land, according to several accounts. What cultivation was like on the sandy, poor-soil terra firme around Santarém city is not clear, but today, people still cultivate house gardens. Chroniclers generally comment on the abundance of food from both plants and aquatic fauna and mention that turtles were collected and maintained in large numbers in corrals.

The archaeological evidence for environment and land use comes from three black-colored garbage middens of the Santarém culture excavated by our project at Tape-rinha, Monte Alegre, and Santarém city. Due to the different extents and site contexts of the excavations, the results at each site are somewhat different. At Caverna da Pedra Pintada, Monte Alegre, midden deposits of the period were excavated in eleven contiguous meter-square units. They were found in the upper stratigraphic levels of the site in and around a large post structure. Ten radiocarbon dates place this deposit in the 300-year period immediately before the first European settlements. The cave occupation was on the fringes of the core Santarém phase occupation areas, a sort of hinterland or hillbilly occupation a two-day walk from the town center at Parizo near Monte Alegre. This group had pottery typical of the Santarém phase but not as elaborate. Potsherds had the basic shapes, typical red paint, and occasional incision, but not the complex modeled ornamentation. At the time, the cave had a small but probably permanent settlement of a single residential group housed in a large, sturdy post-and-thatch structure. The structure might be interpreted as evidence for an interest in security, since the cave itself would presumably have been sufficient shelter.

The biological remains recovered from this deposit were extraordinarily varied. Annual cultigens recovered include several fragments of maize cobs (*Zea mays,* coroico race) and pieces of decorated vessels of *Cucurbita* gourds, species as yet unidentified. The maize could easily have been grown locally, since it is presently grown in quantity in fields only a few kilometers away just outside the rock outcrops. The archaeobotanical evidence for maize is consonant with ethnohistoric accounts of maize cultivation in the confluence area at the time. Manioc, although not yet identified in the food remains, is a likely domesticate, as mentioned above.

Also in the cave midden were abundant fruits, wood, and leaves of many species of trees indigenous to the Lower Amazon. Many of these were already in use during the late Pleistocene occupation. They include Brazil nut, apiranga (*Mouriri apiranga*), pitomba (*Talisia esculenta*), apixuna (*Coccoloba pixuna*), achuá (*Sacoglottis guianensis*), tarumã (*Vitex cymosa*), muruci da mata (*Byrsonima crispa*), jutaí or jatobá (*Hymenaea courbaril, H. oblongifolia, H. parvifolia*), tucumã (*Astrocaryum vulgare*), sacurí (*Attalea microcarpa*), and curuá (*Attalea spectabilis*). Clearly, long-term indigenous exploitation of these species for food and materials did not lead to their extinction. Many of the tree species identified in the Santarém phase occupation but not in the Paleoindian occupation are cultivated, foreign, or floodplain species. These include taperebá (*Spondias mombin*), maracujá do mato (*Passiflora nitida*), cashew (*Anacardium occidentale*), forest cashew (*Anacardium giganteum*), gogo de guariba (*Moutabea chodatiana*), cupixygoba or uxi do campo (*Endopleura uchi*?), itauba (cf. *Silvia itauba*?), mandiocaçu (*Casimirella rupestris*), rabo de arara (*Norantea guianensis*), cedro (*Poupartia amazonica*), cultivated muruci (*Byrsonima crassifolia*), mucujá (*Acrocomia aculeata*), jatá (*Syagrus cocoides*) açaí (*Euterpe oleracea*), moriche (*Mauritia flexuosa*), caraná (*Mauritiella armata*), etc. Some of these species may occasionally have been used by Paleoindians but too rarely to be preserved archaeologically. However, some of these species are valuable orchard trees or trees that are favored when rain forest is cleared. Taperebá and muruci do campo are trees commonly cultivated in local gardens and forest clearings. Mucajá and jatá require forest openings and thrive in the sandy clearings in pasture cut in disturbed rain forest in the Amazon terra firme. Jatá belongs to a group of species concentrated in the drier regions of Brazil (Henderson 1995:128–136). Its presence may be due to the progressive thinning and drying of the local forest as human populations grew and land use intensified. Moriche requires constant moisture, for which the floodplains of the later Holocene would have provided better than the slender Pleistocene plains. Açaí would be considered a cultigen in the Monte Alegre mountains because it has to be planted and watered in areas that have well-drained, seasonally dry soils. In the hills today, the few available springs provide moisture for only a few individual açaí or moriche. Cashew (*Anacardium occidentale*) is not considered indigenous locally but an invader from the littoral. Its presence at this time may be related to extensive clearing for fuel or maize cultivation. The tree invades and proliferates rapidly at the expense of other trees, due to its rapid germination and drought resistance (N. Smith et al. 1992).

In addition to the plants, large numbers of a wide range of faunal species were recovered from the Santarém levels of the cave. These included caiman or jacaré (*Caiman* sp.), caititú or collared peccary (*Tayassu tajacu*), tapir, deer (*Mazama americana,* sometimes juvenile), monkey, many rodents, paca (*Agouti paca*), cutia (*Dasyprocta agouti*), tatú (*Dasypus* sp.), many birds, lizards, river turtles, tartaruga (*Podocnemis expansa*) and tracajá (*P. unifilis*), terrestrial univalves, insects, etc. The majority of specimens, however, are from fish: large armored catfish, tamuatá (*Hoplosternon littorale*), mandí (*Pimelodus armatus*), bacú (*Lithodoras* sp.) and other Doradidae, aruaná (*Osteoglossum bichirrosum*), pirarucú (*Arapaima gigas*), etc.

The late prehistoric opening up of the forest at Monte Alegre is reflected in the stable carbon isotope ratios of archaeological tree specimens, which range several per mil less negative than the late Pleistocene ratios at the site. As human population density and land use intensity increased, the density of the forest decreased, according to the isotope ratios. Now, in the regenerated forest near Pedra Pintada, the range is 2 per mil more negative.

The Santarém city site is an extensive black-soil midden (ca. 4 km²) primarily in and around the port area and the adjacent Aldeia, or native village. Within the midden, geophysical survey and test excavation revealed the presence of apparent house floors with bell-shaped storage pits (locally called bolsões), which also had been uncovered earlier during construction in the Santarém Aldeia. Most of the food remains were excavated from one of the storage pits and adjacent floors. The nature of the remains suggest that they derive from a termination ceremony. In the pit were quantities of broken but highly decorated pottery, numerous small flint grater chips, carbonized tree fruits, and remains of large turtles. Only a few maize kernels were found, but if food processing procedures resembled those of the Contact period people of the area, then maize may have been converted to beer for ceremonies. On the top of the pit were traces of decayed gourd bowls. The few food remains incorporated into the superimposed house floors consisted of tiny fish bones, palm fruits, and grater chips. Despite the possible importance of maize for ceremonial beer consumption, recorded in the chronicles, manioc could have been a useful crop locally, since it can be grown on sandy, poor soil. The large turtles probably can be interpreted as feast food, and the small fish as the faunal supplement to the plant staples of daily subsistence.

The third Santarém phase black-soil midden excavated in the confluence area is the uppermost prehistoric deposit at the archaeological site overlooking the floodplain at Taperinha. There, among rare maize and numerous tree fruits, a wide range of turtle remains, shells of three species of freshwater pearly mussels, bones of numerous species of fish, and bones of tapir (*Tapirus americanus*) were recovered. Larger, deeper black-soil middens occur on the bluff crests above this site but have not yet been excavated.

To summarize, the major site at Santarém city had less diverse biological remains in its refuse than the small hinterland site at Monte Alegre. This contrast suggests that major settlements focused on a narrow range of highly productive, localized resources because their populations were too large and sedentary to rely mainly on resources that were rare, scattered, and slower to reproduce. It is possible that some forest fauna and flora were extinguished locally in this period, but if so, they recovered quickly, since no absent species were noted in the early historic period. The tapir bones at Monte Alegre and Taperinha are evidence that regardless of the extensive, sedentary late prehistoric occupation at Santarém, large forest animals such as tapir had not yet been driven to extinction in the rural areas. The stable isotopes of forest species in rural Pedra Pintada range one or two per mil more negative than those from Santarém City. The isotopic evidence for reduced forest canopies during the Incised and Punctate horizon compared with earlier periods is consonant with the ethnohistoric evidence for savanna patches in the Tapajós confluence area at first Contact (Carvajal 1984) and suggests that the savannas were anthropogenic formations created by human clearing and burning. We need more analyses for the late prehistoric period at the Tapajós confluence, but the more numerous assays from other regions of Greater Amazonia for the time show a similar pattern of increasingly positive ratios (see appendix).

HISTORIC PERIOD

Overall, human impacts on habitats in the Lower Amazon appear to have increased since the Conquest. The population of Amazonia seems to have become more urbanized, while some rural areas have lost population. Locations far from transportation seem to be less occupied now than in prehistoric times and have experienced Postconquest reforestation. The intensity of land use and habitat destruction, however, has tended to increase around towns and cities and in rural parts where industrial plantations, ranches, subsidized agricultural colonization, or timbering operations were established (H. H. Smith 1879; Swales 1993). Throughout the Lower Amazon, widespread environmental change has taken place as a result of the establishment and subsidization of the free-range cattle industry.

In the Tapajós confluence, land deeds and municipal records show that in the late eighteenth and nineteenth centuries A.D., cropping and animal husbandry were carried out at places like Taperinha by Euroamerican owners who took over native land (Hagman family documents and archives of Helcio Amaral at Santarem). During the last half of the nineteenth century, Taperinha in particular underwent deforestation for timber export and for fuel exploitation to support cane liquor production and iron forging for export. Illustrations of the middle of the past century show its uplands devoid of forest vegetation (H. H. Smith 1879). When slavery was abolished in Brazil in 1888, Taperinha was abandoned for industrial use, and the forest has since grown back to a large degree. At Monte Alegre since the Conquest, some upland forests have been decimated by burning to clear pasture (Swales n.d.). Extensive cattle pastures maintained by annual cutting and burning were already in existence at the time of Wallace's visit in the mid-nineteenth century (Wallace 1889). Essentially, Indians were replaced by cattle in the Lower Amazon in the period from about 1750 to 1850, and this change is recorded archaeologically in the replacement of indigenous artifacts and food remains by desiccated cattle dung in the cave at Pedra Pintada.

Some of the trees used for food regularly in both early and late prehistory in Monte Alegre, such as Brazil nut (*Bertholletia excelsa*), are no longer found in the vicinity of the cave because cutting, burning, and pasturing of cattle have eliminated them. Tucumã (*Astrocaryum vulgare*) can be found sporadically in the cave area but has been extinguished by deforestation around Monte Alegre city, where women lament that they cannot make basketry because tucumã is no longer available nearby. Other trees, such as cashew (*Anacardium occidentale*), now grow in extensive groves, apparently because the increasing cutting and burning has opened up habitats that it can invade at the expense of native trees. The species is found by the hundreds in the anthropogenic pastures of denuded sandstone hills near Monte Alegre. Palms that survive cutting and burning, such as sacurí (*Attalea microcarpa*), are still found several kilometers from the cave site, and others that thrive on disturbance, such as curuá (*Attalea spectabilis*) are veritable weeds in the cattle pastures on the poor sandy soils immediately surrounding Monte Alegre city. Both species regenerate stems from beneath the soil surface when the aerial parts are cut or burned off. Several domesticates not found in prehistoric deposits have been introduced to the area, such as species of citrus and mango (*Mangifera indica*).

Today in the confluence area, mature forest is limited to a few forest reserves and isolated areas distant from transportation. Such forests have considerably higher rainfall than deforested areas. For example, the ca. 10,000 ha of mature forest at Taperinha has had very high levels of rainfall a year in recent years, whereas at Monte Alegre City, only 8 km away, annual rain has decreased steadily over the years, with advancing deforestation from subsidized colonization so massive that its extensive quadrate clearings can readily be seen in satellite images. Similarly, forest fauna, such as deer and tapir, now locally extinct at Monte Alegre and Santarém, continue to be abundant in the large private forest refuge at Taperinha.

CONCLUSION

Archaeological and paleoecological data suggest that the human occupation of Lower Amazonia led to progressive transformations of vegetation and soils in the occupied areas. Not all areas were used as intensively as others, and most of the evidence for impacts come from locations at or near occupation sites. Conditions at offsite locations, therefore, may not be a good index of regional occupations (contrary to Athens and Ward 1999). There still seems to be no substitute for conventional regional site survey and excavation to assess the magnitude of human land use and its effects.

One of the major data patterns in the record is the shift from more negative stable carbon isotope ranges in the late Pleistocene to more positive stable carbon isotope ranges during the later Holocene (see appendix). While Pleistocene and early Holocene ratios are consonant with canopied tropical rain forest, ratios postdating the time of Christ in densely settled areas range more than 4 per mil more positive than earlier ratios. According to current knowledge of fractionation of carbon in different types of vegetation, it seems likely that in the later Holocene the forest canopy was opened by clearing and burning in areas near settlements. This progressive pattern seems unlikely to be a climate effect, unless climate modelers are completely wrong. Current models suggest that climate change should have magnified the negativity of vegetation in the Amazon during the late Holocene in comparison with the late Pleistocene. The Holocene forest is supposed to have been more luxuriant and extensive than the Pleistocene forest, which the modelers assume was drier, patchier, and cooler, all characteristics that lead to more positive ratios, not more negative ratios. Since climate-forced Holocene drying of the forest is understood to have been an episodic, not continuous, phenomenon, that could not explain the consistent pattern of greater positivity in Holocene ratios. The only likely explanation at present is that in the Lower Amazon, forest canopy was significantly thinned by human clearing in the more intensively inhabited areas. The conclusion is that the forest remained in a state of moderate disequilibrium throughout the later Holocene. In the vicinity of large or continuous settlements, the tall forest was replaced to some degree by cultivated crops. This change is reflected in the more positive stable carbon isotope ratios of human bones in later prehistoric times. Human ratios reflect vegetation several per mil more positive than in the early Holocene, a pattern interpreted as evidence of greater reliance on the C_4 cultivated plants, such as maize, which is indeed found among the macroarchaeobotanical collections of the period. Thus, although maize presumably came under cultivation in the Amazon by mid-Holocene times, there is no conceivable way it could have been an important food until the last 1,000 years of prehistory, since before then, human bone ratios show few effects of C_4 plants, which require clearings.

Late prehistoric forest disturbance by humans is also the likely cause of the spread of the monotypic terra firme palm forests that are a feature of territory occupied by indigenes today (Anderson 1983; Balée 1989). During the period of greatest forest opening, some tree species originally from the Brazilian littoral, such as cashew, turn up in the interior forest. Humans also, however, planted or encouraged large numbers of useful trees, such as Brazil nut, at their settlements, creating, with time, large valuable groves, many of which survived until the subsidized clearing for cattle in the last three decades. These groves were significant economic resources for Brazil. The majority of the Brazil nuts harvested for commerce in this century are from groves originally established in prehistory. Their elimination by the ill-judged clearing for pasture is a tragic mistake that will haunt the Brazilian economy for many years to come.

Another effect of prehistoric human activities was the deposition of wide areas of black soil, which developed from refuse, around the larger prehistoric settlements. This process began about 3,000 years ago, nearly 1,000 years before the accelerated clearing for intensive annual crop cultivation. The black soil features continue to impact ecology and land use today. The prehistoric black soils are currently the most important Amazonian soils in commercial agriculture. Away from modern settlements, such deposits continue to support diverse and productive ecosystems that are dominated by useful plant species and are rich in game.

Despite these significant effects of prehistoric humans on habitats, the impacts were much less severe than Postconquest ones. For example, no tree or faunal species are known to have become regionally extinct during prehistory. Today, however, some important species are now locally extinct due to habitat destruction or outright extermination. Most ancient Brazil nut groves are gone in the areas cleared for cattle or urbanization, and many animals are locally extinct due to the practices of ranchers, such as profligate burning of vegetation and killing of "vermin." With the diminution of valuable native species, less useful and manageable invaders have proliferated, such as weed trees and inedible pasture herbs. The reasons for the greater magnitude of recent environmental damage are several. Transportation has been mechanized, and Amazonia is more urbanized, so destroyed zones are larger and more deeply impacted by a larger population that has better access to remote areas. In addition, peasant migrants to Amazonia from the northwest are not familiar with local resources, and rural education is negligible. Land tenure practices, furthermore, work against security of ownership and long-term investment. The personnel of federal government agencies and foreign multinational companies, which control increasingly large areas, are even less informed than the new peasants. The centralization of land use, therefore, does not bode well for the future of the ecosystem.

The archaeological and historic record of Amazonia, however, is a potent source of information about the ecosystem, its history of use, and the consequences of different uses. How to improve knowledge of the history of habitats in Amazonia and their alteration by humans? There is need for better integration of archaeological and paleoecological studies. Thus, off-site research also should be carried out in the vicinity of well-studied archaeological sites, instead of primarily in isolated zones outside the major occupation zones. Paleoecological studies need to incorporate more precision and verifiability. Vegetation-based inferences about temperature changes should be tested with oxygen isotope ratios on dated specimens of fauna. Core studies need to be more comprehensive. Cores must always be interpreted against a background of knowledge of plant ecology and the actual pollen rain from a range of current habitats. More types of materials from cores need to be identified and isotopically analyzed. Particularly, the dates and stable isotope ratios of identified pollen, phytoliths, soil organic matter, and macrospecimens need to be compared, to clear up the serious interpretative problems. In archaeological studies, the comprehensive identification of fuel woods from both sites and offsite areas is needed for information about the changing landscape of trees and for comparison with the wood in cores. The history of cultivation of root crops such as manioc also needs to be traced by identifying and dating stemwood from pre-historic charcoals. The presence-absence of the rarer species from prehistoric contexts needs to be refined through comparative research on the morphology of modern and ancient species.

The evidence from land use in history and prehistory in Amazonia for the most part has not been taken into consideration by experts on conservation, ecology, or development in Amazonia. But what could be more relevant to their studies than knowledge about which human activities were the most destructive of habitats in the past and which land uses will be most sustainable in the future? Despite the evidence for greater damage from development by outside interests operating under top-down management and ignorance of local conditions, most government and internationally financed programs still privilege the rich, outsiders, and large hierarchical organizations in allocating ownership, extension, and credit. Today in Amazonia, the smallholders who are the main producers of food and exports and who are generally the most sustainable producers do not have access to such benefits. Furthermore, they are commonly represented as the main agent of destruction, when in fact greater damage is associated with government settlement programs and subsidies. Such scholarly misconceptions and human inequalities need to be addressed in the effort for a better future for the Amazon.

APPENDIX

TABLE 15.1. STABLE CARBON ISOTOPE RATIOS FROM
LATE PLEISTOCENE BIOTA FROM CAVERNA DA PEDRA PINTADA,
MONTE ALEGRE, PARA, BRAZIL

Species	Radiocarbon Years B.P.	$\delta^{13}C$ (‰)	$\delta^{13}C$ Normalized to Wood (‰)
Initial			
Carbonized sacurí seed	11,145 ± 135	−21.6	−23.6
Carbonized tucumã seed	11,110 ± 310	−23.2	−25.2
Carbonized tucumã seed	10,905 ± 295	−20.7	−22.7
Carbonized tucumã seed	10,875 ± 295	−22.5	−24.5
Carbonized tucumã seed	10,655 ± 285	−22.8	−24.8
Carbonized sacurí seed	10,560 ± 60	−21.2	−23.2
Carbonized wood	10,450 ± 60	−23.5	
Carbonized palm seed	10,470 ± 70	−24.9	−26.9
Carbonized sacurí seed	10,410 ± 60	−24.3	−26.3
Solutes from same	10,390 ± 70	−24.9	−26.9
Carbonized sacurí seed	10,392 ± 78	−23.6	−25.6
Carbonized sacurí seed	10,350 ± 70	−25.4	−27.4
Carbonized sacurí seed	10,305 ± 275	−22.7	−24.7
Carbonized sacurí seed	10,275 ± 275	−21.7	−23.7
Carbonized achua seed	10,261 ± 62	−24.8	−26.8
Carbonized sacurí seed	10,250 ± 70	−29.9	−31.9
Average of 16			−25.5
Initial/Early			
Carbonized curuá seed	10,450 ± 60	−23.3	−30.3
Carbonized sacurí seed	10,390 ± 70	−22.1	−29.1
Carbonized curuá seed	10,330 ± 60	−22.5	−29.5
Carbonized sacurí seed	10,320 ± 70	−23.1	−30.1
Carbonized wood	10,300 ± 60	−27.7	
Carbonized wood	10,290 ± 80	−27.6	
Carbonized wood	10,280 ± 70	−26.6	
Carbonized palm seed	10,260 ± 60	−25.8	−27.8
Average of 8			−26.1
Early			
Carbonized palm seed	10,490 ± 80	−23.7	−25.7
Carbonized wood	10,480 ± 70	−24.8	
Solutes of same	10,570 ± 70	−25.2	
Carbonized sacurí seed	10,450 ± 60	−29.9	−31.9
Solutes of same	10,000 ± 60	−30.0	−32.0
Carbonized wood	10,470 ± 70	−28.1	
Solutes of same	10,000 ± 60	−28.9	
Carbonized sacurí seed	10,420 ± 70	−26.2	−28.2
Solutes of same	10,250 ± 70	−25.7	−27.7
Carbonized wood	10,390 ± 60	−26.3	
Solutes of same	10,230 ± 60	−25.2	
Carbonized sacurí seed	10,380 ± 60	−25.8	−27.8
Solutes of same	10,510 ± 60	−25.0	−27.0
Carbonized wood	10,370 ± 70	−26.6	
Solutes of same	10,330 ± 70	−27.8	
Carbonized sacurí seed	10,370 ± 60	−24.8	−26.8
Solutes of same	10,180 ± 60	−24.3	−26.3
Carbonized wood	10,360 ± 60	−23.9	

TABLE 15.1. (Continued)

Species	Radiocarbon Years B.P.	$\delta^{13}C$ (‰)	$\delta^{13}C$ Normalized to Wood (‰)
		Early	
Solutes of same	10,220 ± 60	−26.0	
Carbonized sacurí seed	10,330 ± 70	−25.4	−27.4
Carbonized wood	10,310 ± 70	−29.9	
Solutes of same	10,210 ± 70	−28.6	
Carbonized sacurí seed	10,290 ± 70	−25.7	−27.7
Carbonized sacurí seed	10,290 ± 70	−24.0	−26.0
Solutes of same	10,120 ± 70	−25.8	−27.8
Carbonized jutaí seed	10,261 ± 70	−24.8	−26.8
Carbonized wood	10,260 ± 70	−27.5	
Solutes of same	10,210 ± 60	−27.4	
Carbonized sacurí seed	10,190 ± 50	−27.8	−29.8
Carbonized sacurí seed	10,110 ± 60	−24.7	−26.7
Solutes of same	10,190 ± 60	−25.2	−27.2
Average of 31			−27.4
		Late	
Carbonized jutaí seed	10,683 ± 80	−25.0	−27.0
Carbonized palm seed	10,230 ± 60	−26.8	−28.8
Carbonized palm seed	10,210 ± 60	−27.6	−29.6
Carbonized woody tissue	10,360 ± 50	−12.3	

TABLE 15.2. STABLE CARBON ISOTOPE RATIOS FROM EARLY HOLOCENE BIOTA

Species	Radiocarbon Years B.P.	$\delta^{13}C$
	Preceramic Archaic, Caverna do Gaviaõ, Carajas, Para, Brazil	
Carbonized wood	8,065 ± 360	−26.3
Carbonized wood	7,925 ± 45	−26.6
Carbonized wood	6,905 ± 50	−26.4
Carbonized wood	3,605 ± 160	−25.7
	Pottery Archaic, Taperinha, Para, Brazil	
Carbonized wood	6,860 ± 100	−28.1
Charcoal in pottery sherd	6,590 ± 100	−28.0
Soluble carbon in same	6,640 ± 80	−28.1

TABLE 15.3. STABLE CARBON ISOTOPE RATIOS FROM
MIDDLE AND LATE HOLOCENE BIOTA

Species	Radiocarbon Years B.P.	$\delta^{13}C$ (‰)	$\delta^{13}C$ Normalized to Wood or, for Bone, Normalized for Food (‰)
Formative, Caverna da Pedra Pintada, Monte Alegre, Para, Brazil			
Human cranium apatite	3,563 ± 68	−16.7	−26.7
Carbonized curuá seed	3,390 ± 140	−23.5	−25.5
Carbonized wood	3,725 ± 57	−27.3	
Carbonized wood	3,563 ± 57	−26.5	
Carbonized seed	3,510 ± 90	−26.4	−28.4
Carbonized seed	3,488 ± 64	−26.2	−28.2
Human tooth apatite	3,410 ± 40	−14.4	−24.4
Carbonized wood	3,286 ± 59	−25.3	
Average of 8			−26.5
Early Complex, Igarape de Moura, Para, Brazil			
Carbonized wood	2,395 ± 75	−26.5	
Carbonized wood	2,180 ± 35	−25.6	
Carbonized wood	1,580 ± 75	−26.6	
Early Complex, Ucayali Basin, Peru			
Human bone collagen	c. 2,000	−15.4	−20.4
Human bone collagen	c. 2,000	−15.3	−20.3
Human bone collagen	c. 2,000	−13.6	−18.6
Human bone apatite	c. 1,000	−7.0	−17.0
Human bone collagen	c. 1,000	−14.6	−19.6

TABLE 15.3. (*Continued*)

Species	Radiocarbon Years B.P.	$\delta^{13}C$ (‰)	$\delta^{13}C$ Normalized to Wood or, for Bone, Normalized for Food (‰)
Early Complex, Marajó Island, Para, Brazil			
Human bone collagen	c. 1,400	−17.8	−22.8
Human long bone collagen	c. 1,400	−19.2	−24.2
Human bone apatite	c. 1,400	−12.7	−22.7
Carbonized tucumã seed	1,335 ± 185	−23.8	−25.8
Carbonized wood	1,255 ± 165	−26.5	
Carbonized wood	1,210 ± 50	−24.4	
Carbonized wood	1,120 ± 60	−28.0	
Carbonized wood	1,110 ± 60	−27.6	
Human longbone apatite	c. 1,100	−15.2	−25.2
Human bone apatite	c. 1,100	−17.6	−27.6
Human bone apatite	c. 1,100	−17.8	−27.8
Human longbone collagen	c. 1,100	−18.1	−23.1
Human longbone collagen	c. 1,100	−17.9	−22.9
Human longbone collagen	c. 1,100	−19.6	−24.6
Fish bone collagen	c. 1,100	−23.0	−28.0
Fish bone collagen	c. 1,100	−22.8	−27.8
Human bone apatite	c. 1,100	−12.9	−22.9
Human long bone collagen	c. 1,100	−15.8	−20.8
Carbonized wood	1,080 ± 50	−24.5	
Carbonized wood	1,060 ± 50	−26.2	
Carbonized palm seed	1,030 ± 75	−25.4	−27.4
Carbonized açai seed	1,000 ± 90	−23.9	−25.9
Human bone apatite	c. 1,000	−12.8	−22.8
Human bone apatite	c. 1,000	−12.8	−22.8
Human bone apatite	c. 1,000	−12.7	−22.7
Human bone collagen	c. 1,000	−18.1	−23.1
Human bone collagen	c. 1,000	−16.3	−21.3
Human bone collagen	c. 1,000	−16.6	−21.6
Human bone apatite	c. 1,000	−9.6	−19.6
Human bone collagen	c. 1,000	−15.1	−20.1
Human bone collagen	c. 1,000	−14.3	−19.3
Human bone collagen	c. 1,000	−16.9	−21.9
Carbonized wood	995 ± 220	−24.8	
Carbonized wood	955 ± 195	−26.2	
Carbonized wood	930 ± 190	−23.3	
Carbonized wood	925 ± 75	−25.9	
Carbonized wood	860 ± 110	−27.0	
Fish bone collagen	825 ± 165	−11.0	−16.0
Carbonized wood	755 ± 110	−25.2	
Carbonized palm seed	675 ± 165	−24.4	−26.4
Average of 40			−24.2

(*continued*)

TABLE 15.3. (*Continued*)

Species	Radiocarbon Years B.P.	$\delta^{13}C$ (‰)	$\delta^{13}C$ Normalized to Wood or, for Bone, Normalized for Food (‰)
Contact period peoples, eastern Marajó Island, Para, Brazil			
Human bone apatite	c. 400	−13.1	−23.1
Human bone apatite	c. 400	−11.0	−21.1
Human bone apatite	c. 400	−10.7	−20.7
Human bone apatite	c. 400	−10.8	−20.7
Human bone apatite	c. 400	−11.0	−21.0
Human bone apatite	c. 400	−11.9	−21.0
Human bone apatite	c. 400	−11.9	−21.9
Human bone apatite	c. 400	−12.9	−22.9
Average of 8			−21.6
Late Complex, Caverna da Pedra Pintada, Monte Alegre, Para, Brazil			
Wood	675 ± 60	−25.0	
Carbonized wood	489 ± 56	−25.9	
Carbonized wood	469 ± 65	−24.6	
Wood charcoal	510 ± 56	−28.7	
Wood charcoal	502 ± 56	−25.5	
Wood	430 ± 110	−26.2	
Wood	429 ± 61	−25.5	
Solutes from sacurí seed	422 ± 67	−24.1	−26.1
Solutes from wood	418 ± 60	−24.2	
Tucumã seed	418 ± 56	−24.2	−26.2
Tucumã seed	416 ± 59	−25.5	−27.5
Average of 11			−25.9
Late Complex, Ucayali Basin, Peru			
Anteater bone apatite	c. 500–1,000	−13.7	−23.7
White-tailed deer bone collagen	c. 500–1,000	−22.3	−27.3
White-tailed deer bone collagen	c. 500–1,000	−22.3	−27.3
White-tailed deer bone collagen	c. 500–1,000	−22.4	−27.4
Monkey (*Ateles* sp.) bone apatite	c. 500–1,000	−15.5	−25.5
Monkey bone collagen	c. 500–1,000	−21.5	−26.5
Monkey bone collagen	c. 500–1,000	−21.1	−26.1
Monkey bone collagen	c. 500–1,000	−21.1	−26.1
Monkey bone collagen	c. 500–1,000	−21.4	−26.1
Capybara (*Hydrochoerus hydrochaeris*) bone collagen	c. 500–1,000	−15.9	−20.9
Capybara bone apatite	c. 500–1,000	−8.5	−18.5
Paca (*Cuniculus paca*) bone collagen	c. 500–1,000	−22.2	−27.2

TABLE 15.3. (*Continued*)

Species	Radiocarbon Years B.P.	$\delta^{13}C$ (‰)	$\delta^{13}C$ Normalized to Wood or, for Bone, Normalized for Food (‰)
	Late Complex, Ucayali Basin, Peru		
Paca bone collagen	c. 500–1,000	−21.8	−26.8
Paca bone collagen	c. 500–1,000	−21.3	−26.3
Paca bone collagen	c. 500–1,000	−17.3	−22.3
Peccary (*Tayassu* sp.) bone apatite	c. 500–1,000	−14.1	−24.1
Peccary bone collagen	c. 500–1,000	−21.4	−26.4
Peccary bone collagen	c. 500–1,000	−19.8	−24.8
Rabbit bone collagen	c. 500–1,000	−21.2	−26.2
Bird bone collagen	c. 500–1,000	−22.5	−27.5
Turtle shell apatite	c. 500–1,000	−20.8	−30.8
Fish bone collagen	c. 500–1,000	−23.0	−28.0
Fish bone collagen	c. 500–1,000	−24.8	−29.8
Fish bone collagen	c. 500–1,000	−24.3	−29.3
Fish bones and teeth collagen	c. 500–1,000	−22.5	−27.5
Human bone collagen	c. 500–1,000	−11.7	−16.7
Human bone collagen	c. 500–1,000	−12.1	−17.1
Human bone collagen	c. 500–1,000	−11.8	−16.8
Human bone collagen	c. 500–1,000	−11.7	−16.8
Human bone collagen	c. 500–1,000	−10.3	−15.3
Average of 30			−24.5
	Offsite soil, INPA Forest Reserve, Manaus, Amazonas, Brazil		
Carbonized wood	2,420 ± 75	−26.0	
Carbonized wood	1,080 ± 40	−26.7	
Carbonized wood	675 ± 120	−27.5	

TABLE 15.4. STABLE CARBON ISOTOPE RATIOS FROM
HISTORIC PERIOD BIOTA

Species	Radiocarbon Years B.P.	$\delta^{13}C$ (‰)	$\delta^{13}C$ Normalized to Wood or, for Bone, Normalized for Food (‰)
	17th–18th century, Caverna da Pedra Pintada, Monte Alegre, Para, Brazil		
Tucumã seed	280 ± 150	−21.9	−23.9
Carbonized wood	345 ± 63	−25.0	
Solutes from same	341 ± 56	−22.7	−24.7
Tucumã seed	302 ± 56	−25.4	−27.4
Sacurí seed	300 ± 56	−23.8	−25.8
Sacurí seed	295 ± 61	−26.7	−28.7
Palm seed	288 ± 62	−23.6	−25.7
Tucumã seed	367 ± 62	−26.5	−28.5
Solutes from tucumã seed	230 ± 62	−22.7	−24.7
Sacurí seed	211 ± 57	−27.1	−29.1
Average of 10			−26.4
	20th century, pasture surface, near Lake Arari, Marajó Island, Para, Brazil		
Collagen from cow bone	1980s	−17.6	−22.6
	20th century, uninhabited forest near Caverna da Pedra Pintada, Monte Alegre, Para, Brazil		
Sacurí seed	1990s	−26.6	
Sacurí seed	1990s	−30.7	
Achuá seed	1990s	−28.9	
Tucumã seed	1990s	−28.1	
Sacurí seed	1990s	−25.0	
Tucumã seed	1990s	−25.5	
Cashew seed	1990s	−26.0	
Tucumã seed	1990s	−28.0	
Sacurí seed	1990s	−25.9	
Jutaí seed	1990s	−28.4	
Uxi do Campo seed	1990s	−28.2	
Uxi seed	1990s	−30.7	
Achuá seed	1990s	−28.4	
Sacurí seed	1990s	−27.6	
Miriti seed	1990s	−28.3	
Jatobá seed	1990s	−26.7	
Jutaí seed	1990s	−27.3	
Average of 17		−27.7	

Sources for Appendix: P. Fearnside pers. comm.; Roosevelt 1989, 1991, 1997; Roosevelt et al. 1991, 1996; van der Merwe and Medina 1989, 1991; van der Merwe et al. 1981; van der Merwe pers. comm.; Ambrose pers. comm.; Schoeninger pers. comm.

REFERENCES

Absy, M. L. 1979. *A Palynological Study of Holocene Sediments in the Amazon Basin.* Ph.D. thesis, University of Amsterdam.

———. 1982. Quaternary palynological studies in the Amazon basin. In G. T. Prance, ed., *Biological Diversification in the Tropics,* pp. 67–73. New York: Columbia University Press.

———. 1985. The palynology of Amazonia: The history of forests as revealed by the palynological record. In G. T. Prance and T. Lovejoy, eds., *Key Environments: Amazonia,* pp. 72–82. Oxford: Pergamon Press.

Absy, M. L., and T. van der Hammen. 1976. Some paleoecological data from Rondonia, southern part of the Amazon Basin. *Acta Amazonica* 6: 293–299.

Absy, M. L., A. L. M. Cleef, M. Fournier, L. Martin, M. Servant, A. Sifeddine, M. F. da Silva, F. Soubies, K. Suguio, B. Turcq, and T. van der Hammen. 1991. Mise en évidence de quatre phases d'ouverture de la forêt dense dans le sud-est de l'Amazonia au cours des 60,000 dernières années. *Compte Rendus de l'Académie des Sciences* (Paris) Serie 2, 312: 673–678.

Acuna, C. 1891. Nuevo descubrimiento del gran Río de las Amazonas. *Colleccion de Livros que Tratan de America Raros o Curiosos 2.* Madrid.

Anderson, A. 1983. *The Biology of* Orbignia martiana *(Palmae): A Tropical Dry Forest Dominant in Brazil.* Ph.D. dissertation, Department of Botany, University of Florida, Gainesville.

Athens, S., and J. V. Ward. 1999. The late Quaternary of the western Amazon: Climate, vegetation, and humans. *Antiquity* 73: 287–302.

Balée, W. 1989. The culture of Amazonian forests. In W. Balée and D. A. Posey, eds., *Resource Management in Amazonia: Indigenous and Folk Strategies,* pp. 1–21. Advances in Economic Botany 7. New York: New York Botanical Garden.

Balée, W., and D. Campbell. 1990. Evidence for the successional status of liana forest (Xingu river basin, Amazonian Brazil). *Biotropica* 22(1): 36–47.

Behling, H. 1995. A high-resolution Holocene pollen record from Lago do Pires, SE Brazil: Vegetation, climate, and fire history. *J. Paleolimnology* 14: 253–268.

———. 1996. First report on new evidence for the occurrence of *Podocarpus* and possible human presence at the mouth of the Amazon during the Late-glacial. *Vegetation History and Archaeobotany* 5: 241–246.

Bettendorf, J. F. 1910. *Chronica da Missao dos padres da Companhia de Jesus no estado do Maranaho.* Revista 72(1). Río de Janeiro: Instituto Geografico e Historico.

Bonnefille, R. 1995. A reassessment of the Plio-Pleistocene pollen record of East Africa. In E. S. Vrba, G. H. Denton, T. C. Partridge, and L. H. Buckle, eds., *Paleoclimate and Evolution with Emphasis on Human Origins,* pp. 299–230. New Haven: Yale University Press.

Brochado, J. P. 1980. *The Social Ecology of the Marajoara Culture.* M.A. thesis, Department of Anthropology, University of Illinois, Urbana.

Brown, K. S., Jr., and G. G. Brown. 1992. Habitat alteration and species loss in Brazilian forests. In T. C. Whitmore and J. A. Sayer, eds., *Tropical Deforestation and Species Extinction,* pp.119–142. London: Chapman & Hall.

Brown, K. S., Jr. 1987. Conclusions, synthesis, and alternative hypotheses. In T. C. Whitmore and G. T. Prance, eds. *Biogeography and Quaternary History in Tropical America,* pp. 175–196. Oxford: Clarendon Press.

Bush, M. B., D. Piperno, and P. A. Colinvaux. 1989. A 6,000-year history of Amazonian maize cultivation. *Nature* 340: 303–305.

Campbell, K. E., and D. Frailey. 1984. Holocene flooding and species diversity in southwestern Amazonia. *Quaternary Research* 21: 369–375.

Campbell, D. G., D. C. Daly, G. T. Prance, and U. N. Maciel. 1986. Quantitative ecological inventory of terra firme and varzea tropical forest on the Rio Xingu, Pará, Brazil. *Brittonia* 38: 369–393.

Cardoso da Silva, J. M. 1995. Biogeographic analysis of the South American cerrado avifauna. *Steenstrupia* 21: 49–67.

Carvajal, G. de. 1984. *The Discovery of the Amazon.* New York: American Geographical Society.

Cavalcante, P. B. 1991. *Frutas Comestiveis da Amazonia.* Belem: Editora CEJUP and Museo Paraense Emilio Goeldi.

Chaves, S., and J. Renault-Miskovsky. 1996. Paléoethnologie, paléoenvironnement, et paléoclimatologie du Piaui, Brésil: Apport pollinique de coprolithes humains recueillis dans le gisement préhistorique de "Pedra Furada." *C. R. Academie des Sciences* 322(2a): 1053–1060.

Clapperton, C. M. 1993. Nature of environmental changes in South America at the last glacial maximum. *Palaeogeography, Palaeoclimatology, Palaeoecology* 101: 189–208.

Clay, J. 1988. *Indigenous Peoples and Tropical Forests: Models of Land Use and Management from Latin America.* Cambridge, Mass.: Cultural Survival.

Coetzee, J. A. 1967. *Pollen Analytical Studies in East and Southern Africa: Paleoecology of Africa and of the Surrounding Islands and Antarctica 3.* Capetown: A. A. Balkema.

Colinvaux, P. A., P. E. Oliveira, P. E. Moreno, M. C. Miller, and M. B. Bush. 1996. A long pollen record from lowland Amazonia: Forest and cooling in glacial times. *Science* 274: 85–88.

Denevan, W. 1966. *An Aboriginal Cultural Geography of the Llanos de Mojos de Bolivia.* Ibero-Americana 48. Berkeley: University of California Press.

Dufour, D. 1994. Diet and nutritional status of Amazonia peoples. In A. C. Roosevelt, ed., *Amazonian Indians from Prehistory to the Present,* pp. 151–176. Tucson: University of Arizona Press.

Evans, C. 1950. *The Archeology of the Territory of Amapa, Brazil (Brazilian Guiana).* Ph.D. dissertation, Department of Anthropology, Columbia University, New York.

Falesi, I. C. 1974. Soils of the Brazilian Ama-

zon. In C. Wagley, ed., *Man in the Amazon,* pp. 201–229. Gainesville: University Presses of Florida.

Grace, J., J. Lloyd, J. McIntyre, A. C. Miranda, P. Meir, H. S. Miranda, C. Nobre, J. Moncrief, J. Massheder, Y. Malhi, I. Wright, and J. Gash. 1995. Carbon dioxide uptake by an undisturbed tropical rainforest in southwest Amazonia. *Science* 270: 778–780.

Haberle, S., and M. A. Maslin. 1999. Late Quarternary vegetation and climate change in the Amazon basin based on a 50,000 year pollen record from the Amazon fan, ODP site 932. *Quarternary Research* 51: 27–38.

Haffer, J. 1969. Speciation in Amazonian forest birds. *Science* 169: 131–137.

———. 1993. Times cycle and time's arrow in the history of Amazonia. *Biogeographica* 69(1): 15–45.

Hames, R. B., and W. T. Vickers, eds. 1983. *Adaptive Responses of Native Amazonians.* New York: Academic Press.

Harris, D., ed. 1980. *Human Ecology in Savanna Environments.* New York: Academic Press.

Henderson, Andrew. 1995. *The Palms of the Amazon.* Oxford: Oxford University Press.

Heriarte, M. de. 1964. *Descripcam do Estado do Maranhao, Para, Corupa, e Río das Amazonas, feito por Mauricio de Heriarte, Ouvidor-geral, Provedormor e Auditor, que foi pelo Gobernador D. Pedro de Mello, no Anno 1662.* Facsimile, MSS 5880 and 5879, Osterreichischen National Bibliothek. Graz, Austria: Academische Druck & Verlagsanstalt.

Irion, G. 1984a. Sedimentation and sediments of the Amazonian rivers and evolution of the Amazonian landscapes since Pliocene times. In H. Sioli, ed., *The Amazon: Limnology and Landscape Ecology of a Mighty Tropical River and Its Basin,* pp. 201–214. Dordrecht, Neth.: Dr. W. Junk.

———. 1984b. Clay minerals of Amazonian soils. In H. Sioli, ed., *The Amazon: Limnology and Landscape Ecology of a Mighty Tropical River and Its Basin,* pp. 521–536. Dordrecht, Neth.: Dr. W. Junk.

Jordan, Carl F., ed. 1987. *Amazonian Rain*

Forests: Ecosystem Disturbance and Recovery. New York: Springer-Verlag.

Junk, W. J. 1984. Ecology of the várzea, floodplain of Amazonian whitewater rivers. In H. Sioli, ed., *The Amazon: Limnology and Landscape Ecology of a Mighty Tropical River and Its Basin*, pp. 215–144. Dordrecht, Neth.: Dr. W. Junk.

Kern, D. 1994. *Geoquimica e Pedogeoquimica de Sitos Arqueologicos com Terra Preta na Reserva Florestal de Caxiuama (Portel, Para)*. Ph.D. dissertation, Curso de Pos Graduacao em Geologia e Geoquimica, Geociencias, Universidade Federal do Para.

Ledru, M.-P. 1993. Late Quaternary environment and climatic changes in central Brazil. *Quaternary Research* 39: 90–98.

Liu, K.-B., and P. A. Colinvaux. 1985. Forest changes in the Amazon Basin during the last glacial maximum. *Nature* 318: 556–557.

Lorenzi, H. 1992. *Arvores Brasileiras*. Nova Odessa: Editora Plantarum.

MDME. 1976. *Mapa Geologico*. Santarem, Folha SA-21. Ministerio das Minas e Energia.

Meggers, B. J. 1952. *The Archaeological Sequence on Marajo Island, Brazil, with Special Reference to the Marajoara Culture*. Ph.D. thesis, Department of Anthropology, Columbia University, New York.

———. 1954. Environmental limitation on the development of culture. *American Anthropologist* 56: 801–824.

———. 1971. *Amazonia: Man and Nature in a Counterfeit Paradise*. Chicago: Aldine.

———. 1975. Application of the biological model of diversification to cultural distributions in tropical lowland South America. *Biotropica* 7(3): 141–161.

———. 1977. Vegetational fluctuations and prehistoric cultural adaptation in Amazonia: Some tentative correlations. *World Archaeology* 8(3): 287–303.

———. 1979. Climatic oscillation as a factor in the prehistory of Amazonia. *Antiquity* 44(2): 252–266.

———. 1988. The prehistory of Amazonia. In J. S. Denslow and C. Padoch, eds., *People of the Tropical Forests*, pp. 53–62. Berkeley: University of California Press and

Smithsonian Institution Traveling Exhibition Service.

———. 1996. *Amazonia: Man and Nature in a Counterfeit Paradise*. 2nd ed. Washington, D.C.: Smithsonian Institution Press.

Meggers, B. J., and C. Evans. 1957. *Archaeological Investigations at the Mouth of the Amazon*. Bureau of American Ethnology Bulletin 167. Washington, D.C.: Smithsonian Institution.

Murca Pires, J., and G. T. Prance. 1985. The vegetation types of the Brazilian Amazon. In G. T. Prance and T. C. Lovejoy, eds., *Amazonia: Key Environments*, pp. 109–145. Oxford: Pergamon Press.

Nelson, B., C. Ferreira, M. da Silva, and M. Kawasaki. 1990. Endemism centres, refugia, and botanical collection density in Brazilian Amazon. *Nature* 345: 714–716.

Nimuendaju, K. 1949. Os Tapajo. *Boletim do Museu Paraense Emílio Goeldi* 10: 93–106.

Odum, E. 1971. *Fundamentals of Ecology*. Philadelphia: W. B. Saunders.

Palmatary, H. 1960. *The Archaeology of the Lower Tapajos Valley*. Transactions of the American Philosophical Society n.s. 50(3). Philadelphia.

Phillips, O. L., and A. H. Gentry. 1994. Increasing turnover through time in tropical forests. *Science* 263: 954–957.

Piperno, D. R., and D. M. Pearsall. 1998. *The Origins of Agriculture in the Lowland Tropics*. San Diego, Calif.: Academic Press.

Prance, G. T., ed. 1982. *Biological Diversification in the Tropics*. New York: Columbia University Press.

Prance, G. T., and T. C. Lovejoy, eds. 1985. *Amazonia: Key Environments*. Oxford: Pergamon Press.

Quinn, E. n.d. *The Pottery of the Santarem Culture at the Confluence of the Amazon and Tapajos Rivers, Para, Brazil*. Ph.D. thesis in preparation, Department of Anthropology, University of Illinois, Chicago.

Redford, K. H. 1992. Commentary in K. H. Redford and C. Padoch, eds., *Conservation of Neotropical Forests: Working from Traditional Resource Use*. New York: Columbia University Press.

Roa, P. R. 1980. Algunos espectos de la evolucion sedimentologica y geomorphologica de la llanura aluvial de desborde en el Bajo Llano. *Boletin de la Sociedad Venezolana a Ciencias Exactas y Naturales* 35(139): 31–47.

Roosevelt, A. C. 1980. *Prehistoric Maize and Manioc Subsistence along the Amazon and Orinoco.* New York: Academic Press.

———. 1989. Resource management in Amazonia before the Conquest: Beyond ethnographic projection. In W. Balée and D. A. Posey, eds., *Resource Management in Amazonia: Indigenous and Folk Strategies,* pp. 30–62. Advances in Economic Botany 7. New York: New York Botanical Garden.

———. 1991. *Moundbuilders of the Amazon: Geophysical Archaeology on Marajo Island, Brazil.* San Diego, Calif.: Academic Press.

———. 1993. The rise and fall of the Amazon chiefdoms. In A. C. Taylor and P. Descola, eds., La remontée de l'Amazone: Anthropologie et histoire des societés amazoniennes. *L'Homme* (Paris) 33(126–128): 255–284.

———. 1995a. Educating natural scientists about the environment. In T. A. Arcury and B. R. Johnson, eds., *Anthropological Contributions to Environmental Education. Practicing Anthropology* 17(4): 25–28.

———. 1995b. Early pottery in the Amazon: Twenty years of scholarly obscurity. In W. Barnett and J. Hoopes, eds., *The Emergence of Pottery: Technology and Innovation in Ancient Societies,* pp. 115–131. Washington, D.C.: Smithsonian Institution Press.

———. 1997. *The Excavations at Corozal, Venezuela: Stratigraphy and Ceramic Seriation.* Publications in Anthropology 82. New Haven: Yale University.

———. N.d. *Ancient Amazon: Co-evolution of Humans and the Tropical Rainforest.* Book manuscript.

Roosevelt, A. C., R. Housley, I. M. Imazio da Silveira, S. Maranca, and R. Johnson. 1991. Eighth millennium pottery from a prehistoric shell midden in the Brazilian Amazon. *Science* 254(5038): 1621–1624.

Roosevelt, A. C., M. Lima Costa, C. Lopes Machado, M. Michab, N. Mercier, H. Val-

ladas, J. Feathers, W. Barnett, M. Imazio da Silveira, A. Henderson, J. Silva, B. Chernoff, D. Reese, J. A. Holman, N. Toth, and K. Schick. 1996. Paleoindian cave dwellers in the Amazon: The peopling of the Americas. *Science* 272: 373–384.

Rowe, J. H., ed. and trans. 1952. The Tapajo, by Curt Nimuendaju. *Kroeber Anthropological Society Papers* 6: 1–25.

Sadeck, N. N.d. A decade of rainfall statistics from Monte Alegre, Para, Brazil. Manuscript.

Salati, E. 1985. The climatology and hydrology of Amazonia. In G. T. Prance and T. C. Lovejoy, eds., *Amazonia: Key Environments,* pp. 18–48. Oxford: Pergamon Press.

Salati, E., and J. Marques. 1984. Climatology of the Amazon region. In H. Sioli, ed., *The Amazon: Limnology and Landscape Ecology of a Mighty Tropical River,* pp. 85–127. Dordrecht, Neth.: Dr. W. Junk.

Sales Barbosa, A. 1992. A Tradicao Itaparica: Uma comprehensao ecologica e cultural do povoamento inicial do planalto central Brasileiro. In B. J. Meggers, ed., *Prehistoria Sudamericana: Nuevas Perspectivas,* pp. 145–160. Washington, D.C.: Taraxacum.

Salo, J., R. Kalliola, I. Hakkinen, Y. Makinen, P. Niemela, M. Puhakka, and P. Coley. 1986. River dynamics and the diversity of lowland Amazonian forest. *Nature* 322: 254–258.

Simoes, M. 1969. The Castanheira site: New evidence on the antiquity and history of the Ananatuba phase (Marajo Island, Brazil). *American Antiquity* 34: 402–410.

———. 1981. Coletores-pescadores ceramistas do litoral do Salgado (Para): Nota preliminar. *Boletim do Museu Paraense Emílio Goeldi* n.s. 78: 1–26.

Sioli, H. 1973. Recent human activities in the Amazon region and their ecological effects. In B. J. Meggers, E. S. Ayensu, and W. D. Duckworth, eds., *Tropical Forest Ecosystems in Africa and South America: A Comparative Review,* pp. 321–334. Washington, D.C.: Smithsonian Institution Press.

Skole, D. L., W. H. Chomentowski, W. A.

Salas, and A. D. Nobre. 1994. Physical and human dimensions of deforestation in Amazonia. *Bioscience* 44: 314–322.

Smith, N. H., J. T. Williams, D. L. Plucknett, and J. P. Talbot. 1992. *Tropical Forests and Their Crops*. Ithaca: Cornell University Press.

Smith, H. H. 1879. *Brazil, the Amazons, and the Coast*. New York: Charles Scribner's Sons.

Smith, N. 1980. Anthrosols and human carrying capacity in Amazonia. *Annals of the Association of American Geographers* 70: 533–566.

———. 1999. *The Amazon River Forest*. Oxford: Oxford University Press.

Sombroek, W. 1966. *Amazon Soils: A Reconnaissance of the Soils of the Brazilian Amazon*. Wageningen, Neth.: Centre for Agricultural Publications and Documentation.

Steward, J., and L. Faron. 1959. *Native Peoples of South America*. New York: McGraw-Hill.

Swales, S. 1993. *Agricultural Practices in the Monte Alegre District, Brazil*. M.A. thesis, Geography, University of Illinois, Chicago.

———. N.d. *Agroforestry in the Monte Alegre District, Para, Brazil*. Ph.D. thesis in preparation, Anthropology, University of Illinois, Chicago.

Tieszen, L. L. 1991. Natural variations in the carbon isotope values of plants: Implications for archaeology, ecology, and paleoecology. *J. Archaeological Science* 18: 227–248.

van der Hammen, T. 1963. A palynological study on the Quaternary of British Guiana. *Leidse Geologische Medelingen* 29: 125–180.

van der Hammen, T., and M. L. Absy. 1994. Amazonia during the last glacial. *Palaeography, Palaeoclimatology, Palaeoecology* 109: 247–261.

van der Hammen, T., J. F. Duivenvoorden, J. M. Lips, N. Espejo, and L. E. Urrego. 1992. The late Quaternary of the middle Caqueta River area (Colombian Amazonia). *J. Quaternary Science* 7: 45–55.

van der Merwe, N., and E. Medina. 1989. Photosynthesis and $^{13}C/^{12}C$ ratios in Amazonian rain forests. *Geochimica et Cosmochimica Acta* 53: 1091–1094.

———. 1991. The canopy effect, carbon isotope ratios, and foodwebs in Amazonia. *J. Archaeological Science* 18: 249–259.

van der Merwe, N., A. C. Roosevelt, and J. C. Vogel. 1981. Isotopic evidence for prehistoric subsistence change at Parmana, Venezuela. *Nature* 292(5823): 536–538.

Wallace, A. R. 1889. *A Narrative of Travels on the Amazon and Rio Negro*. 2nd ed. London: Ward, Lock & Co.

Whitmore, T. C. 1990. *An Introduction to Tropical Rain Forests*. Oxford: Clarendon Press.

Whitmore, T. C., and G. T. Prance, eds. 1987. *Biogeography and Quaternary History in Tropical America*. Oxford: Clarendon Press.

Wijmstra, T. A. 1971. *The Palynology of the Guiana Coastal Basin*. Ph.D. thesis, University of Amsterdam.

Wijmstra, T. A., and T. van der Hammen. 1966. Palynological data on the history of tropical savannas in northern South America. *Leidse Geologische Medelingen* 38: 71–83.

D A V I D   L .   L E N T Z

SUMMARY AND CONCLUSIONS

From the preceding essays we can draw a number of conclusions regarding the historical ecology of the Precolumbian Americas. One inescapable conclusion is that the New World is and has been for some time an area of incredible biodiversity, particularly in the Neotropics but also in the varied habitats of temperate North America. Another conclusion, based on an expanding corpus of paleoethnobotanical and archaeological evidence, is that much of the New World was shaped by human influences long before Columbus made landfall. In many areas anthropogenic landscape modifications were significant, often resulting from the use of fire as a game and forest management tool and as an integral component of agriculture. Land use practices in some cases were intensive, sustainable, and highly productive. In other areas land use practices were not sustainable, could be viewed as destructive, and may have brought about ecological disasters.

In essay 2, Hodell, Brenner, and Curtis discuss the kinds of paleoclimatic changes that occurred in Mesoamerica since the last glacial maximum 20,000 years ago. With the shift toward a more humid climate by about 8500 B.P., mesic tropical forests replaced more xeric habitats in Mesoamerica and the Caribbean. The discussion combines evidence from pollen cores, which reflects the ambient pollen rain at certain periods of time, and lacustrine oxygen isotopic evidence from buried ostracods, an approach that is less susceptible to the influences of human activities as is frequently observed with other paleoclimatic approaches. By combining these data sets, it is possible to sort out the effects of climate change from the human influence on vegetation. Hodell, Brenner, and Curtis point out that by about 3000 B.P. there were significant declines in lowland forest taxa accompanied by a greater presence of disturbance taxa in southern Mexico. By the Late Classic period, with estimated populations of 3 million to 14 million Maya in the Petén region of Guatemala (many times greater than the current population), extensive forest removal led to severe soil erosion and concomitant nutrient loss. Exacerbating these long-term trends, droughts in circa A.D. 585 and 862 caused havoc in agricultural systems and, combined with other factors, may even have contributed to the Maya collapse in the late ninth century. Furthermore, prehistoric societies were not just vulnerable to the vagaries of climate change but may even have influenced local climatic regimes. Through hydroengineering, humans disturbed the natural water budget of lakes. Deforestation led to a weakened hydrologic cycle and reduced rainfall in watersheds. Thus, extensive deforestation perhaps contributed to reduced precipitation and increased erosion, thereby reducing the productive capacity of the landscape.

Greller, in essay 3, applies a biogeographer's eye to the vegetation of North and Central America. The plants in this large and diverse geographic area are divided into the Holarctic and Neotropical floristic kingdoms, reflecting their evolutionary heritage dating back to the Lower Cretaceous Epoch (132 million years ago), when the landmass of Pangaea split into Laurasia (the Northern Hemisphere) and Gondwanaland (the Southern Hemisphere). When North America reconnected with South America, these two lineages of plants, isolated for millions of years, were reunited in Central America. Thus we have Holarctic flora, largely temperate in nature, extending to South America along the Central American cordillera but surrounded in the lowlands by Neotropical flora.

The northernmost portions of North America in the circumboreal region are covered by tundra in the Arctic Province and cold-hardy conifer forest in the Canadian Province. The Atlantic North American region extends from the eastern shores of the continent to the eastern slopes of the Rocky Mountains. The Appalachian Province is forested, with conifers predominating in the north and broadleaf trees dominating in the southern reaches. The North American Prairie Province is oak savanna in the east and grassland prairie in the west, following a moisture gradient that declines from east to west. The Rocky Mountain region extends from the Yukon to New Mexico and Texas and includes all of the Great Basin. Altitudinal gradients are influential in this region, with desert grassland and sagebrush in the low elevations and pygmy conifer woodland, oak shrubland, and conifer forests in the highlands. An unusual formation in the Vancouverian Province is the temperate rain forest with its giant conifers, viz., coastal redwoods, sequoia, sitka spruce, and others. The flora of the Madrean region, composed predominantly of both evergreen and deciduous taxa that thrive in humid and perhumid climates, extends from southern California to the southeastern seaboard and into the Sierra Madres in Mexico. The Gulf and Atlantic Coastal Plains Province provided the lush hickory and oak forests and broad alluvial floodplains that were the favored habitat of the prehistoric Mississippian cultures.

The Neotropical Floristic Kingdom reaches its northernmost limits in the southern United States and extends south through the West Indies, lower Mexico, Central America, and into South America. The Mexican xerophytic region, with its hot deserts and abundant succulents, was home to the ancient Hohokam, who were careful to control the permanent water sources for their irrigated fields. The Central Plateau, with its high-elevation grasslands, mesquitals, and scrub forests, was the habitat occupied by the Teotihuacanecos and later the Toltecs and Aztecs, who made this arid land bloom through a variety of hydraulic agricultural approaches. To the southwest lie the Tehuacán and Cuicatlan valleys, also arid lands that were home to some of the earliest farmers of Mesoamerica and subsequently the vaunted Zapotec civilization in Oaxaca, who relied on an extensive irrigation network to practice their agriculture.

The Central American region consists of lowland southern Mexico, Central America, and the West Indies. Plant formations in this region include evergreen rain forest, tropical semideciduous forest, thorn woodland, savanna, and palm communities. The tropical semideciduous forests were home to the ancient Maya, whose splendid civilization was carved from its center.

Essay 4 addresses the topic of domestication in the New World and discusses how human-centered food webs began to develop around 10,000 B.P. The development of these food webs, unique to each region in their full arrays of wild and domesticated components, affected the distribution of species throughout the Americas in five major ways. The first was to extend the range and increase the population of organisms, including host-specific mycorrhizae and nitrogen-fixing species of bacteria, within anthropocentric food webs. Second, the habitat for ruderals and vermin was expanded. Third, habitat for species outside human-centered food webs was reduced. Fourth, the populations of prey species were affected, often reduced. Finally, as the food webs became more agronomically and dietetically efficient, with a greater channeling of energy to human consumption due to improved processing and storage innovations, they supported the expansion of human populations into large, aggregated settlements. The major New World crops that served as the productive foundations of the food webs were squashes (*Cucurbita* spp.), beans (*Phaseolus* spp.), maize (*Zea mays*), chenopodium (*Chenopodium* spp.), potatoes (*Solanum tuberosum*), manioc (*Manihot esculenta*), and numerous tree species, including a variety of palms.

The Basin of Mexico was one of the early centers of mesoamerican civilization. In essay 5, McClung de Tapia provides an assessment of Prehispanic agricultural systems in the basin, which date back to Formative times. She divides the Prehispanic agricultural approaches into two types: temporal and humidity-control mechanisms. Temporal mechanisms are rainfall-dependent and good for a single harvest in one location before shifting to another, with variations such as sloped piedmont cultivation, terrace agriculture, and floodplain cultivation. Humidity-control mechanisms generally permit two harvests per year, and often the same fields can be planted year after year. Approaches of this type common in the basin include floodwater irrigation, drained fields, permanent irrigation, and chinampas (raised fields). All of these systems, which depended on the five major lakes and other permanent water sources of the basin, were in place before the Conquest. They allowed the denizens of the basin, including the inhabitants of Teotihuacán, Tula, Tenochtitlán, and other polities, to convert arid grasslands and scrub forests (as discussed by Greller) into highly productive agricultural land. As McClung de Tapia observes, these diverse agricultural adaptations helped to support the large state organizations that emerged from the Basin of Mexico, yet there is little evidence for a causal link between the initial development of large hydraulic operations and state formation. In other words, the irrigation and drainage systems were developed by localized populations without the aid of centralized control. This is an important distinction, because it refutes the idea that state-level organization is necessary for the implementation of extensive hydraulic engineering projects. Furthermore, McClung de Tapia points out that before the intensive hydraulic systems were developed, Middle Formative inhabitants practiced shifting cultivation in the basin and planted fields too extensively, leading to disastrous episodes of sheet erosion. These events seemed to have encouraged later farmers to turn to floodwater irrigation to make better use of both water and soil resources.

In essay 6, Spencer presents three case studies of Prehispanic, intensified agricultural production that relied on water-control systems. The first study focused on the ancient Purrón Dam site in the Tehuacán Valley, Mexico. The dam was constructed over several centuries and eventually attained a size of 400 m in length and reached a reservoir capacity of 2.6 million m³. The irrigation capability of the structure supported a base population of around 1,000 people and could have produced a surplus in addition to the basic needs of the populace. The dam was abandoned about A.D. 250, apparently for political reasons. Current efforts to irrigate the valley have proven to be far less effective due to poor selection of reservoir locations.

Spencer's second case study comes from the Cañada de Cuicatlán in Oaxaca, Mexico, about 50 km south of the Purrón Dam. From ca. 200 B.C. to A.D. 200 people in the area created a set of stream diversions coupled with aqueducts and irrigation canals that were used to irrigate a major portion of the canyon. At one point the area was controlled by the Zapotecs, who siphoned off agricultural surplus as tribute. Today the land is inefficiently irrigated by the modern Canal de Matamba, which supplies an excess of water for the area it was designed to irrigate, leaving other sectors parched and useless for agriculture.

The third case study is La Tigra, the Prehispanic drained field system in the humid savannas of Barinas, Venezuela. This system served to drain excess water in the rainy season and bring water to the site during periodic droughts. With careful water management, the system permitted two harvests per year versus one harvest achieved by modern farmers without the canal system. In the past, perhaps 20 times as many inhabitants were supported by La Tigra's drained fields as today, and the system probably contributed an additional surplus to the ancient political economy as well.

Spencer presents these cases as potential models for improved land use but warns that planners should avoid large-scale centralized approaches and, instead, promote locally managed, small-scale operations similar to the examples described from Prehispanic times. By providing the details of ancient water-supply projects that seemed to have worked over long periods of time, thus revealing proven technological approaches from the past, archaeologists can play a key role in helping communities improve their productivity.

In essay 7, Dunning and Beach discuss the effects of ancient agricultural and agroforestry practices that led to conditions of both stability and instability in the Maya Lowlands. Pollen evidence from the north-central Petén region of Guatemala revealed that the first arrival of the Maya (2000–1000 B.C.) into the area coincided with extensive primary forest clearance, the appearance of pollen from maize and disturbance taxa, and widespread erosion. This was a pattern repeated throughout much of the ancient Maya realm. The Classic period heirs to this degraded landscape seemed to have learned from the mistakes of their Preclassic forebears and developed, at least in some areas, techniques for conserving soil and water resources while at the same time providing sustenance for a burgeoning human population.

Dunning and Beach present case studies from two areas in the Maya Lowlands: the Petexbatun area in south-central Petén in Guatemala and the Three Rivers area in what is now northwestern Belize. In the Petexbatun area, dams and terraces were built for soil conservation. These constructions seem to have been effective, as evidenced by the use of relict terraces by modern farmers. Comparisons of soil depth in Classic period terraces versus unterraced adjacent land show much greater soil depth in the former. Ancient conservation efforts in the Petexbatun area evidently had positive results, because the gallery forest was well managed, erosion was under control, and local fauna remained abundant throughout the Late Classic period. As the authors point out, this was an "anti-Malthusian example of indigenous, sustainable agriculture in a tropical environment." Studies in the Three Rivers area show similar conservation approaches applied to the lands adjacent to the Classic period civic-ceremonial centers of La Milpa and Dos Hombres. However, the legacy of soil loss from Preclassic times was considerably worse in this area, so their Classic period descendants inherited a more degraded landscape. Despite construction of footslope, cross-channel, and box terraces, erosion remained a problem. Although they attempted to impound runoff, the effects of dry-season desiccation, possibly exacerbated by deforestation, became manifest. The combination of these negative factors led to a "landscape of desperation," with diminished resources and populations out of balance with their food supply. Unfortunately, the Late Classic tragedy described in this essay is being reenacted today in the Petén region. The hope is we can learn from conservation efforts that were successfully employed in the Petexbatun area to avoid future ecological disasters.

In essay 8, Peters paints an insightful picture of agroforestry practices in the Precolumbian Neotropics. Westerners, at least in the past, have had difficulty understanding the complex nature of traditional agricultural and forest management approaches in the tropics. Many have viewed indigenous agroforestry as a destructive, inefficient, and wasteful land use practice. However, upon closer examination, it appears that many indigenous agricultural and silvicultural practices result in carefully managed, conservative, and, some would say, sustainable systems.

Peters outlines three main indigenous silvicultural systems: homegardens, managed fallow systems, and managed forest systems. In the Neotropics, homegardens are heterogeneous patches of useful plants, including annuals, perennials, and tree species, that are intensively managed during occupation and even after abandonment of a house site. Managed fallows are forest segments that are allowed to recover following several years of cultivation. These can be managed as monocyclic or polycyclic fallows, depending on the timing of subsequent clearance for field agriculture or intended utility as an orchard without final harvest cut. Managed forests, which often develop from previously fallowed fields, are subjected to silvicultural treatments that enhance the growth of important economic species. Management activities focus on selective weeding and thinning, protection of desirable volunteers, and planting of useful trees. These activities can decidedly, yet almost imperceptibly, alter the nature and composition of tropical forests. While the data presented in this essay are largely based on ethnographic examples, it seems likely that the silvicultural and forest management practices discussed are at least a reflection of the kinds of forest management activities that were practiced on a wide scale in the Precolumbian Neotropics.

Fritz's contribution (essay 9) examines native farming systems and ecosystems in the Precolumbian Mississippi River valley. She divides the region into three subregions: the Upper, Central, and Lower Mississippi Valley. The central subregion saw the greatest cultural development, at least in terms of plant domestication and population density. Early evidence of cultigens, i.e., sunflower, sumpweed, chenopodium, squash, and maygrass, has been found in this subregion. Although Fritz states that sunflower was domesticated in North America, recent evidence from the San Andrés site in Mexico suggests the crop may have been domesticated elsewhere and introduced into North America. Maize and tobacco were introduced sometime around 2000 B.P. It is unclear whether the first tobacco in the region was *Nicotiana rustica* from South America or *N. quadrivalvis* or *N. multivalvis* from the West Coast of North America. Beans did not arrive until well into the Mississippian period. When the quartet of maize, beans, sunflowers, and squashes did come together as an agronomic unit, the locally developed cultigens fell into disuse. Although Fritz rejects the hypothesis that trade along the Mississippi River was a conduit for the introduction of germ plasm from Mesoamerica and South America, several major cultigens from these regions were introduced into the Mississippi Valley during precontact times, perhaps via the Southwest. In the Upper Mississippi Valley, a kind of "intensification with diversification," with crops grown in ridged fields and corn hills augmented by intensive collection of shellfish and wild plants, occurred during the Oneota period (A.D. 1000–1500). Just prior to European contact in the upper valley, and certainly afterward, bison hunting became a more prominent activity.

One fascinating aspect of prehistory in the Mississippi Valley is the way the inhabitants continued to exploit the bounty of the rivers and forests in spite of possessing a potent array of cultigens. Acorns, nuts, and tree fruits from the verdant woodlands remained prized foodstuffs from Paleoindian to Colonial times. The woodlands adjacent to towns and villages evidently were managed by local inhabitants. Fire, at least in protohistoric times, was used sparingly, principally in areas away from the settlements and mostly for hunting. Anthropogenic ecosystems existed around settlements that paralleled the Mississippi River and its tributaries. The areas of greatest human impact were the larger sites, such as Cahokia, and in the American Bottom that supported a relatively dense population whose heavy demand for wood led to extensive forest clearance during the Mississippian period.

In essay 10, Fish describes the land use practices of the Hohokam, who occupied what is now southern Arizona from A.D. 200 to 1450. They engineered the largest irrigation systems north of Peru in the Salt, Gila, Verde, Hassayampa, Santa Cruz, and lower Colorado river valleys. Through their hydraulic agricultural approaches, they converted more than 13,000 ha of Sonoran Desert thornscrub in the Phoenix basin into well-watered fields. They employed a number of water conservation techniques, including the construction of massive canal networks that drew water from the undammed major rivers, diversionary walls that redirected storm flow onto fields in alluvial fans, terraces (both contour and hillside), and rockpiles that not only retained moisture for domesticated crops but discouraged competing weeds and herbivores as well.

The Hohokam were extremely well adapted to life in a fragile environment; topsoil, water, and fuel had to be carefully conserved. According to archaeological findings, there was no evidence of sheet erosion or salinity buildup (a major concern today), so their soil management practices must have been adequate. To conserve fuel, they employed communal agave roasting pits and efficient pit ovens for cooking, lived in adobe or wattle-and-daub semisubterranean houses, and fired their ceramics at low temperatures. These practices limited the damage to the perennial woody vegetation caused by a dense sedentary population. There was a clear anthropogenic imprint on the landscape, however, mostly in the form of more mesic flora spreading along the canals and other waterways.

Fire, as a human tool for land modification, seems to have been used sparingly. This area is prone to high lightning ignition, so natural fires are a common occurrence. The natural fires probably would have been enough to maintain a diverse mosaic of established perennials and weedy annuals. When the Hohokam did practice intentional burning, it was probably mostly for hunting or in marginal lands well away from their fields. Fuel was a scarce commodity, so the Hohokam tried to guard their wood sources carefully. Also, many useful wild plants, such as cacti, are susceptible to fire damage, so excessive burning would have been counterproductive. As a result of their innovative but conservative subsistence practices, the Hohokam created a long-term sustainable subsistence in a very harsh environment.

In essay 11, Luteyn and Churchill describe an area of rich topographic and biotic diversity, the tropical Andes region. Early migrants into this area were quick to take advantage of varied habitats; some of the most astounding Precolumbian cultural developments flourished in the shadow of the Andes. The mountain chain extends from the Sierra de Santa Marta in Colombia and Venezuela to northern Argentina. Vegetation in this mountainous region varies substantially with elevation, moisture availability, edaphic conditions, and slope exposure.

Starting with the highest elevations and working downward, the páramo vegetation occupies the zone above the forest line (3,000–3,500 m above sea level) and below the snowline (4,400–4,800 m). Luteyn and Churchill subdivide the vegetation type into three zones: true páramo, subpáramo, and superpáramo. There is quite a range in variation in these vegetation subtypes, but they are mostly dominated by bunchgrasses, rosette and cushion plants, xerophytic shrubs, herbaceous plants, and a host of lichens and bryophytes. Small trees (e.g., *Polylepis reticulata*), once common in the region, now grow in protected patches up to 4,100 m in elevation.

The puna, also considered an alpine vegetation type, predominates in altitudes from 3,300 to 5,000 m in the altiplano of Peru and Bolivia. Three puna subdivisions are humid or wet puna, moist puna, and desert puna. These, especially the moist puna, are essentially anthropogenic grasslands that are maintained as such by frequent burning and grazing. Woody species will take hold if the puna is left undisturbed.

The montane forests in South America, although often cleared for agriculture, are valuable sources of timber, fuel wood, and medicinals. Upper montane forests, sometimes called cloud forests, commence from 2,000 to 3,500 m. These are epiphyte-laden, moist forests, often with a canopy as high as 25 m.

Premontane forests are transition zones between the lowland and upper montane forests, with a range in elevation from 1,000 to 2,000 m. They tend to be cooler than the lowland areas but are more nutrient-rich than the upper montane forest. Premontane forests dominate the eastern slopes of the Andes.

On the western side of the Andes lie the Pacific coastal deserts. Robbed of maritime moisture by the cooling Humboldt current, the lowest lands along the coast receive almost no precipitation, while fog banks at higher elevations (300–1,000 m) provide enough moisture for the xerophytic lomas formations. Breaking the monotony of the low deserts are ten permanent rivers fed by Andean snowmelt that support lush green riparian ribbons of phreatophytic trees and shrubs. Both the transdesert river valleys and the lomas formations have been attractive outposts for humans for thousands of years.

Erickson, in essay 12, provides a human-centric perspective on the development of cultural-environmental interactions in the Lake Titicaca Basin of Peru. This perspective emphasizes anthropogenic forces of transformation that create cultural or humanized landscapes designed to meet economic, social, political, and even religious needs. Moreover, the patterning of the built environment is a reflection of cognitive processes of the builders, who operate in a dynamic environment. The processes are historically contingent and not cyclical oscillations that move arbitrarily about some selected norm. Erickson argues that Holocene environmental trajectories in the basin can be truly understood only in terms of human agency because the landscape there has been subjected to human transformative processes for thousands of years.

Folk classifications divide the environs of Lake Titicaca into four categories: the lake and associated wetlands, the seasonally inundated lake plain or pampa, the lower hill slopes (cerro) adjacent to the lake, and the high-altitude puna or grasslands. Each of these zones is managed in a different way, but all are incorporated into an overall subsistence strategy that has been both productive and conservative of scarce resources, viz., soil and water.

Human manipulation of the environment began in Preceramic times and became more elaborate thereafter. Denizens of the basin, although described by some as hardware poor, used simple manual tools, such as the footplow, small hoes, and heavy mallets for clod busting. Nevertheless, through generations of accumulated labor input, they were able to remodel the lands in the basin to produce maximally productive and structurally sound landscape features.

Perhaps the most visible landscape modifications in the basin are the extensive terraces that line the slopes of the valley foothills. Erickson estimates that about 500,000 ha of Precolumbian terraces can be found in the vicinity of the lake today. These terraces serve to deepen soil, control erosion, and act as microclimatic and moisture controls. Raised fields are another major form of landscape modification in the basin. These are useful devices because they help to improve the soil, manage water resources, capture and recycle soil nutrients, improve the microclimate, and provide a source of aquatic products such as fish, reeds, and avifauna in addition to agricultural products from the fields themselves.

Sunken gardens are another agricultural approach widely employed in the Titicaca Basin, with perhaps as much as 256 km² of functioning q'ochas. They are used as fields for potatoes, quinoa, and other crops as well as for growing forage for domesticated camelids. The q'ochas are valuable as aids in water management, nutrient maintenance, forage production, and as a locus for aquaculture. The use of sunken gardens in the basin probably dates back to 1800 B.C. Irrigated pastures, or bofedales, an unusual but ancient feature of the basin, probably originated as early as 2000 B.C. These canal networks are extensive and were designed primarily to provide dry-season pasturage for the large herds of llamas and especially alpacas maintained in the area.

Other miscellaneous artificial landscape features include modified rivers and streams, artificial canals, improved springs, causeways, roads and paths, ponds, reservoirs, walls, stone piles, corrals, gardens, cemeteries, burial towers, temples, shrines, and of course residences. It is easy to understand why Erickson maintains a human-centric perspective; the inhabitants of the Lake Titicaca Basin constructed, modified, and managed the area for their own benefit, creating an anthropogenic ecosystem that completely dominated the landscape. What is impressive about this area is the longevity of the system; it supported substantial human populations through several millennia of dramatic climatic fluctuations and remains effectively in place today.

In essay 13, D'Altroy describes land use in the Andes region as a reflection of its vertically compressed topography. He outlines five major production zones influenced by local climate, water availability, topography, and elevation and how each was exploited according to resource potential, human technology, and suitability of domesticates. First, moving from west to east, is the Pacific coastal plain, in areas below 150 m where marine and riparian zones once teemed with fish and shellfish (modern pollutants have curtailed productivity in recent times). Rivers fed by Andean snowmelt provided irrigation water for crops such as maize, lúcuma, cucurbits, gourds, and cotton. The second zone, the yungas, ranged from 300 to 2,300 m on both eastern and western slopes of the Andes. This was a highly productive zone used to grow maize, tropical fruits, peppers, and the all-important coca. Two of the products of this zone, maize (consumed mostly in the form of maize beer) and coca, were essential to the Inka political economy. The third, the quechua zone, extended from 3,100 to 3,500 m. These inter-Andean valleys were loci for drained fields, terraces, and irrigated lands planted with maize, beans, quinoa, and various tuber crops. Above 3,500 m, maize normally will not grow because of frost, but hardier crops, like quinoa, legumes, and certain tubers, can be coaxed from the ground. The fourth zone, the puna and lower páramo, is an alpine tundra extensively used for the herding of Andean camelids, the llama and the alpaca. Although some crops can be grown there, such as frost-resistant potatoes, it is a marginal area for agriculture. The fifth zone is the montaña, or montane forest zone, on the eastern slopes of the Andes above the floor of the Amazon floodplain and the rain forest proper. This zone, with its fertile valley bottoms and terraced hillsides, is used to produce maize, coca, tropical fruits, and many other warm-weather crops. Throughout much of prehistory in the region, a common extractive socioeconomic strategy entailed the development of intensified agricultural systems in as many of these zones as possible by local groups to maximize access to resources and maintain economic viability and independence. A variety of redistribution networks were developed to insure availability of products from the different production zones as a hedge against the periodic shortfalls that ensued from climatic irregularities. In the final act of Precolumbian political-ecological hegemony, the late prehistoric Inka unified these extractive systems over large areas and effectively managed the agropastoral exploitation approaches that completely transformed the intermontane valleys and slopes of the Andes and Pacific coastal areas into anthropogenic landscapes.

The flora of the South American lowland tropics is one of the most diverse in the world. In essay 14, Daly and Mitchell provide a classification of the region's vegetation using the phytochorion concept. They have essentially divided the phytochoria into three groups: humid forests, tropical dry forests, and savannas.

Humid forests are the dominant vegetation in the Chocó, the Magdalena and the northwestern Caribbean coast, the Guayana Highlands, the Guianas and eastern Amazonia, central and western Amazonia, the southern Amazonian transition zone, and the Atlantic forest complex. Humid forests exist in an environment of high rainfall, 1,700 to 9,000 mm, that is evenly distributed throughout the year. In general they have high alpha-diversity, aseasonal climates, high canopies (up to 40 m with emergents up to 65 m), and understories of coarse herbs, vines, and slender trees. Variation in soils, geological parent material, physical barriers such as rivers and mountains, and water tables create phytogeographically distinct zones.

Tropical dry forests are found in areas with less available moisture or at least where annual precipitation is irregularly distributed with extended periods of dryness. The southern Brazilian region, the Chaco, the Caatinga, and the Coastal Cordillera of Venezuela are regions covered with tropical dry forests. They tend to receive less than 1,600 mm of annual precipitation with one or two dry seasons per year. Generally, the trees are shorter in stature and the species richness is less than in humid forests, yet dry forests are often high in endemism. Wind dispersal of seeds and pollen is more common, and there is a greater diversity of vines and epiphytes in dry forests than in humid forests.

The cerrado, the llanos of Venezuela and Colombia, the Llanos de Moxos, and the campo rupestre are all examples of South American savannas. Essentially they are tropical grasslands that range from treeless areas dominated by sedges and grasses to dense woodlands with understories dominated by grasses. Most have a severe dry season with frequent fires or floods. Many areas, such as the Roraima-Rupununi are species rich, with considerable endemism. At least some of these savanna formations, such as the Llanos de Moxos, have been strongly influenced by anthropogenic forces, especially repeated burning.

In essay 15, Roosevelt summarizes her recent work in the eastern Amazon Basin, or the Lower Amazon, as she refers to it. Her data, based on archaeological, paleoethnobotanical, and stable carbon isotope analyses, come from the Tapajos-Amazon confluence at Santarém–Monte Alegre. In terms of paleoenvironmental conditions, Roosevelt, contrary to traditional interpretation, hypothesizes that during the late Pleistocene and extending to late Holocene times, most of the Lower Amazon was covered by a closed-canopy, oligarchic tropical rain forest, and the presence of extensive savannas resulting from a Pleistocene desiccation, described by others, was not to be found. When savannas do appear in the paleoenvironmental record in the late Holocene, they do so as a result of anthropogenic factors, not climatic causes.

The late prehistoric Santarém culture, known largely from three habitation sites at Taperinha, Monte Alegre, and Santarém city, relied on maize, probably manioc, palms and other fruit-bearing trees, terrestrial and aquatic vertebrates, and shellfish for their sustenance. Through an agricultural approach based largely on shifting cultivation as seen in their Tapajo-culture descendants, the precontact inhabitants of the Tapajos confluence cleared land and burned forests around their settlements, often creating degraded landscapes such as savannas or monotypic stands of palms and other successional trees. It was estimated that the site core of Santarém was approximately 4 km^2 in area, indicating a substantial settlement. Other human impacts include the raising of flooded land through construction of mounds and raised fields and the formation of black soil from layer upon layer of organically rich trash deposits. These areas are fertile agricultural zones today.

Since the Conquest, the Tapajos region has undergone substantial population increases. Obviously, with larger modern populations the impact on the region is much greater.

Roosevelt concludes that beginning in the late Holocene, forests in the Lower Amazon adjacent to human settlements were in a state of disequilibrium due to agriculture and other extractive activities. She points to a need for more regional surveys that include teams of archaeologists and paleoethnobotanists to gain a real understanding of what transpired in the precontact Amazon Basin. The human impact prior to the arrival of Europeans seems to have been significant in localized areas, but a true understanding of the anthropogenic effects in the Amazon Basin as a whole must await further study.

One issue that should be addressed is the concept of wilderness in the New World before Contact. Some have argued that there was no such thing as a pristine landscape in the Precolumbian Americas (e.g., Cronon 1995). Without a doubt, the impact of Precolumbian Americans on the landscape has been largely underestimated, but the extent of that impact varied substantially from region to region, depending on the underlying resilience and fragility of the local biota, the technological wherewithal of the inhabitants, the level of trophic interaction, the duration of human occupation, and population density. If we use Denevan's population estimates (see table 1.2), which are higher than most, we see a density of 1.4 persons/km^2 for South America and only 0.2 persons/km^2 for North America (excluding Mexico and Central America). At face value, these densities are quite low, and when humans, especially hunters and gatherers, occupy an area at low densities, they tend to mimic other forces of natural disturbance. Moreover, the Precolumbians were not randomly distributed throughout the Americas; populations were dense in some areas and sparse in others. For example, almost two thirds of the South American population was centered in the Andes mountains. Much of the lowland population was clustered along the Amazon River and other major rivers. Thus, the ecological impact of the Precolumbian Americans was strongly felt in areas most intensively settled, viz., the intermontane valleys and altiplano of the Andes, the river valleys of the Peruvian and Atacama deserts, the lower Amazon River basin, the central Mississippi Valley (especially the American Bottom), parts of eastern and southwestern North America, the central Oaxaca, Tehuacán, and Cuicatlan valleys in Mexico, most of the Maya Lowlands and adjacent portions of Central America, and certain islands in the Caribbean, such as Hispaniola. Outside these areas, populations were low, with a concomitant less humanized environmental imprint. Yet even hunter-gatherers equipped with nothing more than stone tools and fire can subtly change an environment by encouraging and protecting valuable species, eliminating those that are not, and otherwise changing the frequency of disturbance.

The focus of this book has been to provide an overview of Holocene vegetation in the Americas as it might have existed without the influence of humans, to examine how Precolumbians interacted with the environment, and to assess the nature of that interaction. The goal has been to move toward some understanding of the past to help us better prepare for the future. Many of the indigenous land use practices of the Precolumbians were sustainable and conservative and represent the culmination of thousands of years of empirical experimentation. Now that the increases in crop yields from improved cultivars, resulting in the Green Revolution of the last half-century, are showing signs of diminishing returns (Mann 1999), alternative land use strategies, especially those employed successfully in the past, should be explored and implemented in cooperation with local agriculturalists to improve productivity and conserve scarce resources. Grassroots participation coupled with appropriate technology is essential if rural development projects are to be successful (Bocco 1991; Pawluk et al. 1992). We can also learn about the mistakes of the past, which are richly detailed above. Notwithstanding indications of earlier calamities, it seems that the errors of the ancients, such as farming on steep slopes without terracing and deforesting large tracts of land at once, are being repeated throughout the hemisphere, giving new meaning to the Maya concept of cyclical time. These serve as good examples of what not to do; unsustainable practices should be avoided, and knowing their end result can help modern managers prevent them from recurring. Efforts based on applied indigenous knowledge, if combined with population control measures, can help make marginal areas in the Americas productive and self-supporting once again. Thus the footprints of the Precolumbian Americans can be a blueprint for the future.

REFERENCES

Bocco, G. 1991. Traditional knowledge for soil conservation in central Mexico. *J. Soil and Water Conservation* 46(5): 346–348.

Cronon, W. 1995. The trouble with wilderness; or, getting back to the wrong nature. In W. Cronon, ed., *Uncommon Ground: Toward Reinventing Nature*, pp. 69–90. New York: W. W. Norton.

Mann, C. C. 1999. Crop scientists seek a new revolution. *Science* 283(5400): 310–314.

Pawluk, R. R., J. A. Sandor, and J. A. Tabor. 1992. The role of indigenous soil knowledge in agricultural development. *J. Soil and Water Conservation* 47(4): 298–302.

INDEX

For plants, see both scientific and common names.

Cleyera, 62
Clidemia, 295
Cliffrose, 53
Climate: of Basin of Mexico, 124–126; of Chaco, 429; at end of Pleistocene, 90; and food webs, 90; of Maya Lowlands, 181–183. *See also* Climate change; Rainfall
Climate change: causes of, 28–30; in early to middle Holocene, 22–25, 29; and insolation forcing, 28–29; in last Ice Age, 17–20; in late Holocene, 25–28, 29; and Maya impact on Yucatán Peninsula, 30–32; mechanisms of, in circum-Caribbean and linkages to other regions, 28–30; modern climate of Middle America and Intra-Americas Sea, 14–20; natural sources of information on, 13–14; and neoenvironmental determinism, 315–316; in northern American tropics and subtropics since last Ice Age, 13–33, 493–494; and ocean-atmosphere system variability, 29–30; paleoclimate history, 17–28; Pleistocene-Holocene transition, 20–22; and solar variability, 28; and volcanic eruptions, 30
Cloud forests, 288–289
Clusia, 287, 289, 298, 299, 399, 401, 415
Clusiaceae, 280, 287, 296, 303
Cnestidium rufescens, 80
Coahuila scrub oak, 54
Coastal Plain Province of the Northeast of Mexico (Tamaulipan Province), 65–67
Cobble features, 257, 260–266
Coca, 100, 361, 366, 371, 501
Coccoloba, 73, 78, 110, 298
Coccoloba ovata, 409
Coccoloba pixuna, 473
Cochambamba Valley, Bolivia, 372, 383
Cochlospermaceae, 69, 297
Cochlospermum vitifolium, 69, 297, 437
Coconuts, 73
Cocoyol, 108
Coffea arabica, 206
Coffee, 68, 212
Cohune, 103, 108
Coldenia, 67
Coleogyne ramosissima, 49, 50
Coles Creek culture, 238–239
Collared peccary, 474
Colocasia esculenta, 206
Colombia: archaeological evidence for early plant use in, 91; Chocó region of, 283, 297, 403; deforestation in, 304; domesticated plants in, 99, 101, 103; forests in, 304; Lla-

nos of, 396, 436, 438; vegetation in, 290–291, 294–295. *See also* Amazonia
Colorado Plateau, 47, 50
Columbia Plateau, 47
Columellia oblonga, 295
Columelliaceae, 295
Comandra, 46
Comarostaphylis arbutoides, 62
Combretaceae, 189, 297–298
Combretum leprosum, 441
Commelina, 109
Commelinaceae, 109, 300
Commiphora leptophloeos, 427
Compressed settlement pattern, 364
Condalia, 65, 67
Condalia hookeri, 66, 67
Conifers, 42–51, 82, 291
Connaraceae, 80
Convolvulaceae, 99, 109, 295, 298
Cooking techniques, 258, 261
Copaifera, 415
Copaifera lansdorfii, 436
Copaifera martii, 410
Copaifera officinalis, 437
Copaifera pubiflora, 414
Copal, 211–212
Copán, 197
Copernicia, 430
Copernicia alba, 430, 439, 441
Copernicia prunifera, 428
Copernicia tectorum, 437
Copiapoa, 302
Copiapoa cinerea var. haseltoniana, 302
Cordia, 79, 79–80, 110, 295, 297, 299, 300
Cordia alliodora, 69
Cordia elaeagnoides, 69
Cordilleran (Rocky Mountain) Region, 47–50
Coriaria, 296
Coriaria thymifolia, 295
Coriariaceae, 295, 296
Corn. *See* Maize; Milpa
Cornus, 45
Cornus disciflora, 55
Cornus excelsa, 55
Cornus florida, 58, 60
Corozo, 103, 108
Corrals in Lake Titicaca Basin, 345
Cortaderia, 292, 293
Costa Rica: homegardens in, 205–206; in late Holocene, 27; vegetation of, 39, 62, 72, 77, 290–291; Younger Dryas period in, 20–22
Cotton, 73, 91, 101, 104, 142, 156, 301, 358, 371, 501